Organic Reactions

Organic Reactions

VOLUME 26

JOHN WILEY & SONS, INC.
NEW YORK · CHICHESTER · BRISBANE · TORONTO

Published by John Wiley & Sons. Inc.

Library of Congress Catalog Card Number 42–20265

ISBN 0-471-05209-4

Printed in the United States of America

10 9 8 7 6 5 4 3 2 1

PREFACE TO THE SERIES

In the course of nearly every program of research in organic chemistry the investigator finds it necessary to use several of the better-known synthetic reactions. To discover the optimum conditions for the application of even the most familiar one to a compound not previously subjected to the reaction often requires an extensive search of the literature; even then a series of experiments may be necessary. When the results of the investigation are published, the synthesis, which may have required months of work, is usually described without comment. The background of knowledge and experience gained in the literature search and experimentation is thus lost to those who subsequently have occasion to apply the general method. The student of preparative organic chemistry faces similar difficulties. The textbooks and laboratory manuals furnish numerous examples of the application of various syntheses, but only rarely do they convey an accurate conception of the scope and usefulness of the processes.

For many years American organic chemists have discussed these problems. The plan of compiling critical discussions of the more important reactions thus was evolved. The volumes of *Organic Reactions* are collections of chapters each devoted to a single reaction, or a definite phase of a reaction, of wide applicability. The authors have had experience with the processes surveyed. The subjects are presented from the preparative viewpoint, and particular attention is given to limitations, interfering influences, effects of structure, and the selection of experimental techniques. Each chapter includes several detailed procedures illustrating the significant modifications of the method. Most of these procedures have been found satisfactory by the author or one of the editors, but unlike those in *Organic Syntheses* they have not been subjected to careful testing in two or more laboratories.

Each chapter contains tables that include all the examples of the reaction under consideration that the author has been able to find. It is inevitable, however, that in the search of the literature some examples will be missed, especially when the reaction is used as one step in an extended synthesis. Nevertheless, the investigator will be able to use the tables and their accompanying bibliographies in place of most or all of the literature search so often required.

Because of the systematic arrangement of the material in the chapters and the entries in the tables, users of the books will be able to find information desired by reference to the table of contents of the appropriate chapter. In the interest of economy the entries in the indices have been kept to a minimum, and, in particular, the compounds listed in the tables are not repeated in the indices.

The success of this publication, which will appear periodically, depends upon the cooperation of organic chemists and their willingness to devote time and effort to the preparation of the chapters. They have manifested their interest already by the almost unanimous acceptance of invitations to contribute to the work. The editors will welcome their continued interest and their suggestions for improvements in *Organic Reactions.*

Chemists who are considering the preparation of a manuscript for submission to Organic Reactions are urged to write either secretary before they begin work.

CONTENTS

CHAPTER 1

HETEROATOM-FACILITATED LITHIATIONS

Heinz W. Gschwend and Herman R. Rodriguez

Pharmaceuticals Division, Ciba-Geigy Corporation, Summit, New Jersey

CONTENTS

ACKNOWLEDGMENT

We wish to acknowledge the support of Dr. Neville Finch, whose encouragement and patience with our extensive efforts were indispensable for a timely completion of our endeavors. It is also a privilege to express our thanks to Mrs. Dorothy Alves, Ms. Donna Kreiss, and Ms. Marybeth McAlister for their meticulous typing of a demanding manuscript.

INTRODUCTION

Some 25 years have elapsed since the topic of metalation reactions was reviewed by Gilman and Morton.[1] The intervening years have been notable for intensive explorations in this area, in part because many organolithium reagents are now commercially available. Specifically, research efforts have been characterized by the discovery of new functional groups that promote metalation, elaboration of novel heterocyclic and olefinic substrates as metalatable species, recognition of new types of lithiating agents, and the continuation of efforts to define accurately the mechanism of metalation. Accordingly, heteroatom-facilitated lithiation has become recognized as an increasingly important tool, not only in the elaboration of carbocyclic aromatic and heteroaromatic systems, but also in synthetic aliphatic chemistry. A few recent reviews have covered the

[1] H. Gilman and J. W. Morton, Jr., *Org. Reactions*, **8**, 258 (1954).

topic in a more limited[2,3] or less specific sense.[4–6] It is the purpose of this chapter to survey and classify the vast accumulation of heteroatom-facilitated lithiations recorded since the first coverage in *Organic Reactions*.[1]

As outlined by Gilman and Morton, the terms "metalation" in general and "lithiation" in particular denote any replacement of a hydrogen atom by metal or lithium. In this review, however, lithiation is defined as the exchange of a hydrogen atom attached to an sp^2-hybridized carbon atom by lithium to form a covalent lithium-carbon bond (Eq. 1). More specifically, discussion is limited to those metalations that, through the influence of a heteroatom, are characterized by rate enhancement and regioselectivity. In fact, lithiation reactions of this type are noted for an extraordinarily high degree of regioselectivity, metalation generally occurring on the sp^2-carbon atom closest to the heteroatom. Based on the relative position of the heteroatom, such lithiations are conveniently classified into two principal categories: alpha and beta (*ortho*) lithiations.

$$\text{>C—H} \xrightarrow{\text{RLi}} \text{>C—Li} \qquad \text{(Eq. 1)}$$

In *alpha lithiations* the metalating agent deprotonates the sp^2-carbon atom *alpha* to the heteroatom to form a carbon–lithium bond (Eqs. 2 and 3). This sp^2-carbon atom may be part of an olefinic or heteroaromatic π system (Eqs. 2 and 3, X = heteroatom, Y = heteroatom or C_{sp^2}).

$$\text{X–C(=Y)–H} \xrightarrow{\text{RLi}} \text{X–C(=Y)–Li} \qquad \text{(Eq. 2)}$$

$$\text{(ring with X, Y)–H} \xrightarrow{\text{RLi}} \text{(ring with X, Y)–Li} \qquad \text{(Eq. 3)}$$

In *beta lithiations* the metalating agent is directed to deprotonate the sp^2-carbon atom *beta* to the heteroatom-containing substituent (Eqs. 4 and 5, X = heteroatom and $n = 0$–2). The sp^2-carbon atom can be part of an aromatic (Eq. 4) or an olefinic (Eq. 5) π system. It should be noted

[2] D. W. Slocum and D. I. Sugerman, *Adv. Chem. Ser.*, **130,** 222 (1974).

[3] E. M. Kaiser and D. W. Slocum in *Organic Reactive Intermediates*, S. P. McManus, Ed., Academic Press, New York, 1973, Chap. 5.

[4] J. M. Mallan and R. L. Bebb, *Chem. Rev.*, **69,** 693 (1969).

[5] U. Schöllkopf in *Houben-Weyl Methoden der Organischen Chemie*, Vol. 13/1, Georg Thieme Verlag, Stuttgart, 1970, p. 93.

[6] B. J. Wakefield, *The Chemistry of Organolithium Compounds*, Pergamon Press, New York, 1974, p. 38.

that the designation "*ortho* metalation" is used specifically for the beta metalation of carbocyclic aromatic systems.

$(C)_n$ X —RLi→ $(C)_n$ X, H → Li (Eq. 4)

$(C)_n$ X —RLi→ $(C)_n$ X, H → Li (Eq. 5)

This chapter surveys all systems in which alpha and beta lithiations have been observed, with the exception of ferrocenes.[2,3,7]

MECHANISM

Although numerous studies on the mechanism of heteroatom-facilitated lithiations have been undertaken, particularly those of alkyl aryl ethers, most of the current views are based more on speculation and hypothesis than on data. A major reason for this situation lies in the properties and/or characteristics of the traditional lithiating agents, *i.e.*, lithium alkyls and lithium aryls. These include high but variable reactivity and complexity (depending on solvent, concentration, and temperature) and instability.[6,8] In recent years an additional dimension has been added through the utilization of new types of lithiating agents, particularly *n*-butyllithium/amine complexes and lithium dialkylamides.

An understanding of the mechanism of metalation can be facilitated by the recognition of the existence of *two* distinct types of metalating agents. Accordingly, before the actual mechanism is discussed, attention should be brought to the individual characteristics of these lithiating agents. The first type, lithium alkyls and lithium aryls, are oligomers of varying complexity in solution (see p. 8). In addition they are electron-deficient species,[9,10] and therefore Lewis acids which can coordinate with Lewis bases such as ethers and amines[6,8] with consequent depolymerization to varying extents. It is important to note that kinetically these reagents become more basic as the aggregate size diminishes; therefore, tetrahydrofuran is the solvent of choice for generating reactive species (see

[7] D. W. Slocum and B. P. Koonsvitsky, *J. Org. Chem.*, **41,** 3664 (1976).

[8] U. Schöllkopf in *Houben-Weyl Methoden der Organischen Chemie,* Vol. 13/1, Georg Thieme Verlag, Stuttgart, 1970, p. 7.

[9] C. T. Viswanathan and C. A. Wilkie, *J. Organomet. Chem.*, **54,** 1 (1973).

[10] T. L. Brown, D. W. Dickerhoof, and D. A. Bafus, *J. Am. Chem. Soc.*, **84,** 1371 (1962).

accompanying tabulation). The pK_a of n-butane is 45–50,[11–13] with that of other alkanes typically falling in the range 42–60.[6,11–13]

Lithiating Agent	Solvent	State of Aggregation	Reference
n-BuLi	hydrocarbon	hexameric	14
	ether	tetrameric	14
	tetrahydrofuran	dimeric (solvated)	15
n-BuLi/TMEDA*	hydrocarbon	monomeric	16
t-BuLi	hydrocarbon	tetrameric	17, 9
	tetrahydrofuran	dimeric (solvated)	15
C_6H_5Li	ether	dimeric	4, 18
	tetrahydrofuran	dimeric	19, 20

The distinguishing characteristic of the second type of lithiating agent is its negligible extent of Lewis-acid character relative to uncomplexed lithium alkyls and lithium aryls. Included in this category are n-butyllithium/amine complexes and lithium dialkylamides, the former being monomeric in hydrocarbon solution, whereas some evidence exists that the latter are dimeric aggregates in solution.[21] Another distinctive property of the lithium dialkylamides is their decreased thermodynamic basicity relative to the lithium alkyls, the pK_a's of secondary amines being approximately 30.[5]

Although an apparent contradiction in terms, it has often been noted[22–25] that lithium dialkylamides are generally more effective metalating agents than the thermodynamically more basic lithium alkyls and lithium aryls particularly vis-à-vis substrates with pK_a values <30. The phenomenon of an increased kinetic basicity of heteroatom bases in

* Throughout the text TMEDA is taken to stand for tetramethylethylenediamine.

[11] D. J. Cram, *Fundamentals of Carbanion Chemistry*, Academic Press, New York, 1965.

[12] K. P. Butin, I. P. Beletskaya, A. N. Kashin, and O. A. Reutov, *J. Organomet. Chem.*, **10,** 197 (1967).

[13] H. F. Ebel in *Houben-Weyl Methoden der Organischen Chemie*, Vol. 13/1, Georg Thieme Verlag, Stuttgart, 1970, p. 32.

[14] H. L. Lewis and T. L. Brown, *J. Am. Chem. Soc.*, **92,** 4664 (1970).

[15] F. A. Settle, M. Haggerty, and J. F. Eastham, *J. Am. Chem. Soc.*, **86,** 2076 (1964).

[16] A. W. Langer, Jr., *N.Y. Acad. Sci.*, **27,** 741 (1965).

[17] D. A. Shirley, T. E. Harmon, and C. F. Cheng, *J. Organomet. Chem.*, **69,** 327 (1974).

[18] D. A. Shirley, M. J. Danzig, and F. C. Canter, *J. Am. Chem. Soc.*, **75,** 3278 (1953).

[19] D. W. Slocum and C. A. Jennings, *J. Org. Chem.*, **41,** 3653 (1976).

[20] V. Ramanathan and R. Levine, *J. Org. Chem.*, **27,** 1667 (1962).

[21] R. Huisgen, H. König, and N. Bleeker, *Chem. Ber.*, **92,** 424 (1959).

[22] R. Huisgen and J. Sauer, *Chem. Ber.*, **92,** 192 (1959).

[23] H. W. Gschwend, Ciba-Geigy Pharmaceutical Co., Summit, New Jersey, unpublished results.

[24] H. R. Rodriguez, Ciba-Geigy Pharmaceutical Co., Summit, New Jersey, unpublished results.

[25] R. Huisgen and J. Sauer, *Angew. Chem.*, **72,** 91 (1960).

general and lithium amides in particular may be rationalized by the availability of a free pair of electrons, which permits the formation of a four-membered transition state (Eq. 6). The intermediacy of a free carbanion is thus avoided.[22]

$$\text{C=C–H}\cdots\text{N}(\text{R})_2\text{–Li} \longrightarrow \text{C=C–Li} + \text{HNR}_2 \qquad \text{(Eq. 6)}$$

It should be recognized that the deprotonation of an unsaturated carbon acid, such as benzene or ethylene (pK_a 36.5–37.0),[11–13] is thermodynamically feasible with lithium alkyls, as the ΔpK_a is greater than 5. Yet these metalations are exceedingly slow even in ethereal solvents; *e.g.*, benzene shows negligible lithiation with *n*-butyllithium in hexane after 3 hours at room temperature.[1,26] Therefore, one phenomenon that must be explained by any mechanism concerning heteroatom-facilitated lithiation is the great rate enhancement; *e.g.*, anisole metalates to the extent of 30% with *n*-butyllithium in ether in 2 hours.[27] In addition the high degree of regioselectivity must be accounted for.

An example of rate enhancement in an intermolecular sense is the lithiation of benzene with *n*-butyllithium coordinated with TMEDA (3 hours/25°) *vs.* that of *n*-butyllithium alone (*vide supra*).[16,26] The function of the TMEDA is the depolymerization of the normally hexameric *n*-butyllithium to the kinetically more reactive monomer via coordination of the nitrogen atoms of the bidentate ligand with the lithium atom. Analogously, as a working hypothesis it has been assumed for many years that the initial step in heteroatom-facilitated lithiation reactions is the coordination of the electron-deficient metalating agent with the nonbonding electrons in the substrate heteroatom (with attendant depolymerization).[28] This coordination is then followed by a protophilic attack of the carbanionic portion of the lithiating agent on the adjacent hydrogen atom, leading to the metalated product.[29] Although this simplistic view of the mechanism is quite useful in explaining and predicting numerous experimental observations, other factors, particularly the inductive effects of heteroatoms or substituents, can play important roles in these reactions. It is the contention of these authors that the hydrocarbon acidity can become the sole determinant and that a new mechanism is indicated.

[26] M. D. Rausch and D. J. Ciappenelli, *J. Organomet. Chem.*, **10,** 127 (1967).
[27] R. A. Ellison and F. N. Kotsonis, *J. Org. Chem.*, **38,** 4192 (1973).
[28] J. D. Roberts and D. Y. Curtin, *J. Am. Chem. Soc.*, **68,** 1658 (1946).
[29] D. Bryce-Smith, *J. Chem., Soc.*, **1954,** 1079.

Accordingly, it is postulated that there are, in fact, *two* limiting mechanisms:

1. "coordination only" mechanism
2. "acid–base" mechanism

Between these extremes there is a continuous spectrum of cases in which both effects simultaneously contribute in varying degrees to the observed phenomena. A full explication of these concepts is best presented through illustration.

Beta (*Ortho*) Lithiations

The best example of a "coordination only" mechanism is the *ortho* lithiation of N,N-dialkylbenzylamines.[30] Despite the fact that the benzylic methylene group (neglecting any effect of the basic nitrogen) has an inductive effect on the *ortho* position which actually decreases its acidity, lithiation occurs exclusively at position 2 as well as far more rapidly relative to benzene.[31,32] Thus it has been assumed that the initial step in this reaction is the coordination of the metalating agent with the lone pair of the basic nitrogen atom. The nearest available proton in the *ortho* position then suffers a protophilic attack, leading to the internally chelated and isolable organolithium species (Eq. 7).[2,3,9] One may consider this an intramolecular version of the lithiation of benzene with the *n*-butyllithium/TMEDA complex alluded to earlier. This view is further supported by the observation that even with excess metalating agent only a monolithio species is formed, since the chelate no longer possesses Lewis-base character.[30,33]

$C_6H_5CH_2N(CH_3)_2$ $\underset{}{\overset{RLi}{\rightleftharpoons}}$ [complex: $CH_2N(CH_3)_2$ coordinated to Li; H, Li, R] $\longrightarrow$ [chelate: $CH_2N(CH_3)_2$, Li] (Eq. 7)

Most mechanistic studies of the *ortho* lithiation have been carried out with alkyl aryl ethers, where a combination of the two mechanistic determinants, *i.e.*, coordinative and inductive effects, is operative. Nmr data have shown that lithiation in hydrocarbon solvents results in the initial disruption of the oligomeric alkyllithium to yield a reactive complex with the substrate.[34] The stoichiometry of this complex depends on

[30] F. N. Jones, M. F. Zinn, and C. R. Hauser, *J. Org. Chem.*, **28,** 663 (1963).

[31] A. I. Shatenshtein, *Tetrahedron*, **18,** 95 (1962).

[32] R. Huisgen, W. Mack, K. Herbig, N. Ott, and E. Anneser, *Chem. Ber.*, **93,** 412 (1960).

[33] F. N. Jones, R. L. Vaulx, and C. R. Hauser, *J. Org. Chem.*, **28,** 3461 (1963).

[34] R. A. Ellison and F. N. Kotsonis, *Tetrahedron*, **29,** 805 (1973).

the structure of the ether. For anisole the whole process is generally and simplistically pictured in Eq. 8.[35] In the intermediate coordinated species the carbon–lithium bond of the metalating agent and the carbon–hydrogen bond of the substrate are polarized to a greater extent, thus rendering the proton to be removed more acidic by induction. The second step, namely the actual deprotonation (protophilic attack), is rate determining. Hydrogen–deuterium exchange studies clearly show the expected isotope effect (k_H/k_D) to be in the range 6–8.[36]

$$C_6H_5OCH_3 + RLi \underset{}{\overset{\text{Fast}}{\rightleftharpoons}} [CH_3O(\cdots Li-R)C_6H_4-H] \xrightarrow{\text{Slow}} CH_3O\cdots Li-C_6H_4 \quad \text{(Eq. 8)}$$

A pivotal example is the lithiation of *p*-methoxy-N,N-dimethylbenzylamine.[19] The acidity of the hydrogen atom at position 3 should be distinctly higher than at position 2 because of the inductive effect of the methoxyl group.[13,31,32] Nevertheless, metalation with a lithiating agent of high Lewis-acid character (*n*-butyllithium) occurs exclusively at position 2 as a result of the higher coordinative capacity of the basic nitrogen ("coordination only" mechanism). Alternatively, with the monomeric *n*-butyllithium/TMEDA complex whose Lewis-acid character has been considerably diminished via coordination (*vide supra*), the most *acidic* proton at position 3 is removed selectively ("acid–base" mechanism).[19]

Alpha Lithiations

For alpha metalations of π-excessive heterocycles, particularly those of thiophene, the precoordination of the metalating agent with the heteroatom has also been invoked as the initial step of the mechanism.[37,38] However, it has been demonstrated that thiophene, like benzene, cannot be lithiated readily in hexane solution by *n*-butyllithium[23,39,40] but that lithiation does occur rapidly in ethereal solvents (see p. 35). This evidence suggests that the sulfur atom in thiophene does not in fact act as

[35] R. A. Finnegan and J. W. Altschuld, *J. Organomet. Chem.*, **9**, 193 (1967).
[36] D. A. Shirley and J. P. Hendrix, *J. Organomet. Chem.*, **11**, 217 (1968).
[37] D. A. Shirley and K. R. Barton, *Tetrahedron*, **22**, 515 (1966).
[38] S. Gronowitz, *Ark. Kemi*, **13**, 295 (1958) [*C.A.*, **53**, 15056d (1959)].
[39] C. G. Screttas and J. F. Eastham, *J. Am. Chem. Soc.*, **87**, 3276 (1965).
[40] D. J. Chadwick and C. Willbe, *J. Chem. Soc., Perkin Trans. I*, **1977**, 887.

a Lewis base, as does the oxygen atom in anisole, and therefore is unable to decrease the state of aggregation of the hexameric *n*-butyllithium (see p. 8) with consequent increase in its activity. Therefore the rate enhancement and regioselectivity of most alpha metalations can be attributed merely to the inherently higher acidity of the substrates. Consequently, alpha metalations can generally be considered to proceed by the "acid–base" mechanism.

There are few accurate pK_a measurements available for these substrates (*e.g.*, trichloroethylene, $pK_a = 18$,[12] thiophene (alpha position), $pK_a \approx 30$[23]); however, ample evidence from base-catalyzed hydrogen exchange rates indicates that the kinetic acidity of the alpha position (C_{sp^2}–H) is dramatically enhanced by the adjacent electronegative atom relative to ethylene or benzene (Eq. 9). This observation is corroborated by the fact that 2-thienyllithium can be generated to the extent of 50% with lithium diisopropylamide, a considerably weaker base than *n*-butyllithium.[23] It has similarly been established that the hydrogen exchange rate in the alpha position is greater by a factor of 2.5×10^5 than that in the beta position.[41,42] Again, this observation explains the regioselectivity of the metalation without invoking precoordination.

$$\text{C=C–H} \xrightarrow{K_1} \text{C=C}^{\ominus} \qquad \text{X–C=C–H} \xrightarrow{K_2} \text{X–C=C}^{\ominus} \qquad \text{(Eq. 9)}$$

$$K_1 \ll K_2$$

X = heteroatom

Another example supportive of the "acid–base" mechanism hypothesis is the lithiation of pyrazoles. If precoordination were indeed an important factor, one would assume that N-substituted pyrazoles would lead to lithiation in the 3 position via coordination with the unshared pair of electrons of the nitrogen in the 2 position. Metalation, however, occurs exclusively at the more acidic 5 position (see p. 23).

In synopsis, it can be stated that lithiating agents with little or no Lewis-acid character generally lead to thermodynamic products derived from deprotonation at the most acidic position. In contrast, coordinatively unsaturated metalating agents, such as *n*-butyllithium in hydrocarbons or ethers, produce kinetic products derived from deprotonation at the atom closest to the most effective ligand.

[41] A. I. Shatenshtein, A. G. Kamrad, I. O. Shapiro, Y. I. Ranneva, and E. N. Zvyagintseva, *Dokl. Akad. Nauk. SSSR*, **168**, 364 (1966) [*C.A.*, **65**, 8695g (1966)].

[42] J. A. Elvidge, J. R. Jones, C. O'Brien, E. A. Evans, and H. C. Sheppard, *Adv. Heterocycl. Chem.*, **16**, 1 (1974).

SCOPE AND LIMITATIONS

Nature and Reactivity of the Lithiated Species

Few reports describe the physical properties or the relative reactivity of alpha- or beta-lithiated species. The most complete information is available on *o*-lithio-N,N-dimethylbenzylamine, which can actually be isolated as a pure solid.[9] In this case there is good evidence for strong intra- and/or intermolecular lithium–nitrogen interactions[9,43] and, that by virtue of the latter, a solvent- and concentration-dependent aggregation occurs. Because of this type of internal chelation and/or the acidifying inductive effect of electron-withdrawing directors, heteroatom-stabilized lithioorganics appear to be both somewhat less reactive[23,24,44] and weaker Lewis acids. The decrease in these parameters is dependent on the nature and strength of the internal chelation and inductive effects. It is this very decrease in Lewis–acid character that permits the isolation of such lithiated reagents, notably *o*-lithio-N,N,-dimethylbenzylamine, free of ethereal solvents.[9]

Influence of Other Substituents

In carbocyclic aromatic, heteroaromatic, and olefinic systems, the presence or absence of substituents other than the one arbitrarily considered to be the directing functionality is of considerable importance, since several factors in the lithiation of a particular substrate will be affected. These can be classified into the following categories: effects on rates, regioselectivity, and compatibility.

Effects on Rates. Substituents clearly affect the rates of lithiation in a particular π system. These effects are largely inductive in nature and are reflected in an increased or decreased kinetic acidity of a particular proton. Although various substrates readily lend themselves to kinetic studies under carefully controlled conditions, the rate-enhancing effects of substituents have been quantitatively studied only for bromobenzene derivatives.[32] These studies on the kinetics of aryne formation from substituted bromobenzenes, in which the *ortho* lithiation is considered to be the rate-determining step, probably represent the best basis for the assessment of these substituent effects. By applying the data to *ortho* lithiations in general (X = any *ortho*-directing group), the effect of the substituent R on the rate of metalation should decrease as indicated in Eqs. 10–12.

[43] D. W. Slocum and P. L. Gierer, *Chem. Commun.*, **1971**, 305.
[44] P. Beak and B. Siegel, *J. Am. Chem. Soc.*, **96**, 6803 (1974).

Effect of R on a *meta* position:

$k_m(R)$

$R = F > CF_3, Cl^* > OCH_3, H > CH_3 > N(CH_3)_2$

(Eq. 10)

(Eq. 11)

Effect of R on an *ortho* position:

$k_o(R)$

$R = F > OCH_3 > Cl, CF_3^* > N(CH_3)_2 > H > CH_3$

(Eq. 12)

One interesting aspect of these data is the unusual position occupied by the methoxyl and dimethylamino groups in 1,3-disubstituted benzenes. Although both groups exhibit a smaller acidifying effect on an *ortho* position than does chlorine or the trifluoromethyl group,[31] their rate-enhancing effect, particularly that of the methoxyl group, is greater than that shown by the trifluoromethyl group. This phenomenon is ascribed to the enhanced coordinative involvement of these two moieties,[32] as discussed in the mechanistic section. Accordingly, in 1,3-disubstituted benzenes the rate enhancement of R is determined not only by its inductive effect but also by its coordinative potential.

These relative directing effects are corroborated in several other systems, as judged by the rough estimates of the degree of *ortho* lithiation determined by deuteration. For example, from the studies of N,N-dimethylbenzylamines,[24,45] N-methylbenzamides,[23,24] and 2-aryloxazolines,[46] it can be stated that a chlorine in a *para* position has a rate-enhancing effect, whereas a methoxyl group has a small, if any, rate-decreasing effect relative to hydrogen. These effects run parallel to the relative rate factors *f* of the potassium amide–catalyzed hydrogen–deuterium exchange in liquid ammonia for substituted benzenes.[31,47]

In alpha lithiations the situation is rather similar. Any substituent in a position beta to the facilitating heteroatom increases the rate of metalation if it exerts an inductive effect that increases the acidity of the alpha

*The inductive effect of the halogens decreases from fluorine to iodine.[22] Although chlorine is not included in Huisgen's studies,[32] it is arbitrarily put on the same level with the trifluoromethyl group.

[45] K. P. Klein and C. R. Hauser, *J. Org. Chem.* **32,** 1479 (1967).

[46] H. W. Gschwend and A. Hamdan, *J. Org. Chem.*, **40,** 2008 (1975).

[47] G. E. Hall, R. Piccolini, and J. D. Roberts, *J. Am. Chem. Soc.*, **77,** 4540 (1955).

proton. The reverse inductive effect, of course, has the opposite effect. This is true for both olefins and π-excessive, five-membered heterocycles. Specifically, this rate-enhancing or -decreasing effect is shown, for example, in the thiophene series: 3-bromothiophene > thiophene > 3-methylthiophene.[48]

Regioselectivity. Since a large number of functional groups and atoms possess an inherent directing effect on the metalation of π-systems, their presence in a substrate, together with the dominant directing group, can have a pronounced effect on the regioselectivity of the metalation. For benzenoid systems, when a substituent (Y) is present in either a 1,2 or 1,4 relationship to the dominant *ortho*-directing group or atom (X), the general rule is that the higher degree of lithiation occurs *ortho* to the more powerful directing group. The strength of the director, however, is interdependent on the nature of the lithiating agent utilized (see p. 49 and pp. 77–80).

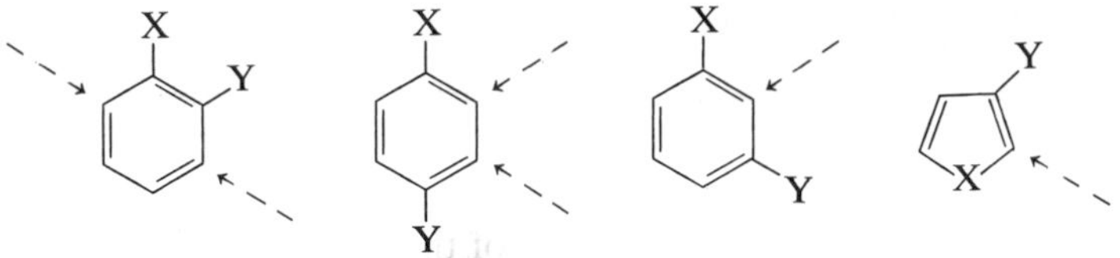

X Y 1,2 X Y 1,4 X Y 1,3 Y X 1,3

When X and Y are in a 1,3 relationship, even if the directing effect of Y is weak, lithiation occurs predominantly or exclusively in the position *ortho* to both the beta-directing group X and the substituent Y. The degree of regioselectivity depends on the relative inductive and coordinative capacities of Y. Known exceptions to this phenomenon are those in which $X = Y = CF_3$[49,50] and $X = Y = CONHC_4H_9$.[24,51] Analogously, in π-excessive five-membered heterocycles, the general effect of Y in the 3 position (studied most extensively in the thiophene series)[52] is to provide selective or specific lithiation in the 2 position.

Whereas most substituents exert some beta-directing effect, some groups cannot coordinate, and these influence the regioselectivity in a different sense. For Y = alkyl, for instance, metalation occurs preferably *not* in the position *ortho* to both substituents in 1,3-disubstituted benzenoid systems but, rather *para* to Y, whereas in π-excessive, five-membered heterocycles position 5 is preferred.

[48] S. Gronowitz, *Ark. Kemi*, **7,** 361 (1954) [*C.A.*, **49,** 13216b (1955)].
[49] K. D. Bartle, G. Hallas, and J. D. Hepworth, *Org. Magn. Reson.*, **5,** 479 (1973).
[50] D. A. Shirley, J. R. Johnson, Jr., and J. P. Hendrix, *J. Organomet. Chem.*, **11,** 209 (1968).
[51] B. H. Bhide, *Chem. Ind. (London)*, **1974,** 75.
[52] S. Gronowitz, *Adv. Heterocycl. Chem.*, **1,** 73 (1963).

Compatibility. As a generalization, any substituent that does not react with the lithiating agent used to effect a metalation is considered compatible. Most of the knowledge concerning such compatibility is based on data available for benzene and thiophene systems. It should be emphasized that for any particular example the compatibility of a substituent with the lithiation conditions depends both on the facility of metalation of the substrate and on the lithiating agent itself. Whether or not a given substituent is compatible, therefore, is often a question of competing reactions, *i.e.*, nucleophilic attack of the lithiating agent on the substituent *vs.* the actual metalation. Accordingly, it should be noted that substituents often considered incompatible in the metalation of carbocyclic aromatic systems are quite compatible in the more facile alpha lithiations. In addition the use of the non-nucleophilic lithium dialkylamides renders compatibility to certain electrophilic substrates in isolated instances. The following summary lists those functional groups that are generally considered incompatible, except where noted. These substituents conveniently fall into three categories: electrophiles, acidic groups, and halogens.

Electrophiles. Included in this category are nitro groups, aldehydes, ketones, acids, esters, nitriles, and, in most instances, primary and tertiary amides.[53,54] Low-temperature lithiations of *m*-chlorobenzonitrile,[24] hindered tertiary benzamides,[55,56] and β-aminoacrylic acid derivatives,[57a–c] however, are known.

Acidic Groups. This category encompasses alkylmercapto,[1,58,59] alkyl sulfinyl, alkylsulfonyl,[60] and *ortho* alkyl groups. The acidic character of these groups usually precludes their presence in substrates. However, the alkylmercapto groups are perfectly acceptable in heterocyclic or olefinic systems. Alkyl substituents are generally quite compatible under the usual lithiating conditions. If, however, such alkyl groups, particularly methyl, are located in an *ortho* position with respect to the directing group, deprotonation of the alkyl group can be either a side reaction, as for *o*-alkoxytoluenes[61,62] or *o*-toluidines,[63] or the exclusive pathway, as for

[53] C. J. Upton and P. Beak, *J. Org. Chem.*, **40,** 1094 (1975).
[54] W. H. Puterbaugh and C. R. Hauser, *J. Org. Chem.*, **29,** 853 (1964).
[55] P. Beak, G. R. Brubaker, and R. F. Farney, *J. Am. Chem. Soc.*, **98,** 3621 (1976).
[56] P. Beak and R. Brown, *J. Org. Chem.*, **42,** 1823 (1977).
[57a] R. R. Schmidt and J. Talbiersky, *Angew. Chem., Int. Ed. Engl.*, **15,** 171 (1976).
[57b] R. R. Schmidt and J. Talbiersky, *Angew. Chem., Int. Ed. Engl.*, **17,** 205 (1978).
[57c] R. R. Schmidt and J. Talbiersky, *Angew. Chem., Int. Ed. Engl.*, **16,** 853 (1977).
[58] D. A. Shirley and B. J. Reeves, *J. Organomet. Chem.*, **16,** 1 (1969).
[59] D. Seebach, *Angew. Chem., Int. Ed. Engl.*, **8,** 639 (1969).
[60] M. Julia and J. M. Paris, *Tetrahedron Lett.*, **1973,** 4833.
[61] R. L. Letsinger and A. W. Schnizer, *J. Org. Chem.*, **16,** 869 (1951).
[62] T. E. Harmon and D. A. Shirley, *J. Org. Chem.*, **39,** 3164 (1974).
[63] R. E. Ludt, G. P. Crowther, and C. R. Hauser, *J. Org. Chem.*, **35,** 1288 (1970).

o-toluamides,[64,65] *o*-tolylsulfonamides,[66] *o*-tolyloxazolines,[46] and *o*-methylbenzylamines[67] (Eq. 13). This type of reaction is synthetically rather useful but, since it lies beyond the defined scope of this chapter, it is not discussed in detail.

$$\text{o-}X\text{-}C_6H_4\text{-}CH_3 \xrightarrow{RLi} \text{o-}X\text{-}C_6H_4\text{-}CH_2Li \text{ (X}\cdots\text{Li chelate)} \qquad \text{(Eq. 13)}$$

Halogens. Whereas metal–halogen exchange of chlorine and fluorine usually is not a problem, aryne formation can become the predominant reaction if the lithium enters *ortho* or beta with respect to the halogen. The use of lithium amides generally increases the rate of aryne formation.[25] This eliminative pathway can be suppressed at low temperatures. In bromine- or iodine-containing substrates metal–halogen exchange is usually the preferred reaction. In the readily lithiatable heterocyclic systems, however, even in the presence of bromine and, in a few cases, iodine metalation proceeds normally, and little if any halogen–metal exchange is observed.[48,68] A few exceptions are known in which metalation is preferred in bromo-substituted benzenoid systems.[69]

Alpha Lithiation

Alpha metalations are deprotonations of olefinic, aromatic, or other π systems at the sp^2-hybridized carbon that is alpha to a heteroatom X, as illustrated by Eq. 14. Alpha lithiations occur primarily in families of

$$Y{=}C(H)(X) \xrightarrow{RLi} Y{=}C(Li)(X) \qquad Y = {>}C, O, S, {-}N; \quad X = OR, SR, NRR', \text{halogen} \qquad \text{(Eq. 14)}$$

five-membered heterocycles that have one or more heteroatoms. More recently, the metalation of simple heterosubstituted olefins has been developed into a synthetically useful tool (*vide infra*). One common feature of alpha lithiations is the rapidity with which they proceed in comparison to beta lithiations.

Alpha Activators and Their Relative Activating Ability

The inductive effect of an alpha activator greatly increases the acidity of the adjacent carbon–hydrogen bond, thus making alpha lithiations very

[64] R. L. Vaulx, W. H. Puterbaugh, and C. R. Hauser, *J. Org. Chem.*, **29,** 3514 (1964).
[65] J. J. Fitt and H. W. Gschwend, *J. Org. Chem.*, **41,** 4029 (1976).
[66] H. Watanabe and C. R. Hauser, *J. Org. Chem.*, **33,** 4278 (1968).
[67] R. L. Vaulx, F. N. Jones, and C. R. Hauser, *J. Org. Chem.*, **29,** 1387 (1964).
[68] N. Gjøs and S. Gronowitz, *Acta. Chem. Scand.*, **25,** 2596 (1971).
[69] W. E. Truce and M. F. Amos, *J. Am. Chem. Soc.*, **73,** 3013 (1951).

facile. The following heteroatoms are known to lead to alpha metalations: nitrogen, oxygen, sulfur, selenium, tellurium, fluorine, chlorine, and bromine. In the series of five-membered heterocycles—thiophene, furan, and N-alkylpyrroles—competitive metalation studies have revealed that under thermodynamic conditions the rate of metalation is greatest with sulfur, resulting in the following rank order: sulfur > oxygen > N-alkyl.[70] The fact that sulfur is the best alpha activator is explained by the $(d\text{–}\sigma)$ overlap, which apparently outweighs the inductive effect of the more electronegative oxygen and nitrogen atoms.[42,70] It appears that this $(d-\sigma)$ overlap is a major factor in the enhancement of the thermodynamic stability of such anions adjacent to sulfur.[71,72] Conversely, the more rapid lithiation of furan under kinetic conditions points to the superior ability of oxygen to act as a ligand.[40] Accordingly, under kinetic conditions the order becomes oxygen > sulfur > N-alkyl.

By inference from the available experimental facts the following additional conclusions can be drawn:

1. Alpha metalations occur more readily on a carbon bearing two heteroatoms (Eq. 14, X and Y = heteroatom).
2. The rate of alpha metalation increases with an increasing number of heteroatoms present in five-membered heterocycles.[71,73]
3. Alpha metalations of olefins occur most readily, usually at temperatures below −70°. The relative activating potency of the heteroatom under thermodynamic conditions appears to be halogen > sulfur > oxygen > nitrogen.

Nitrogen as an Alpha-Activating Atom

Enamines. The alpha lithiation of enamines has received only scant attention, being limited to special cases. N,N-Diethyl-3-(1-pyrrolidinyl)acrylamide, for instance, can be metalated at −115° to produce the corresponding α-lithio species.[57a] The latter reacts with methyl iodide, for instance, to give the analogous crotonamide in high yield. The facility of this metalation is probably a result of both the alpha-activating ability of the pyrrolidine group and the acidifying and directing effects of the beta-directing carboxamide function. It should be noted that at higher temperatures 1,4 addition of the lithiating agent prevails.[57a]

Alpha lithiation of ethyl 3-(1-pyrrolidinyl)acrylate proceeds equally well at low temperatures.[57b] However, a different picture emerges on the

[70] D. A. Shirley and J. C. Goan, *J. Organomet. Chem.*, **2,** 304 (1964).

[71] R. A. Olofson, J. M. Landesberg, K. N. Houk, and J. S. Michelman, *J. Am. Chem. Soc.*, **88,** 4265 (1966).

[72] R. A. Coburn, J. M. Landesberg, D. S. Kemp, and R. A. Olofson, *Tetrahedron*, **26,** 685 (1970).

[73] A. C. Rochat and R. A. Olofson, *Tetrahedron Lett.*, **1969,** 3377.

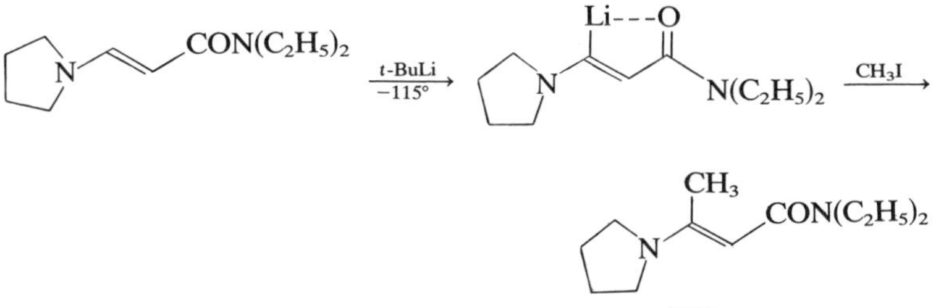

(95%)

metalation of 3-(1-pyrrolidinyl)acrylonitrile. Reaction with lithium diisopropylamide at temperatures lower than −105° results in the kinetically controlled deprotonation of the expected position, *i.e.*, alpha to the enamine group, to give the lithio species **1**. When warmed above −100°, however, this intermediate (**1**) slowly rearranges to the thermodynamically more stable beta-lithiated isomer **2a**, which is mesomeric with the imine anion **2b**. Both lithiated species **1** and **2a,b** can be reacted selectively with electrophiles to produce, after treatment with methyl iodide, *e.g.*, 3-(1-pyrrolidinyl)crotonitrile and 2-methyl-3-(1-pyrrolidinyl)-acrylonitrile, respectively.[57c]

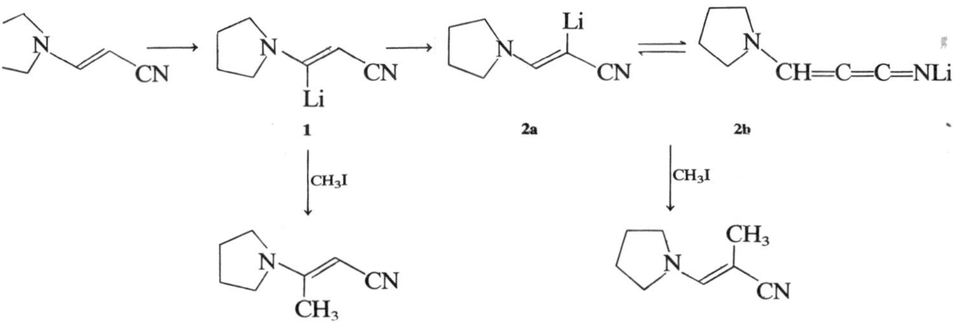

Deuterium incorporation has shown that the dihydropyridine **3** undergoes alpha metalation.[74]

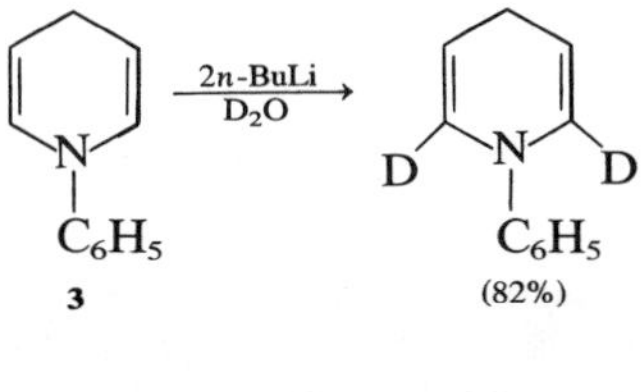

(82%)

[74] D. M. Stout, T. Takaya, and A. I. Meyers, *J. Org. Chem.*, **40**, 563 (1975).

Vinyl Isocyanides. Functionally, vinyl isocyanides may be considered derivatives of enamines. The strong acidifying character of the isocyanide group on sp^3 carbons is well known.[75] Apparently, this electron-withdrawing effect is also sufficiently operative on sp^2 carbons to permit alpha lithiation of certain types of vinyl isocyanides. Styryl isocyanide, for instance, is readily deprotonated at low temperatures to give α-lithio styryl isocyanide.[76] This reacts readily with various electrophiles; reaction with carbon dioxide, for example, gives a high yield of lithio α-isocyanocinnamate. With aldehydes and ketones, *e.g.*, acetone, the initial adduct is in equilibrium with the lithiated oxazoline (see p. 31), which in turn can further react to give a diadduct. The ratio of mono- and diadduct appears to depend on the character of the carbonyl reactant.[76] In view of the limited thermal stability of vinyl isocyanides, these reactions are clearly only of theoretical interest.

$$C_6H_5CH{=}CHNC \xrightarrow[-110^\circ]{n\text{-BuLi}} C_6H_5CH{=}C(Li)NC \xrightarrow{CO_2} C_6H_5CH{=}C(CO_2Li)NC$$

(*cis/trans*) (94%)

Formamides and Thioformamides. Deprotonation of dialkylformamides is essentially instantaneous even at $-100°$,[77] and there seem to be very few limitations to the nature of the N-alkyl substituent (Eq. 15). Alpha-lithiated formamides represent a novel type of acyl anion equivalent or, more specifically, a highly reactive one-carbon synthon with the oxidation state of cyanide. The synthetic potential of lithio dialkylformamides is documented by their excellent reactivity with numerous electrophiles.[77,78]

$$HC(=X)NR_1R_2 \xrightarrow{LDA} LiC(=X)NR_1R_2 \qquad X = O, S \qquad \text{(Eq. 15)}$$

The deprotonation of dimethylformamide can be carried out with lithium diisopropylamide at $-78°$; subsequent reaction with pivalaldehyde, for example, gives a good yield of the hydroxyamide **4**.[78] A variant of this technique using N-methoxymethyl substituents gives access

[75] U. Schöllkopf, *Angew. Chem., Int. Ed. Engl.*, **16,** 339 (1977).

[76] U. Schöllkopf, D. Stafforst, and R. Jentsch, *Justus Liebigs Ann. Chem.*, **1977,** 1167.

[77] D. Seebach, W. Lubosch, and D. Enders, *Chem. Ber.*, **109,** 1309 (1976).

[78] B. Banhidai and U. Schöllkopf, *Angew. Chem., Int. Ed. Engl.*, **12,** 836 (1973).

to adducts such as **5** and **6**, which upon deprotection give either secondary or primary amides.[79] An interesting facet of this reaction is that the lithiated species need not be preformed, for a 1:1 mixture of dimethylformamide and a substrate, *e.g.*, cyclohexanone, can be treated with lithium diisopropylamide at −78° to form the corresponding hydroxyamide in 62% yield.[78] This observation certainly attests to the kinetic acidity of the formamide proton relative to those of cyclohexanone.

$$HCON(CH_3)_2 \xrightarrow[2.\ t\text{-}C_4H_9CHO]{1.\ LDA} t\text{-}C_4H_9CH(OH)CON(CH_3)_2$$

4 (76%)

$$HCONR_1R_2 \xrightarrow[2.\ (C_6H_5)_2CO]{1.\ LDA} (C_6H_5)_2C(OH)CONR_1R_2$$

5 $R_1 = CH_3$, $R_2 = CH_2OCH_3$ (85%)
6 $R_1 = R_2 = CH_2OCH_3$ (88%)

The reported lithiation of N,N-dimethylthioformamide[77,80] is carried out at −100° to form the desired anion within 3 minutes. Reaction of the anion with methyl benzoate gives an 85% yield of the corresponding α-ketothioamide.[77]

Pyrroles and Indoles. Compared with the oxygen and sulfur isosteres, furan and thiophene, pyrroles are lithiated only poorly in their alpha position. In fact, pyrrole itself undergoes only deprotonation of the nitrogen atom, even with excess reagent.[81] N-Alkyl- and N-arylpyrroles, however, do undergo the expected alpha lithiation, but higher temperatures or longer reaction times are needed. Markedly increased rates of metalation can be achieved, however, by the use of reactive alkyllithium/TMEDA complexes.[40,68,82] This is illustrated by the transformation of N-methylpyrrole into N-methylpyrrole-2-carboxylic acid in good yield,[68] whereas the same reaction with *n*-butyllithium alone provides only a 42% yield of the acid.[81] With excess *n*-butyllithium and TMEDA products derived from 2,5-dilithio-N-methylpyrrole can be isolated, whereas in the absence of TMEDA the 2,4-dilithio species predominates.[40] N-Arylpyrroles can also be considered as tertiary anilines and, in fact, products of both alpha *and ortho* lithiation are isolated, as illustrated by the formation of the tricyclic ketone **7**.[81]

The metalation of indoles follows a similar pattern. N-Unsubstituted indoles are transformed into their lithium salts only, whereas N-alkylindoles are metalated in their alpha positions.[83] The rate of lithiation

[79] U. Schöllkopf and H. Beckhaus, *Angew. Chem., Int. Ed. Engl.*, **15,** 293 (1976).

[80] D. Enders and D. Seebach, *Angew. Chem., Int. Ed. Engl.*, **12,** 1014 (1973).

[81] D. A. Shirley, B. H. Gross, and P. A. Roussel, *J. Org. Chem.*, **20,** 225 (1955).

[82] J. W. F. Wasley, Ciba-Geigy Pharmaceutical Co., Summit, New Jersey, personal communication.

[83] D. A. Shirley and P. A. Roussel, *J. Am. Chem. Soc.*, **75,** 375 (1953).

EtLi/TMEDA, CO_2 — N-CH_3 pyrrole → 1-methylpyrrole-2-CO_2H (70%)

n-BuLi, CO_2 — N-C_6H_5 pyrrole → 1-phenylpyrrole-2-CO_2H (14%) + **7** (15%)

is moderate; however, as for the metalation of pyrroles, the rate can be enhanced by using tetrahydrofuran as solvent.[84] The modest reactivity of indoles, particularly in comparison with benzofuran, is documented by the results observed in the lithiation of 5-methoxy-1-methylindole. After reaction with pyridine-2-carboxaldehyde, all three possible products (**8**–**10**) are formed, indicating that the *ortho*-directing effect of the methoxyl group is as strong as the alpha-activating effect of the indole nitrogen.[84]

1. t-BuLi
2. 2-Pyridinecarboxaldehyde

8 + **9** + **10**

(74%, total) (5:1:4)

R = CH(OH)-(2-pyridyl)

The alpha lithiation of indoles can be facilitated by using methoxymethyl[85] or arylsulfonyl[84] protecting groups, *i.e.*, functionalities that have a beta-directing influence of their own. The cumulative directing effect of both the indole nitrogen and the protecting group is now sufficient, for instance, to lead exclusively to alpha lithiation, with 5-methoxy-1-(phenylsulfonyl)indole producing the 2-substituted derivative

[84] R. J. Sundberg and R. L. Parton, *J. Org. Chem.*, **41,** 163 (1976).
[85] R. J. Sundberg and H. R. Russell, *J. Org. Chem.*, **38,** 3324 (1973).

11 in very good yield.[84] The presence of the methoxymethyl group has a similar rate-enhancing effect, as illustrated by the high yields achieved in the preparation of the acid **12**.[85] A potential advantage in the use of protected indoles lies, of course, in the possibility of removing the protecting group.

CH_3O ... N–$SO_2C_6H_5$ → (1. *t*-BuLi; 2. 2-Pyridinecarboxaldehyde) → CH_3O ... –CH(OH)–(2-pyridyl), N–$SO_2C_6H_5$

11 (84%)

N–CH_2OCH_3 indole → (*t*-BuLi) → 2-Li, N–CH_2–OCH_3 → (CO_2) → 2-CO_2H, N–CH_2OCH_3

12 (80%)

Pyrazoles. 1-Substituted pyrazoles are metalated in the 5 position. 3-Methyl-1-propylpyrazole, for instance, is alpha lithiated rapidly and essentially quantitatively to produce the carbinol **13**.[86] Although N-unsubstituted pyrazoles are reported to undergo metalation in the 5 position, the yields are poor.[87] The generation of a 1,2 dianion may be responsible for this phenomenon. Again, an alternative is to use an

CH_3 pyrazole N–C_3H_7-*n* → (1. *n*-BuLi; 2. C_6H_5CHO) → CH_3 pyrazole 5-$CH(OH)C_6H_5$, N–C_3H_7-*n*

13 (95%)

appropriate protecting group. The tetrahydropyranyl ether **14** is metalated under mild conditions, allowing very good yields of the 5-substituted products, as documented by the isolation of the thioether **15**.[23]

Two interesting side reactions can be observed in the metalation of certain pyrazoles. The first is a relatively facile deprotonation of the

[86] D. E. Butler and S. M. Alexander, *J. Org. Chem.*, **37,** 215 (1972).

[87] R. Hüttel and M. E. Schön, *Justus Liebigs Ann. Chem.*, **625,** 55 (1959).

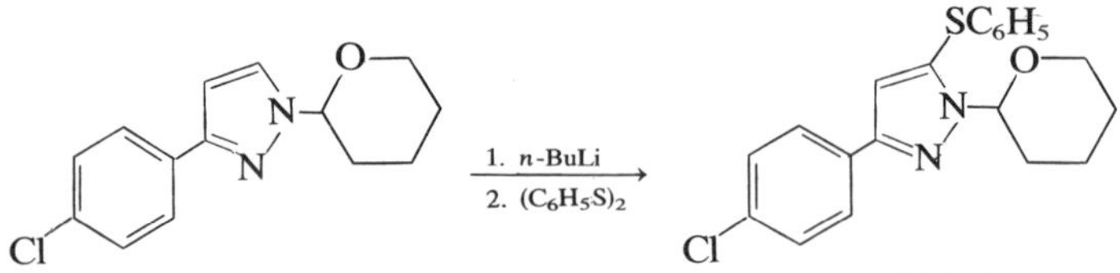

N-alkyl substituent, particularly when it is a methyl group. Lithiation of 1,3-dimethylpyrazole, for instance, results not only in the formation of the expected alpha-metalated heterocycle **16**, but also appreciable amounts of the species **17**. Subsequent reaction with benzaldehyde yields the corresponding carbinols.[86] It appears that the two anions are interconvertible. It should therefore be possible to run such reactions under conditions permitting the trapping of only the kinetic deprotonation product. The second example pertains to 1-phenylpyrazoles, where, aside from the normal alpha lithiation, *ortho* metalation of the phenyl ring is observed because of chelation with the imine nitrogen, giving rise to two acids **18** and **19**.[88] This type of side reaction becomes predominant when organomagnesium halides are utilized as metalating agents.[89]

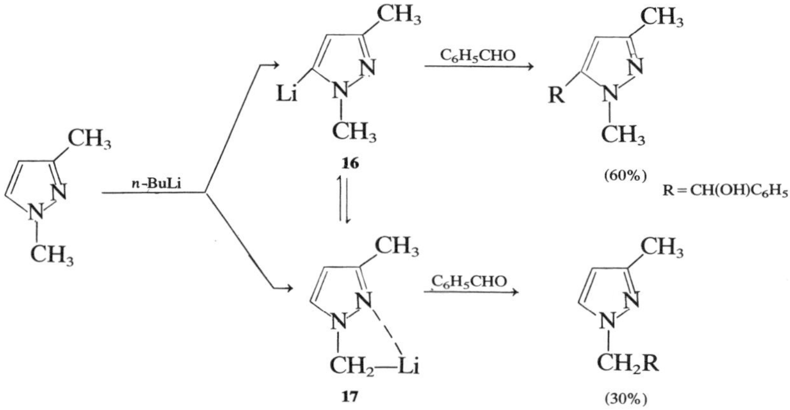

Imidazoles and Condensed Imidazoles. As with other π-excessive, five-membered heterocycles with two heteroatoms in a 1,3 arrangement, lithiation of 1-substituted imidazoles occurs between the nitrogen atoms when the position is available. If the 2 position is occupied by an alkyl group, deprotonation may occur either at the 5 position or on the alkyl

[88] P. W. Alley and D. A. Shirley, *J. Am. Chem. Soc.*, **80,** 6271 (1958).

[89] A. Marxer and M. Siegrist, *Helv. Chem. Acta*, **57,** 1988 (1974).

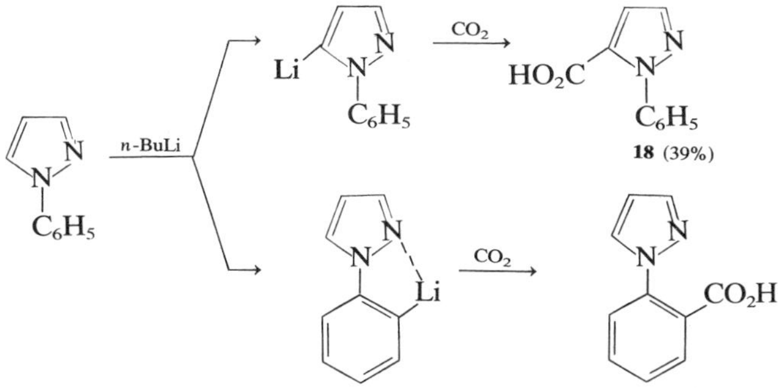

18 (39%)

19 (10%)

group, depending on its nature. Although 1,2-dimethylimidazole is reported to lithiate at position 5,[90] a more recent investigation clearly demonstrates that the products of such reactions are derived exclusively from deprotonation of the relatively acidic 2-methyl group,[91] as illustrated by the isolation of the carbinol **20**.

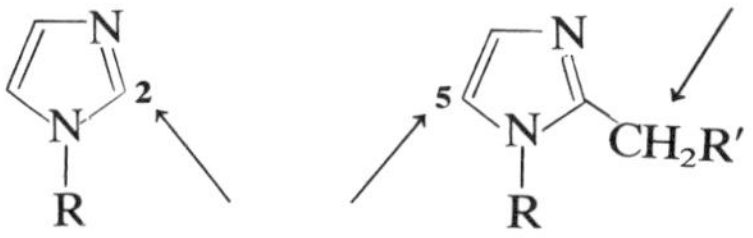

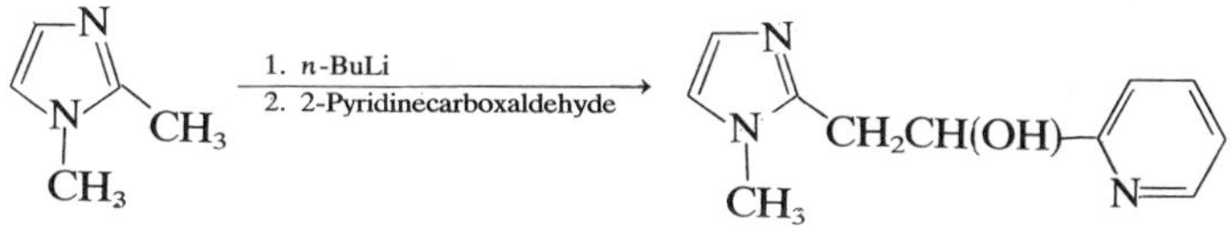

1. *n*-BuLi
2. 2-Pyridinecarboxaldehyde

20 (66%)

Alpha lithiation between the nitrogen atoms is extremely facile, occurring even in the presence of a bromine substituent otherwise liable to undergo halogen–metal exchange. This is illustrated by the preparation of the carbinol **21**.[92]

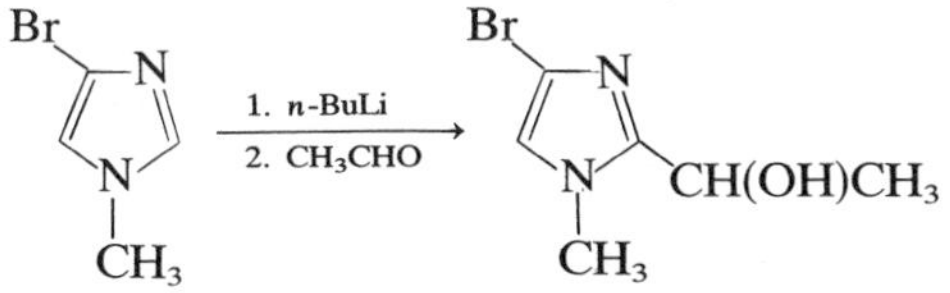

1. *n*-BuLi
2. CH_3CHO

21 (50%)

[90] B. A. Tertov, V. V. Burykin, and I. D. Sadekov, *Khim Geterotsikl. Soedin.*, **1969,** 560 [*C.A.*, **71,** 124328y (1969)].

[91] F. Vinick, Ciba-Geigy Pharmaceutical Co., Summit, New Jersey, unpublished results.

[92] D. S. Noyce and G. T. Stowe, *J. Org. Chem.*, **38,** 3762 (1973).

Lithiation of 1-substituted benzimidazoles proceeds analogously with great facility and leads to 2-lithiated species. 1-Methylbenzimidazole can thus be silylated in high yield to give **22**.[93]

1. n-BuLi
2. $(CH_3)_3SiCl$

$Si(CH_3)_3$ CH_3 CH_3

22 (91%)

Imidazo[1,2-*a*]pyridine, a condensed imidazole with a blocked 2 position, undergoes alpha metalation as expected to produce, after quenching with dimethylformamide, the aldehyde **23**.[94]

1. n-BuLi
2. DMF

CHO

23 (24%)

Triazoles and Tetrazoles. 1,2,3-Triazoles[95] and 1,2,4-triazoles[96] are lithiated with extreme rapidity in their respective alpha positions. It should be noted, however, that 5-lithio-1,2,3-triazoles are only stable at low temperatures. Above −40° they display a tendency to undergo a cycloreversion with loss of nitrogen and formation of an N-substituted ketenimine anion.[97] Nevertheless, low-temperature lithiation of 1-phenyl-1,2,3-triazole followed by reaction with methyl iodide gives an excellent yield of the 5-methyl derivative **24**.[95] Under similar conditions, followed by quenching with benzophenone, 1-benzyl-3-phenyl-1,2,4-triazole gives an equally high yield of the alcohol **25**.[96]

1. n-BuLi
2. CH_3I

C_6H_5 C_6H_5 CH_3

24 (94%)

$CH_2C_6H_5$ $CH_2C_6H_5$

1. n-BuLi
2. $(C_6H_5)_2CO$

$C(C_6H_5)_2OH$

C_6H_5 C_6H_5

25 (92%)

[93] P. Jutzi and W. Sakriss, *Chem. Ber.*, **106,** 2815 (1973).
[94] W. W. Paudler and H. G. Shin, *J. Org. Chem.*, **33,** 1638 (1968).
[95] R. Raap, *Can. J. Chem.*, **49,** 1792 (1971).
[96] H. Behringer and R. Ramert, *Justus Liebigs Ann. Chem.*, **1975,** 1264.
[97] U. Schöllkopf and I. Hoppe, *Justus Liebigs Ann. Chem.*, **1974,** 1655.

Tetrazoles are probably among the most readily lithiatable heterocycles because they contain the maximum number of four electronegative ring atoms.[73] The acidity of the only available ring hydrogen must be appreciable. N-Methyltetrazole, for instance, can thus be converted to the thiol **26**.[98]

n-BuLi, "S" →

CH_3 CH_3 (—SH)

26 (67%)

Pyridines, Condensed Pyridines, and Pyridine N-Oxides. Alpha metalation of pyridines and condensed pyridines is not a practical reaction, largely because these compounds are susceptible to nucleophilic attack. However, what may appear to be a direct lithiation can be achieved through use of lithium diisopropylamide and hexamethylphosphoramide as a cosolvent. Under these conditions only dimeric products (2,2′-bipyridines) can be isolated.[99] Comparatively stable 2-lithiopyridines (or quinolines) are, nevertheless, readily available via halogen–metal exchange.[100]

Direct metalation of pyridine N-oxides with *n*-butyllithium is feasible, although the yields are generally modest to poor.* This is accounted for by the fact that addition competes with alpha deprotonation.[101] Although the lithiation proceeds alpha to the nitrogen atom, one has to assume a good deal of *ortho*-directing effect on the part of the oxygen atom. Thus a comparison with the *ortho* lithiation of phenol[1] (an equally poor process) may be suggested. The reactivity of the 2-lithio derivatives is not high enough to permit reactions with epoxides.[102] However, it is not uncommon for products derived from 2-lithiopyridine N-oxides to be highly electrophilic and therefore subject to additions of unreacted metalated species to the remaining azomethine linkage.[101] The reaction thus has only a limited synthetic potential.

Pyrimidines. Unlike imidazoles, pyrimidines apparently are not metalated by alkyllithiums in the 2 position, *i.e.*, between the nitrogen atoms, because of the known propensity of the pyrimidine nucleus to

*The use of lithium diisopropylamide gives better results (personal communication from R. A. Lyle, North Texas State University).

[98] R. Raap, *Can. J. Chem.*, **49,** 2139 (1971).
[99] A. J. Clarke, S. McNamara, and O. Meth-Cohn, *Tetrahedron Lett.*, **1974,** 2373.
[100] R. G. Jones and H. Gilman, *Org. Reactions*, **6,** 339 (1951).
[101] R. A. Abramovitch, R. J. Coutts, and E. M. Smith, *J. Org. Chem.*, **37,** 3584 (1972).
[102] R. A. Abramovitch and E. E. Knaus, *J. Heterocycl. Chem.*, **12,** 683 (1975).

suffer nucleophilic attack at one of the azomethine linkages by nucleophilic organometallic reagents. One report, however, describes a successful metalation in the 4 position of 5-methylpyrimidine by using the non-nucleophilic lithium diisopropylamide. On quenching with benzophenone a modest but nonetheless significant yield of the carbinol **27** is isolated.[99]

1. LDA
2. $(C_6H_5)_2CO$

CH_3 $C(C_6H_5)_2OH$

27 (30%)

Oxygen as an Alpha-Activating Atom

Alkyl Vinyl Ethers and Allenic Ethers. The synthetic potential of alpha-metalated vinyl ethers is substantial because they represent acyl anion equivalents, a type of synthon that has only recently become available (see p. 90). Although numerous enol ethers have been metalated,[103] methyl vinyl ether has been explored most thoroughly.[104] It is readily lithiated by *t*-butyllithium in tetrahydrofuran at low temperatures, reacting with essentially any electrophile to give good to excellent yields of the primary products. These in turn may then be hydrolyzed to ketones and elaborated further. The conversion of methyl vinyl ether into the ketone **28** is but one example of this useful type of reaction.[104] α-Methoxyvinyllithium undergoes 1,2 addition exclusively on α,β-unsaturated systems,[104] but 1,4 addition can be effected by conversion to the corresponding copperlithium reagent.[105,106]

OCH_3 —(*t*-BuLi, −65°)→ Li, OCH_3 —(C_6H_5CN)→ COC_6H_5, OCH_3

28 (70%)

An interesting and potentially useful variant of this reaction is the *in situ* preparation of an enol ether via tautomerization of a protected allylic alcohol. Thus the tetrahydropyranyl ether of allyl alcohol is isomerized to the enol ether **29**, which in turn is readily metalated to the lithio species **30**. Subsequent reaction with methyl iodide produces a very good yield of the enol ether **31**.[107] It is likely that the oxygen of the protecting group of **30** provides additional chelation and stability to such reagents.

[103] U. Schöllkopf and P. Hänssle, *Justus Liebigs Ann. Chem.*, **763,** 208 (1972).

[104] J. E. Baldwin, G. A. Höfle, and O. W. Lever, Jr., *J. Am. Chem. Soc.*, **96,** 7125 (1974).

[105] C. G. Chavdarian and C. H. Heathcock, *J. Am. Chem. Soc.*, **97,** 3822 (1975).

[106] R. K. Boeckman, Jr., K. J. Bruza, J. E. Baldwin, and O. W. Lever, Jr., *J. Chem. Soc., Chem. Commun.*, **1975,** 519.

[107] J. Hartmann, M. Stähle, and M. Schlosser, *Synthesis*, **1974,** 888.

$$CH_2{=}CHCH_2OTHP \xrightarrow{t\text{-}C_4H_9OK} CH_3CH{=}CHOTHP\ (\mathbf{29}) \xrightarrow{s\text{-BuLi}} \mathbf{30} \xrightarrow{CH_3I} \mathbf{31}\ (83\%)$$

THP =

29 **30** **31** (83%)

Metalation of enol ether systems was first realized with alkoxyallenes.[108a] If the higher metalation temperature in the reaction between methoxyallene and *n*-butyllithium is any indication, it would appear that the resulting anion **32** has a greater stability than simple lithiated vinyl ethers. The product yields are generally good[108a] to excellent,[23] as illustrated by the formation of the allenic ether **33**.[108a] The accessibility and facility of metalation of allenic ethers, combined with their high reactivity toward a broad spectrum of electrophiles, make species like **32** very valuable synthons indeed.

$$CH_2{=}C{=}CHOCH_3 \xrightarrow{n\text{-BuLi}} CH_2{=}C{=}C(OCH_3)Li\ (\mathbf{32}) \xrightarrow{C_6H_5CH_2Br} CH_2{=}C{=}C(OCH_3)CH_2C_6H_5$$

32

33 (64%)

Interestingly, metalation of hindered allenic ethers, such as *t*-butoxyallene, with lithium dicyclohexylamide occurs predominantly at the gamma position (*ca.* 80%).[108b] Similarly, lithiation of alpha-substituted allenic ethers with *n*-butyllithium takes place at the gamma position.[108b]

Higher alkoxycumulenes are apparently metalated equally well. Treatment of 1-methoxy-4-methyl-1,2,3-pentatriene with *n*-butyllithium provides the lithiated derivative **34**,[109] which can also be prepared from 1,4-dimethoxy-4-methyl-2-pentyne via *in situ* elimination and metalation.[110,111] It reacts with electrophiles such as cyclohexanone to give the expected products, *e.g.*, the alcohol **35**, in good yields.

[108a] S. Hoff. L. Brandsma, and J. F. Arens, *Rec. Trav. Chim. Pays-Bas*, **87,** 916 (1968).

[108b] J. C. Clinet and G. Linstrumelle, *Tetrahedron Lett.*, **1978,** 1137.

[109] H. A. Selling, J. A. Rompes, P. P. Montijn, S. Hoff. J. H. Van Boom, L. Brandsma, and J. F. Arens, *Rec. Trav. Chim. Pays-Bas*, **88,** 119 (1969).

[110] R. Mantione, A. Alves, P. P. Montijn, H. J. T. Bos, and L. Brandsma, *Tetrahedron Lett.*, **1969,** 2483.

[111] R. Mantione and A. Alves, *Receuil*, **89,** 97 (1970).

$(CH_3)_2C{=}C{=}C{=}CHOCH_3 \xrightarrow{n\text{-BuLi}} (CH_3)_2C{=}C{=}C{=}C(OCH_3)Li$ (**34**) $\xleftarrow{2\,n\text{-BuLi}} (CH_3)_2C(OCH_3)C{\equiv}CCH_2OCH_3$

34 $\xrightarrow{\text{Cyclohexanone}}$ $(CH_3)_2C{=}C{=}C{=}C(OCH_3)$–C(OH) (cyclohexyl)

35 (90%)

Furans and Condensed Furans. The alpha metalation of furans can be compared with the lithiation of vinyl ethers, since furan is itself a cyclic divinyl ether. Accordingly, the reaction of furan with *n*-butyllithium, although considerably less facile, does proceed in high yield in refluxing ether, as shown by the isolation of the carbinol **36**.[112] Dilithiation of furan can be achieved efficiently with *n*-butyllithium/TMEDA in refluxing hexane, as indicated by the isolation (90%) of dimethyl furan-2,5-dicarboxylate.[40] With a substituent in the 2 position, metalation occurs equally well in the free 5 position. As is discussed in more detail in regard to the thiophene series, a 3 substituent with a beta-directing capacity permits a regioselective metalation in the 2 position. Furan-3-carboxaldehyde ethylene ketal, for instance, leads to the 2-lithio species **37**, which can be quenched with various electrophiles to produce, after hydrolysis, 2-substituted-3-carboxaldehydes[113] such as the boronic acid **38**.[114,115]

Furan $\xrightarrow{1.\ n\text{-BuLi};\ 2.\ C_6H_5CHO}$ 2-furyl–$CH(OH)C_6H_5$

36 (98%)

Furan-3-carboxaldehyde ethylene ketal $\xrightarrow{n\text{-BuLi}}$ **37** (2-Li) $\xrightarrow{(n\text{-}C_4H_9O)_3B;\ H_3O^+}$ 3-CHO-2-$B(OH)_2$-furan

37 **38** (36%)

Condensed furans such as benzofuran undergo alpha metalation in the 2 position rather than *ortho* lithiation in the 7 position, which, at least in principle, could be considered an alternate pathway. The fact that alpha metalation is achieved far more readily than *ortho* metalation at any of

[112] V. Ramanathan and R. Levine, *J. Org. Chem.*, **27,** 1216 (1962).

[113] M. C. Fournie-Zaluski and C. Chatain-Cathaud, *Bull. Soc. Chim. Fr.*, **1974,** 1571.

[114] S. Gronowitz and U. Michael, *Ark. Kemi*, **32,** 283 (1970) [*C.A.*, **73,** 109824e (1970)].

[115] D. Florentin, B. P. Roques, and M. C. Fournie-Zaluski, *Bull. Soc. Chim. Fr.*, **1976,** 1999.

the three possible sites (4, 6, or 7) of 5-methoxybenzofuran is documented by the reaction with *n*-butyllithium, followed by ethylene oxide, to obtain only the 2-substituted ethanol **39**.[116] The result is additional evidence that alpha metalation proceeds more readily in furans than pyrroles, for the analogous 1-methyl-5-methoxyindole gives a mixture of products,[84] as indicated on p. 22.

CH_3O 4 2 6 7 O → (1. *n*-BuLi; 2. ethylene oxide) CH_3O O $-(CH_2)_2OH$

39 (62%)

Isoxazoles. In contrast to the isothiazoles (p. 39) the lithiation of isoxazoles does not lead to the expected products. Still, metalation does occur next to the more electronegative oxygen atom when the 3 position is occupied, as illustrated for 3,4-diphenylisoxazole, but the lifetime of the lithiated species is very short and it suffers fragmentation into benzonitrile and a deprotonated phenyl ketene which can be trapped with electrophiles.[117]

C_6H_5, C_6H_5 (N, O) → (*n*-BuLi, −60°) C_6H_5, C_6H_5 (N, O, Li) ⟶ $C_6H_5CN + C_6H_5C{=}C{=}O\ Li^+$

Oxazoles and Oxazolines. Oxazoles can be readily lithiated in the 2 position. Unlike their nitrogen analogs, the imidazoles, the ambireactive character of their lithiated species indicates that they are in equilibrium with the enolate of a β-ketoisonitrile. Lithiation of 4,5-diphenyloxazole, for instance, produces the carbinol **40** on reaction with benzaldehyde, whereas reaction with chlorotrimethylsilane leads exclusively to the O-silyl ether **41**.[118]

2-Oxazolines can be metalated in a similar manner, leading to an equilibrium mixture of alpha-lithiated species and isonitrile.[119,120] Lithiated 4,5-diphenyl-2-oxazoline, for example, is acylated on oxygen with acetic anhydride, whereas reaction with benzaldehyde produces the dihydro derivative of **40**.[120]

[116] P. Cagniant and G. Kirsch, *C. R. Acad. Sci., Paris, Ser, C,* **281,** 111 (1975).

[117] U. Schöllkopf and I. Hoppe, *Angew. Chem., Int. Ed. Engl.,* **14,** 765 (1975).

[118] R. Schroeder, U. Schöllkopf, E. Blume, and I. Hoppe, *Justus Liebigs Ann. Chem.,* **1975,** 533.

[119] A. I. Meyers and E. W. Collington, *J. Am. Chem. Soc.,* **92,** 6676 (1970).

[120] U. Schöllkopf, F. Gerhart, I. Hoppe, R. Harms, K. Hantke, K. H. Scheunemann, E. Eilers, and E. Blume, *Justus Liebigs Ann. Chem.,* **1976,** 183.

$$\text{2,3-}(C_6H_5)_2\text{-oxazole} \xrightarrow{n\text{-BuLi}} \text{2-Li-4,5-}(C_6H_5)_2\text{-oxazole} \rightleftharpoons C_6H_5C(OLi){=}C(C_6H_5)N{=}C$$

$$\xrightarrow{C_6H_5CHO} \text{4,5-}(C_6H_5)_2\text{-oxazol-2-yl}{-}CH(OH)C_6H_5 \quad \textbf{40}\ (67\%)$$

$$\xrightarrow{(CH_3)_3SiCl} C_6H_5C(OSi(CH_3)_3){=}C(C_6H_5)N{=}C \quad \textbf{41}\ (85\%)$$

Sulfur as an Alpha-Activating Atom

Vinyl Sulfides and Allenic Thioethers. Vinyl sulfides undergo rapid alpha metalation with lithium alkyls under carefully controlled conditions, and thus provide yet another acyl anion equivalent.[121] However, a possible side reaction in these alpha lithiations is a Michael-type addition of the metalating agent to the olefinic bond, leading to a saturated carbanion stabilized by the (sp^3–d) overlap.[59,122] The reaction is fairly general with respect to the nature of the substitution on sulfur; *i.e.*, both alkyl and aryl groups are permissible. Allyl or benzyl sulfides are excluded, however, because of their known propensity for deprotonation of the sp^3-bound hydrogen, as is the case with allyl vinyl sulfide.[123] The metalation of phenyl vinyl sulfide can be effected readily with lithium diisopropylamide,[124,125] as well as with *n*-butyllithium/TMEDA,[126] without incurring addition of the metalating agent (see pp. 98–99). The vinyl sulfide anions react readily with aldehydes, ketones, epoxides, and alkyl halides. For example, alkylation of lithiated ethyl vinyl sulfide with *n*-octyl bromide, followed by mercuric ion-catalyzed hydrolysis of the primary product gives 2-decanone in excellent yield.[121]

$$CH_2{=}CHSC_2H_5 \xrightarrow[-78^\circ]{s\text{-BuLi}} CH_2{=}C(Li)SC_2H_5 \xrightarrow{n\text{-}C_8H_{17}Br} CH_2{=}C(SC_2H_5)C_8H_{17}\text{-}n \xrightarrow{Hg^{2+}} CH_3COC_8H_{17}\text{-}n \ (90\%)$$

An interesting situation arises with 2-alkoxy-1-(alkylthio)ethylenes where alpha lithiation is, in principle, possible at either of the two

[121] K. Oshima, K. Shimozi, H. Takahashi, H. Yamamoto, and H. Nozaki, *J. Am. Chem. Soc.*, **95,** 2694 (1973).

[122] W. E. Parham and R. F. Motter, *J. Am. Chem. Soc.*, **81,** 2146 (1959).

[123] K. Oshima, H. Takahashi, H. Yamamoto, and H. Nozaki, *J. Am. Chem. Soc.*, **95,** 2693 (1973).

[124] R. C. Cookson and P. J. Parsons, *J. Chem. Soc., Chem. Commun.*, **1976,** 990.

[125] B. Harirchian and P. Magnus, *J. Chem. Soc., Chem. Commun.*, **1977,** 522.

[126] R. C. Cookson, University of Southampton, England, personal communication.

sp^2-carbon atoms. Paralleling the observations made in the rates of lithiation of thiophene and furan under thermodynamic conditions, metalation of 2-ethoxy-1-(pentylthio)ethylene proceeds exclusively at the position alpha to sulfur.[127] The reactivity of the resulting anion appears to be comparable to that of the simple α-alkylthiovinyllithium species, and reaction with various electrophiles is possible. As shown, reaction with ethylene oxide produces the alcohol **42** in good yield.[127] The ethoxy group quite likely provides an additional beta-directing or at least an acidifying effect. One aspect worthy of note in this reaction is the apparent absence of elimination of lithium alkoxide. Similarly, lithiated (Z)-1,2-bis(diethylthio)ethylene[128] is stable at low temperatures, whereas the correspondingly metalated 1,2-bis(phenylthio)ethylenes (Z or E) undergo elimination at higher temperatures.[129]

C_2H_5O / SC_5H_{11}-*n* $\xrightarrow[-70°]{t\text{-BuLi}}$ C_2H_5O- - -Li / SC_5H_{11}-*n* $\xrightarrow{\text{ethylene oxide}}$ C_2H_5O $(CH_2)_2OH$ / SC_5H_{11}-*n*

42 (60%)

In contrast to the vinyl sulfides discussed thus far, the amine **43** is unique in that the beta-directing nitrogen ligand serves to depolymerize the metalating agent and to chelate the resulting alpha-lithiated species. Despite the stability of the anion at ice-bath temperature, it is still reactive enough to undergo addition to a variety of electrophiles, as illustrated by the reaction with dimethyl disulfide to produce the basic ketene-S-acetal **44**.[23]

$(CH_3)_2N$ / SC_6H_5 $\xrightarrow{n\text{-BuLi}}$ $(CH_3)_2N$- - -Li / SC_6H_5 $\xrightarrow{(CH_3S)_2}$ $(CH_3)_2N$ SCH_3 / SC_6H_5

43 **44** (90%)

The alpha lithiation of allenic thioethers is facile, but in contrast to the oxygen congeners (see p. 29), the reactivity of the metalated species is more complex because of an increased stability of the propargylic anions. Thus metalation of the allene **45** followed by reaction with formaldehyde leads to a mixture of the two alcohols **46** and **47**.[130] This ambireactive character can be changed in favor of the products derived from alpha-lithiated allenic thioethers by using lithium amide and ammonia as the metalating system.[131a,b]

[127] I. Vlattas, L. DellaVecchia, and A. O. Lee, *J. Am. Chem. Soc.*, **98,** 2008 (1976).
[128] R. R. Schmidt and B. Schmid, *Tetrahedron Lett.*, **1977,** 3583.
[129] W. E. Parham and P. L. Stright, *J. Am. Chem. Soc.*, **78,** 4783 (1956).
[130] A. Schaap, L. Brandsma, and J. F. Arens, *Rec. Trav. Chim. Pays-Bas*, **84,** 1200 (1965).
[131a] L. Brandsma, C. Jonker, and M. H. Berg, *Rec. Trav. Chim. Pays-Bas*, **84,** 560 (1965).
[131b] J. H. vanBoom, L. Brandsma, and J. F. Arens, *Rec. Trav. Chim. Pays-Bas*, **85,** 580 (1966).

$$CH_3CH{=}C{=}CHSC_2H_5 \ (\mathbf{45}) \xrightarrow[2.\ CH_2O]{1.\ CH_3Li} CH_3CH{=}C{=}C(SC_2H_5)CH_2OH \ (\mathbf{46}) + CH_3CH(CH_2OH)C{\equiv}CSC_2H_5 \ (\mathbf{47})$$

2 : 3

Vinyl Sulfoxides. Alpha metalation of alkenyl aryl sulfoxides is particularly facile and occurs at very low temperatures with lithium diisopropylamide. One noteworthy feature observed in the lithiation of E/Z mixtures of such substrates is the preferential formation of E-products which must be accounted for by an isomerization of the intermediate anion. Thus lithiation of a 1 : 1 mixture of (E)- and (Z)-3-methoxypropenyl phenyl sulfoxide followed by alkylation with methyl iodide leads almost exclusively to the E-product.[132] The yield of products is even higher in alkenyl 2-pyridyl sulfoxides. It appears that the metalation of propenyl 2-pyridyl sulfoxide is greatly facilitated by the presence of the pyridine nitrogen, which serves as a ligand to stabilize the resulting species via a five-membered chelate. The methylated product, 1-methylpropenyl 2-pyridyl sulfoxide, is isolated in almost quantitative yield.[132]

(E) $C_6H_5S(O)CH{=}CHCH_2OCH_3$ $\xrightarrow{LDA}$ $C_6H_5S(O)C(Li{\cdots}OCH_3)=CHCH_2$ $\rightleftharpoons$ $C_6H_5SCH{=}C{=}CHOCH_3$ (OLi on S) $\rightleftharpoons$ $C_6H_5S(O)C(Li){=}CHCH_2OCH_3$ $\xleftarrow{LDA}$ (Z) $C_6H_5S(O)CH{=}CHCH_2OCH_3$

$\xrightarrow{CH_3I}$ $C_6H_5S(O)C(CH_3){=}CHCH_2OCH_3$ (73%)

[132] H. Okamura, Y. Mitsuhira, M. Miura, and H. Takei, *Chem. Letters*, **1978**, 517.

LDA CH_3I

CH_3 Li CH_3 CH_3 CH_3

(96%)

Thiophenes and Condensed Thiophenes. Thiophene is metalated very readily, and under appropriate conditions even 2,5-dilithiothiophene can be generated.[40] Competitive lithiations between furan and thiophene have revealed[4,41,133] that under thermodynamic conditions the product ratio is 96:4 in favor of thiophene, whereas under kinetic conditions furan is metalated faster[40] (see p. 18). The calculated relative activities for the two heterocycles indicate that the alpha position in thiophene is 500 times more reactive than that in furan.[41] In addition, the greater stability of 2-thienyllithium is explained by *d*-orbital participation.

The lithiation of thiophene and numerous substituted derivatives has been extensively studied and reviewed.[38,40,42,134] From the wealth of data available the following generalizations can be made for the lithiation of thiophene derivatives:

1. Because of the great facility with which thiophene itself can be metalated in its alpha position, 2-substituted derivatives in general give exclusive metalation in the 5 position. Two exceptions are discussed (p. 77).
2. A 3-substituted thiophene whose 3 substituent has any beta-directing ability (either inductive or coordinative) gives rise to exclusive or predominant 2 lithiation.
3. 3-Substituted thiophenes whose substituent has *no* beta-directing capability, such as 3-alkyl or arylthiophenes, are lithiated in both the 2 and predominantly in the 5 positions, depending on the degree of steric hindrance.
4. 3,4-Disubstituted thiophenes are lithiated beta to the stronger directing group.
5. Metalation of 2,5-disubstituted thiophenes is governed by the relative effectiveness of the beta-directing substituents (see p. 43).

The most intriguing aspect of the alpha metalation of thiophenes is the sheer inability of almost any beta-directing group to compete with the

[133] Y. L. Gol'dfarb and Y. L. Danywshevskii, *J. Gen. Chem.* (*USSR*) (*Engl. Transl.*), **31,** 3410 (1961).
[134] D. W. Slocum and P. L. Gierer, *J. Org. Chem.*, **41,** 3668 (1976).

alpha lithiation. Although the sulfonamide, sulfone, and dimethylaminomethyl groups are among the strongest beta directors, when present as 2 substituents in thiophenes, their effect is completely overshadowed by a selective alpha lithiation in the 5 position, as illustrated by the exclusive formation of the products **48**,[135] **49**,[69] and **50**.[134] In the presence of excess *n*-butyllithium, however, 3,5-dilithio species can be generated. For example, the diacid **51** can be isolated after an exhaustive metalation of *t*-butyl 2-thienyl sulfone.[136] In fact, the only beta-directing functionality capable of competing with an alpha activator is the imine moiety inherent in pyridine nuclei[137,138] and oxazolines.[139] These results are discussed in more detail in the competitive metalations section (p. 77).

S $SO_2N(C_2H_5)_2$ $\xrightarrow[CO_2]{n\text{-BuLi}}$ HO_2C S $SO_2N(C_2H_5)_2$

48 (82%)

S $SO_2C_6H_5$ $\xrightarrow[CO_2]{n\text{-BuLi}}$ HO_2C S $SO_2C_6H_5$

49 (42%)

S $CH_2N(CH_3)_2$ $\xrightarrow[D_2O]{n\text{-BuLi}}$ D S $CH_2N(CH_3)_2$

50 (88%)

S $SO_2C_4H_9$-*t* $\xrightarrow[CO_2]{2\ n\text{-BuLi}}$ HO_2C S $SO_2C_4H_9$-*t* (CO_2H)

51 (68%)

The regioselectivity provided by a 3 substituent with beta-directing ability, however, weak it may be, is quite striking. With the exception of 3-(alkylseleno)thiophenes,[140] alpha metalation in the 2 position is the predominant or exclusive process. 3-Alkoxy substituents provide a very high degree of regioselectivity and, in contrast to the carbocyclic aromatic systems, even a bulky *t*-butoxy group gives access to a high yield of the

[135] D. W. Slocum and P. L. Gierer, *J. Org. Chem.*, **38,** 4189 (1973).

[136] F. M. Stoyanovich and B. P. Fedorov, *Khim. Geterotsikl. Soedin.*, **1967,** 823 [*C.A.*, **69,** 77038a (1968)].

[137] T. Kauffmann, E. Wienhofer, and A. Wöltermann, *Angew. Chem., Int. Ed. Engl.*, **10,** 741 (1971).

[138] T. Kauffmann and A. Mitschker, *Tetrahedron Lett.*, **1973,** 4039.

[139] I. Vlattas and L. DellaVecchia, *J. Org. Chem.*, **42,** 2649 (1977).

[140] Y. L. Gol'dfarb, V. P. Litvinov, and A. Sukiasyan, *Izv. Akad. Nauk SSSR, Ser. Khim.*, **1971,** 1296 [*C.A.*, **75,** 140592t (1971)].

2-methylated product **52**.[141] The dimethylaminomethyl group gives regioselective metalation in the 2 position, as evidenced by the isolation of the aldehyde **53**.[43,134] A methoxymethyl group in the 3 position has a similar effect.[43,134,142]

OC_4H_9-*t* ; 1. *n*-BuLi, 2. $(CH_3O)_2SO_2$ → OC_4H_9-*t*, CH_3

52 (87%)

$CH_2N(CH_3)_2$; 1. *n*-BuLi, 2. DMF → $CH_2N(CH_3)_2$, CHO

53 (75%)

Such selective metalations are utilized in the preparation of condensed thiophenes, *e.g.*, the acid **54**. The reported overall yield in this multistep transformation is impressive.[143] 3,3′-Thiodithiophene can be dilithiated in the 2 and 2′ positions to give, after copper-catalyzed coupling, the trithienyl system **55**.[144]

1. *n*-BuLi, 2. "S" → SLi ; 1. $BrCH_2CO_2CH_3$, 2. H_3O^+ → CO_2H

54 (94%)

2 *n*-BuLi → Li Li ; Cu^+ →

55 (52%)

3-Alkyl- or -arylthiophenes give varying amounts of 2- and predominantly 5-metalated products, depending on the steric and/or inductive effects of these groups. These effects are best illustrated by comparing the relative amounts of lithiation in the respective positions, as determined by the isolation of various products.[20,145,146]

[141] H. J. Jakobsen and S. O. Lawesson, *Tetrahedron*, **21,** 3331 (1965).

[142] K. Y. Tserng and L. Bauer, *J. Org. Chem.*, **40,** 172 (1975).

[143] S. Gronowitz and B. Persson, *Acta Chem. Scand.*, **21,** 812 (1967).

[144] F. DeJong and M. J. Janssen, *J. Org. Chem.*, **36,** 1645 (1971).

[145] S. Gronowitz, B. Cederlund, and A. B. Hörnfeldt, *Chem. Scr.*, **5,** 217 (1974) [*C.A.*, **81,** 49013t (1974)].

[146] N. Gjøs and S. Gronowitz, *Ark. Kemi*, **30,** 225 (1968) [*C.A.*, **70,** 78043f (1969)].

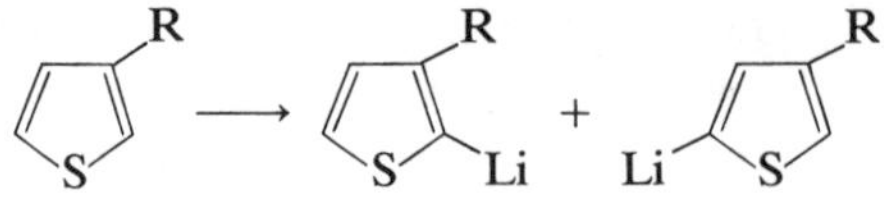

2-Li		5-Li	R	Yield (%)
22	:	78	CH_3	87
—		100	C_4H_9-*t*	72
44	:	56	C_6H_5	72

3,4-Disubstituted thiophenes are lithiated in the alpha position that is beta to the more effective directing group. Thus 3-*t*-butoxy-4-methylthiophene gives exclusively the 2-substituted product **56**.[147] Likewise, the intermediate **57**, derived from 3-*t*-butoxy-4-lithiothiophene and dimethylformamide, is metalated selectively next to the more strongly beta-directing dimethylaminomethyl group leading, after reaction with dimethylformamide, to the dialdehyde **58**,[148]

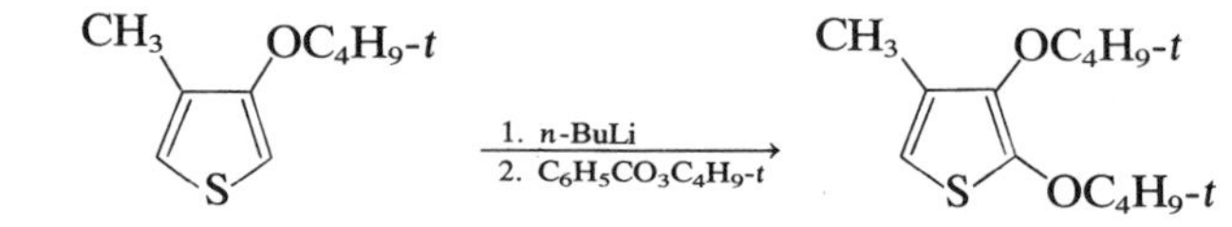

1. *n*-BuLi
2. $C_6H_5CO_3C_4H_9$-*t*

56 (79%)

1. *n*-BuLi
2. DMF

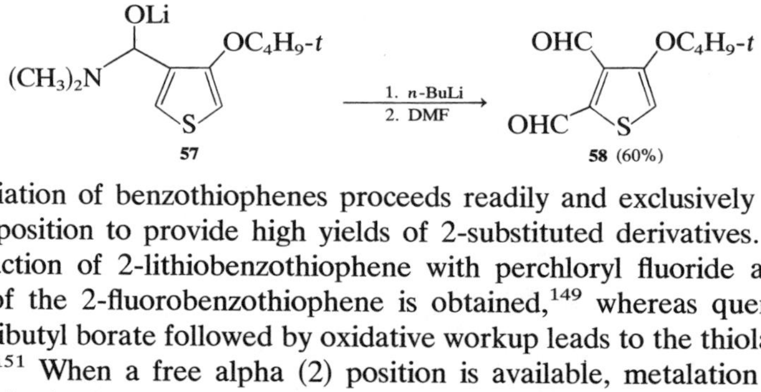

57

58 (60%)

Lithiation of benzothiophenes proceeds readily and exclusively at the alpha position to provide high yields of 2-substituted derivatives. Thus, on reaction of 2-lithiobenzothiophene with perchloryl fluoride a good yield of the 2-fluorobenzothiophene is obtained,[149] whereas quenching with tributyl borate followed by oxidative workup leads to the thiolactone **59**.[150,151] When a free alpha (2) position is available, metalation never occurs in the benzene ring *ortho* to the sulfur atom, as is the case with dibenzothiophene (see p. 69).[1,152] The intermediacy of a 2,7-dilithio species, however, has been observed with excess metalating agent.[40] 3-Substituted benzothiophenes are metalated equally well, as documented by the quantitative deuteration of the amine **60**.[153] In

[147] J. Z. Mortensen, B. Hedegaard, and S. O. Lawesson, *Tetrahedron*, **27**, 3839 (1971).

[148] C. Paulmier, J. Morel, D. Semard, and P. Pastour, *Bull. Soc. Chim. Fr.*, **1973**, 2434.

[149] R. D. Schultz, D. D. Taft, J. P. O'Brien, J. L. Shea, and H. M. Mork, *J. Org. Chem.*, **28**, 1420 (1963).

[150] W. C. Lumma, G. A. Dutra, and C. A. Voeker, *J. Org. Chem.*, **35**, 3442 (1970).

[151] R. P. Dickinson and B. Iddon, *J. Chem. Soc., C*, **1970**, 1926.

[152] H. Gilman and D. L. Esmay, *J. Am. Chem. Soc.*, **76**, 5786 (1954).

[153] T. R. Bosin and R. B. Rogers, *J. Labelled Compd.*, **9**, 395 (1973) [*C.A.*, **80**, 47761r (1974)].

RLi → –Li; $FClO_3$ → –F (70%); $(n\text{-}C_4H_9O)_3B$, H_2O_2 → =O **59** (76%)

$(CH_2)_2N(CH_3)_2$ **60** → n-BuLi, D_2O → $(CH_2)_2N(CH_3)_2$, –D (100%)

contrast to N,N-dimethylphenethylamine (p. 52), no benzylic deprotonation is observed.

Isothiazoles. The activating influence of the two heteroatoms in isothiazoles permits a facile lithiation in the 5 position at very low temperatures, making this reaction preparatively useful, as demonstrated by the formation of isothiazole-5-carboxaldehyde.[154] However, at higher temperatures metalation can be complicated by fragmentations that occur either via nucleophilic attack of the lithiating agent at sulfur[155] or by other mechanisms.[154,155]

1. n-BuLi/−70°, 2. DMF → CHO (75%)

Thiazoles and Benzothiazoles. Thiazoles have two possible sites for metalation, namely, the 2 and the 5 positions. As with imidazoles, metalation in the 2 position is favored for derivatives lacking a 2 substituent, as illustrated for the preparation of thiazole-2-carboxylic acid.[156] 2-Substituted thiazoles such as 2-chlorothiazole,[157] undergo metalation exclusively in the 5-position. If the 2 substituent is a methyl group, however, lithiation in the 5 position, while still the major pathway, is accompanied by a lateral deprotonation of the relatively acidic methyl group.[158,159] This side reaction normally amounts to as much as 10% of the product mixture, as evinced by the two alkylated products **61** and

[154] M. P. L. Caton, D. H. Jones, R. Slack, and K. R. H. Wooldridge, *J. Chem. Soc.*, **1964,** 446.
[155] R. G. Micetich, *Can. J. Chem.*, **48,** 2006 (1970).
[156] J. Metzger and B. Koether, *Bull. Soc. Chim. Fr.*, **1953,** 708.
[157] D. S. Noyce and S. A. Fike, *J. Org. Chem.*, **38,** 3316 (1973).
[158] J. Crousier and J. Metzger, *Bull. Soc. Chim. Fr.*, **1967,** 4134.
[159] G. Knaus and A. I. Meyers, *J. Org. Chem.*, **39,** 1192 (1974).

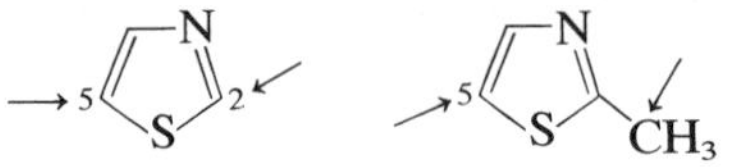

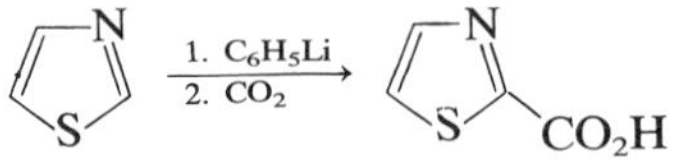

(40%)

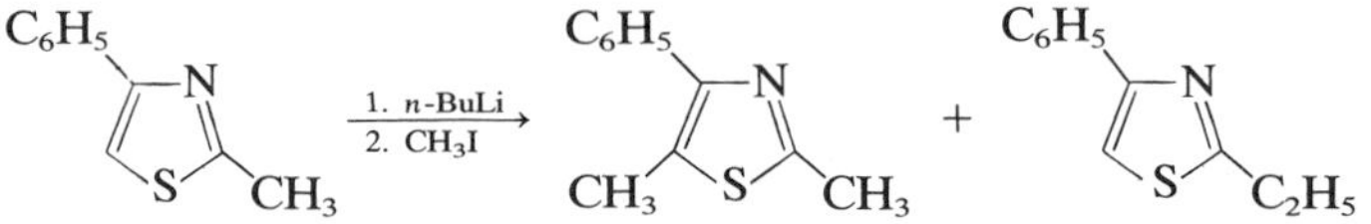

61 (95%) 62 (5%)

62.[159] In comparison to the corresponding imidazoles, this side reaction is less pronounced, as would be expected from the higher alpha-activating ability of the sulfur atom.

Benzothiazole undergoes rapid metalation in the 2 position, as illustrated by the formation of the silane **63.**[160]

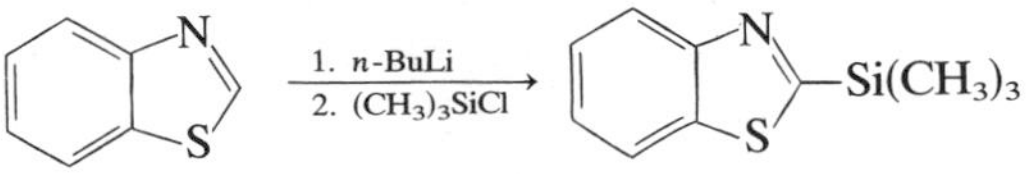

63 (77%)

Selenium and Tellurium as Alpha-Activating Atoms

Vinyl Selenides. In comparison to the vinyl sulfides (p. 32) phenyl vinyl selenide can react with metalating agents via three different pathways. In diethyl ether, *n*-butyllithium adds in a Michael-type fashion (A), whereas in tetrahydrofuran attack is at selenium (B), followed by cleavage of the phenyl-Se bond. It is only through the use of lithium diisopropylamide that alpha lithiation can be achieved (C) as illustrated by the isolation of 1-methyleneundecyl phenyl selenide.[161] Tetraselafulvalene on the other hand, which is a more complex cyclic vinyl selenide, can be alpha metalated with *n*-butyllithium.[162]

Selenophenes and Tellurophenes. Like their oxygen and sulfur analogs, selenophene[163] and tellurophene can be alpha-metalated effectively to give, for example, 2-methyltellurophene in very good yield.[164] In unsymmetrically substituted derivatives the directing influence of an ether

[160] F. H. Pinkerton and S. F. Thames, *J. Heterocycl. Chem.*, **8,** 257 (1971).

[161] M. Sevrin, J. N. Denis, and A. Krief, *Angew. Chem., Int. Ed. Engl.*, **17,** 527 (1978).

[162] D. C. Green, *J. Chem. Soc., Chem. Commun.*, **1977,** 161.

[163] Y. K. Yur'ev and N. K. Sadovaya, *Zh. Obshch. Khim.*, **34,** 1803 (1964) [*C.A.*, **61,** 8258f (1964)].

[164] F. Fringuelli and A. Taticchi, *J. Chem. Soc., Perkin Trans. I*, **1972,** 199.

$n\text{-}C_4H_9CH_2CH(Li)SeC_6H_5$ ⟵("A", n-BuLi) $CH_2{=}CHSeC_6H_5$ ⟶("B", n-BuLi) $CH_2{=}CHSeC_4H_9\text{-}n$ + C_6H_5Li

$CH_2{=}CHSeC_6H_5$ ⟶("C", LDA) $CH_2{=}C(Li)SeC_6H_5$ ⟶($n\text{-}C_{10}H_{21}Br$) $CH_2{=}C(SeC_6H_5)C_{10}H_{21}\text{-}n$ (70%)

functionality provides for regioselectivity in very much the same way as for the thiophenes, as illustrated by the formation of 3-methoxyselenophene-2-carboxaldehyde.[165]

Tellurophene ⟶(1. n-BuLi; 2. $(CH_3O)_2SO_2$) 2-methyltellurophene (75%)

3-Methoxyselenophene ⟶(1. n-BuLi; 2. DMF) 3-methoxyselenophene-2-carboxaldehyde (50%)

Halogens as Alpha-Activating Atoms

Haloalkenes. The alpha metalation of haloalkenes, or chloroalkenes in particular, has been extensively studied and reviewed.[166] Unlike all other alpha-lithiated species reported thus far, α-lithio haloalkenes can add to electrophilic substrates (E) as well as undergo an alpha elimination of lithium halide to form carbenoid species (Eq. 16). The stability of

$(CH_3)_2C{=}CHCl$ ⟶(RLi) $(CH_3)_2C{=}C(Cl)Li$ ⟶(E) $(CH_3)_2C{=}C(Cl)E$; ⟶(−LiCl) $(CH_3)_2C{=}C{:}$ (Eq. 16)

α-lithio haloalkenes is generally lower than that of other alpha-metalated species, and their generation must often be carried out at temperatures

[165] G. Henrio, J. Morel, and P. Pastour, *Bull. Soc. Chim. Fr.*, **1976**, 265.

[166] V. G. Köbrich, A. Akhtar, F. Ansari, W. E. Breckoff, H. Büttner, W. Drischel, R. H. Fischer, K. Flory, H. Fröhlich, W. Goyert, H. Heinemann, I. Hornke, H. R. Merkle, H. Trapp, and W. Zündorf, *Angew. Chem., Int. Ed. Engl.*, **6**, 41 (1967).

below −100°. Vinyl chloride, for instance, can be metalated and carboxylated quantitatively at −115° to produce α-chloroacrylic acid.[167,168] The presence of one or two phenyl substituents tends to increase the stability of the alpha-lithiated species. Thus β-chlorostyrene can be metalated at −80°, and carboxylation gives a high yield of (E)-2-chlorocinnamic acid.[169] Even more stable is the trisubstituted species **64**, which can be prepared at −43°.[170] Carboxylation at that temperature leads to formation of the acid **66**. At higher temperatures, such as −12°, the lithio derivative **65** undergoes a Fritsch–Buttenberg–Wiechell rearrangement which is quite general for various *gem*-diaryl derivatives of this type.[166,170,171]

$C_6H_5CH{=}CHCl$ —(1. *n*-BuLi/−80°; 2. CO_2)→ $C_6H_5CH{=}C(Cl)CO_2H$ (90%)

64 $Ar_2C{=}CHCl$ —(*n*-BuLi, −43°)→ **65** $Ar_2C{=}C(Cl)Li$ —(CO_2)→ **66** $Ar_2C{=}C(Cl)CO_2H$ (88%); —(−12°)→ Ar—C≡C—Ar (61%)

Ar = p-ClC_6H_4

As with other appropriately 1,2-disubstituted olefins, beta elimination can also be a problem. For example, lithiation of (Z)-1,2-dichloroethylene followed by carboxylation leads via elimination of lithium chloride to chloropropiolic acid. It should be noted, however, that the (E) isomer gives the normal product of alpha lithiation, namely (E)-2,3-dichloroacrylic acid, in essentially quantitative yield.[167,168]

$ClCH{=}CHCl$ —(1. *n*-BuLi/−100°; 2. CO_2)→ $ClCH{=}C(Cl)CO_2H$ (99%)

The alpha metalation of fluoroalkenes[172,173] and bromoalkenes,[174a,b] although feasible, has received only scant attention.

[167] G. Köbrich and K. Flory, *Chem. Ber.*, **99,** 1773 (1966).

[168] G. Köbrich and K. Flory, *Tetrahedron Lett.*, **1964,** 1137.

[169] M. Schlosser and V. Ladenberger, *Chem. Ber.*, **100,** 3893 (1967).

[170] G. Köbrich, H. Trapp, and I. Hornke, *Tetrahedron Lett.*, **1964,** 1131.

[171] G. Köbrich and H. Trapp, *Chem. Ber.*, **99,** 680 (1966).

[172] F. G. Drakesmith, R. D. Richardson, O. J. Stewart, and P. Tarrant, *J. Org. Chem.*, **33,** 286 (1968).

[173] F. G. Drakesmith, O. J. Stewart, and P. Tarrant, *J. Org. Chem.*, **33,** 472 (1968).

[174a] J. Ficini and J. C. Depezay, *Tetrahedron Lett.*, **1968,** 937.

[174b] K. S. Y. Lau and M. Schlosser, *J. Org. Chem.*, **43,** 1595 (1978).

Beta (*Ortho*) Lithiation

As defined in the introduction, beta (*ortho*) lithiation consists in the replacement of a C_{sp^2}-bound hydrogen atom by lithium at the position beta to a functional group with nonbonding electrons. The designation "ortho lithiation," which is a type of beta lithiation, is used in this chapter exclusively to denote the deprotonation of the position adjacent to the directing atom or functional group attached to carbocyclic aromatic systems. The reaction itself is characterized by its high degree of regioselectivity, as with very few exceptions no other carbon atoms are deprotonated. The reaction is shown in Eqs. 17 and 18, where X represents a heteroatom and n may vary from 0 to 2. Metalations of this type are observed with a wide variety of groups in which X may be nitrogen, oxygen, sulfur, halogen, selenium, or phosphorus.

$(CH_2)_n$ X, H $\xrightarrow{RLi}$ $(CH_2)_n$ X, Li + RH beta lithiation (Eq. 17)

$(CH_2)_n$ X, H $\xrightarrow{RLi}$ $(CH_2)_n$ X, Li + RH *ortho* lithiation (Eq. 18)

Beta (*Ortho*)-Directing Groups and Their Relative Directing Ability

An attempt to classify all directing groups based on their inductive electron-withdrawing or electron-donating and coordinative properties leads to the following categories:

Coordination Only. Heteroatoms that are attached to a π system (carbocyclic aromatic, heteroaromatic, olefinic) by one or at the most two saturated carbon atoms can lead to a beta metalation only via the coordinative mechanism if the effect of the saturated carbon(s) is considered neutral. The most effective groups are basic amines, *e.g.*, $\geq CNR_1R_2$, $—CR_1(O^-)NR_2R_3$, and $\geq CCH_2NR_1R_2$. Considerably less efficient as *ortho* directors are alkoxides, ethers, acetals, and ketals, *e.g.*, $\geq CO^-$, $\geq COR$, and $>C(OR)_2$. The latter groups, however, provide excellent beta-directing effects in the lithiation of more readily metalated systems, such as π-excessive, five-membered heterocycles.

Electron-Withdrawing Groups. Electron-withdrawing groups with little or no coordinative capacity are effective beta directors because of their ability to markedly increase the hydrocarbon acidity of adjacent positions. Groups that have shown beta-directing properties include $\rangle$CO, CF_3, F, Cl, Br, and I.

Electron-Withdrawing Groups with Moderate Coordination Potential. A common feature of these functionalities is a moderate capacity for coordination with the lithiating agent. Their inherent effect on the beta position is electron withdrawing by induction. On coordination with the metalating agent this effect is enhanced. Such groups are, in order of decreasing effect: SO_2-aryl, $SO_2NR_1R_2 > OR > SR > SeR > NR_1R_2$.

Electron-Withdrawing Groups with Pronounced Coordination Potential. These functional groups exhibit the ideal features for facile beta metalations, *i.e.*, a high capacity for effective coordination with the lithiating agent via the nonbonding pair of electrons of their sp^2-hybridized nitrogen atoms, and the ability to inductively acidify adjacent hydrogens through their electron-withdrawing properties. Some of these functionalities are —$CONR_1R_2$, 2-oxazolines, 2-pyridines, and $\rangle C{=}NNR_1R_2$.

Each individual group has its own distinct beta-directing ability. Thus the establishment of a ranking system is not only of theoretical but certainly of considerable practical interest for π systems bearing more than one directing group. Various research groups have addressed themselves to this problem.[1,2,19,23,24] Their results, combined with additional inferential interpretations, permit the following generalizations:

1. Under kinetic lithiation conditions, the strongest beta-directing groups combine both an electron-withdrawing effect *and* the properties of a good ligand.
2. Among the groups in which the directing heteroatom is separated from the π system by a saturated carbon atom, thus providing no electron-withdrawing effect, a basic nitrogen atom is the most powerful director.
3. In the presence of non-Lewis-acid metalating agents, the rank order of the directing groups is determined largely by their inductive or acidifying effect.

In the lithiation of carbocyclic aromatic systems with coordinatively unsaturated metalating agents, the following order of beta-directing potential can be established. The list includes only those groups that either have been studied in competitive experiments or for which assumptions can be made based on the wealth of data available:

$SO_2NR_1R_2$,* SO_2-aryl[23] > 2-oxazolines > CONHR, CSNHR[2,24] > $CH_2N(CH_3)_2$
> $CR(O^-)CH_2N(CH_3)_2$[175] > OCH_3 > O-aryl > NH-aryl[176]
> S-aryl[177,178] > NR-aryl > $N(CH_3)_2$ > $CR_1R_2O^-$

(The lack of comparative data precludes an exact ranking of the halogens.)

Nitrogen as a Beta (*Ortho*)-Directing Atom

Mono-, Di-, and Triarylamines. The nitrogen atoms in N,N-disubstituted anilines rate among the poorest *ortho*-directing groups. This observation may be attributed to the fact that their nitrogen lone pairs are not readily available for coordination with the lithiating agent because they are strongly engaged in the resonance of the π system to which they are attached. N,N-Dimethylaniline is lithiated in refluxing hexane in 55% yield, as shown by the isolation of an *ortho*-substituted product.[179,180] The inferiority of a tertiary aromatic amine as a directing group is clearly exemplified by the metalation of N,N-dimethyl-*p*-anisidine (**67**), which is lithiated exclusively (85%) *ortho* to the methoxyl group.[19] The observation that the *meta* isomer is lithiated in the position *ortho* to both directing groups (**68**, 80%) underscores the fact that the anilino nitrogen nevertheless possesses a distinct, albeit weak, *ortho*-directing ability.[19] The use of *n*-butyllithium/TMEDA greatly facilitates the otherwise sluggish metalation of anilines, as evinced by the lithiation of N,N-dimethyl-*p*-toluidine (80%).[63] The predominant deprotonation of the methyl group in N,N-dimethyl-*o*-toluidine is characteristic of this general alternative pathway (see p. 16).[63]

$N(CH_3)_2$ Li OCH_3

67

$N(CH_3)_2$ Li OCH_3

68

No successful *ortho* lithiations of N-alkyl or N-unsubstituted anilines have been reported. One way to achieve the latter goal (see p. 48) is

[175] D. W. Slocum and W. Ackermann, *Chem. Commun.*, **1974,** 968.
[176] D. A. Shirley and J. C. Liu, *J. Org. Chem.*, **25,** 1189 (1960).
[177] G. Cauquil, A. Casadevall, and E. Casadevall, *Bull. Soc. Chim. Fr.*, **1960,** 1049.
[178] R. D. Nelson, *Iowa State Coll. J. Sci.*, **27,** 229 (1953) [*C.A.*, **48,** 2069e (1954)].
[179] A. R. Lepley, W. A. Khan, A. B. Giumanini, and A. G. Giumanini, *J. Org. Chem.*, **31,** 2047 (1966).
[180] D. W. Slocum, G. Bock, and C. A. Jennings, *Tetrahedron Lett.*, **1970,** 3443.
*See Note on page 105.

exemplified by the dilithiation of pivalanilide where the oxygen or nitrogen atom of the monodeprotonated species can serve as a ligand for the metalating agent. The method is general and the reaction of the dilithio species proceeds in very good yield, as illustrated by the isolation of *o*-(methylthio)pivalanilide.[181a]

$NHCOC_4H_9$-*t* —(2 *n*-BuLi)→ C_4H_9-*t*, N, O, Li, Li —($(CH_3S)_2$)→ $NHCOC_4H_9$-*t*, SCH_3 (78%)

An example of the lithiation of a diarylamine is that of 2-anilinonaphthalene, which occurs predominantly in the 3 position of the naphthalene nucleus.[181b] Conversely, 1-anilinonaphthalene is metalated almost exclusively in the 8 position to give, after treatment with carbon dioxide, good yields of the products **69** and **70**.[181b] Thus, when there is a choice between a four- or five-membered cyclometalated intermediate, the latter is clearly favored, and lithiation takes place in a position formally gamma rather than beta with respect to the nitrogen ligand.

HNC_6H_5 —(1. *n*-BuLi; 2. CO_2)→ HO_2C HNC_6H_5 **69** (56%) + O, N, C_6H_5 **70** (25%)

13, H N, 1, S, 6 **71** —(1. *n*-BuLi; 2. CO_2)→ HO_2C, H N, S **72** (80%)

In the more complex system **71,** where there is a choice of three metalatable sites, namely positions 1, 6, and 13, lithiation at carbon 1 via a five-membered cyclometalated species is again clearly the preferred pathway. Carboxylation produces the acid **72** in excellent yield.[182]

Another interesting observation was made with the phenothiazines.

[181a] W. Fuhrer and H. W. Gschwend, *J. Org. Chem.*, **44,** 1133 (1979).

[181b] N. S. Narasimhan and A. C. Ranade, *Indian J. Chem.*, **7,** 538 (1969).

[182] D. A. Shirley and J. C. Gilmer, *J. Org. Chem.*, **27,** 4421 (1962).

N-Unsubstituted derivatives are lithiated more readily and regioselectively than are the N-alkylated homologs. Lithiation of N-methylphenothiazine leads to a mixture of about equal amounts of products **73** and **74**,[177,183a] indicating the approximate equivalence of sulfur and nitrogen as *ortho*-directing atoms. However, phenothiazine itself is lithiated exclusively *ortho* to nitrogen to give **76**.[178] If it is indeed the relative unavailability of the nitrogen lone pair for chelation with the lithiating agent that makes the tertiary nitrogen of N-methylphenothiazine a poorer *ortho* director, one rationale is the following: As the formal negative charge of the monoanion generated by N-deprotonation of phenothiazine is delocalized over the entire π-system (**75**), the nonbonding pair of electrons of the nitrogen atom, now coplanar with the tricyclic system, becomes available for chelation with the lithiating agent. This phenomenon is not limited to phenothiazines, but has also been observed with benzo- and dibenzophenothiazines where the exclusive site of metalation is *ortho* to the (NH) group.[176,182,183b]

S N CH_3 —RLi / CO_2→ S N CH_3 CO_2H + CO_2H S N CH_3

73 (14%) **74** (16%)

S N H —2 *n*-BuLi→ ⊖ N ---Li —$(C_6H_5)_2CO$→ S N H $C(C_6H_5)_2OH$

75 **76** (70%)

The unusual *meta* lithiation of triphenylamine was explained in terms of steric hindrance by Gilman and Morton[1] in the latest review on this subject. An alternate rationalization may also be the unavailability of the nitrogen lone pair for coordination. A reexamination of these results confirms the preponderance of a rather sluggish *meta* metalation.[23]

α-Lithio(N-alkylidene)arylamines (Aryl Isocyanides). As discussed previously (p. 20), the activating influence of the isocyanide group promotes alpha lithiation at very low temperatures, with little or no attendant addition of the lithiating agent to the electrophilic isocyanide carbon. In contrast, the isocyanide group is only a very poor *ortho* director, and therefore the initial step in the attempted *ortho* lithiation of phenyl isocyanide with *t*-butyllithium is addition of the metalating agent

[183a] G. Cauquil, M. A. Casadevall, and E. Casadevall, *C. R. Acad. Sci. Paris, Ser. C*, **243**, 590 (1956).
[183b] D. A. Shirley and W. E. Tatum, *J. Org. Chem.*, **25**, 2238 (1960).

to form α-lithio(N-neopentylidene)aniline. The latter in turn can now be metalated in its *ortho* position by a second equivalent of *t*-butyllithium to form the dilithio species, α,2-dilithio(N-neopentylidene)aniline.[184] Formally, therefore, one is not dealing with an *ortho* lithiation of an arylisocyanide but rather with an *ortho* metalation of an α-lithio(N-alkylidene)arylamine. It should be noted that the metalation conditions (*t*-butyllithium/TMEDA) are critical, and in the absence of the chelating amine no ring deprotonation occurs.[184] This finding is indicative of the weak *ortho*-directing influence of the intermediate metallo aldimine group. Nevertheless, under the conditions indicated generation of the dilithio species proceeds in high yield, as illustrated by the reaction products with methyl iodide and dimethylgermanium dichloride.[184]

NC; *t*-BuLi, −78° → N=C(C_4H_9-*t*)Li; *t*-BuLi, TMEDA → N=C(C_4H_9-*t*)Li, Li; CH_3I → N=C(C_4H_9-*t*)CH_3, CH_3 (92%)

$(CH_3)_2GeCl_2$ → N, C_4H_9-*t*, Ge, CH_3 CH_3 (68%)

From a synthetic point of view the reaction permits in effect an *ortho* lithiation of a primary aromatic amine since the initial products (ketimines) can readily be hydrolyzed to the respective anilines. It should be pointed out that generally both carbon–lithium bonds in the dilithio species react with the electrophile, which may prove to be a limitation in certain instances.

Aralkylamines. The dialkylaminomethyl group is the most powerful non-electron-withdrawing *ortho* director (see p. 43). This group has served to metalate variously substituted benzene rings,[30,33,45] naphthalenes,[185] thiophenes,[134] isoxazoles,[186] and vinyl sulfides[23] (Eq. 19). Although the nitrogen substituents may be varied, the dimethylamino

[184] H. M. Walborsky and P. Ronman, *J. Org. Chem.*, **43,** 731 (1978).

[185] Z. Horii, Y. Matsumoto, and T. Momose, *Chem. Pharm. Bull.* (*Tokyo*), **19,** 1245 (1971).

[186] J. Garner, G. A. Howarth, W. Hoyle, S. M. Roberts, and H. Suschitzky, *J. Chem. Soc.*, *Perkin Trans. I*, **1976,** 994.

function has received the broadest attention. Nevertheless, other cases have been studied, e.g., benzylmorpholine.[187]

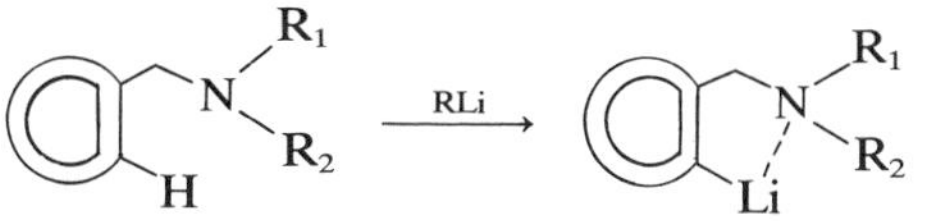

(Eq. 19)

Even the metalation of secondary amines is feasible, proceeding to give good yields of products. The use of *n*-butyllithium/TMEDA, however, is essential to attain appreciable amounts of the required dilithio species **77a,b**.[188] In N-benzylaniline the superiority of the $CH_2NR_1R_2$ group over the anilino function with respect to their *ortho*-directing abilities (see p. 45) is documented by the isolation of the alcohol **78a**.[188] The benzylic position may be fully substituted without affecting the degree of *ortho* lithiation,[33] or it may even be joined with the nitrogen atom to form a ring, as in **79**.[24] The excellent yield of **80** formed after reaction of the lithiated species with methyl iodide clearly indicates the broad applicability of this reaction.

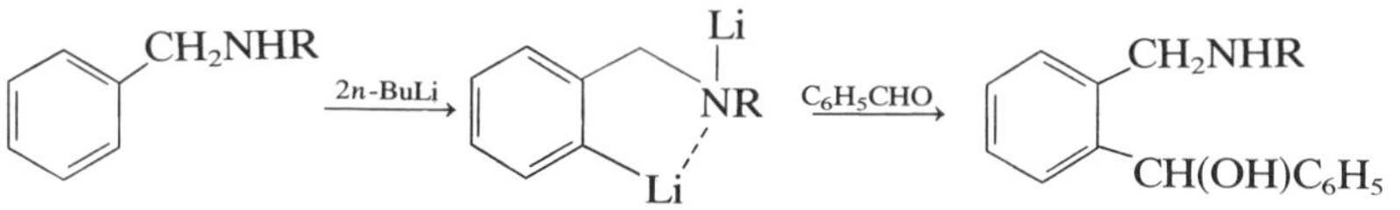

77 a $R = C_6H_5$ b $R = CH_3$

78 a (75%) b (63%)

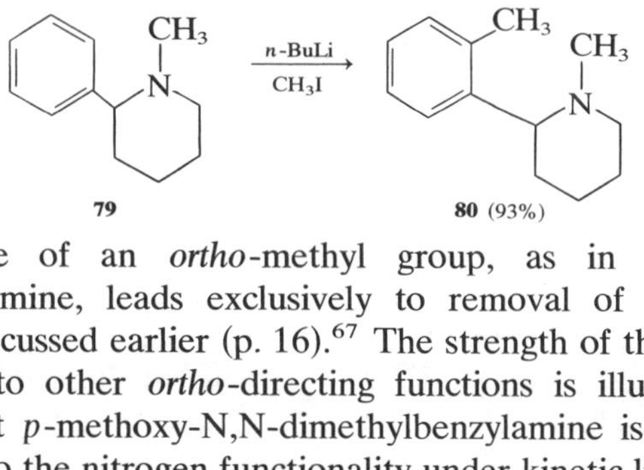

79

80 (93%)

The presence of an *ortho*-methyl group, as in *o*-methyl-N,N-dimethylbenzylamine, leads exclusively to removal of the sp^3-bound hydrogen, as discussed earlier (p. 16).[67] The strength of the benzylamino group relative to other *ortho*-directing functions is illustrated by the observation that *p*-methoxy-N,N-dimethylbenzylamine is metalated exclusively *ortho* to the nitrogen functionality under kinetic lithiation conditions to give the alcohol **81** after quenching with benzophenone.[19,45,180] An interesting reversal of this result can be achieved by lithiating with the monomeric *n*-butyllithium/TMEDA complex. In this case the major (thermodynamic) product **82** is derived from lithiation *ortho* to the methoxyl group, with only a minor amount (7%) of **81** being formed.[19]

[187] R. Coombs, W. J. Houlihan, J. Nadelson, and E. I. Takesue, *J. Med. Chem.*, **14**, 1072 (1971).
[188] R. E. Ludt and C. R. Hauser, *J. Org. Chem.*, **36**, 1607 (1971).

By analogy with the anisidines, *m*-methoxy-N,N-dimethylbenzylamine is lithiated exclusively in the position *ortho* to both directing groups.[19,45,189] Further insight into the chelating ability of nitrogen and oxygen with organolithium compounds is provided by *o*-methoxy-N,N-dimethylbenzylamine. The predominant lithiation *ortho* to the methoxyl group (58% yield of **84**) is postulated to occur because the substrate serves as a bidentate ligand to convert, as does TMEDA, the tetrameric *n*-butyllithium (in ether) into the monomeric species **83**. This in turn would then serve as the actual lithiating agent and thus, like the *n*-butyllithium/TMEDA complex, preferentially lead to metalation at the most acidic position, that is, *ortho* to the methoxyl group.[19] Alternatively, of course, it could be postulated that the nitrogen lone pair in **83** is no longer available for coordination.

$CH_2N(CH_3)_2$ / CH_3O / $C(C_6H_5)_2OH$ ← *n*-BuLi — $CH_2N(CH_3)_2$ / CH_3O — *n*-BuLi/TMEDA, $(C_6H_5)_2CO$ →

81 (80%)

$CH_2N(CH_3)_2$ / CH_3O / $C(C_6H_5)_2OH$

82 (55%)

$CH_2N(CH_3)_2$ / OCH_3 — *n*-BuLi → $N(CH_3)_2$ / Li—Bu-*n* / O / CH_3 / $O(C_2H_5)_2$ → $CH_2N(CH_3)_2$ / OCH_3 / Li

83 **84**

Benzylamines with halogen substituents (F, Cl, CF_3) in an *ortho* or *para* position are lithiated *ortho* to the nitrogen director to give, *e.g.*, the metalated species **85** and **86**.[23,24] For the *meta*-trifluoromethyl derivative **87** lithiation occurs as expected in the position *ortho* to both substituents (**87**).[45] *meta*-Fluoro and *meta*-chloro derivatives lead to benzyne formation, and are therefore not suitable as substrates under the conditions reported.[45] However, at lower temperatures successful lithiations may be anticipated.

The facile alpha deprotonation of the thiophene nucleus precludes a successful competition of the dimethylaminomethyl group. However, by blocking the 5 position, as in **88**, lithiation proceeds normally beta to the

[189] N. S. Narasimhan and B. H. Bhide, *Tetrahedron Lett.*, **1968**, 4159.

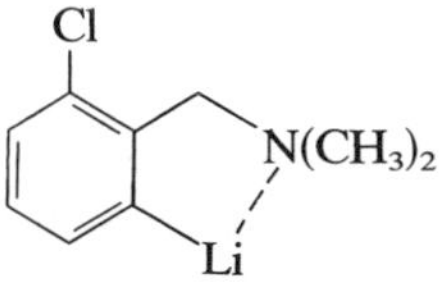

Cl N(CH$_3$)$_2$ Li **85**   Cl N(CH$_3$)$_2$ Li **86**   N(CH$_3$)$_2$ Li CF$_3$ **87**

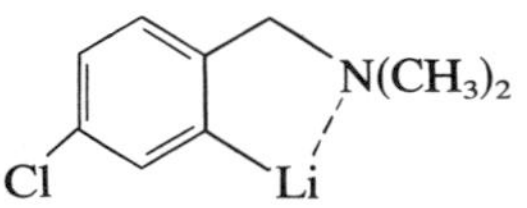

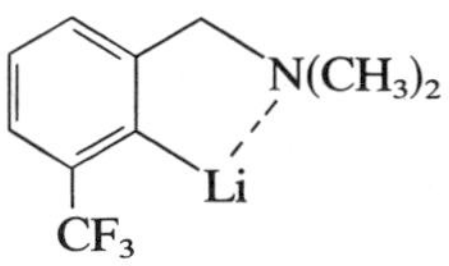

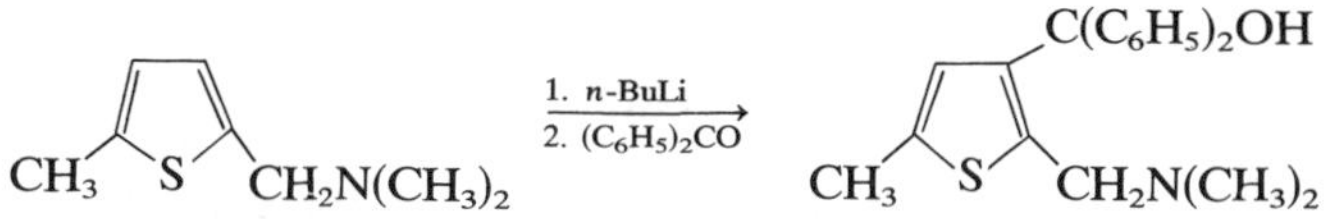

CH$_3$ S CH$_2$N(CH$_3$)$_2$ **88** $\xrightarrow[\text{2. } (C_6H_5)_2CO]{\text{1. } n\text{-BuLi}}$ C(C$_6$H$_5$)$_2$OH CH$_3$ S CH$_2$N(CH$_3$)$_2$ **89** (65%)

nitrogen function, *i.e.*, in the 3 position, to give the carbinol **89** in good yield.[43,134]

In the 1-substituted naphthalene **90** lithiation occurs with surprising selectivity in the 8 position[185,190] to give the alcohol **91** with only a minor amount of the 2-substituted isomer.[191] This finding parallels the results with 1-alkoxynaphthalenes[192] (p. 63) and 1-anilinonaphthalenes[181] (p. 46). In the 2-substituted naphthalene **92** metalation occurs both in the 1 and 3 positions, as reflected by the isolation of approximately equal amounts of products **93** and **94**.[185,191]

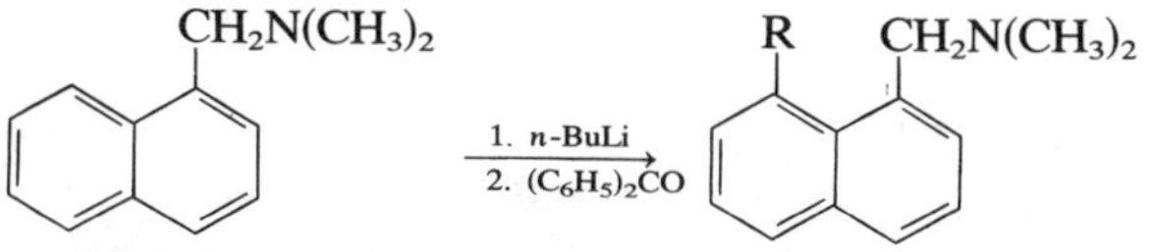

CH$_2$N(CH$_3$)$_2$ **90** $\xrightarrow[\text{2. } (C_6H_5)_2CO]{\text{1. } n\text{-BuLi}}$ R CH$_2$N(CH$_3$)$_2$ **91** (73%)   R = C(C$_6$H$_5$)$_2$OH

CH$_2$N(CH$_3$)$_2$ **92** $\xrightarrow[\text{2. } (C_6H_5)_2CO]{\text{1. } n\text{-BuLi}}$ R CH$_2$N(CH$_3$)$_2$ **93** (34%) + CH$_2$N(CH$_3$)$_2$ R **94** (45%)

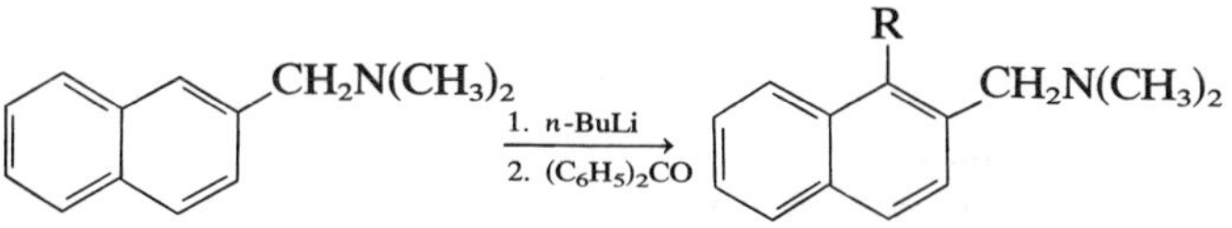

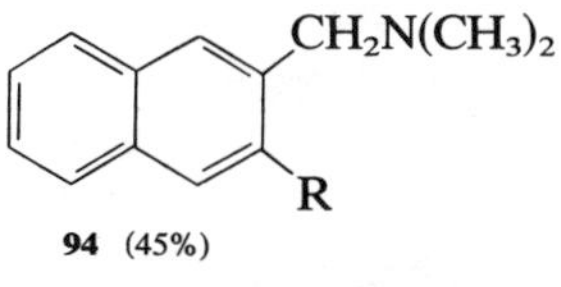

[190] G. Van Koten, A. J. Leusink, and J. G. Noltes, *J. Organomet. Chem.*, **84,** 117 (1975).
[191] R. L. Gay and C. R. Hauser, *J. Am. Chem. Soc.*, **89,** 2297 (1967).
[192] D. A. Shirley and C. F. Cheng, *J. Organomet. Chem.*, **20,** 251 (1969).

Arylethylamines. The phenethylamine **95a** can be lithiated in minimal yield in the *ortho* position,[193,194] although there are conflicting reports.[195] The predominant pathway is benzylic deprotonation followed by beta elimination. By blocking the benzylic position, as in **95b**, *ortho* metalation can be achieved in poor yield.[195] A more interesting and potentially rather useful way to circumvent the elimination problem is illustrated by the lithiation of the phenethanolamine **96**, in which the alkoxide prevents benzylic deprotonation. Thus *ortho*-substituted derivatives such as the silane **97**, can be isolated in moderate to good yields.[175]

$C_6H_5CR_2CH_2N(CH_3)_2$ ⟶ (ortho-Li, CR_2, $N(CH_3)_2$ chelate)

95 **a** R=H
b R=CH_3

$C_6H_5CH(OH)CH_2N(CH_3)_2$ $\xrightarrow{n\text{-BuLi}}$ (OLi, ortho-Li, $N(CH_3)_2$ chelate) $\xrightarrow{(CH_3)_3SiCl}$ $o\text{-}(CH_3)_3SiC_6H_4CH(OH)CH_2N(CH_3)_2$

96

97 (61%)

Moving the basic nitrogen still further from the aromatic nucleus as in N,N-dimethyl-3-phenylpropylamine prevents *ortho* metalation, and benzylic deprotonation becomes the exclusive pathway of the reaction.[19,195] This is not unexpected in view of the fact that the lithium at the benzylic position can form the favored five-membered ring through chelation with the basic nitrogen.

α-Alkoxidobenzylamines. α-Alkoxidobenzylamines of type **98,** obtained by the addition of an organolithium reagent to a tertiary amide, can be successfully used as *ortho*-directing groups. The sequence, as outlined in Eq. 20, permits a one-pot transformation of an organolithium reagent and a tertiary carboxamide into an *ortho*-substituted aryl ketone **100**. The basic nitrogen in the tetrahedral intermediate **98** is assumed to be responsible for directing the subsequent lithiation into the *ortho* position to produce the dilithiated species **99**. The process is another

[193] N. S. Narasimhan and A. C. Ranade, *Tetrahedron Lett.*, **1966,** 603.
[194] D. W. Slocum, T. R. Engelmann, and C. A. Jennings, *Aust. J. Chem.*, **21,** 2319 (1968).
[195] R. L. Vaulx, F. N. Jones, and C. R. Hauser, *J. Org. Chem.* **30,** 58 (1965).

example of *ortho* metalation of an α-*gem*-disubstituted benzylamine. The versatility of the reaction is indicated by the transformation of *p*-chloro-N,N-dimethylbenzamide into 4-chloro-2-methylacetophenone and to the benzophenone **101**.[196] The latter example is quite interesting in that there are two possible sites for *ortho* metalation, namely, in the chlorinated and in the unsubstituted phenyl rings. The known rate-enhancing effect of the *para*-chloro substituent (p. 14) provides good regioselectivity, and no detectable lithiation is observed in the unsubstituted ring.[196]

C_6H_5Li + $RCON(CH_3)_2$ or $C_6H_5CON(CH_3)_2$ + RLi → LiO(R)C[$N(CH_3)_2$]–C_6H_5 (**98**) $\xrightarrow{n\text{-BuLi}}$ **99** (ortho-Li, coordinated to N) $\xrightarrow{E}$ 2-E-C_6H_4COR (**100**) (Eq. 20)

4-Cl-$C_6H_4CON(CH_3)_2$ $\xrightarrow{1.\ CH_3Li;\ n\text{-BuLi}\quad 2.\ CH_3I}$ 4-chloro-2-methylacetophenone (81%)

4-Cl-$C_6H_4CON(CH_3)_2$ $\xrightarrow{1.\ C_6H_5Li;\ n\text{-BuLi}\quad 2.\ (CH_3S)_2}$ 4-chloro-2-(methylthio)benzophenone, **101** (60%)

Lithiation of the selenophene intermediate **102** exemplifies the superiority of an α-alkoxidoamine over a *t*-butoxy group (under kinetic conditions) in terms of their beta-directing ability, as illustrated by the formation of 4-*t*-butoxyselenophene-2,3-dicarboxaldehyde.[148]

4-*t*-C_4H_9O-3-[$CH(OLi)N(CH_3)_2$]selenophene (**102**) $\xrightarrow{1.\ n\text{-BuLi}\quad 2.\ DMF}$ 4-*t*-C_4H_9O-selenophene-2,3-(CHO)$_2$ (50%)

[196] L. Barsky, H. W. Gschwend, J. McKenna, and H. R. Rodriguez, *J. Org. Chem.*, **41,** 3651 (1976).

Arylcarboxamides and Thioamides. Deprotonated secondary carboxamides unlike their carboxylate analogs,[197] are essentially inert toward nucleophilic attack by organolithium reagents.[54] In addition, they possess not only electron-withdrawing, but also chelating characteristics. The combination of these properties has led to the development of the selective *ortho* functionalization of arylcarboxamides by lithiation (Eq. 21).

CONHR, H —R'Li→ OLi, NR, H —R'Li→ OLi, NR, Li (Eq. 21)

Primary benzamides fail to undergo *ortho* lithiation even with an excess of *n*-butyllithium.[54] Certain tertiary amides, however, can successfully be metalated, particularly if the metalation is facilitated by additional factors, such as the influence of an alpha activator, as in N,N-diethyl-3-thiophenecarboxamide[134] or β-aminoacrylamides,[57a] or by the careful selection of conditions.[55,56] For instance, N,N-diethylbenzamide can be lithiated with *s*-butyllithium/TMEDA at low temperatures and subsequently treated with various electrophiles. Reaction with methyl iodide, for example, gives a good yield of N,N-diethyl-*o*-toluamide.[56] In general, however, tertiary benzamides suffer nucleophilic attack by the lithiating agent[54,196] (see p. 53) for the practical use made of this fact), and it is the secondary arylcarboxamides that are synthetically more useful. The nature of the nitrogen substituent does not appear to influence the outcome of the lithiation to any noticeable extent, and both alkyl or aryl groups are feasible. The use of the *t*-butyl group permits a later transformation into primary amides or nitriles.[24] There are few substituents (see p. 16) that affect this rather general reaction in a negative manner, and the directing potency of the carboxamide group virtually assures regioselective metalation in its *ortho* position. The reactions generally proceed in good yield, as illustrated by the lithiation of N-methylbenzamide and the isolation of the carbinol **103**.[54,198]

$C_6H_5CON(C_2H_5)_2$ —1. *s*-BuLi/TMEDA/−78°; 2. CH_3I→ o-$CH_3C_6H_4CON(C_2H_5)_2$ (75%)

$C_6H_5CONHCH_3$ —1. 2 *n*-BuLi; 2. $(C_6H_5)_2CO$→ o-$[C(C_6H_5)_2OH]C_6H_4CONHCH_3$

103 (81%)

[197] M. J. Jorgenson, *Org. Reactions*, **18,** 1 (1970).

[198] C. L. Mao, I. T. Barnish, and C. R. Hauser, *J. Heterocycl. Chem.*, **6,** 475 (1969).

It should be noted that certain primary products are relatively unstable, particularly those derived from addition of the lithiated amide to aldehydes, ketones, nitriles, and epoxides. The tendency of these adducts to form cyclic products is rather high, and only under carefully controlled conditions can the intermediates be isolated.[199] This may in fact be one of the reasons for the relatively modest yields often cited for isolation of primary products. Alternatively, if the resulting heterocycles are desired, the method is well suited for obtaining bicyclic products, such as phthalides **104**,[200] phthalimidines **105**,[24] **106**,[201] or isocoumarins **107**.[202] The first (**104**) and last examples (**107**) also illustrate the superiority of the carboxamide function over a fluorine and a methoxyl group as an *ortho* director. Numerous additional studies have confirmed this fact.[19]

$CONHC_6H_5$, F; 1. *n*-BuLi 2. N-Methylpiperidone → O, O, F, N, CH_3

104 (45%)

$CONHC_4H_9$-*t*; 1. *n*-BuLi 2. DMF → O, NC_4H_9-*t*, OH

105 (86%)

$CONHCH_3$; 1. *n*-BuLi 2. C_6H_5CN → O, NCH_3, NH_2, C_6H_5

106 (62%)

OCH_3, CH_3O, $CONHCH_3$, CH_3O; 1. *n*-BuLi 2. Propylene oxide → CH_3O, O, CH_3O, O, CH_3O, CH_3

107 (35%)

[199] Sandoz-Wander Inc., U.S. Pat. 3,878,215 (1975) [*C.A.*, **83**, 79099j (1975)].

[200] A. Marxer, H. R. Rodriguez, J. M. McKenna, and H. M. Tsai, *J. Org. Chem.*, **40**, 1427 (1975).

[201] H. Watanabe, C. L. Mao, I. T. Barnish, and C. R. Hauser, *J. Org. Chem.*, **34**, 919 (1969).

[202] N. S. Narasimhan and R. S. Mali, *Chem. Ind. (London)*, **1975**, 519.

For N-alkyl-*o*-toluamides the exclusive pathway of the lithiation reaction is again deprotonation of the relatively acidic *ortho*-methyl group[64] (see p. 16).

Secondary thiobenzamides can also be utilized for regioselective *ortho* metalation.[65] The formation of the thioether **108** illustrates this reaction. Although the tendency of primary adducts of this type to form cyclic products is apparently no less than that of the carboxamides, the thioamide function may nevertheless prove useful for further synthetic elaboration via sulfur alkylation and extrusion.[65]

$$p\text{-ClC}_6\text{H}_4\text{CSNHCH}_3 \xrightarrow[2.\ (C_6H_5S)_2]{1.\ n\text{-BuLi}} \mathbf{108}$$

108 (91%)

2-Aryloxazolines and 2-Aryloxazines. The azomethine linkage has long been recognized as an excellent ligand.[203] This property, combined with a strong electron-withdrawing effect, makes this functionality one of the most powerful beta directors known to date. 4,4-Dimethyl-2-oxazolines and 1,3-oxazines constitute synthetically useful directing groups that contain this common structural element. The *ortho* lithiation of 2-aryloxazolines proceeds readily even at low temperatures. The *ortho*-metalated intermediates react with numerous electrophiles.[46,204–206] For example, the method can be used to prepare 2,6-dideuterated benzoic acids.[204] For the *para*-methoxyl derivative **109** lithiation occurs, not unexpectedly, regioselectively in the position *ortho* to the oxazoline ring, as the aldehyde **110** is the only product isolated.[46] In fact, the directing capacity of the oxazoline group is so high that it can compete successfully

$$\mathbf{109} \xrightarrow{n\text{-BuLi}} \text{(ortho-Li intermediate)} \xrightarrow{\text{DMF}} \mathbf{110}$$

109

110 (70%)

[203] M. I. Bruce, *Angew. Chem., Int. Ed. Engl.*, **16,** 73 (1977).

[204] A. I. Meyers and E. D. Mihelich, *J. Org. Chem.*, **40,** 3158 (1975).

[205] A. Padwa, A. Ku, A. Mazzu, and S. I. Wetmore, Jr., *J. Am. Chem. Soc.*, **98,** 1048 (1976).

[206] A. I. Meyers and E. D. Mihelich, *Angew. Chem., Int. Ed. Engl.*, **15,** 270 (1976).

with the alpha metalation of the thiophene nucleus, one of the more readily lithiated substrates known.[139] Moreover, despite the susceptibility of the pyridine nucleus to attack by nucleophilic metalating agents, the oxazoline derived from pyridine-4-carboxylic acid can successfully be lithiated in the 3 position.[207]

Thus far the following limitations of this reaction have been observed: *ortho*-methoxyl[208a,b] (**111a**) and *ortho*-fluoro derivatives[208c] (**111b**) suffer clean nucleophilic displacement of the methoxyl or fluoro groups in preference to *ortho* metalation, whereas *ortho*-tolyl derivatives undergo the familiar deprotonation at the methyl group,[46] quite analogous to the *o*-toluamides (p. 16).[64] It should be noted that in certain cases nucleophilic substitution can be suppressed in favor of *ortho* lithiation by the use of *n*-butyllithium/TMEDA.[23] Interestingly, treatment of 2,2′-*m*-phenylenebis(4,4-dimethyloxazoline) with alkyl or aryl lithium reagents leads to addition of the metalating agent to the aromatic ring.[208d] Normal *ortho* lithiation at the 2 position, however, can be effected with lithium diisopropylamide/TMEDA.

O (CH$_3$)$_2$ N Y —RLi or (RMgX)→ O (CH$_3$)$_2$ N R

111 a: Y = OCH$_3$
b: Y = F

Arylcarbimines. The synthetic utility of carbimines as *ortho*-directing groups is limited by the propensity of the azomethine linkage in such substrates to suffer attack by a nucleophilic lithiating agent or to facilitate alpha deprotonation.[209,210] Nevertheless, in systems where neither of these two side reactions can or does occur, the C=N group again proves to be a very powerful *ortho*-directing functionality.

Imines of alkyl aryl ketones clearly undergo deprotonation of the rather acidic alpha position. Nevertheless, *ortho* metalation can also be observed,[211] although it is of little synthetic value. For aryl carboxaldimines, which lack acidic alpha protons, nucleophilic attack is the generally observed reaction, even with N-*t*-butylimines.[24] In special cases, such as in the cyclohexylimine **112**, where *ortho* metalation is facilitated by the

[207] A. I. Meyers and R. A. Gabel, *Tetrahedron Lett.*, **1978,** 227.
[208a] A. I. Meyers and E. D. Mihelich, *J. Am. Chem. Soc.*, **97,** 7383 (1975).
[208b] A. I. Meyers, R. Gabel, and E. D. Mihelich, *J. Org. Chem.*, **43,** 1372 (1978).
[208c] A. I. Meyers and B. E. Williams, *Tetrahedron Lett.*, **1978,** 223.
[208d] T. D. Harris, B. Neuschwander, and V. Boekelheide, *J. Org. Chem.*, **43,** 727 (1978).
[209] G. Stork and S. R. Dowd, *J. Am. Chem. Soc.*, **85,** 2178 (1963).
[210] G. Wittig, H. D. Frommeld, and P. Suchanek, *Angew. Chem., Int. Ed. Engl.*, **2,** 683 (1963).
[211] G. Wittig and H. Reiff, *Angew. Chem., Int. Ed. Engl.*, **7,** 7 (1968).

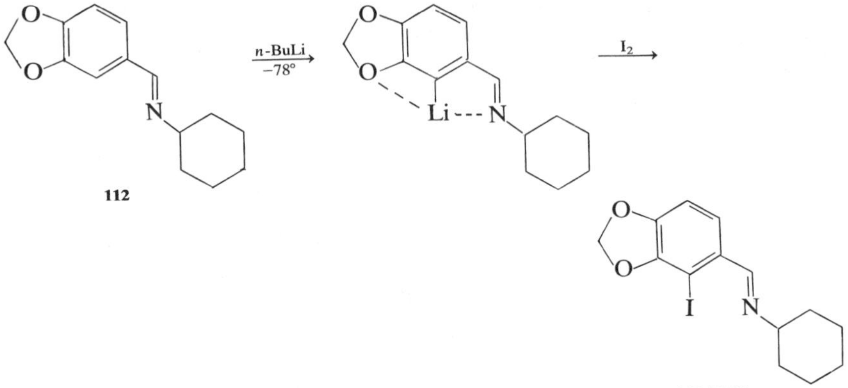

$\xrightarrow[-78^\circ]{n\text{-BuLi}}$ $\xrightarrow{I_2}$

112 **113** (60%)

additional directing effect of an ether function, good yields of *ortho*-substituted products such as **113** are observed.[212]

2-Arylpyridines. Although pyridines generally are susceptible to nucleophilic attack by the metalating agent on the azomethine linkage, some can serve as powerful *ortho*-directing groups. Metalation of the arylpyridine **114** followed by deuteration indicates that the kinetic product **115** arises from an *ortho* lithiation and slowly equilibrates to the thermodynamic product **116**.[213,214] When viewed in the light of the rather facile lateral deprotonation of picolines,[215] the successful *ortho* lithiation of **114** clearly underscores the marked directing ability of the pyridine nucleus. In 2-(2-thienyl)quinoline the *ortho*-directing ability of the quinoline nitrogen leads to predominant metalation in the 3 position of

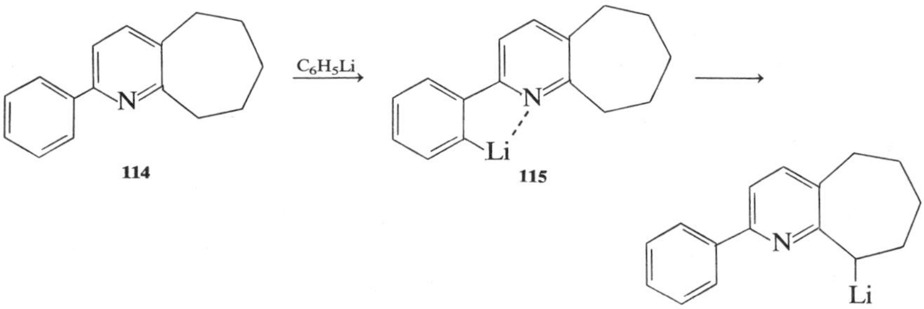

$\xrightarrow{C_6H_5Li}$ $\longrightarrow$

114 **115** **116**

[212] F. E. Ziegler and K. W. Fowler, *J. Org. Chem.*, **41,** 1564 (1976).

[213] J. Epsztajn, A. Bieniek, and J. Z. Brzezinski, *Bull. Acad. Pol. Sci., Ser. Sci. Chim.* **23,** 917 (1975) [*C.A.*, **85,** 46792t (1976)].

[214] J. Epsztajn, W. E. Hahn, and J. Z. Brzezinski, *Rocz. Chem.*, **49,** 123 (1975) [*C.A.*, **83,** 58604e (1975)].

[215] C. Osuch and R. Levine, *J. Am. Chem. Soc.*, **78,** 1723 (1956).

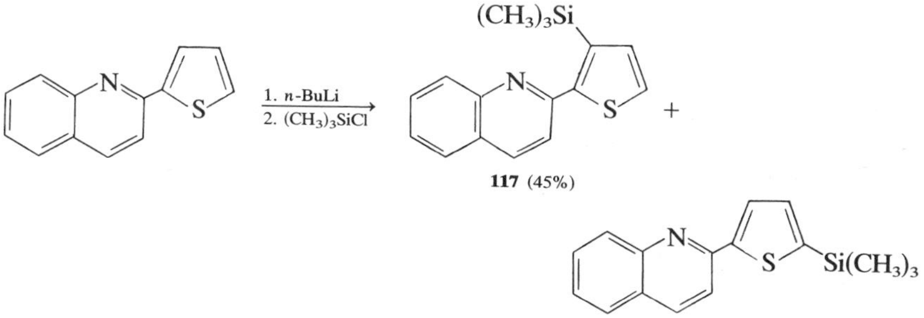

117 (45%)

118 (8%)

the thiophene nucleus and only a small amount of alpha metalation, as reflected by the isolation of the silanes **117** and **118**.[138] The significance of this result becomes apparent in the discussion of the competitive *ortho* *vs.* alpha lithiations (p. 77).

Hydrazones. Although the N,N-dialkylhydrazones of alkyl aryl ketones suffer from the same alpha deprotonation as imines, a good degree of *ortho* metalation is nevertheless observed.[23] Thus metalation of the hydrazone **119** followed by reaction with dimethyl disulfide produces the bisthioether **120** in good yield.[23] N,N-Dimethylhydrazones of diaryl ketones, on the other hand, appear to be promising substrates for *ortho* metalations, as indicated by the formation of the thioether **121**.[23]

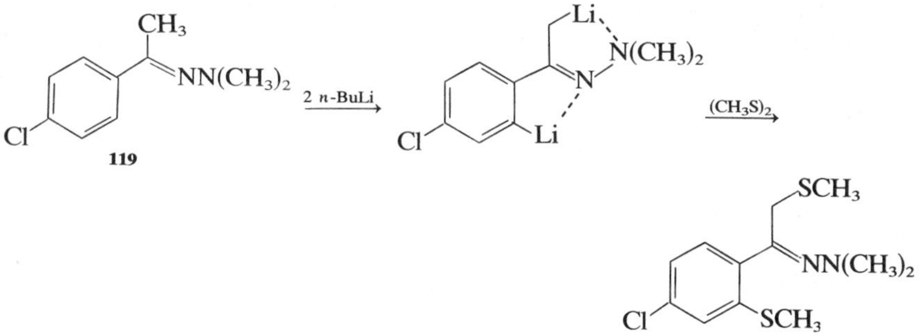

119

120 (60%)

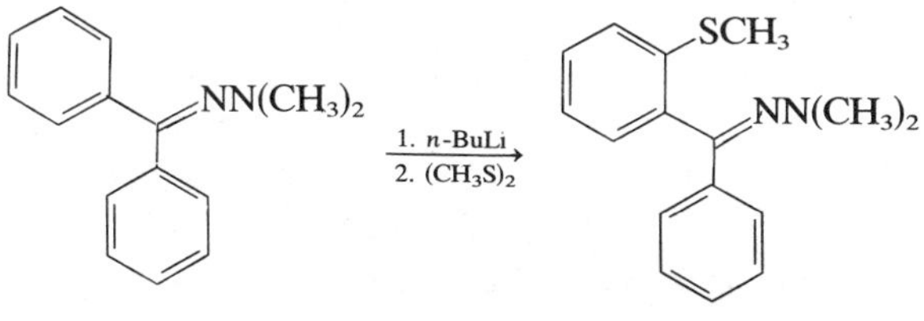

121 (50%)

3-Arylpyrazoles. 1-Alkyl-3-arylpyrazoles incorporate the same functional array of nitrogen atoms as hydrazones. Unlike the diaryl hydrazones, however, such pyrazoles offer two additional sites of deprotonation, both of which are attacked,[23] namely, the 5 position of the pyrazole nucleus and the N-alkyl substituent (see p. 24).[86] Both side reactions can be suppressed by a bulky tertiary nitrogen substituent, such as the methoxyisopropyl protecting group in the pyrazole **122**. Metalation now proceeds predominantly at the *ortho* position. After removal of the protecting group the ketone **123** can be isolated in fair yield.[23]

$NC(CH_3)_2OCH_3$ — 1. s-BuLi; 2. $ArCON(CH_3)_2$; 3. H_3O^+ → NH, COAr, Cl; $Ar = o\text{-}ClC_6H_4$

122 **123** (40%)

2-Arylimidazolines. 2-Arylimidazolines are metalated in their *ortho* position in a manner reminiscent of the 2-aryloxazolines. Thus lithiation of 2-phenylimidazoline followed by reaction with *p*-chlorobenzaldehyde produces the alcohol **124**.[216]

HN, N — 1. 2 n-BuLi; 2. $p\text{-}ClC_6H_4CHO$ → HN, N, $CH(OH)C_6H_4Cl\text{-}p$

124

Nitriles. The use of a nitrile as an *ortho* director is limited because of its highly electrophilic character. One report describes the low temperature alpha metalation of 3-cyanothiophene, which evidently is kinetically favored over nucleophilic addition.[217] The predominant formation of 3-cyano-2-thiophenecarboxylic acid attests to the directing influence of the cyano group. An interesting example is *m*-chlorobenzonitrile. The basicity of the non-nucleophilic lithium tetramethylpiperidide is apparently sufficient to metalate the position *ortho* to both substituents. The stability of the lithiated species at low temperatures permits reaction with electrophiles. For example, the thioether **125** can be isolated in moderate yield.[24]

CN, Cl — 1. LTMP; 2. $(CH_3S)_2$ → CN, SCH_3, Cl

125 (32%)

[216] W. J. Houlihan, U.S. Pat. 3,932,450 (1976) [*C.A.* **84,** 135662x (1976)].

[217] S. Gronowitz and B. Eriksson, *Ark. Kemi,* **21,** 335 (1963) [*C.A.*, **59,** 13918f (1963)].

1. LDA; 2. D_2O

126 (37%)

For α,β-unsaturated nitriles beta metalation can be achieved under carefully selected conditions, as shown by the deuteration of the dihydropyridine **126**.[218] The analogous lithiation of β-aminoacrylonitriles is more complex (see p. 19).[57c]

Oxygen as a Beta (*Ortho*)-Directing Atom

Alkyl Vinyl Ethers. Lithiations in the beta position of vinyl ethers, unassisted by additional directing groups, have not been reported. However, when the anion is further activated by the presence of an adjacent chlorine, such as in (E)-2-chlorovinyl ethyl ether, metalation occurs at low temperatures, as indicated by the quantitative conversion to the acid **127**.[174a] Conversely, when the same conditions are applied to (Z)-2-chlorovinyl ethyl ether, the corresponding acid is produced in only 40% yield.[174a] This may be a result of both the absence of a chelating effect by the ether oxygen and the fact that the lithiated species can undergo a *trans* beta elimination. Similar observations were made with the corresponding (E/Z)-2-bromovinyl ethyl ethers.[174b]

n-BuLi, −100°; CO_2

127 (100%)

Alkyl Aryl Ethers. Alkyl aryl ethers are by far the most widely studied group of compounds, not only in terms of the scope of the metalation, but also with respect to the mechanism of the reaction. A recent study gives a systematic account of the relative *ortho*-directing ability of the methoxyl group in relation to other functionalities.[19]

Most lithiations of alkyl aryl ethers are carried out in diethyl ether as solvent, largely because the original reports call for it.[1] More recently, however, it was realized that, as with diaryl ethers, the use of tetrahydrofuran accelerates the metalation considerably and greatly improves the yields of products. Of the numerous examples studied in carbocyclic aromatic systems, a few deserve special attention. 1,3-Dimethoxybenzene illustrates a characteristic common to most other

[218] R. R. Schmidt and G. Berger, *Chem. Ber.*, **109**, 2936 (1976).

directing groups, namely, the highly selective lithiation in the position *ortho* to both methoxyl groups. The extent of metalation in that position is 96–97%.[36,50] As illustrated, reaction with 1,6-dibromohexane gives an excellent yield of the diarylhexane **128**.[219] Interesting aspects of the metalation of the bromoether **129** are its regioselectivity and, quite obviously, the absence of any appreciable metal–halogen exchange[220,221] (p. 17). The use of phenyllithium, rather than *n*-butyllithium, in the preparation of **130** should be noted. Another example of regioselectivity is the lithiation of chlorotrimethoxybenzene (**131**), which is extremely facile, being complete within 3 minutes at −70°. The major product **132** is derived from metalation between the methoxyl group and the chlorine, with formation of only a minor amount of the isomer **133**.[222,223] Although

1. C_6H_5Li; 2. $Br(CH_2)_6Br$ → **128** (93%)

129 — 1. C_6H_5Li; 2. Cyclopentanone → **130** (33%)

131 — 1. *n*-BuLi; 2. CO_2 → **132** (79%) + **133** (16%)

1. *n*-BuLi; 2. D_2O → **134** (94%)

[219] H. Lettré and A. Jahn, *Chem. Ber.*, **85,** 346 (1952).
[220] G. I. Feutrill, R. N. Mirrington, and R. J. Nichols, *Aust. J. Chem.*, **26,** 345 (1973).
[221] K. Yamada, H. Yazawa, D. Uemura, M. Toda, and Y. Hirata, *Tetrahedron*, **25,** 3509 (1969).
[222] W. Schäfer and R. Leute, *Chem. Ber.*, **99,** 1632 (1966).
[223] W. Schäfer, R. Leute, and H. Schlude, *Chem. Ber.*, **104,** 3211 (1971).

chlorine is known to be a less effective *ortho* director for coordinatively unsaturated metalating agents than the methoxyl group, the apparent inconsistency can be accounted for by the greater *acidifying* effect of the chlorine,[31,32] since for each lithiatable position the necessity for coordination of the metalating agent is satisfied by either *single* methoxyl group. Metalation of 1,2,3-trimethoxybenzene in tetrahydrofuran proceeds rapidly to produce the deuterated product **134** in excellent yield.[224]

The side reaction common to most *ortho* lithiations, namely, the deprotonation of an *ortho*-methyl group, is also apparent in *o*-methylanisole, albeit to a lesser extent[1,61] (see p. 16). Cyclic alkyl aryl ethers such as **135** can be metalated as expected in the position *ortho* to the oxygen function to give upon carbonation the acid **136** in respectable yield.[225]

1. *n*-BuLi
2. CO_2

135

CO_2H

136 (73%)

The controversy concerning the lithiation of 1-methoxynaphthalene was apparently caused not only by divergent interpretations but also by differing reaction conditions.[226,227] The conditions indeed play an important role, particularly with respect to the lithiating agent and its origin. A thorough investigation of the various factors affecting the ratio of isomeric products indicates that under carefully controlled conditions either one of the isomers **137** and **138** can be obtained essentially as the exclusive product.[192] These results imply that metalation in the 2 position, which is achieved exclusively with *n*-butyllithium/TMEDA, is strictly an acid–base reaction (see p. 10), whereas in hydrocarbon solvents the precoordination of the lithiating agents leads to the more favorable five-membered cyclometalated intermediate. The regioselective metalation in the 8 position can formally be compared to the recently reported

OCH_3 1. RLi 2. CO_2/CH_2N_2 → OCH_3 CO_2CH_3 (**137**) + CH_3O_2C OCH_3 (**138**)

	137 : 138	
R = *n*-Bu/ether, hexane	73 : 27	(28% total)
n-Bu/hexane, TMEDA	>99.3 : <0.3	(60% total)
t-Bu/cyclohexane	1 : 99	(35% total)

[224] G. Schill and E. Logemann, *Chem. Ber.*, **106,** 2910 (1973).
[225] H. Christensen, *Synth. Commun.*, **4,** 1 (1974).
[226] R. A. Barnes and L. J. Nehmsmann, *J. Org. Chem.*, **27,** 1939 (1962).
[227] B. M. Graybill and D. A. Shirley, *J. Org. Chem.*, **31,** 1221 (1966).

ortho lithiation of α-silyloxystyrenes in which, in fact, an enol ether serves as the directing group.[228] Here again, the ring size of the cyclometalated intermediate is five rather than four as for anisoles. The analogous observation is made in the case of 1-anilinonaphthalenes (p. 46). The metalation of 2-methoxynaphthalenes also depends on the conditions. In tetrahydrofuran the 3-lithiated species is formed almost exclusively, as indicated by the isolation of the imine **139**, whereas other conditions lead to a mixture of 1- and 3-substituted products with the latter predominating.[17,229,230]

OCH_3 — 1. n-BuLi, 2. C_6H_5CN → OCH_3, C_6H_5, NH

139 (90%)

In five-membered, π-excessive heterocycles the alkoxy group can provide regioselectivity for alpha metalations as outlined earlier (p. 37). In the absence of any free alpha protons, as in 2-methoxy-5-methylthiophene, the beta-directing effect of the ether function leads to metalation in the 3 position to give the carboxylic acid **140**.[231] In the isoxazole **141** metalation facilitated by the methoxyl group occurs in the only available position, and the iodo derivative **142** is obtained in high yield.[232]

CH_3, S, OCH_3 — 1. n-BuLi, 2. CO_2 → CO_2H, CH_3, S, OCH_3

140 (50%)

C_6H_5, N, O, OCH_3 — 1. n-BuLi, 2. I_2 → C_6H_5, I, N, O, OCH_3

141 **142** (88%)

Alkoxyalkyl Aryl Ethers. Free phenols can be metalated in their *ortho* position,[1,233] but the yields of products are too low to be of preparative interest. Considerably more valuable is the metalation of protected phenols such as methoxymethyl or tetrahydropyranyl ethers. The yields in

[228] J. Klein and A. Medlik-Balan, *J. Org. Chem.*, **41,** 3307 (1976).
[229] N. S. Narasimhan and R. S. Mali, *Tetrahedron*, **31,** 1005 (1975).
[230] N. S. Narasimhan and M. V. Paradkar, *Indian J. Chem.*, **7,** 536 (1969).
[231] J. Sicé, *J. Am. Chem. Soc.*, **75,** 3697 (1953).
[232] R. G. Micetich and C. G. Chin, *Can. J. Chem.*, **48,** 1371 (1970).
[233] L. Santucci and H. Gilman, *J. Am. Chem. Soc.*, **80,** 4537 (1958).

the metalation of these derivatives are generally good to excellent, as illustrated by the preparation of the aldehyde **143**[234] using the methoxymethyl protecting group, and the alcohol **144**,[235] where the tetrahydropyranyl ether is utilized.

Cl OCH$_2$OCH$_3$ → 1. *n*-BuLi/TMEDA 2. DMF → Cl OCH$_2$OCH$_3$ CHO

143 (85%)

O O → 1. *n*-BuLi 2. O 3. H_3O^+ → OH (CH$_2$)$_2$OH

144 (70%)

Diaryl Ethers and Condensed Diaryl Ethers. The ability of *n*-butyllithium to deprotonate the *ortho* position of dibenzofuran in particular and diaryl ethers in general was recognized as early as 1934.[1] This observation constitutes the discovery of the heteroatom-facilitated metalations discussed in this chapter. Competitive experiments show that the ethers have a greater *ortho*-directing effect than either sulfides or amines.[1] This differential directing ability is best documented by the lithiation of phenoxathiine, in which monometalation occurs exclusively adjacent to oxygen. Even dimetalation, using excess *n*-butyllithium, takes place predominantly in the two positions *ortho* to oxygen and only to a smaller extent at position 1.[236,237] Likewise, N-ethylphenoxazine is metalated adjacent to oxygen exclusively to give the acid **145** (p. 66).[238]

The lithiation of diaryl ethers is used extensively to prepare not only specifically *ortho*-substituted derivatives, but also condensed tricyclic systems. Dimetalation of diphenyl ether, for example, followed by reaction with a dichlorosilane gives reasonable yields of phenoxasilins[239–242] such as **146** or, with tetrachlorosilane, the spirosilane **147**.[239]

In asymmetrically substituted diaryl ethers the rate-increasing or -decreasing influence of substituents can be used to provide regioselective

[234] H. Christensen, *Synth. Commun.*, **5**, 65 (1975).
[235] R. Stern, J. English, Jr., and H. G. Cassidy, *J. Am. Chem. Soc.*, **79**, 5797 (1957).
[236] H. Gilman and S. H. Eidt, *J. Am. Chem. Soc.*, **78**, 2633 (1956).
[237] S. H. Eidt, *Iowa State Coll. J. Sci.*, **31**, 397 (1957) [*C.A.*, **51**, 14729i (1957)].
[238] H. Gilman and L. O. Moore, *J. Am. Chem. Soc.*, **80**, 2195 (1958).
[239] K. Oita and H. Gilman, *J. Am. Chem. Soc.*, **79**, 339 (1957).
[240] H. Gilman and W. J. Trepka, *J. Org. Chem.*, **26**, 5202 (1961).
[241] H. Gilman and W. J. Trepka, *J. Org. Chem.*, **27**, 1418 (1962).
[242] H. Gilman and D. Miles, *J. Org. Chem.*, **23**, 1363 (1958).

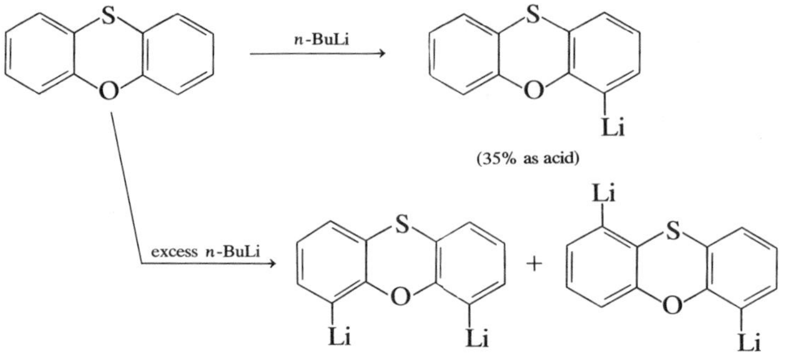

n-BuLi

(35% as acid)

excess n-BuLi

(35% as diacid) (9% as diacid)

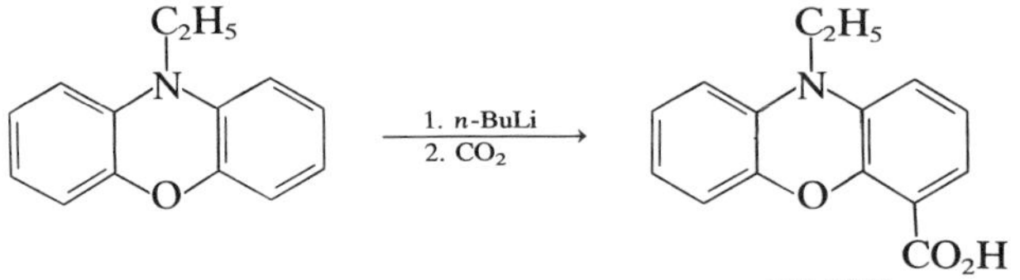

1. n-BuLi
2. CO_2

145 (35%)

2 n-BuLi

$(C_6H_5CH_2)_2SiCl_2$ $SiCl_4$

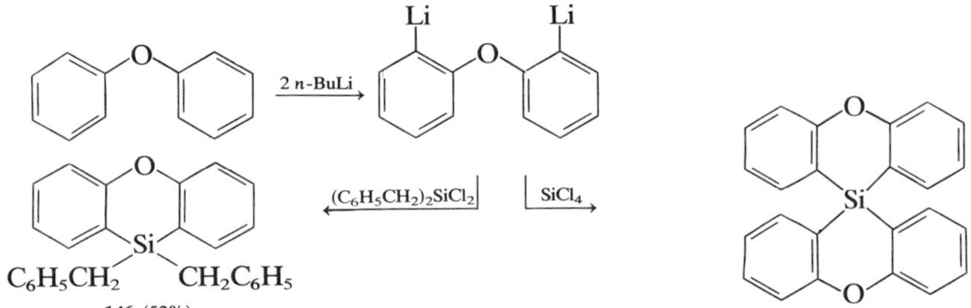

146 (52%) **147** (25%)

ortho lithiation. Thus *p*-chlorophenyl phenyl ether is lithiated in the substituted ring to give the silane **148**.[243]

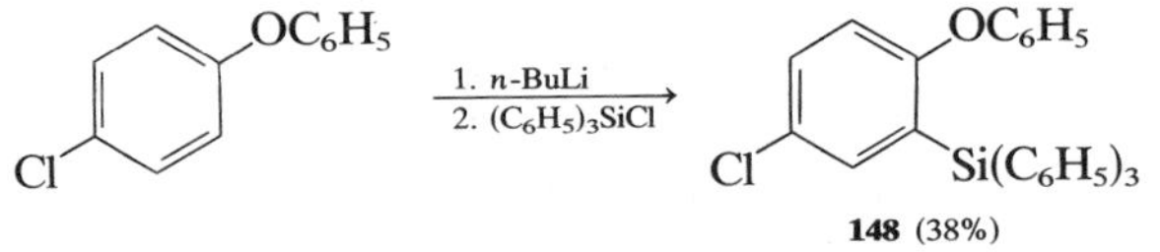

1. n-BuLi
2. $(C_6H_5)_3SiCl$

148 (38%)

Alkyl Aralkyl Ethers and Aralkyl Alcohols. *Ortho* lithiation of alkyl benzyl ethers is generally not feasible because of their well-known propensity to undergo benzylic deprotonation, followed by Wittig rearrangement.[244a] Nevertheless, alkoxymethyl groups possess a proven beta-directing effect and are extensively used to provide regioselectivity in the

[243] K. Oita and H. Gilman, *J. Org. Chem.*, **21,** 1009 (1956).
[244a] G. Wittig, P. Davis, and G. Koenig, *Chem. Ber.*, **84,** 627 (1951).

alpha metalation of π-excessive, five-membered heterocycles. The most commonly encountered functionalities are the methoxymethyl group and the acetals and ketals of carbonyl compounds. Thus 3-(methoxymethyl)thiophene is metalated exclusively in the 2 position to give a high yield of the acid **149**.[142] This is in distinct contrast to the random metalation of 3-methylthiophene (p. 38). Furan-3-carboxaldehyde ethylene ketal is not only metalated more rapidly than furan itself but also regioselectively in the 2 position, as illustrated by the isolation of the ketone **150**.[113] For 1-(methoxymethyl)indole the methoxymethyl group provides a more rapid and clean lithiation in the 2 position than is observed for 1-alkylindoles[85] (p. 23).

CH_2OCH_3 ; n-BuLi ; OCH_3 ; S ; Li ; CO_2 ; CH_2OCH_3 ; CO_2H

149 (89%)

O ; O ; 1. n-BuLi 2. $C_6H_5CON(CH_3)_2$ 3. H_3O^+ ; CHO ; COC_6H_5

150 (61%)

In carbocyclic aromatic systems, where benzylic ethers can generally not serve as *ortho*-directing groups, benzylic alcohols may be used instead. The hydroxyalkyl group is by itself a poor *ortho* director, as evinced by the low-yielding metalation of benzyl alcohol.[1] Under forcing conditions, however, lithiation can be effected remarkably well.[244b] In addition, it can be successfully utilized to provide a desired regioselectivity in more complex aromatic templates with additional *ortho*-directing functionalities. In the octahydrophenanthrene system **151**, for example, lithiation proceeds with a high degree of regioselectivity in the position *ortho* to both the methoxyl and hydroxyalkyl groups to produce, after carboxylation, the tetracyclic lactone **152**.[245]

OCH_3 ; OH ; H ; 2 n-BuLi ; CH_3O ; Li ; OLi ; H ; CO_2 ; CH_3O ; O ; O ; H

151 **152** (80%)

[244b] N. Meyer and D. Seebach, *Angew. Chem., Int. Ed. Engl.*, **17,** 522 (1978).
[245] M. Uemura, S. Tokuyama, and T. Sakan, *Chem. Lett.*, **1975,** 1195.

Arylcarboxylic Acids, Esters, and Diaryl Ketones. In contrast to arylcarboxamides, the direct *ortho* lithiation of arylcarboxylic acids is generally not feasible because of the increased electrophilicity of the carboxylate group. Thus treatment of lithium benzoate with an additional equivalent of an organolithium reagent leads to addition rather than metalation, providing useful ketone syntheses.[197] However, lithiations in the position *ortho* to the carbonyl groups of acids and ketones are reported in a few special cases. For example, the use of the non-nucleophilic lithium dialkylamides is successful with thiophene-3-carboxylic acid.[246] Alternatively, metalation can be achieved even with alkyllithium reagents if, because of additional activation, the rate of deprotonation is faster than the rate of addition of the lithiating agent to the carboxylate carbonyl group. For example, the bromo compound **153** is prepared in good yield from 3-methylisothiazole-4-carboxylic acid.[154] Similar reactions, albeit of little synthetic value because of self-condensations, are reported for arylcarboxylic esters and diaryl ketones.[53] For example, the phthalan **154** can be obtained in high yield.[53]

CH_3 CO_2H N S —1. 2 *n*-BuLi/−70°; 2. Br_2→ CH_3 CO_2H N S Br

153 (52%)

$(C_6H_5)_2CO$ —LTMP, −78°→ C_6H_5 O Li + $(C_6H_5)_2CO$ ⟶ C_6H_5 C_6H_5 O C_6H_5 OH

154 (80%)

Sulphur as a Beta (*Ortho*)-Directing Atom

Alkyl Aryl Sulfides. The ability of a thioether to facilitate *ortho* lithiation is intermediate between that of ethers and anilines. The lithiation of thioanisole is, in contrast to that of anisole, more complex because of an equilibration of the kinetic ring metalation product and the thermodynamic sidechain-metalated species (Eq. 22).[1,58] Hence this *ortho*

SCH_3 —*n*-BuLi→ SCH_3 Li ⇌ SCH_2Li (Eq. 22)

[246] G. M. Davies and P. S. Davies, *Tetrahedron Lett.*, **1972,** 3507.

functionalization is of little preparative value. However, the lateral deprotonation, which can be made the exclusive reaction, has received considerable attention.[247] The reason for this competitive reaction must be the additional anion stabilization by the *d* shells of sulfur.[59] Such a stabilizing effect is clearly not available for anisole.

Thioethers, however, do provide high regioselectivity in the alpha metalation of π-excessive, five-membered heterocycles. Evidently, in such substrates deprotonation of the alkylmercapto group is not a problem, possibly because of the greater thermodynamic stability of the alpha-metalated species. 3-(Methylthio)thiophene is lithiated regioselectively in the 2 position to give a good yield of the corresponding acid.[248] For the isoxazole **155** the thioether function allows rapid *ortho* lithiation to produce the carboxylic acid **156** in good yield.[232]

SCH_3 (thiophene) —[1. *n*-BuLi; 2. CO_2]→ SCH_3, CO_2H (thiophene) (70%)

C_6H_5, SC_2H_5 (isoxazole, N, O) **155** —[1. *n*-BuLi; 2. CO_2]→ C_6H_5, CO_2H, SC_2H_5 (isoxazole, N, O) **156** (73%)

Diaryl Sulfides and Condensed Diaryl Sulfides. Diaryl or condensed diaryl sulfides can be metalated without occurrence of the side reaction reported for alkyl aryl sulfides. Dibenzothiophene is lithiated as expected, *ortho* to sulfur.[1,152] The use of tetrahydrofuran as solvent greatly facilitates this reaction, as indicated by the good yield of 4-bromodibenzothiophene.[152,249] The reported yields for the lithiation of thianthrene seem rather low, as exemplified by the preparation of the boronic acid **157**.[250] It is conceivable that the use of tetrahydrofuran could be equally beneficial here.

As already discussed, the metalation of N-alkylphenothiazines leads to approximately equal amounts of products derived from lithiation *ortho* to both sulfur and nitrogen,[177] whereas in phenoxathiins the oxygen is clearly the dominant directing group.[236,237]

The lithiation of diphenyl sulfide is reported to give a moderate yield of products derived from *ortho* metalation.[1,251] However, a more recent

[247] E. J. Corey and D. Seebach, *J. Org. Chem.*, **31,** 4097 (1966).
[248] S. Gronowitz, *Ark. Kemi*, **13,** 269 (1958) [*C.A.*, **53,** 15056e (1959)].
[249] E. B. McCall, A. J. Neale, and T. J. Rawlings, *J. Chem. Soc.*, **1962,** 4900.
[250] H. Gilman and D. R. Swayampati, *J. Am. Chem. Soc.*, **79,** 208 (1957).
[251] C. Eaborn and J. A. Sperry, *J. Chem. Soc.*, **1961,** 4921.

1. n-BuLi
2. Br_2

(77%)

1. n-BuLi
2. $(n\text{-}C_4H_9O)_3B$

$B(OH)_2$

157 (22%)

investigation indicates that the crude metalation products are 9:1 mixtures of *ortho* and *meta* isomers.[252]

Sulfones. The potential of the sulfone group as an *ortho* director is limited to diaryl and *t*-alkyl aryl sulfones for reasons outlined earlier (p. 16). An additional limiting factor is the propensity of aryl sulfones, in particular naphthyl sulfones, to undergo conjugate addition of the lithiating agent.[253] Nevertheless, the sulfone is a powerful directing group, and clean *ortho* metalation can be achieved at low temperatures. Thus *t*-butyl 1-naphthyl sulfone is metalated exclusively at the 2 position at −70° to give the acid **158**, whereas conjugate addition occurs at the reflux temperature of ether, leading to the *trans*-substituted dihydronaphthalene **159**.[253] The increased directing effect as compared with the ether function is evinced by the dilithiation of phenoxathiin-10,10-dioxide, which is metalated exclusively at the positions *ortho* to the sulfonyl group to produce the diacid **160**.[236] This is in contrast to phenoxathiin itself, where the ether function is the dominant directing group.

A unique example is the metalation of *m*-bromophenyl phenyl sulfone, which occurs at the position *ortho* to both directing groups, leading to the

$SO_2C_4H_9\text{-}t$ CO_2H

158 (47%)

1. n-BuLi/−70°
2. CO_2

$SO_2C_4H_9\text{-}t$

n-BuLi/34°

H $SO_2C_4H_9\text{-}t$ H $C_4H_9\text{-}n$

159 (57%)

[252] F. D. Bailey and R. Taylor, *J. Chem. Soc., B*, **1971,** 1446.

[253] F. M. Stoyanovich, R. G. Karpenko, and Y. L. Gol'dfarb, *Tetrahedron*, **27,** 433 (1971).

1. 2 n-BuLi
2. CO_2

160 (52%)

1. n-BuLi
2. CO_2

161 (54%)

acid **161**.[69] The example is remarkable insofar as there is neither any appreciable metal–halogen exchange nor any benzyne formation. The *para* isomer is also metalated *ortho* to the sulfone with less than 2% metal–halogen exchange.[69] These results attest not only to the powerful directing influence of the sulfone group but also to the thermodynamic stability of the resulting anion.

Arylsulfonamides. Secondary and tertiary sulfonamides are the strongest *ortho*-directing groups known to date, at least in carbocyclic aromatic systems.[2,19] Unlike the tertiary arylcarboxamides,[56] tertiary arylsulfonamides can be lithiated without any special conditions, *e.g.*, N,N-dimethylbenzenesulfonamide is metalated cleanly in its *ortho* position, and the imine **162** can be isolated in very good yield.[254] For secondary sulfonamides two molar equivalents of an alkyl lithium are necessary to achieve nuclear metalation. The reactions are rapid and the lithiated species may be treated with a variety of substrates to give, quite generally, high yields of products such as the allylic alcohol **163**.[24] 1-Sulfamoylnaphthalenes are metalated in the 8 position predominantly,[255]

1. n-BuLi
2. C_6H_5CN

162 (79%)

1. n-BuLi
2. $CH_2{=}CHCHO$

163 (>80%)

[254] H. Watanabe, R. A. Schwarz, C. R. Hauser, J. Lewis, and D. W. Slocum, *Can. J. Chem.*, **47,** 1543 (1969).

[255] J. G. Lombardino, *J. Org. Chem.*, **36,** 1843 (1971).

whereas the 2 isomers give largely the 1-lithiated intermediates.[255] Because of the low yields of isolated products, the possibility of additional regioisomers, particularly in view of the facts stated about the 1-naphthyl sulfones (p. 70),[253] cannot be excluded. A recent study on the competitive lithiation of N-*t*-butyl-N-methyl-*p*-benzenesulfonamide and 2-(*p*-chlorophenyl)-4,4-dimethyl-2-oxazoline reveals that the sulfonamide substrate is lithiated almost exclusively.[23]

One side reaction deserving elaboration pertains to an apparently quite general rearrangement observed with lithiated arylsulfonamides of N-substituted anilines[256–259] leading to N-substituted-*o*-(arylsulfonyl)anilines. The mechanism of this rearrangement is now rigorously established.[257] The initial step is the normal *ortho* lithiation, for example, of the sulfonamide **164** to give **165**, which is stable at −70°. Above −20° a rate-determining transmetalation occurs, giving **166** which then rearranges to the sulfone **167**.[256,257] The intramolecular character of

164 →(n-BuLi, −70°) 165 →(>−20°) 166 → 167

168 →(n-BuLi) 169 → 170

[256] S. J. Shafer and W. D. Closson, *J. Org. Chem.*, **40,** 889 (1975).
[257] D. Hellwinkel and M. Supp, *Tetrahedron Lett.*, **1975,** 1499.
[258] D. Hellwinkel and M. Supp, *Angew. Chem., Int. Ed. Engl.*, **13,** 270 (1974).
[259] D. Hellwinkel and M. Supp, *Chem. Ber.*, **109,** 3749 (1976).

the transmetalation is documented by the stability of the cyclic derivative **168**. Only upon addition of a second equivalent of *n*-butyllithium is **168** metalated to the dilithio species **169**. This then rearranges to produce, after workup, the tricyclic sulfone **170** in good yield.[257]

The preferential deprotonation of an *ortho*-methyl group rather than nuclear metalation is also observed with arylsulfonamides (p. 16).[66]

Halogens as Beta (*Ortho*)-Directing Atoms

Aryl Fluorides and Vinyl Fluorides. Fluorine is quite an effective *ortho* director largely because of its inductive capacity rather than any coordinating effect. Of all the halogens fluorine exerts the strongest acidifying effect on *ortho* positions.[31,32] Pentafluorobenzene, for instance, exhibits a pK_a of 23 *vs.* 30.5 for pentachlorobenzene and 37 for benzene itself.[12] Although the products derived from metalation of such substrates, namely, 2-lithiofluorobenzenes, can undergo elimination to form benzynes, they may nevertheless be prepared efficiently and in a stable form at temperatures below −50°. Fluorobenzene itself can thus be lithiated in tetrahydrofuran solution, and upon carboxylation *o*-fluorobenzoic acid is obtained in 60% yield.[260] Lithiation of 1-fluoronaphthalene leads after carboxylation to 1-fluoro-2-naphthoic acid,[260] whereas the 2 isomer gives approximately equal amounts of 2-fluoro-1-naphthoic acid and 3-fluoro-2-naphthoic acid.[261] The ease of metalation, as well as the stability of the metalated species, is somewhat greater with polyfluorinated benzenes.[6] 1,2,3,4-Tetrafluorobenzene, for example, is lithiated readily even at −70°. Upon reaction with mercuric chloride the diarylmercury derivative **171** is isolated in high yield.[262] In terms of relative directing abilities toward coordinatively unsaturated metalating agents, fluorine ranks below the methoxyl group, since *p*-fluoroanisole is lithiated exclusively *ortho* to the ether function.[19] Fluorine also serves as a beta director for the metalation of cyclic olefins. The nonafluorocyclohexene **172**, for instance, is lithiated readily to produce after carboxylation the corresponding acid in very good yield.[263]

$$C_6H_5F \xrightarrow[2.\ CO_2]{1.\ n\text{-BuLi}} o\text{-}FC_6H_4CO_2H \quad (60\%)$$

[260] H. Gilman and T. S. Soddy, *J. Org. Chem.*, **22,** 1715 (1957).
[261] T. H. Kinstle and J. P. Bechner, *J. Organomet. Chem.*, **22,** 497 (1970).
[262] C. Tamborski and E. J. Soloski, *J. Organomet. Chem.*, **17,** 185 (1969).
[263] S. F. Campbell, R. Stephens, and J. C. Tatlow, *Chem. Commun.*, **1967,** 151.

1. n-BuLi/−70°
2. $HgCl_2$

171 (90%)

172

1. CH_3Li/−70°
2. CO_2

(77%)

Aryl Chlorides and Vinyl Chlorides. The metalation of chlorinated aromatics is complicated not only by their propensity for benzyne formation but also by the possibility of halogen–metal exchange.[6,100] Although the *ortho* metalation of chlorobenzene is not practical because of the rapid formation of benzyne, in polychlorinated aromatics the inductive effect of neighboring halogen atoms is sufficient to provide some stability to the lithiated species. Metal–halogen exchange can often be suppressed and even eliminated by using appropriate metalating agents. A good example is the lithiation of 1,2,3,4-tetrachlorobenzene, where either the product of *ortho* lithiation (**173**) or chlorine–lithium exchange (**174**) can be obtained almost exclusively.[264]

1. RLi
2. $(CH_3)_3SiCl$

	173	**174**
R = t-C_4H_9	(0%)	(100%)
R = CH_3	(93%)	(7%)

Metal–halogen exchange is apparently not a problem in the lithiation of 2,3,6-trichloropyridine, which metalates quite selectively to give the 4-picolyl derivative **175**.[265] In the lithiation of 3-chloro-4,5-dihydrofuran the beta-directing influence of the chlorine facilitates the alpha metalation of the dihydrofuran system. The methylated product **176** can be obtained in respectable yield.[266]

[264] I. Haiduc and H. Gilman, *Chem. Ind.* (*London*), **1968,** 1278.
[265] N. J. Foulger and B. J. Wakefield, *J. Organomet. Chem.*, **69,** 161 (1974).
[266] M. Schlosser, B. Schaub, B. Spahic, and G. Sleiter, *Helv. Chim. Acta*, **56,** 2166 (1973).

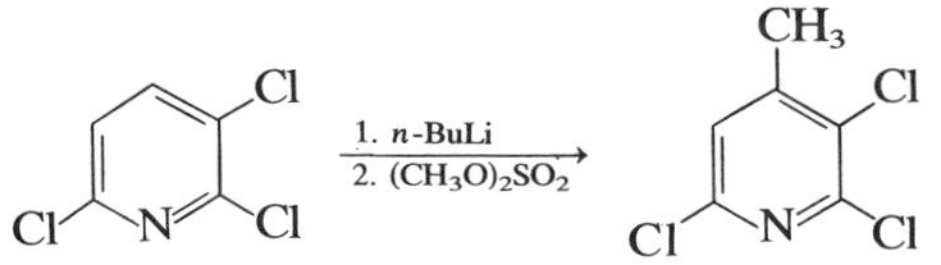

1. n-BuLi
2. $(CH_3O)_2SO_2$

175 (45%)

1. n-BuLi
2. CH_3I

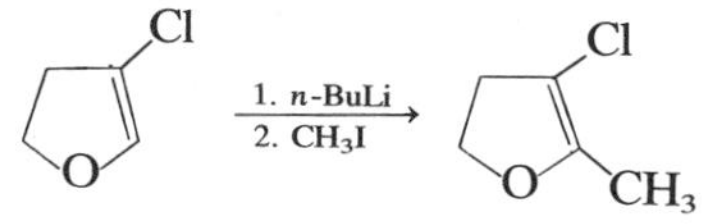

176 (74%)

Aryl Bromides. Whereas metal–halogen exchange is only a minor side reaction with chlorinated aromatics, it becomes with relatively few exceptions almost the exclusive pathway in the attempted *ortho* lithiation of bromobenzenes with alkyllithiums. Nevertheless, bromine, like other halogen atoms, does have a pronounced *ortho*-directing capacity, as evinced by the rapid formation of *ortho*-lithiated intermediates when lithium dialkylamides are used. However, except for isolated instances in which special stabilizing effects are operative, these intermediates are of no preparative value for electrophilic reactions, because rapid aryne formation occurs.[32] The exchange and elimination problems are generally only minor in substrates bearing a powerful *ortho*- or alpha-directing group. In the metalation of 3-bromothiophene the inductive effect of bromine provides excellent regioselectivity, in that 3-bromo-2-thiophenecarboxylic acid is almost the exclusive product.[48] Another example of this type is the lithiation of 3-bromophenyl phenyl sulfone, which occurs regioselectively between both groups (p. 71).[69]

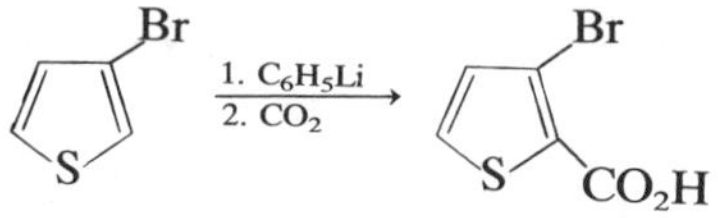

1. C_6H_5Li
2. CO_2

(72%)

Aryl Iodides. Lithiation at the position *ortho* to an iodo substituent in preference to metal–halogen exchange is only observed in readily metalated sulfur-containing heterocycles. By analogy with the bromo analog cited earlier, 3-iodothiophene can be metalated to produce after carboxylation 3-iodo-2-thiophenecarboxylic acid as the major product.[68] Likewise, 4-iodo-3-methylisothiazole is converted into the corresponding 5-carboxaldehyde **177** in respectable yield.[154]

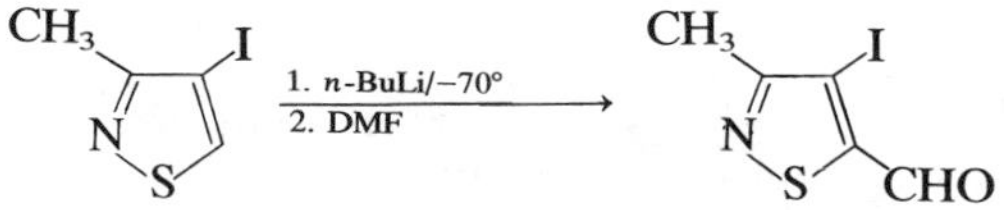

1. n-BuLi/−70°
2. DMF

177 (68%)

(Trifluoromethyl)benzenes. The trifluoromethyl group, generally considered a pseudo-halogen, exerts a moderate *ortho*-acidifying effect. It appears to be a considerably weaker director than the methoxyl group for coordinatively unsaturated metalating agents, since *p*-(trifluoromethyl)anisole is lithiated exclusively in the 2 position by *n*-butyllithium.[19] Mono- or bis(trifluoromethyl)benzenes are lithiated quite readily,[49,50] and the corresponding carboxylic acids are produced in good to excellent yield. However, it should be noted that the trifluoromethyl group is an exception in that, unlike most other *ortho*-directing groups, the bis-1,3-disubstituted derivatives, *e.g.*, *m*-bis(trifluoromethyl)benzene, do not metalate preferentially in the 2 position, as indicated by the isolation of the two acids **178** and **179**.[49,267–269] The reasons for this unusual result are not clear.

CF_3 CF_3 $\xrightarrow[\text{2. } CO_2]{\text{1. } n\text{-BuLi}}$ CF_3 CF_3 CO_2H + CF_3 CO_2H CF_3

178 **179**

(3:2) (85% total)

Other Beta (*Ortho*)-Directing Groups

Arylphosphine Oxides and Imides. Although triphenylphosphine is reported to undergo an anomalous *meta* metalation,[1] arylphosphine oxides and imides can be lithiated in the expected *ortho* position. Thus, phenylbis(3-thienyl)phosphine oxide is dimetalated regioselectively with *n*-butyllithium in the 2 and 2′ positions of the thiophene rings[270] to give, after reaction with ethyl benzoate, the tricyclic alcohol **180**. Metalation of the phosphine imide **181** occurs without halogen–metal exchange in the position *ortho* to phosphorus to give the silane **182**.[271] The ease of metalation can be ascribed to both the ligand effect of the phosphinimidyl group and its inductive effect on the *ortho* position.[271] The use of phenyllithium in such metalations appears rather critical, however. The scope of the phosphine oxide and imide metalations is limited to the triaryl derivatives, since the alkyl group is preferentially deprotonated in alkyl diarylphosphine imides. [271,272]

[267] P. Aeberli and W. J. Houlihan, *J. Organomet. Chem.*, **67,** 321 (1974).
[268] W. J. Houlihan, U.S. Pat. 3,751,491 (1973) [*C.A.*, **79,** 91787g (1973)].
[269] W. J. Houlihan, U.S. Pat. 3,825,594 (1974) [*C.A.*, **81,** 91232p (1974)].
[270] J. P. Lampin and F. Mathey, *J. Organomet. Chem.*, **71,** 239 (1974).
[271] C. G. Stuckwisch, *J. Org. Chem.*, **41,** 1173 (1976).
[272] D. J. Peterson and J. H. Collins, *J. Org. Chem.*, **31,** 2373 (1966).

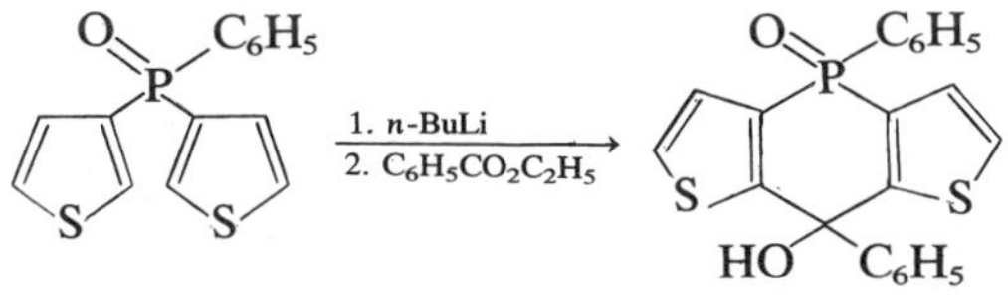

1. n-BuLi
2. $C_6H_5CO_2C_2H_5$

180 (68%)

1. C_6H_5Li
2. $(CH_3)_3SiCl$

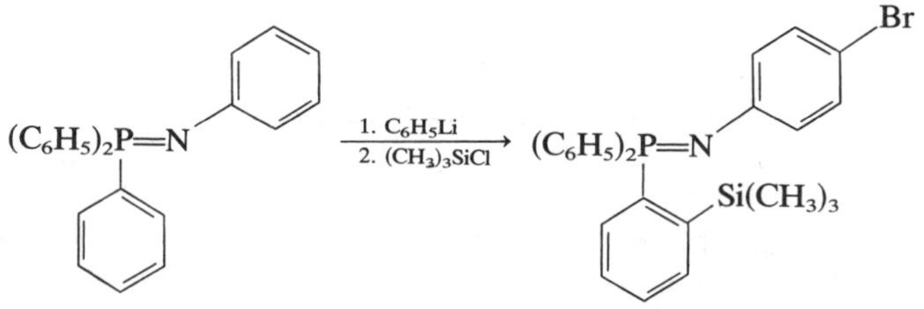

181

182 (50%)

Selenides. As with its sulfur isostere, the *ortho*-directing potential of selenium is confined to diaryl selenides and the alkyl selenides of π-excessive five-membered heterocycles. In the heterocyclic examples the very modest directing ability of the selenide group provides minimal regioselectivity, as illustrated by the ratio of the acids **183** and **184**.[140] The only reported *ortho* metalation of a diarylselenide is that of dibenzoselenophene. After carboxylation an essentially quantitative yield (96%) of the 4-carboxylic acid was obtained.[273]

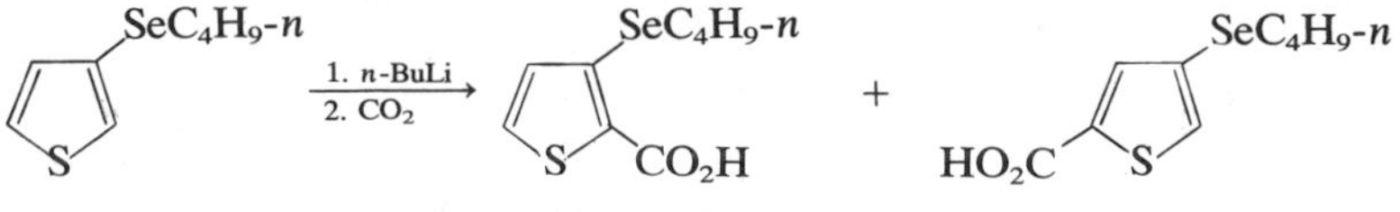

1. n-BuLi
2. CO_2

183

184

(6:4) (79% total)

Competitive Beta *vs.* Alpha Lithiation

The presence of both a beta- and an alpha-directing group within the same molecule provides for some interesting possibilities. Whereas it is accepted that alpha metalations generally proceed with greater facility than beta lithiations, specific cases have been studied in which both reactions can successfully compete. In fact, a judicious choice of conditions can allow either one of the two processes to become dominant. The following examples are illustrative of the general concept.

2-(2′-Thienyl)pyridine offers two potential sites for lithiation: the 3 position (invoking the beta-directing effect of the pyridine nitrogen) and the 5 position (alpha lithiation). By the appropriate combination of

[273] W. J. Burlant and E. S. Gould, *J. Am. Chem. Soc.*, **76,** 5775 (1954).

solvent, temperature, and metalating agent, either of these two positions can be lithiated predominantly, as documented by the formation of the silanes **187** and **188**.[138] It appears that the beta metalation is kinetically controlled because the species **185** slowly equilibrates to the thermodynamically more stable **186** under the reaction conditions. Evidently, under kinetic conditions *n*-butyllithium (tetrameric in ether[14]) preferentially coordinates with the pyridine nitrogen; abstraction of the nearest proton then leads to **185**. By contrast, in tetrahydrofuran, *n*-butyllithium, (a solvated dimeric species[15]) acts more as a base than as a Lewis acid (see p. 11), thus abstracting the most acidic proton in the alpha position of the thiophene nucleus. The fact that lithium diisopropylamide in ether produces essentially the same result is consistent with this rationale, as it displays, unlike tetrameric *n*-butyllithium, only negligible Lewis-acid character.

RLi → **185** + **186** $\xrightarrow{(CH_3)_3SiCl}$ **187** + **188**

RLi	Solvent	Temperature (°C)	**187** Yields (%)	**188** Yields (%)
n-BuLi	THF	0	4	93
LDA	ether	0	1	74
n-BuLi	ether	0	62	13

A similar observation is made in the lithiation of the oxazoline **189**. The major product in tetrahydrofuran is derived from alpha lithiation (**191**), whereas the use of ether leads almost exclusively to beta lithiation and subsequently to the alcohol **190**.[139] This result is the basis for the generalization pointed out earlier (p. 36) that in the thiophene system oxazolines are the strongest beta-directing groups known.

Another example is 1-*t*-butyl-3-(*p*-chlorophenyl)pyrazole. Again, two different metalation sites are possible, but this time in two separate rings of the molecule. The first and evidently more reactive site is the 5 position of the pyrazole ring; the alternative site is the *ortho* position of the phenyl ring, for which the imine group of the pyrazole ring acts as the

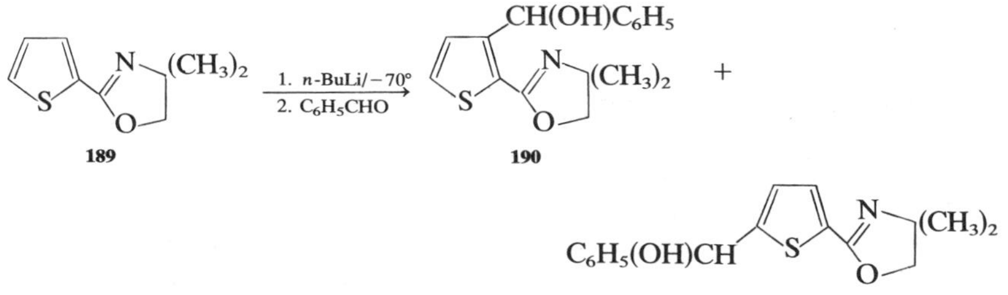

Solvent	190 Yields (%)	191 Yields (%)
THF	36	55
ether	91	4

ortho director. Here, the choice of appropriate conditions permits making either reaction dominant: in tetrahydrofuran the products are almost exclusively derived from the alpha-lithiated species **192**, whereas in ether the product arising from the *ortho*-lithio derivative **193** dominates.[23]

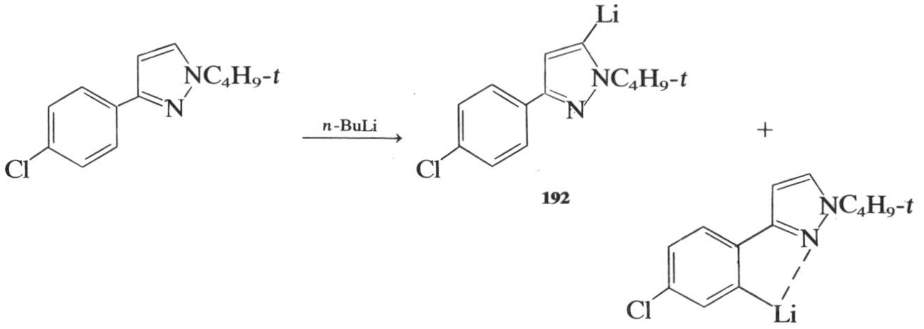

It is certainly no coincidence that in all these examples the azomethine linkage is the director that can successfully compete with the generally facile alpha lithiation process. As alluded to in the thiophene section (p. 36), this appears to be the only group capable of inducing beta lithiation in the presence of an alpha activator. The unique character of the azomethine linkage is apparently the combination of an appreciable electron-withdrawing effect with the pronounced coordinative capacity of the nonbonding orbital of the imine nitrogen.

One additional interesting facet of the competing alpha *vs.* beta lithiation is the relative reactivity of these carbon–lithium bonds toward electrophiles. Subtle differences can be turned into synthetic advantages, particularly in reactions with only moderately active electrophilic substrates. One such example is illustrated by the doubly protected arylpyrazole **194**, which is metalated both in its alpha and its *ortho* position.

Because of the lesser reactivity of the lithiated pyrazole compared to the lithiated phenyl ring, a selective reaction with a tertiary amide is possible. The reactivity of the lithiated pyrazole in the dilithio species **195** is diminished, not only for electronic reasons, but additionally because of the combined chelating effect provided by the three ether oxygens. Accordingly, the product **196** of this reaction is the one derived from *ortho* lithiation.[23] It should be noted that more reactive substrates, such as deuterium oxide or dimethyl disulfide, do indeed lead to products arising from reaction at both sites.[23]

$CH_3O(CH_3)_2COCH_2$ ... $NC(CH_3)_2OCH_3$, Cl

194

$\xrightarrow{2\ s\text{-BuLi}}$

CH_3 CH_3, OCH_3, O, Li, OCH_3, N, N, CH_3, CH_3, Cl, Li

195

$\xrightarrow[H_3O^+]{o\text{-ClC}_6H_4CON(CH_3)_2}$

$HOCH_2$, NH, N, O, Cl, Cl

196 (64%)

The Substrate

The number of substrates that react with alpha- or beta-lithiated species is legion. The highly nucleophilic character of such organometallics provides for a reactivity and versatility approaching that of normal alkyl- and aryllithium[6] or Grignard reagents. Thus, any electrophilic center in a given substrate apt to react with a Grignard reagent is very likely to react with an alpha- or beta-lithiated species. A rather detailed discussion of the behavior of a broad spectrum of electrophilic substrates toward organolithium compounds in general has been presented.[6] The following survey is organized according to the nature of the bonds formed.

C–D/T Bonds. Deuteration (or tritiation) with D_2O (T_2O) or ROD proceeds readily and can be used as a diagnostic tool for the extent and site of lithiation,[23,24] except for metalations with lithium dialkylamides, where D incorporation may be lower.*

C–C Bonds. Formation of this particular bond is by far the most common, and a wide variety of substrates containing electrophilic functionalities of various oxidation states centered at carbon participate in this reaction. These include alkyl halides or sulfates, epoxides, aldehydes, ketones, carboxylate salts, carboxylic acid halides, anhydrides, esters, amides, nitriles, isocyanates, isothiocyanates, alicyclic and heterocyclic imines, as well as Michael acceptors, *e.g.*, the enol ethers of 1,3-diketones and β-amino nitroethylenes.

Alkylation of lithiated species proceeds best with *primary halides* or *sulfates*, in particular with methyl iodide[46,95] or dimethyl sulfate.[274] Yields with secondary halides are often lower because of their decreased reactivity and/or tendency for elimination. Alkylations with *allylic* or *benzylic halides* are generally more successful with the somewhat less basic alpha-lithiated species,[20,275] since deprotonation of these alkylating agents may occur, *e.g.*, benzyl bromide leading to stilbene.[23,112] *Epoxides* generally are well suited for the introduction of $-CH_2CHROH$.[116] Ethylene oxide works especially well in this regard. *Aldehydes* never fail to react with metalated species, and in fact can be used as a label to follow the progress of a lithiation reaction (p. 95).[23,112] *Ketones* react readily. However, with enolizable ketones the yields are often considerably lower than those of the corresponding reactions with aldehydes.[79] Lower temperatures or a change to a less polar solvent can be advantageous on occasion.[23,24,276] Reaction with *carbon dioxide* produces carboxylic acids. This substrate has often been used, particularly in the older literature, to determine the site and the extent of lithiations.[1,277] In view of the possible further reactions of lithium carboxylates,[197] particularly at room temperature, the yields of acid do not always reflect the degree of lithiation. On the other hand, *lithium carboxylates* can be useful substrates for the preparation of ketones.[177,197] *Anhydrides* and *acid halides* are with few exceptions poor substrates for the preparation of ketones. However, the direct conversion of an organolithium species into an ester is feasible under certain conditions by utilizing alkyl chloroformates.[141] Similarly, the use of dialkylcarbamoyl halides leads to tertiary carboxamides.[24] Generally, the use of *esters* to prepare ketones

* Personal communication from B. M. Trost, University of Wisconsin, Madison, Wisconsin.

[274] E. N. Karaulova, D. S. Meilanova, and G. D. Gal'pern, *Dokl. Akad. Nauk SSSR*, **123**, 99 (1958) [*C.A.*, **53**, 5229f (1959)].

[275] H. J. Jakobsen, E. H. Larsen, and S. O. Lawesson, *Tetrahedron*, **19**, 1867 (1963).

[276] J. D. Buhler. *J. Org. Chem.*, **38**, 904 (1973).

[277] C. Tamborski amd E. J. Soloski, *J. Org. Chem.*, **31**, 743 (1966).

has the disadvantage of overreaction.[6] Under certain conditions, however, the desired products may be obtained in good yield.[270] The reaction of lithioorganics with *dimethylformamide*[234] or N-methylformanilide[278] is exceedingly clean and useful for the preparation of aldehydes. Reactions with other *tertiary amides*, particularly those with no acidic alpha hydrogens, lead to the corresponding ketones.[113] The electrophilicity of tertiary amides is, however, not always high enough to make this a general reaction. *Isocyanates* are highly reactive substrates that produce carboxamides in high yield.[279] The addition to *isothiocyanates* proceeds equally well.[46] The reaction with *aldimines* gives access to secondary amines.[102] Alpha deprotonation of the substrate, however, does limit the generality of this reaction.[209,210] The facile addition to the *azomethine* linkage of pyridines,[99] pyrimidines,[280] isoquinolines,[99] quinolines,[83,281] and quinoxalines[282] gives access to the correspondingly substituted heterocycles. The initial products are oxidized readily to the heteroaromatic systems. *Nitriles*, particularly aromatic nitriles, react well to give the corresponding imines,[254] which in turn can be hydrolyzed to the respective ketones.[283] It should be noted, however, that aliphatic nitriles have a tendency to be deprotonated.[112] The *enol ethers* of 1,3-*diketones* may undergo addition/elimination to provide access to beta-substituted enones.[284] In analogous fashion 2-(*dialkylamino*)*nitroethylenes* react with lithiated species to yield beta-substituted nitroolefins.[285]

C–N Bonds. The types of substrates capable of forming this bond are currently very limited. The direct introduction of an amino group is best achieved with methoxylamine,[286] although yields are usually only poor to moderate. A similar conversion is feasible with phenyl azide.[287]

C–O Bonds. Direct oxygenation, although not a general reaction, can be accomplished.[288] The desired conversion, however, is more readily achieved via boronic acids or borate esters as intermediates.[150] The use of alkyl peroxybenzoates permits the direct introduction of alkoxyl groups.[289]

[278] M. B. Groen, H. Schadenberg, and H. Wynberg, *J. Org. Chem.*, **36,** 2797 (1971).

[279] D. A. Shirley and M. D. Cameron, *J. Am. Chem. Soc.*, **74,** 664 (1952).

[280] A. Mitschker, U. Brandl, and T. Kauffmann, *Tetrahedron Lett.*, **1974,** 2343.

[281] D. A. Shirley and M. D. Cameron, *J. Am. Chem. Soc.*, **72,** 2788 (1949).

[282] T. Kauffmann, B. Muke, R. Otter, and D. Tigler, *Angew. Chem., Int. Ed. Engl.*, **14,** 714 (1975).

[283] C. Paulmier, J. Morel, and P. Pastour, *Bull. Soc. Chim. Fr.*, **1969,** 2511.

[284] T. Kametani, H. Nemoto, M. Takeuchi, S. Hibino, and K. Fukumoto, *Chem. Pharm. Bull.* (*Tokyo*), **24,** 1354 (1976).

[285] T. Severin, D. Scheel, and P. Adhikary, *Chem. Ber.*, **102,** 2966 (1969).

[286] H. Gilman and R. K. Ingham, *J. Am. Chem. Soc.*, **75,** 4843 (1953).

[287] B. A. Tertov, V. V. Burykin, and A. V. Koblik, *Khim. Geterotsikl. Soedin.*, **1972,** 1552. [*C.A.* **78,** 58308h (1973)].

[288] H. Gilman and D. L. Esmay, *J. Am. Chem. Soc.*, **76,** 5787 (1954).

[289] A. B. Hörnfeldt and S. Gronowitz, *Ark. Kemi*, **21,** 239 (1963) [*C.A.*, **59,** 13917b (1963)].

C–S Bonds. The reaction of an organolithium reagent with a diaryl or dialkyl disulfide is quite clean and gives high yields. In fact, dimethyl disulfide can be used successfully as a label for following the extent of metalations (p. 95).[65,134,196] Elemental sulfur can be used to produce lithium thiolates which may either be alkylated *in situ*[143] or worked up directly to yield the corresponding mercaptans.[275,290] Sulfinic acids, on the other hand, are accessible by reaction of the lithiated precursors with sulfur dioxide.[112,163]

C–Se Bonds. The reaction of lithioorganics with diselenides corresponds to their reaction with disulfides, and the analogous products are obtained.[291] The same products are also accessible by treatment with elemental selenium, followed by *in situ* alkylation.[292]

C–Halogen Bonds. Synthetic procedures for the preparation of all carbon–halogen bonds are available. For example, the use of perchloryl fluoride with lithioorganic species leads directly to the corresponding fluorinated compounds.[149] N-Chlorosuccinimide may be used to produce chloro derivatives.[204] Similar results are obtained with hexachloroethane.[293] Reaction of these organometallics with bromine,[98,249] 1,2-dibromoalkanes,[293] or *p*-toluenesulfonyl bromide[134] produces the brominated derivatives. Treatment of lithioorganics with iodine is frequently a remarkably good reaction, often producing the corresponding iodides in high yield.[46,149,294] The use of diiodomethane leads to the same iodinated compounds.[293]

C–B Bonds. Reaction with trialkyl borates produces boronic acids in high yields after acid hydrolysis.[114] These in turn may serve as precursors for phenols (treatment with CuOAc), chlorides (CuCl), bromides (CuBr), or iodides (CuI).[2] Boronic acid amides are accessible directly by treatment of lithioorganic species with bis(dialkylamino)boron halides.[295]

C–P Bonds. Reaction of lithioorganics with phosphorus trihalides leads directly to trisubstituted phosphines.[296] Use of phosphorus oxychloride, on the other hand, permits formation of the analogous phosphine oxides directly.[296]

[290] Y. L. Gol'dfarb, M. A. Kalik, and M. L. Kirmalova, *Izv. Akad. Nauk SSSR, Ser. Khim.*, **1964,** 1675 [*C.A.*, **61,** 16032c (1964)].

[291] S. Gronowitz and T. Frejd, *Acta Chem. Scand., Ser. B*, **30,** 439 (1976).

[292] E. Niwa, H. Aoki, H. Tanaka, K. Munakata, and M. Namiki, *Chem. Ber.*, **99,** 3215 (1966).

[293] R. L. Gay, T. F. Crimmins, and C. R. Hauser, *Chem. Ind.* (*London*), **1966,** 1635.

[294] P. Jordens, G. Rawson, and H. Wynberg, *J. Chem. Soc., C*, **1970,** 273.

[295] B. Wrackmeyer and H. Nöth, *Chem. Ber.*, **109,** 1075 (1976).

[296] E. Niwa, H. Aoki, H. Tanaka, and K. Munakata, *Chem. Ber.*, **99,** 712 (1966).

C–Si Bonds. Chlorosilanes are excellent substrates for producing tetrasubstituted silanes.[93,297,298]

C–Metal Bonds. Since the carbon-lithium bond of organolithium reagents is considerably more reactive than the carbon-metal bond of most other metals, unidirectional lithium-metal exchange can be readily achieved. These new organometallic reagents are generally considerably more stable. Moreover, their reactivity is of a softer nature, and therefore they may be utilized for reaction with substrates generally not suited for direct treatment with lithiated species, *e.g.*, acid halides or anhydrides, and allylic or benzylic halides. In the preparation of the new organometallics the source of the less electropositive metal is almost always the halide, although other ligands can be used as well. The following list encompasses some examples representative of the numerous possibilities for forming new carbon–metal bonds. A more comprehensive account of their formation and reactivity is available.[6]

C–Fe Organoiron derivatives are accessible via $Fe(CO)_5$.[299]

C–Co These derivatives are available through the use of $CoCl_2$.[300]

C–Ni Organonickel compounds can be generated from Ni^{2+}–phosphine complexes.[301]

C–Cu Organocopper derivatives or organolithio cuprates are readily available by treatment of lithiated species with cuprous halides.[105,190] Their use in organic synthesis has been reviewed.[302] Reaction of organolithium reagents with cupric chloride produces coupled compounds.[144]

C–Pd $PdCl_2$ or its complexes with phosphines or sulfides are generally used to convert (C–Li) into (C–Pd) bonds.[303]

C–Ag The halides of Ag^+ or Ag^{2+}, when reacted with lithiated species, yield either organosilver derivatives[304] or lithioorganoargentates.[305]

C–Sn Organotin compounds are readily prepared from organotin halides of the general formula R_mSnCl_n, where $m+n=4$, and $R=$organic residue.[306]

[297] S. F. Thames and H C. Odom, Jr., *J. Heterocycl. Chem.*, **3,** 490 (1966).

[298] S. F. Thames and J. E. McCleskey, *J. Heterocycl. Chem.*, **4,** 146 (1967).

[299] I. Rhee, Y. Hirota, M. Ryang, and S. Tsutsumi, *Bull. Chem. Soc. Jpn.*, **43,** 947 (1970).

[300] A. C. Cope and R. N. Gourley, *J. Organomet. Chem.*, **8,** 527 (1967).

[301] L. L. Garber and C. H. Brubaker, Jr., *J. Am. Chem. Soc.*, **90,** 309 (1968).

[302] G. H. Posner, *Org. Reactions*, **19,** 1 (1972).

[303] G. Longini, P. Fantucci, P. Chini, and F. Canziani, *J. Organomet. Chem.*, **39,** 413 (1972).

[304] A. J. Leusink, G. Van Koten, and J. G. Noltes, *J. Organomet. Chem.*, **56,** 379 (1973).

[305] A. J. Leusink, G. Van Koten, J. W. Marsman, and J. G. Noltes, *J. Organomet. Chem.*, **55,** 419 (1973).

[306] H. J. Kroth, H. Schumann, H. G. Kuivila, C. D. Schaeffer, and J. J. Zuckerman, *J. Am. Chem. Soc.*, **97,** 1754 (1975).

C–Pt Dialkyl sulfide complexes of $PtCl_2$ serve to form carbon–platinum complexes.[303]

C–Au The formation of carbon–gold bonds can be achieved by the use of Au^+ complexes such as $R_3P \cdot AuCl$.[307]

C–Hg The reaction between lithiated species and mercuric chloride is very general, leading either to R_2Hg or RHgCl, depending on the stoichiometry of the reaction.[262]

Ortho Interactions

The expected primary products from the reaction of an *ortho*-lithiated species with certain substrates are either not isolable or of only limited stability. The products actually obtained are either derived from a subsequent intramolecular nucleophilic attack of the newly introduced fragment on the electrophilic center of the *ortho*-directing group or, conversely, by a nucleophilic attack of the *ortho*-directing group on the electrophilic center of the new *ortho* substituent.

Thus, *ortho*-directing groups fall into one of two categories:

Electrophiles: CONHR, CSNHR, 2-oxazolines, 2-imidazolines, $CR(O^-)NR_1R_2$ (after hydrolysis to a carbonyl group).
Nucleophiles: CONHR, CSNHR, 2-imidazolines, SO_2NHR, CH_2NHR, CH_2OLi.

Among newly introduced *ortho* groups, the following are apt to undergo intramolecular interactions:

Electrophiles: CHO, $CONR_1R_2$, COR, CO_2H
Nucleophiles: CR_1R_2OH, $CH_2CR_1R_2OH$, C(R)=NH, CR_1R_2NHR.

In addition to the examples illustrated on p. 46 (compound **70**) and p. 55 (compounds **104–107**), the following examples are representative of these types of intramolecular interactions:

$CSNHCH_3$ / Cl —(1. *n*-BuLi, 2. DMF)→ $HNCH_3$, S, CHO, Cl → NCH_3, S, OH, Cl (Ref. 65)

[307] G. Van Koten and J. G. Noltes, *J. Am. Chem. Soc.*, **98,** 5393 (1976).

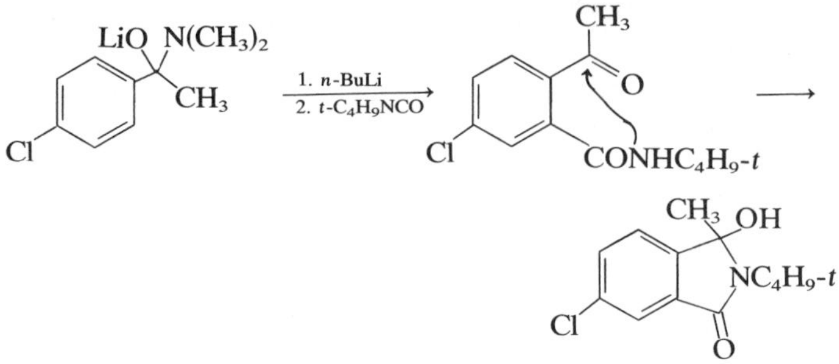

(Ref. 196)

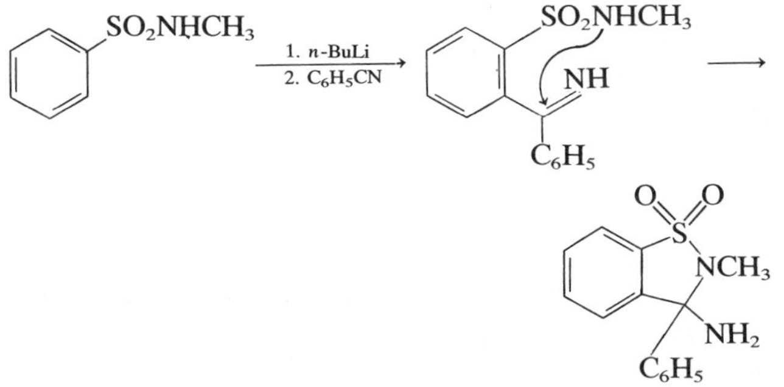

(Ref. 201)

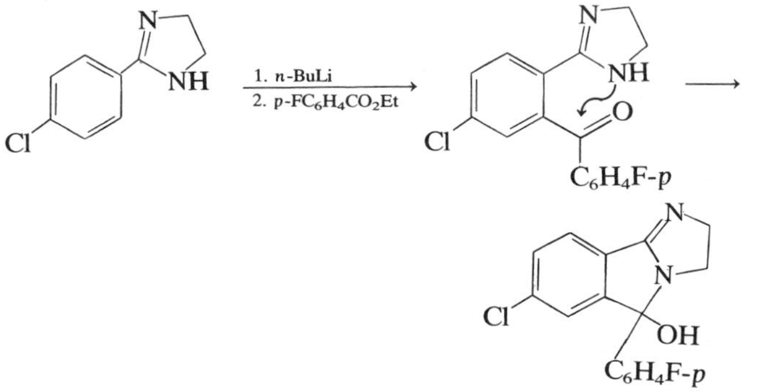

(Ref. 308)

SYNTHETIC UTILITY

The multifaceted nature of heteroatom-facilitated lithiations, encompassing carbocyclic and heterocyclic aromatic as well as olefinic systems, is suggestive of broad synthetic applicability. Moreover, the high degree of regioselectivity and relative facility of this type of reaction, combined with the high reactivity of the new organolithium species toward a variety of electrophilic substrates, contribute to the attractiveness of heteroatom-facilitated lithiations as a synthetic tool. The purposes of the following

[308] W. J. Houlihan, U.S. Pat. 3,932,450 (1976) [*C.A.*, **84**, 135662x (1976)].

discussion are, first, to underscore and pinpoint the distinct advantages of directed lithiations over other methods in the substitution of various aromatic systems; second, to put in perspective the role of lithiations in the development of modern synthetic methodology, in particular, of acyl anion equivalents; and third, to give a general idea of the synthetic potential of selected directing groups.

Carbocyclic Aromatic Systems

The most generally applicable reaction in synthetic benzenoid chemistry is electrophilic substitution. In monosubstituted benzenes this type of reaction produces either a *mixture* of *ortho* and *para* isomers or *meta* isomers, depending on the inductive and resonance effects of the original substituent.[309,310] Whereas only certain *ortho* functionalizations can be achieved via Claisen-type rearrangements,[311] heteroatom-facilitated metalation is the only regioselective and direct, hence practical, synthetic procedure for making accessible *almost any ortho*-substituted benzene derivative. This method is capable of producing novel funtionalities *ortho* to phenolic ethers, benzylamines, and more importantly, carboxylic acids and their derivatives, ketones, sulfones, and sulfonamides. The latter grouping is significant insofar as electrophilic substitution *ortho* to strongly electron-withdrawing groups is essentially impossible. The available alternatives, *i.e.*, nucleophilic substitution of either the appropriate diazonium salts or strongly activated leaving groups,[312] are limited by the availability of the respective starting materials as well as by the nature of the nucleophiles.

For 1,3-disubstituted benzenes electrophilic reactions generally lead to a variety of products depending on the character of the substituents. In contrast, with few exceptions (p. 15) heteroatom-facilitated lithiations permit the regioselective preparation of 1,2,3-trisubstituted compounds. For example, whereas 1,3-dimethoxybenzene undergoes electrophilic substitution in the 4 position,[313] lithiation occurs exclusively between the ether functions.[314] Alternatively, 1,2,3-trisubstituted benzenes are readily available by lithiating 1,2-disubstituted compounds, metalation occurring *ortho* to the more effective directing group (p. 15). In turn the trisubstituted derivatives can be utilized to prepare 1,2,3,4-tetrasubstituted compounds, *etc.*

[309] G. A. Olah, *Acc. Chem. Res.*, **4,** 240 (1971).
[310] D. E. Pearson and C. A. Buehler, *Synthesis*, **1971,** 455.
[311] S. J. Rhoads and N. R. Raulins, *Org. Reactions*, **22,** 1 (1974).
[312] J. F. Bunnett and R. E. Zahler, *Chem. Rev.*, **49,** 273 (1951).
[313] N. P. Buu-Hoï, *Justus Liebigs Ann. Chem.*, **556,** 1 (1944).
[314] K. H. Boltze and H. D. Dell, *Justus Liebigs Ann. Chem.*, **709,** 63 (1967).

For 1,4-disubstituted benzenes, electrophilic substitution may still occur in either of the two *ortho* positions unless the character of the respective substituents is sufficiently different. This is again a situation where *ortho* lithiation provides a viable alternative. For example, *p*-chlorotoluene is acylated under Friedel-Crafts conditions by benzoyl chloride in almost equal amounts at the 2 (56%) and 3 positions (44%), making this a preparatively impractical reaction.[23] In sharp constrast, *p*-chloro-N,N-dimethylbenzylamine is lithiated regiospecifically and almost quantitatively in the 2 position.[23,24,45] Electrophilic substitution of *p*-methylanisole occurs preferentially adjacent to the methoxyl group,[315] as does lithiation,[61] providing access to a 2-substituted-4-methylanisole. In contrast, lithiation and subsequent functionalization of *p*-methoxy-N,N-dimethylbenzylamine take place regioselectively *ortho* to the aminoalkyl group,[19] which in turn can be converted to a methyl group, thus leading to a 3-substituted 4-methylanisole.

Lithiation of naphthalene systems is also feasible, the most useful aspect probably being the accessibility of 1,8-disubstituted derivatives, whose availability by other routes requires more elaborate synthetic sequences.[316] Metalation in the *peri* (8) position of 1-substituted naphthalenes is particularly favored with 1-anilinonaphthalene,[181] 1-[N,N-(dimethylamino)methyl]naphthalene,[191] and under special conditions, 1-methoxynaphthalene.[192]

It should be noted that certain lithiated species obtained by heteroatom-facilitated metalation can also be prepared either by metal–halogen exchange of the corresponding bromo derivatives[317–319] or by destannylation with *n*-butyllithium.[320–322] Although the direct lithiation is usually simpler, more economical, and far more versatile, there are substrates that are incompatible with alkyllithiums even at −78° (p. 16) or whose *ortho*-lithiated species self-condense at this temperature. In these instances metal–halogen exchange can occasionally be utilized to good advantage because of the extremely low temperatures at which the exchange can be effected, *e.g.*, −100°.[323,324]

Ortho metalations with metals other than lithium have been studied

[315] B. Jones, *J. Chem. Soc.*, **1941,** 267.

[316] P. L. Letsinger, J. A. Gilpin, and W. J. Vullo, *J. Org. Chem.*, **27,** 672 (1962).

[317] A. I. Meyers, D. L. Temple, D. Haidukewych, and E. D. Mihelich, *J. Org. Chem.*, **39,** 2787 (1974).

[318] W. E. Parham and Y. A. Sayed, *J. Org. Chem.*, **39,** 2051 (1974).

[319] W. E. Parham and D. C. Egberg, *J. Org. Chem.*, **37,** 1545 (1972).

[320] D. Seyferth and M. A. Weiner, *J. Am. Chem., Soc.*, **83,** 3583 (1961).

[321] E. J. Corey and R. H. Wollenberg, *J. Org. Chem.*, **40,** 2265 (1975).

[322] R. H. Wollenberg, K. Albizati, and R. Peries, *J. Am. Chem. Soc.*, **99,** 7365 (1977).

[323] W. E. Parham and Y. A. Sayed, *J. Org. Chem.*, **39,** 2053 (1974).

[324] W. E. Parham and L. D. Jones, *J. Org. Chem.*, **41,** 1187 (1976).

extensively and used for synthetic purposes.[203,325–328] However, the economy and high reactivity of lithioorganics as well as their lack of toxicity (compare thallium[326] or mercury[329]), clearly make heteroatom-facilitated lithiations a more generally acceptable and practical method.

Heterocyclic Aromatic Systems

The uses of alpha lithiation in heteroaromatic chemistry are legion. It would be redundant to reelaborate the metalation of every heterocyclic system, as much of the utility of this methodology is covered implicitly in the appropriate sections (pp. 17–42). However, a few particularly noteworthy aspects should be emphasized.

For most five-membered heterocycles with one, two, or three heteroatoms, the two-step sequence of lithiation followed by reaction with an electrophile gives access to regioselectively substituted derivatives, many of which differ in the position of the newly introduced functionality from those obtained by direct electrophilic attack on the heterocycle. For example, both pyrazoles[330] and isothiazoles[331] undergo electrophilic substitution on carbon at their respective 4 positions, whereas they are lithiated at position 5 (pp. 23 and 39). In addition, imidazoles,[332] oxazoles,[333] and thiazoles[334] suffer electrophilic attack at either the 5 or a combination of both the 4 and 5 positions, whereas metalation occurs exclusively between the two heteroatoms (pp. 24, 31, and 39). Although few reports exist on the electrophilic substitution of 1,2,3-triazoles, there is evidence that the reaction takes place at the 4 position,[335] whereas lithiation is known to occur at position 5 (p. 26).

Benzimidazoles[332] and benzothiazoles[334] present a different picture in that they undergo electrophilic substitution in the benzenoid portion of the molecule, whereas metalation again occurs between the two heteroatoms (pp. 24 and 39). Another noteworthy example is that of indoles, whose behavior toward electrophiles parallels that of simple enamines.[336] By contrast, lithiation of N-substituted indoles occurs regioselectively at position 2 (p. 22).

[325] H. P. Abicht and K. Issleib, *Z. Chem.*, **17,** 1 (1977).
[326] E. C. Taylor and A. McKillop, *Acc. Chem. Res.*, **3,** 338 (1970).
[327] J. Dehard and M. Pfeffer, *Coord. Chem. Rev.*, **18,** 327 (1976).
[328] G. W. Parshall, *Acc. Chem. Res.*, **3,** 139 (1970).
[329] W. Kitching, *Organomet. Chem. Rev.*, **3,** 35 (1968).
[330] G. Marino, *Adv. Heterocycl. Chem.* **13,** 235 (1971).
[331] R. Slack and K. R. H. Woolridge, *Adv. Heterocycl. Chem.*, **4,** 107 (1965).
[332] E. S. Schipper and A. R. Day, *Heterocycl. Compd.*, **5,** 194 (1957).
[333] R. Lakhan and B. Ternai, *Adv. Heterocycl. Chem.*, **17,** 100 (1974).
[334] J. M. Sprague and A. H. Land, *Heterocycl. Compd.*, **5,** 484 (1957).
[335] T. L. Gilchrist and G. E. Gymer, *Adv. Heterocycl. Chem.*, **16,** 33 (1974).
[336] W. A. Remers, *Heterocycl. Compd.*, **25,** 70 (1972).

Heterosubstituted Olefins—Acyl Anion Equivalents

The principle of reversing the polarity of synthons has received considerable attention during the past decade.[337,338] In particular, it was the development of acyl anion equivalents which opened up numerous new possibilities in synthetic organic chemistry. Although metalated dithianes were first introduced to the armamentarium of acyl anion equivalents,[339] the discovery of the alpha lithiation of ethyl vinyl ether[103] considerably broadened the scope of this synthetic tool. Additionally, the facile metalation of other alkyl vinyl ethers, alkyl or aryl vinyl sulfides, and allenic ethers further extended the accessibility of acyl anion equivalents from commercially or readily available materials.

The highly nucleophilic character of these lithiated species permits reactions with a wide variety of electrophiles. Characteristically, like other organolithium reagents these acyl anion equivalents add 1,2 to most α,β-unsaturated systems. Conjugate addition, however, can be achieved via conversion into the corresponding organolithium cuprates, a topic covered in an earlier volume of this series.[302] Although this aspect has not been exploited to any great extent with the alpha- and beta-lithiated species discussed in this chapter, the possibility nevertheless exists.[105] A representative list of these acyl anion equivalents follows:

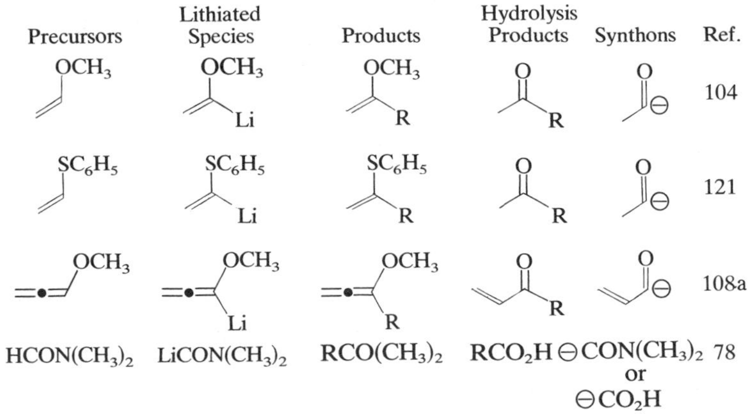

Precursors	Lithiated Species	Products	Hydrolysis Products	Synthons	Ref.
OCH_3	OCH_3, Li	OCH_3, R	O, R	O, ⊖	104
SC_6H_5	SC_6H_5, Li	SC_6H_5, R	O, R	O, ⊖	121
OCH_3	OCH_3, Li	OCH_3, R	O, R	O, ⊖	108a
$HCON(CH_3)_2$	$LiCON(CH_3)_2$	$RCO(CH_3)_2$	RCO_2H	$\ominus CON(CH_3)_2$ or $\ominus CO_2H$	78

Synthetic Potential of *Ortho*-Directing Groups

Of all the *ortho*-directing groups two types stand out as being particularly useful for subsequent elaborations, *i.e.*, dialkylaminoalkyls and carboxylic acid derivatives. The N,N-dialkylaminoalkyl functionality is an

[337] D. Seebach, *Synthesis*, **1969,** 17.

[338] O. W. Lever, Jr., *Tetrahedron*, **32,** 1943 (1976).

[339] E. J. Corey and D. Seebach, *Angew. Chem., Int. Ed. Engl.*, **4,** 1075 (1965).

exceedingly useful synthetic handle, since it can readily be converted to either a reactive halide or quaternary salt suitable for further transformations. Alternatively, the basic functionality can be hydrogenolyzed to a simple alkyl substituent.

$$\text{C}_6\text{H}_5\text{CH}_2\text{N(CH}_3)_2 \xrightarrow[\text{2. Electrophile}]{\text{1. RLi}} o\text{-E-C}_6\text{H}_4\text{CH}_2\text{N(CH}_3)_2 \xrightarrow{\text{Reagent}} o\text{-E-C}_6\text{H}_4\text{CH}_2\text{X}$$

Reagent	X
H_2/catalyst	H
$ClCO_2R$	Cl
BrCN	Br
CH_3I	$N(CH_3)_3^+I^-$

Of the carboxylic acid derivatives only the secondary amides, thioamides, and 2-oxazolines are noteworthy for their synthetic potential. Among the secondary amides the N-*t*-butylcarboxamides offer the most versatility for further transformations, as they can be cleaved either to primary amides with anhydrous acid or to nitriles with acidic dehydrating agents,[24] as well as completely hydrolyzed to carboxylic acids. The particular asset characteristic of thioamides is their potential for mild carbon–carbon bond formation via S-alkylation followed by sulfur extrusion.[340] The 2-oxazolines readily lend themselves to transfunctionalization, as they are convertible not only to aldehydes or ketones[341–343] but also to acids and esters.[317]

$$\text{C}_6\text{H}_5\text{X} \xrightarrow[\text{2. Electrophile}]{\text{1. RLi}} o\text{-E-C}_6\text{H}_4\text{X} \xrightarrow{\text{Reagent}} o\text{-E-C}_6\text{H}_4\text{Y}$$

X	Y	Reagent	Ref.
2-oxazoline	CHO	$CH_3I/NaBH_4$	342, 343
	COR	CH_3I/RMgX	341
	CO_2H, CO_2R	H_3O^+/ROH_2^+	317
$CONHC_4H_9$-*t*	$CONH_2$	CF_3CO_2H	24
	CN	$POCl_3$	24
CSNHR	RNH–C(CH₃)=CH–C(=O)R	1. $R'COCH_2Br$ 2. R_3P	340

[340] M. Roth, P. Dubs, E. Götschi, and A. Eschenmoser, *Helv. Chim. Acta*, **54,** 710 (1971).

[341] A. I. Meyers and E. M. Smith, *J. Org. Chem.*, **37,** 4289 (1972).

[342] I. C. Nordin, *J. Heterocycl. Chem.*, **3,** 531 (1966).

[343] A. I. Meyers, A. Nabeya, H. W. Adickes, I. R. Politzer, G. R. Malone, A. C. Kovelesky, R. L. Nolen, and R. C. Portnoy, *J. Org. Chem.*, **38,** 36 (1973).

Other beta-directing groups with some synthetic potential are masked carbonyl functionalities, in particular α-alkoxidoaralkylamines and the acetals and ketals of aldehydes and ketones. Upon hydrolysis they collapse to the respective carbonyl groups that are well suited to further synthetic elaboration. The α-alkoxidoaralkylamines, generated by the addition of an organometallic to the carbonyl group of a tertiary amide, appear to have a reasonably general applicability, as they may serve to functionalize the *ortho* or beta positions of both carbocyclic and heterocyclic aromatic systems (p. 52). In contrast, the directing effect exerted by acetals and ketals is modest, restricting their synthetic utility to substrates in which lithiation is facilitated by the activating effect of an additional heteroatom (situated in a 1,3-relationship with respect to the acetal); the masked 3-acylated five-membered heterocycles (p. 66) are an example.

As alluded to earlier (p. 85), the intramolecular interaction of *ortho* directors or their transformation products with the newly introduced functionality can give access to numerous bicyclic systems. Such interactions are illustrated on p. 37 (compounds **54** and **55**) and pp. 85–86. Another particularly cogent example is the synthesis of 4-phenyl-1H-2,3-benzoxazine from N,N-dimethylbenzylamine in a four-step sequence.[344] This conversion not only demonstrates the elaboration of bicyclic systems, but also underscores the concept of transfunctionalization of the original directing group.

$CH_2N(CH_3)_2$ → (1. *n*-BuLi; 2. C_6H_5CN; 3. H_3O^+) → $CH_2N(CH_3)_2$, $C(=O)C_6H_5$ → (1. NH_2OH; 2. CH_3I) → $CH_2\overset{+}{N}(CH_3)_3\ I^-$, $C(=NOH)C_6H_5$ → (*n*-BuLi) → O, N, C_6H_5 (51%)

[344] I. T. Barnish and C. R. Hauser, *J. Org. Chem.*, **33**, 1372 (1968).

EXPERIMENTAL CONSIDERATIONS

The Lithiating Agent

Numerous organolithium reagents are now available from commercial sources, generally as filtered solutions packed in bottles capped with serum stoppers or, for larger quantities, in metal cylinders.* If *n*-butyllithium is not available, it can be conveniently prepared from 1-chloro- or 1-bromobutane and lithium metal following detailed published directions.[1,4,6] The *n*-butyllithium/TMEDA complex in hexane[225] or ether[19] and lithium diisopropylamide (p. 98) are generally prepared *in situ*. The quality of fresh commercial solutions of *n*-butyllithium in hexane, the most commonly used lithiating agent, is generally good, and the indicated assay is accurate to within a ±5 % range. Older preparations, particularly those with considerable sedimentation, or reagents prepared *in situ* should be assayed according to methods reviewed previously.[4,8,345,346] We favor the direct titration with *s*-butyl alcohol employing the charge-transfer complex formed between the alkyllithium and 2,2′-biquinoline.[347] In hydrocarbon solvents most of the commonly used lithiating agents, in particular *n*-butyllithium, can be stored at room temperature under an inert atmosphere (nitrogen or argon) for several months. Storage at lower temperatures is not essential and may even prove to be disadvantageous, since the titers may vary because of crystallization or the possible condensation of moisture.

The Solvent

The most commonly used solvents for lithiation reactions are diethyl ether, tetrahydrofuran, and hexane. Other ethers, such as dimethoxyethane,[348] and hydrocarbons, such as cyclohexane[17] or benzene,[24] are used occasionally. As with other reactions involving organometallic reagents, all solvents should be dry.

One limitation in the use of ethereal solvents is their reaction with the lithiating agent itself. The rather rapid deprotonation of tetrahydrofuran by *n*-butyllithium leads to butane, ethylene, and the lithium enolate of acetaldehyde.[349–352] Even diethyl ether is slowly decomposed by *n*-

* U.S. suppliers: Aldrich Chemical Co., Alpha Inorganics, Inc., Columbia Organics, Foote Mineral, Lithium Corporation of America, Tridom Chemical, Inc.

In Europe: Fluka Ltd.

345 A. G. Davies, *Ann. Rep. Prog. Chem., Sec. B*, **64,** 219 (1967) [*C.A.*, **70,** 96836w (1969)].

346 W. G. Kofron and L. M. Baclawski, *J. Org. Chem.*, **41,** 1879 (1976).

347 S. C. Watson and J. F. Eastham, *J. Organomet. Chem.*, **9,** 165 (1967).

348 M. Schlosser, *J. Organomet. Chem.*, **8,** 9 (1967).

349 R. B. Bates, L. M. Kroposki, and D. E. Potter, *J. Org. Chem.*, **37,** 560 (1972).

350 H. Gilman and G. L. Schwebke, *J. Organomet. Chem.*, **4,** 483 (1965).

351 S. C. Honeycutt, *J. Organomet. Chem.*, **29,** 1 (1971).

352 M. E. Jung and R. B. Blum, *Tetrahedron Lett.*, **1977,** 3791.

butyllithium,[350,353] the half-life in this solvent being 153 hours at 25°.[354] A compilation of the stability of various lithiating agents in ethereal solvents is available.[4,6,8]

Selection of Experimental Conditions

To determine the appropriate experimental conditions for an unfamiliar substrate, several factors must be considered. These include the number and types (alpha or beta) of potential sites of metalation, compatibility of additional functional groups with the lithiating agent, solubility of the substrate, and solvent dependence of any subsequent reactions.

If there is a single lithiatable site, the choice of conditions, *i.e.*, metalating agent, solvent, temperature, etc., is fairly straightforward. A suitable starting point may be a search for relevant examples such as those compiled in the tabular survey. However, it should be noted that many of the older procedures have to be viewed in the light of current experimental and theoretical knowledge. For example, the more reactive *n*-butyllithium/TMEDA in ether or hexane (p. 21) may be used to considerable advantage for rate enhancement of some of the more sluggish *ortho* lithiations, notably of those substrates activated by electron-withdrawing groups with moderate coordination potential (p. 44). With few exceptions, *e.g.*, pyrroles, indoles, and enamines, most other *ortho* and alpha lithiations are sufficiently rapid with *n*-butyllithium in ether to obviate the need for additives or accelerators. Still, if a lower temperature or a shorter duration is desirable, the *n*-butyllithium/TMEDA complex can potentially be advantageous. On occasion *n*-butyllithium in tetrahydrofuran can have a similar salutary effect on both alpha and *ortho* metalations. However, caution should be exercised when considering the use of this potential expedient because of the indicated instability of this combination of solvent and metalating agent.

Any solvent choice should also take into consideration the solubility characteristics of the substrate. It should be noted, however, that on occasion intractable starting materials dissolve on metalation.[23,24] Alternatively, soluble substrates with acidic protons, *e.g.*, secondary arylcarboxamides, may precipitate on initial deprotonation and redissolve on lithiation.[23,24] The nature of the solvent also affects the reactivity of the metalated species. For instance, it appears that alkylations are best performed in tetrahydrofuran. The use of ether or hexane may also be successful, however, if prior to the addition of the alkylating agent, an equimolecular amount of hexamethylphosphoramide is added to the lithiated species. On the other hand, tetrahydrofuran is often the least

[353] A. Maercker and W. Theysohn, *Justus Liebigs Ann. Chem.*, **747,** 70 (1971).
[354] H. Gilman and B. J. Gaj, *J. Org. Chem.*, **22,** 1165 (1957).

suitable solvent for reactions of metalated species with enolizable substrates such as ketones.

For those substrates that contain multiple lithiatable sites, the selection of experimental conditions becomes more exacting, since the success or failure of the desired metalation depends on competitive rates. If the competition is between a "coordination-only" group and either an alpha activator or an electron-withdrawing group (other than with pronounced coordination potential) (p. 44), the use of *n*-butyllithium/TMEDA in ether or hexane and in many instances *n*-butyllithium/tetrahydrofuran promotes lithiation adjacent to the inductively acidifying functionality, thus leading to the thermodynamic product. For this purpose lithium dialkylamides can also be used if the substrate site is sufficiently acidic ($pK_a < 30$). They are especially suitable for substrates containing functional groups susceptible to either nucleophilic attack or metal–halogen exchange. In contrast, utilization of a coordinatively unsaturated, Lewis-acid-type metalating agent, such as *n*-butyllithium in ether or hexane, should favor lithiation *ortho* or beta to the more effective coordinating group, hence yielding the kinetic product. If there are *several* sites available *ortho* or beta to the dominant directing group, the position actually metalated is the one rendered most acidic by the inductive effect of additional substituents (for examples see compounds **101**, p. 53, and **148**, p. 66). It should be noted that low temperatures also favor kinetic products.

Analysis of the competition between other combinations of functionalities is no longer simple. Therefore the factors governing the facilitating ability of each functional group should be evaluated in the light of the discussion in preceding sections (pp. 7 and 77) before the choice of appropriate experimental conditions is made. Recent studies have been addressed to the problem of predicting metalation sites from ^{13}C–H bond couplings.[355]

Typical Procedure for an Exploratory Lithiation

The following general experimental procedure may serve as a guideline for the lithiation of a novel substrate. A three-necked, round-bottomed flask equipped with thermometer, serum stopper and dropping funnel, nitrogen inlet, and a magnetic stirring bar is purged with a stream of dry nitrogen. It is then charged with a solution of the substrate in the appropriate dry solvent, normally in a concentration of 3–10 mL/mmol. The solution of the lithiating agent is placed in the dropping funnel, usually 1.1 molar equivalents if a monolithio species is to be generated. The constant stream of nitrogen may then be replaced by a balloon filled

[355] E. B. Pedersen, *J. Chem. Soc., Perkin Trans. II*, **1977**, 473.

with nitrogen, thus keeping the reaction system under a positive nitrogen pressure. In this manner the accumulation of traces of oxygen and moisture carried with the nitrogen can be avoided. Alternatively, a mercury bubbler system may be used. The solution is adjusted to the desired temperature. The metalating agent is then added from the dropping funnel. To monitor the progress of the metalation it is best to withdraw aliquots by means of a syringe and to quench these under an atmosphere of nitrogen at 0° with any of the following substrates: methyl alcohol-*d*, deuterium oxide, dimethyl disulfide, or acetaldehyde. If the proton to be removed is easily detectable in the nmr spectrum, both qualitatively and quantitatively, CH_3OD or D_2O are the reactants of choice, since the amount of deuterated product can be assayed (except with lithium amides as metalating agents; see p. 80). In other cases the use of acetaldehyde or dimethyl disulfide is advisable, particularly if the new methyl signals of the product are well defined and recognizable in the nmr spectrum. An additional advantage of all these "label reactants" is their relative volatility as well as the volatility of the addition products with the lithiating agent (*e.g.*, *n*-butyllithium) itself. Once the degree of lithiation reaches a satisfactory level, the actual reactant can be added. The procedure of the aqueous workup is generally determined by the nature of the expected products.

EXPERIMENTAL PROCEDURES

Unless indicated otherwise, the following experiments were carried out in a three-necked flask equipped with a magnetic stirring bar, dropping funnel, thermometer, and nitrogen inlet.

3-(*p*-Chlorophenyl)-1-(2-tetrahydropyranyl)pyrazole-5-methanol [Typical Alpha Lithiation of a 5-Membered Nitrogen Heterocycle (Pyrazole) with *n*-Butyllithium in Tetrahydrofuran].[23] A solution of 5.25 g (20 mmol) of 3-(*p*-chlorophenyl)-1-(2-tetrahydropyranyl)pyrazole in 100 mL of tetrahydrofuran was cooled under an atmosphere of nitrogen in an ice bath, and 15 mL (24 mmol) of a 1.6 *M* solution of *n*-butyllithium in hexane was added dropwise. After the addition the mixture was stirred at 0–5° for 85 minutes. Then 1.6 g of powdered and dried (at 100°/0.1 mm for 1 hour) paraformaldehyde was added in one portion, and the reaction mixture was allowed to stir at room temperature for 16 hours. The reaction mixture was diluted with ether and washed with basic brine. The organic layer was dried over sodium sulfate and evaporated under reduced pressure. The residue of 5.9 g was crystallized from ether–hexane to yield 4.2 g (72.5%) of product, mp 95–97°; ir (Nujol) 3150–3330 cm^{-1}; nmr ($CDCl_3$) δ 1.4–2.5 (m, 6*H*), 3.2–4.2

(m, 3*H*, 1 exch), 4.65 (s, 2*H*), 5.48 (dd, 1*H*), 6.48 (s, 1*H*), and 7.2–7.8 (AB quartet, 4*H*).

3-O-Methyl-17α-(α-methoxyvinyl)estra-3,17β-diol (Preparation of α-Methoxyvinyllithium via Alpha Lithiation with *t*-Butyllithium in Tetrahydrofuran at Low Temperatures).[104,356] To a solution of 1.1 g (19.2 mmol) of methyl vinyl ether in dry tetrahydrofuran, cooled to −65°, was added under an atmosphere of nitrogen 7.5 mL (12 mmol) of a 1.6 *M* solution of *t*-butyllithium in pentane. After removal of the cooling bath, the yellow precipitate redissolved and the solution became colorless between −5 and 0°. This solution, now containing 12 mmol of α-methoxyvinyllithium, was cooled again to −60°. Then 1.15 g (4 mmol) of estrone methyl ether in 20 mL of dry tetrahydrofuran was added, and the mixture was allowed to warm to 0° over a period of 0.5 hour. The mixture was then quenched with aqueous ammonium chloride and extracted with ether. The organic layer was dried over anhydrous magnesium sulfate, and most of the ether was removed under reduced pressure. Addition of 50 mL of hexane induced crystallization of the product, which was collected in 74% yield (1.02 g), mp 144–146°; ir (CCl_4) 3650, 1665, and 1620 cm^{-1}; nmr (CCl_4) δ 0.90 (s, 3*H*), 1.0–2.4 and 2.6–3.0 (m, 16*H*), 3.54 (s, 3*H*), 3.70 (s, 3*H*), 4.02 and 4.15 (AB quartet, $J = 2.5$ Hz, 2*H*), 6.5–6.8 and 7.05–7.3 (m, 3*H*).

α,α-Diphenyl-2-furylmethanol (Alpha Lithiation of Furan with *n*-Butyllithium in Tetrahydrofuran or Ether).[23] To an ice-cooled solution of 4.08 g (60 mmol) of furan in 60 mL of dry tetrahydrofuran was added dropwise 34.4 mL (55 mmol) of a 1.6 *M* solution of *n*-butyllithium in hexane under an atmosphere of nitrogen. After the addition the solution was allowed to stir at ice-bath temperature for 3 hours. Then a solution of 9.1 g of benzophenone (50 mmol) in 30 mL of dry tetrahydrofuran was added at a rapid rate, and the reaction mixture was stirred at room temperature for 2 hours. The mixture was poured onto brine, the aqueous layer was reextracted with ether, and the organic phases were dried over sodium sulfate. After evaporation of the solvents the residue of 12.8 g was crystallized from ether–hexane to give a total of 12.45 g of analytically pure product (98.8% based on benzophenone), mp 86–87°; ir (Nujol) 3380 cm^{-1}; nmr ($CDCl_3$) δ 3.5 (s, 1*H*), 5.9 (d, $J = 3$ Hz, 1*H*), 6.25 (dd, $J = 3$ and 2 Hz, 1*H*), and 7–7.7 (m, 11*H*).

Alternatively, the reaction can be carried out in ether (−20° during the addition of *n*-butyllithium, then reflux for 4 hours) to give the same product in 98% yield.[112]

[356] J. E. Baldwin, O. W. Lever, Jr., and N. R. Tzodikov, *J. Org. Chem.*, **41**, 2312 (1976).

1-Hydroxycyclohexanecarboxylic Acid Dimethylamide (Alpha Lithiation of Dimethylformamide with Lithium Diisopropylamide and Reaction with an Enolizable Ketone).[23,78] A solution of 2.92 g (40 mmol) of dimethylformamide and 3.92 g (40 mmol) of cyclohexanone in 100 mL of dry ether–tetrahydrofuran (4 : 1) was cooled under an atmosphere of nitrogen to −78°. In a separate three-necked flask a solution of lithium diisopropylamide (44 mmol) [prepared from 6.16 mL (44 mmol) of diisopropylamine in 17 mL of ether and 27.5 mL of a 1.6 *M* solution of *n*-butyllithium in hexane (44 mmol)] was also precooled to −70° under nitrogen. The solution of lithium diisopropylamide was then withdrawn by means of a syringe via a serum stopper and transferred, again through a serum stopper, into the flask containing the substrates. The reaction mixture was then kept at −70° for 6 hours. Subsequently, the temperature was raised to 20°, the reaction was quenched with an aqueous solution of ammonium chloride, and the layers were separated. After reextraction of the aqueous layer with ether and washing of the organic layers with brine, the ethereal solution of the product was dried over sodium sulfate and evaporated. The oily residue, containing both product and unreacted (because of enolization) cyclohexanone, was kept under high vacuum at 70° to remove the unreacted ketone. The solidified colorless residue was recrystallized from cyclohexane to yield 3.1 g (45.5%) of 1-hydroxycyclohexanecarboxylic acid dimethylamide as colorless needles, mp 105°; ir (Nujol) 3340 and 1600 cm^{-1}; nmr ($CDCl_3$) δ 1.2–2.1 (m, 10*H*), 3.14 (s, 6*H*), and 4.24 (s, 1*H*). A similar procedure gave a higher yield (62%).[78]

3-Hydroxy-2-(phenylthio)-1-octene (Alpha Lithiation of Phenyl Vinyl Sulfide in Tetrahydrofuran).[124,126]

A. With Lithium Diisopropylamide.* A solution of freshly distilled phenyl vinyl sulfide (1.36 g, 10 mmol) in 10 mL of dry tetrahydrofuran was added dropwise and under nitrogen to a cold (−60°) solution of lithium diisopropylamide (10 mL) [prepared from diisopropylamine (1.01 g) and *n*-butyllithium (7.69 mL of a 1.3 *M* solution in hexane)] and hexamethylphosphoramide (3 mL). After this solution was stirred at −60° for 30 minutes, 1.0 g (10 mmol) of *n*-hexanal in 5 mL of dry tetrahydrofuran was added dropwise. After an additional 30 minutes the cooling bath was removed, and the mixture was allowed to reach room temperature. The reaction mixture was then poured into water and extracted with two 10-mL portions of ether. The ethereal extracts were combined and

* Lithiation of phenyl vinyl sulfide with lithium diisopropylamide proceeds efficiently under these conditions, but the yields of adducts obtained are variable. The use of *n*-butyllithium/TMEDA appears to give more consistent results.

dried over anhydrous sodium sulfate. After evaporation of the solvent the crude product was obtained as an oil, which was distilled to give 1.81 g (76%) of 3-hydroxy-2-(phenylthio)-1-octene, bp 120° (0.1 mm); nmr ($CDCl_3$) δ 1.4 (m, 12*H*), 4.2 (t, $J = 6$Hz, 1*H*), 4.95 (s, 1*H*), 5.5 (m, 1*H*), and 7.4 (m, 5*H*).

B. With *n*-Butyllithium/TMEDA* Procedure A was repeated by adding the phenyl vinyl sulfide to a cold solution (−70°) of TMEDA (1.16 g, 10 mmol) and *n*-butyllithium (7.69 mL of a 1.3 *M* solution in hexane, 10 mmol) in 30 mL of dry tetrahydrofuran. The yield of distilled product was 2.0 g (84%).

2-Thiophenethiol. The preparation of this compound in 65-70% yield by alpha lithiation of thiophene and reaction with elemental sulfur is described in *Organic Syntheses.*[357]

Benzo[*b*]thiophen-2(3H)-one (Alpha Lithiation of Benzothiophene with *n*-Butyllithium in Ether and Reaction with Trialkylborate–Hydrogen Peroxide).[150] To a solution of 26.8 g (0.2 mol) of benzothiophene in 100 mL of anhydrous ether was added 130 mL of a 1.6 *M* solution of *n*-butyllithium in hexane (0.208 mol). The mixture was refluxed for 45 minutes, and the resulting solution of 2-lithiobenzothiophene was cooled in a dry ice–acetone bath and treated with 64.5 g (0.28 mol) of tri-*n*-butyl borate. The resulting gelatinous precipitate was hydrolyzed with 200 mL of 1 *N* hydrochloric acid, initially at 0°, then for 1 hour at room temperature. The two layers were separated, and the aqueous phase was reextracted with ether. The combined organic layers were then extracted with 200 mL of 1 *N* sodium hydroxide, and the basic aqueous layer was backwashed with ether. Acidification of the aqueous layer with ice-cold 3 *N* hydrochloric acid gave a pink-yellow, odorous precipitate of crude boronic acid, which was filtered and washed with water. The crude boronic acid was dissolved in a small volume of ether with stirring and treated with 96 mL of 10% hydrogen peroxide containing 2 mL of saturated aqueous sodium carbonate solution. The resulting mixture was refluxed for 1 hour and then stirred at room temperature. The layers were separated, and the aqueous phase was reextracted with ether. The combined organic phase was washed with water until the ferrous ion test no longer indicated the presence of hydrogen peroxide. Finally the ether was washed with brine, dried over magnesium sulfate, and evaporated under vacuum to give 21.7 g (72%) of

* See footnote on p. 98.

[357] E. Jones and I. M. Moodie, *Org. Syntheses*, **50,** 104 (1970).

crystalline product, mp 32–34°. Recrystallization from hot aqueous methanol and cooling to −20° gave pale-yellow needles, mp 34–35°; ir (CCl_4) 1723, 1595, and 1460 cm^{-1}; nmr ($CDCl_3$) δ 3.92 (s, 2*H*), and 7.24 (m, 4*H*).

2-Chloro-3,3-diphenylacrylic acid (Alpha Lithiation of a Chloroalkene with *n*-Butyllithium in Tetrahydrofuran).[358] To a solution of 6.45 g (30 mmol) of 1-chloro-2,2-diphenylethylene in 50 mL of tetrahydrofuran, cooled to −71°, was added over a 1-hour period 23 mL of a 1.32 *M* solution of *n*-butyllithium in ether (30 mmol). At the onset of the addition the color of the reaction mixture was pink, which then changed slowly to yellow and finally to a light brown. The mixture was poured onto solid powdered carbon dioxide which had been covered with dry ether. After the addition of water the ether–tetrahydrofuran mixture was removed on the rotary evaporator. The basic aqueous layer was then extracted successively with ether, acidified with an excess of dilute sulfuric acid, and extracted again with ether. The combined layer was dried over calcium chloride, filtered, and evaporated. The crude acid so obtained (6.43 g, 83%, mp 130–133°) was recrystallized from cyclohexane to give 6.20 g of 2-chloro-3,3-diphenylacrylic acid, mp 136°.

Evaporation of the neutral layer gave 0.8 g of crystals (mp 45–47°) consisting of a mixture of starting material (240 mg, 3.7%) and diphenylacetylene (560 mg, 10%).

An analogous lithiation in ether as solvent resulted in the recovery (99%) of starting material.[358]

[6-(Dimethylamino)-*m*-tolyl]diphenylmethanol. The preparation of this compound in 49–57% yield by *ortho* lithiation of N,N-dimethyl-*p*-toluidine with *n*-butyllithium–TMEDA is described in *Organic Syntheses.*[359]

3-Chloro-6-[(dimethylamino)methyl]-2′-fluorobenzophenone (*Ortho* Lithiation of a Tertiary Benzylamine with *n*-Butyllithium in Ether).[23] A solution of 119 g (0.7 mol) of *p*-chloro-N,N-dimethylbenzylamine in 1.7 L of anhydrous ether was cooled to 2–4° in an ice bath. Then 480 mL (0.77 mol) of a 1.6 *M* solution of *n*-butyllithium in hexane was added at such a rate that the temperature did not exceed 4°. The reaction mixture was allowed to stir under a positive nitrogen pressure (balloon) at ice-bath temperature for 12 hours. Subsequently a solution of 85 g (0.7 mol) of *o*-fluorobenzonitrile in 350 mL of ether was added, and the

[358] G. Köbrich and H. Trapp, *Chem. Ber.*, **99**, 670 (1966).

[359] J. V. Hay and T. M. Harris, *Org. Syntheses*, **53**, 56 (1973).

mixture was stirred at 25° for an additional 12 hours. After the flask was again cooled in an ice bath, 1.3 L of 5 *N* hydrochloric acid was added, carefully at the beginning, and the biphasic mixture was then refluxed for 40 minutes. The two layers were separated in a separatory funnel, the ethereal layer was reextracted with 200 mL of dilute hydrochloric acid, and the aqueous layers were combined. After the addition of ice, the acidic layer was brought to pH 11 with 30% aqueous sodium hydroxide solution. The oily product was then extracted into methylene chloride (two extractions), and the organic layer was separated and dried over sodium sulfate. The solvent was removed under vacuum, and the residue was crystallized from hexane (600 mL)/ether (50 mL) to give a total of 160.5 g (79%) (three crops) of analytically pure product, mp 89°; ir (Nujol) 1653 cm^{-1}; nmr ($CDCl_3$) δ 1.97 (s, 6*H*), 2.36 (s, 2*H*), and 7.0–7.8 (m, 7*H*).

2-Acetyl-5-chlorobenzaldehyde (One-Pot Transformation of a Tertiary Benzamide into an *Ortho*-Functionalized Acetophenone by Addition of Methyllithium, *Ortho* Metalation with *n*-Butyllithium in Tetrahydrofuran, and Reaction with Dimethylformamide).[196] A solution of 10 g (53.6 mmol) of *p*-chloro-N,N-dimethylbenzamide in 100 mL of dry tetrahydrofuran was cooled to −78°. Then 31.2 mL of a 2.0 *M* solution of methyllithium in ether (59 mmol) was added dropwise. After the mixture was stirred at that temperature for 1 hour, 368 mL (59 mmol) of a 1.6 *M* solution of *n*-butyllithium in hexane was added dropwise. The mixture was then stirred at 20–25° for 16 hours. The solution of the dilithio species thus formed was cooled in an ice bath and allowed to react with a solution of 4.4 g (59 mmol) of dimethylformamide in 20 mL of dry tetrahydrofuran. The reaction mixture was stirred at 0° for 0.5 hours, and then the temperature was raised to 25° and maintained for 1 hour. The mixture was quenched with 50 mL of 3 N hydrochloric acid and extracted with ether. The combined extracts were washed with water and dried over sodium sulfate. After evaporation of the solvents under vacuum the crude orange oil (6.5 g) was distilled [165–170° (0.7 mm)] to give 5.6 g (56%) of the product as a colorless oil; ir (film) 1688 cm^{-1}; nmr ($CDCl_3$) δ 10.1 (s, 1*H*), 7.3–8.0 (m, 3*H*), and 2.68 (s, 3*H*).

2-*t*-Butyl-3-hydroxyphthalimidine (*Ortho* Lithiation of a Secondary Benzamide in Tetrahydrofuran with 2 Equivalents of *n*-Butyllithium).[23,24] A solution of 5.32 g (30 mmol) of N-*t*-butylbenzamide in 100 mL of dry tetrahydrofuran was cooled in an ice bath under an atmosphere of nitrogen. To this solution was added 41.3 mL (66 mmol) of a 1.6 *M* solution of *n*-butyllithium in hexane at such a rate that the internal temperature did not exceed 10°. The generation of the monoanion is markedly exothermic. After the addition the mixture was allowed to stir at ice-bath temperature for 4 hours. The generation of the desired dianion is evident by the formation of a thick white precipitate. Subsequently, 2.9 g (3.1 mL, 40 mmol) of dry dimethylformamide was added neat, and the mixture was allowed to stir for 2.5 hours at room temperature. The reaction was quenched with ice and excess 2 *N* hydrochloric acid, and the aqueous phase was reextracted with ether. The organic layers were washed with brine and dried over anhydrous sodium sulfate. The thick, viscous oil obtained after evaporation of the solvent was dissolved in hot hexane, whereupon the product crystallized to give 5.7 g, mp 105–125°. Recrystallization from hot toluene gave 5.05 g (82%) of pure 2-*t*-butyl-3-hydroxyphthalimidine, mp 132–135°; ir (Nujol) 3205 and 1655 cm^{-1}; nmr (DMSO-d_6) δ 1.58 (s, 9*H*), 6.0 (d, $J=$ 10 Hz, 1*H*), 6.34 (d, $J=10$ Hz, 1*H* exch), and 7.53 (s, 4*H*).

2-[*p*-Methoxyphenyl-2-(phenylthio)]-4,4-dimethyl-2-oxazoline (*Ortho* Lithiation of an Aryloxazoline Using *n*-Butyllithium in Ether and Reaction with a Diaryl Disulfide).[46] A solution of 3.07 g (15 mmol) of 2-(*p*-methoxyphenyl)-4,4-dimethyl-2-oxazoline in 65 mL of dry ether was cooled in an ice bath under an atmosphere of nitrogen. Then 10.3 mL (16.5 mmol) of a 1.6 *M* solution of *n*-butyllithium in hexane was added dropwise, and the mixture was stirred at this temperature for 4 hours. A solution of 3.65 g (16.5 mmol) of diphenyl disulfide in 30 mL of dry ether was added, and the reaction mixture was stirred for 16 hours at ambient temperature. The reaction was quenched with ice water, and the ethereal layer was washed with dilute sodium hydroxide solution to separate the thiophenol and finally with brine. The organic layer was dried over sodium sulfate and evaporated to produce a residue of 5.6 g. The thioether was crystallized from ether–hexane to give 4.2 g (89%), mp 51–53°; ir (Nujol) 1630 and 1590 cm^{-1}; nmr ($CDCl_3$) δ 1.4 (s, 6*H*), 3.55 (s, 3*H*), 4.02 (s, 2*H*), 6.35 (d, $J=3$ Hz, 1*H*), 6.52 (dd, $J=8$ and 2 Hz, 1*H*), 7.2–7.7 (m, 5*H*), and 7.75 (d, $J=8$ Hz, 1*H*).

3-Chloro-2-(methoxymethoxy)benzaldehyde (*Ortho* Lithiation of a Protected Phenol with *n*-Butyllithium/TMEDA in Hexane and Reaction with Dimethylformamide).[234] To a solution of 261 mL of 1.92 *M n*-butyllithium in hexane (0.5 mol) was added 86.3 g (0.5 mol) of TMEDA. After the mixture had cooled in an ice bath, 86.3 g (0.5 mol) of *o*-chlorophenyl methoxymethyl ether was added during 30 minutes, with the temperature maintained between 0–5°. The mixture was stirred for an additional 30 minutes at this temperature. The yellow slurry was transferred under an atmosphere of nitrogen through a polyethylene tube into an addition funnel, from which it was added over a 25-minute period to a solution of 43.8 g (0.6 mol) of dimethylformamide in 470 mL of xylene. During the addition the reaction mixture was stirred vigorously, and the temperature was maintained at 0–5° by external cooling. Stirring at this temperature was continued for an additional hour, after which the reaction mixture was transferred slowly through a polyethylene tube into a stirred mixture of 37% hydrochloric acid (190 mL) and crushed ice (900 g). The temperature was watched carefully and not permitted to rise above 5°. Stirring at 0–5° was continued for 20 minutes after the transfer. The aqueous phase was then separated and discarded, and the organic layer was washed with cold dilute hydrochloric acid (1 *N*, 200 mL) and with brine (200 mL). The organic layer was stirred for 20 minutes with a solution of sodium bisulfite (52 g, 0.5 mol) in water (120 mL) mixed with ice (150 g). The aqueous solution was separated and kept cold. This extraction scheme was repeated twice, each time with half of the initial quantities. A solution of sodium hydroxide (45 g) was added slowly to the combined bisulfite extracts with stirring and cooling to maintain the temperature below 10° while the pH was adjusted to 11 toward the end. The crystals were collected by suction, washed with water, and air-dried overnight. The yield of product was 85.4 g (65%), mp 38–40°. Recrystallization of a sample from *n*-hexane afforded pure material, mp 40–41°; nmr ($CDCl_3$) δ 3.56 (s, 3*H*), 5.16 (s, 2*H*), 6.95–7.35 (m, 1*H*), 7.5–7.9 (m, 2*H*), and 10.25 (s, 1*H*).

N-*t*-Butyl-*o*-formyl-N-methylbenzenesulfonamide (*Ortho* Lithiation of a Tertiary Arylsulfonamide with *n*-Butyllithium in Ether).[24] A solution of 200 g (0.88 mol) of N-*t*-butyl-N-methylbenzenesulfonamide in 2.1 L of anhydrous ether was cooled to −70° under an atmosphere of nitrogen. At that temperature the sulfonamide precipitated partially. Then 570 mL of a 1.6 *M* solution of *n*-butyllithium in hexane (0.91 mol) was added dropwise. The mixture was warmed to 0° and kept at that temperature for 1 hour. Subsequently, a solution of dimethylformamide (76 g, 1 mol) in 200 mL of anhydrous ether was added dropwise. After an additional hour at ice-bath temperature, the reaction was quenched with water, and the ether was separated and dried over sodium sulfate. The solvent was evaporated under reduced pressure, and the residue was triturated with cold isopropyl alcohol to give 183 g (81%) of white crystalline product, mp 69–71°; ir (Nujol) 1690 cm^{-1}; nmr ($CDCl_3$) δ 1.20 (s, 9*H*), 2.72 (s, 3*H*), 7.2–7.8 (*m*, 4*H*), and 10.54 (s, 1*H*).

TABULAR SURVEY

The information in the following tables is an extension of that reviewed previously,[1] covering the literature to the end of December 1977. Additional significant findings reported through August 1978 are included as well. Although the attempt has been made to present an exhaustive list of *successful* examples of this reaction, difficulties in searching the literature make it likely that some references were overlooked. The arrangement of the tabular survey parallels the text; *i.e.*, the main division is according to the *type* of lithiation. Alpha lithiations are presented first, followed by beta (*ortho*) lithiations. Within each type the next determinant is the heteroatom of the directing functionality, the tables being arranged in the order of the increasing atomic number of the heteroatom. The final arrangement is then determined by the specific directing functionality.

Within each table the compounds lithiated are listed according to the increasing number of carbon atoms. If a compound contains more than one director, it can be found in each table appropriate for the specific director. The conditions stated are those of the lithiation itself and not of subsequent reactions with electrophiles. To avoid ambiguity temperature ranges below 0°C are expressed, for example, as: −70° to −30°. If a reference contains more than one set of lithiation conditions for a particular compound, these data are fully presented only if it was deemed significant. When there is more than one reference for a given compound–substrate combination, the information tabulated is considered to describe the optimum lithiation conditions reported. This reference is listed first, and the remaining references are arranged in numerical order. In the substrate column, aside from the electrophile, other reagents are listed only if the initial intermediate is further transformed, *e.g.*, oxidized, hydrolyzed, *etc.* In naming products attempts have been made to conform to *Chemical Abstracts* nomenclature (Vols. **56–65**). Yields, which are indicated by a dash (–), are not specified in the reference(s) cited. Lithiation conditions or substrates, marked by an asterisk (*), are to be taken as an interpretation of the authors because the original report was not explicit.

Standard abbreviations used throughout the tables are the following:

BuLi	*n*-butyllithium
LDA	lithium diisopropylamide
LDCA	lithium dicyclohexylamide
LTMP	lithium tetramethylpiperidide
DMF	dimethylformamide
DMA	dimethylacetamide

THF	tetrahydrofuran
HMPA	hexamethylphosphoramide
ether	diethyl ether
pet. ether	petroleum ether
Ac	acetyl
Et	ethyl
THP	tetrahydropyranyl

*Under thermodynamic conditions, tertiary amides are more powerful *ortho* directors than sulfonamides and oxazolines (personal communication from P. Beak, University of Illinois, Urbana, Illinois).

TABLE I. ENAMINES (ALPHA)

Formula	Compound Lithiated	Conditions	Substrate	Product and Yield (%)	Refs.
$C_7H_8N_2$	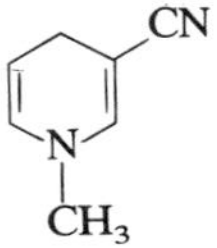	LDA/THF/ −80°/20 hr	CH_3OD	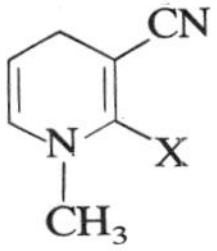 X = D (65)	218
		”	FSO_2OCH_3	$X = CH_3$ (77)	218
	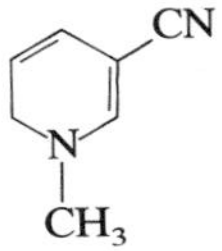	LDA/THF/ −80°/1.5 hr	CH_3OD	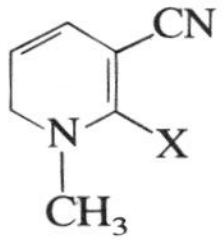 X = D (54)	218
		”	FSO_2OCH_3	$X = CH_3$ (72)	218
$C_7H_{10}N_2$	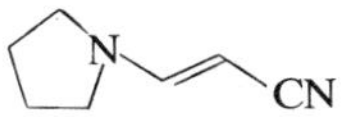	LDA/THF, pentane/ −113°/40 min	CH_3OD	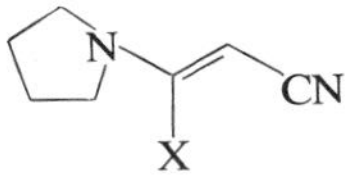 X = D (95)	57c
		”	CH_3I	$X = CH_3$ (74)	57c
		”	C_2H_5I	$X = C_2H_5$ (35)	57c
		”	C_6H_5CHO	$X = CH(OH)C_6H_5$[a] (42)	57c
$C_9H_{15}NO_2$	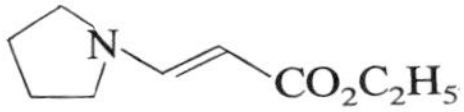 $CO_2C_2H_5$	t-BuLi/THF/ −113°/15 min	$C_2H_5CO_2CH{=}CHN$ 	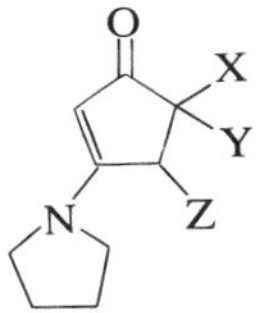 $X = CO_2C_2H_5$, Y = H, Z = N  (41)	57b

		”	$C_6H_5CH{=}CHCO_2C_4H_9\text{-}t$		$X = CO_2C_4H_9\text{-}t$, $Y = H$, $Z = C_6H_5$ (67)	57b
		”	$C_6H_5COCH{=}CHN$ 		$X = COC_6H_5$, $Y = H$, $Z = N$  (49)	57b
		”	$C_6H_5CH{=}C(C_6H_5)CN$		$X = Z = C_6H_5$, $Y = CN$ (32)	57b
		”	$C_6H_5COCH{=}CHCOC_6H_5$		$X = Z = COC_6H_5$, $Y = H$ (25)	57b
		”	CH_3CHO	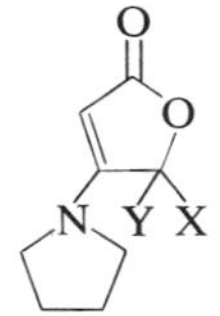 O, O, N, Y, X	$X = CH_3$, $Y = H$ (49)	57b
		”	$HCO_2C_2H_5$		$X = OC_2H_5$, $Y = H$ (52)	57b
		”	$(CO_2CH_3)_2$		$X = CO_2CH_3$, $Y = OCH_3$ (58)	57b
		”	C_6H_5CHO		$X = C_6H_5$, $Y = H$ (51)	57b
		”	$C_6H_5CO_2CH_3$		$X = C_6H_5$, $Y = OCH_3$ (43)	57b
$C_{11}H_{11}N$	 N, C_6H_5	BuLi/THF/0°/ 4.5 hr	D_2O	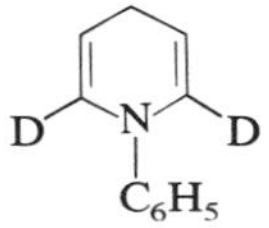 D, N, D, C_6H_5	(82)	74

TABLE I. ENAMINES (ALPHA) (*Continued*)

Formula	Compound Lithiated	Conditions	Substrate	Product and Yield (%)	Refs.
$C_{11}H_{20}N_2O$	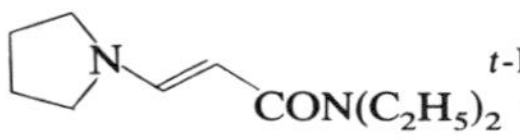 1-pyrrolidinyl-CH=CH-$CON(C_2H_5)_2$	t-BuLi/THF/ −115°	CH_3OD	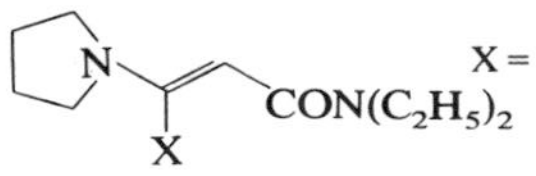 1-pyrrolidinyl-C(X)=CH-$CON(C_2H_5)_2$, X = D (100)	57a
		”	CH_3I	X = CH_3 (95)	57a
		”	C_2H_5I	X = C_2H_5 (60)	57a
		”	$C_6H_5CO_2CH_3$	X = COC_6H_5 (95)	57a
		”	p-$CH_3C_6H_4CO_2CH_3$	X = $COC_6H_4CH_3$-p (60)	57a
		t-BuLi/THF/ −120°/0.5 hr	p-$O_2NC_6H_4CHO$	X = $CH(OH)C_6H_4NO_2$-p (60)	604
		”	C_6H_5CHO	X = $CH(OH)C_6H_5$ (61)	604
		”	fluorenone	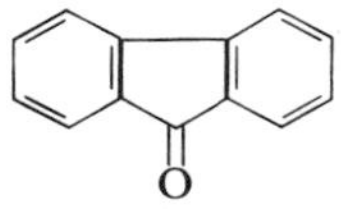 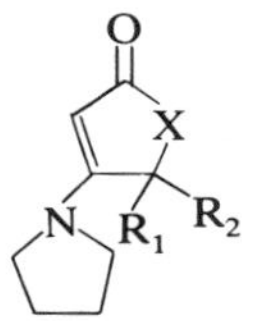 4-(1-pyrrolidinyl)-5-R_1-5-R_2-X-cyclopent-3-en-2-one: X = O, R_1R_2 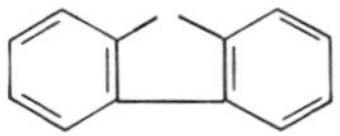 = fluorenylidene (56)	604
		”	$(C_6H_5)_2CO$	X = O, $R_1 = R_2 = C_6H_5$ (45)	604
		”	$C_6H_5CH{=}NC_6H_5$	X = NC_6H_5, R_1 = H, $R_2 = C_6H_5$ (26)	604
		”	p-$(CH_3)_2NC_6H_4CH{=}NC_6H_5$	X = NC_6H_5, R_1 = H, $R_2 = NC_6H_4N(CH_3)_2$-p (21)	604

[a] This compound was contaminated with the ring isomer.

Note: References 360–607 are on pp. 355–360.

TABLE II. Vinyl Isocyanides (Alpha)

Formula	Compound Lithiated	Conditions	Substrate	Product and Yield (%)	Refs.
C_7H_9N	(cyclopentylidene)methyl isocyanide [structure: cyclopentane=CH–NC]	BuLi/THF, ether, pet. ether/−110°/ 0.5 hr	$(CH_3)_3SiCl$	α-(Trimethylsilyl)$\Delta^{1,\alpha}$-cyclopentane isocyanide (63)	76
C_9H_7N	$C_6H_5CH{=}CHNC$	BuLi/THF, ether, pet. ether/-110°/ 0.5 hr	CO_2	$C_6H_5CH{=}C(NC)X$ $X = CO_2Li$ (95)	76
		"	CH_3I	$X = CH_3$ (75)	76
		"	$ClCO_2C_2H_5$	$X = CO_2C_2H_5$ (70)	76
		"	$(CH_3)_3SiCl$	$X = Si(CH_3)_3$ (53)	76
		"	C_6H_5COCl	$X = COC_6H_5$ (94)	76
		"	CH_3COCH_3	$C_6H_5CH{=}$(oxazoline ring, N=C–C(OH)RR′, O) $R = R' = CH_3$ (77)	76
		"	C_6H_5CHO	$R = H$, $R' = C_6H_5$ (36)	
		"	$(C_6H_5)_2CO$	$R = R' = C_6H_5$ (—)	76
$C_{10}H_9N$	$C_6H_5(CH_3)C{=}CHNC$	BuLi/THF, ether, pet. ether/−110°/ 0.5 hr	$(CH_3)_3SiCl$	β-Methyl-α-(trimethylsilyl)cinnamylisocyanide (78)	76

TABLE III. Formamides and Thioformamides (Alpha)

Formula	Compound Lithiated	Conditions	Substrate	Product and Yield (%)		Refs.
C_3H_7NO	$HCON(CH_3)_2$	LDA/THF/ether, −78°	t-C_4H_9CHO	$(CH_3)_2NCOX$	$X = CH(OH)C_4H_9$-t (76)	78
		"	Cyclohexanone		$X = HO$-(cyclohexyl) (62)	78
		"	C_6H_5CHO		$X = CH(OH)C_6H_5$ (45)	78
		"	$C_6H_5CH{=}CHCHO$		$X = CH(OH)CH{=}CHC_6H_5$ (48)	78
		"	$(C_6H_5)_2CO$		$X = C(C_6H_5)_2OH$ (85)	78
		"	Estra-1,3,5(10)-triene-17-one	17-Hydroxy-3-methoxy-N,N-dimethyl-1,3,5(10)-estratriene-17-carboxamide (72)		78
C_3H_7NS	$HCSN(CH_3)_2$	LDA/THF/-100°/ 3 min	CH_3I	$(CH_3)_2NCSX$	$X = CH_3$ (50)	77, 80
		LDA/THF, ether, HMPA/-100°/ 3 min	C_2H_5I		$X = C_2H_5$ (70)	77
		LDA/THF/-100° 3 min	C_2H_5CHO		$X = CH(OH)C_2H_5$ (80)	77
		"	CH_3COCH_3		$X = C(CH_3)_2OH$ (85)	77, 80
		"	Cyclohexenone		$X = HO$-(cyclohex-2-enyl) (50)	77
		"	Cyclohexanone		$X = HO$-(cyclohexyl) (65)	77, 80
		"	C_6H_5CHO		$X = CH(OH)C_6H_5$ (75)	77, 80

		"	$C_6H_5COCH_3$		$X = C(CH_3)(OH)C_6H_5$ (65)	77, 80
		"	$C_6H_5CO_2CH_3$		$X = COC_6H_5$ (85)	77, 80
		"	$(C_6H_5)_2CO$		$X = C(C_6H_5)_2OH$ (85)	80
$C_4H_9NO_2$	$HCON(CH_3)CH_2OCH_3$	LDA/THF/-75°/ 3 hr	Cyclohexanone	$CH_3OCH_2(CH_3)NCOX$	$X = HO\text{–}C_6H_{10}$ (ring) (39)	79
		"	C_6H_5CHO		$X = CH(OH)C_6H_5$ (76)	79
		"	$C_6H_5CH{=}CHCHO$		$X = CH(OH)CH{=}CHC_6H_5$ (40)	79
		"	$(C_6H_5)_2CO$		$X = C(C_6H_5)_2OH$ (85)	79
$C_5H_{11}NO_3$	$HCON(CH_2OCH_3)_2$	LDA/THF/−75°/ 3 hr	Cyclohexanone	$(CH_3OCH_2)_2NCOX$	$X = HO\text{–}C_6H_{10}$ (ring) (44)	79
		"	C_6H_5CHO		$X = CH(OH)C_6H_5$ (74)	79
		"	$(C_6H_5)_2CO$		$X = C(C_6H_5)_2OH$ (88)	79
$C_5H_{11}NS$	$HCSN(C_2H_5)_2$	LDA/THF/−100°/ 3 min	C_6H_5CHO	N,N-Diethylthiomandelamide (86)		77
$C_7H_{15}NO$	$HCON(C_3H_7\text{-}i)_2$	t-BuLi/THF, ether, pentane/−95°	D_2O	$(i\text{-}C_3H_7)_2NCOX$	$X = D$ (70)	569, 578, 579
		LDA/THF/−78°/ 5 min	CH_3I		$X = CH_3$ (20)	579
		t-BuLi/THF, ether, pentane/−95°	C_2H_5CHO		$X = CH(OH)C_2H_5$ (62)	569, 578, 579
		"	CH_3COCH_3		$X = C(CH_3)_2OH$ (81)	569, 578, 579
		"	Cyclohexanone		$X = HO\text{–}C_6H_{10}$ (ring) (83)	569
		"	C_6H_5CHO		$X = CH(OH)C_6H_5$ (80)	569, 578, 579
			$C_6H_5CH_2Br$		$X = CH(C_6H_5)CH_2C_6H_5$ (68)	569

TABLE III. FORMAMIDES AND THIOFORMAMIDES (ALPHA) (*Continued*)

Formula	Compound Lithiated	Conditions	Substrate	Product and Yield (%)	Refs.
$C_7H_{15}NO$ (*Contd.*)	$HCON(C_3H_7\text{-}i)_2$	"	$C_6H_5CO_2C_2H_5$	$X = COC_6H_5$ (70)	569, 578, 579
		"	$C_6H_5CH{=}CHCHO$	$X = CH(OH)CH{=}CHC_6H_5$ (68)	569, 578
		LDA/THF/−78°/ 5 min	$(C_6H_5)_2CO$	$X = C(C_6H_5)_2OH$ (92)	578, 579
		t-BuLi/THF, ether, pentane/−95°/ 45 min	"	$X = C(C_6H_5)_2OH$ (85)	569
$C_9H_{19}NS$	$HCSN(C_4H_9\text{-}n)_2$	LDA/THF/−100°/ 3 min	C_6H_5CHO	N,N-Dibutylthiomandelamide (71)	77
		"	$(C_6H_{11})_2CO$	N,N-Dibutyl-2,2-dicyclohexyl-2-hydroxythioacetamide	77
$C_{13}H_{11}NS$	$HCSN(C_6H_5)_2$	LDA/THF/−100°/ 3 min	C_6H_5CHO	N,N-Diphenylthiomandelamide (57)	77

Note: References 360–607 are on pp. 355–360.

TABLE IV. Pyrroles (Alpha)

Formula	Compound Lithiated	Conditions	Substrate	Product and Yield (%)	Refs.
C_5H_7N	N—CH_3	EtLi/ether, TMEDA/ reflux/1 hr	CuCl	N(—CH_3)—X; X = N(—CH_3) (8)	68
		BuLi/ether, TMEDA/ reflux/1 hr	I_2	X = I (30)	68
		"	CO_2	X = CO_2H (70)	68, 81
		BuLi/ether/reflux/ "overnight"	$CH_2=CHCH_2Cl$	X = $CH_2CH=CH_2$ (20)	580
		BuLi/ether	$ClB(CH_3)N(CH_3)_2$	X = $B(CH_3)N(CH_3)_2$ (37)	295
		"	$ClB[N(CH_3)_2]_2$	X = $B[N(CH_3)_2]_2$ (45)	295
		"	Cl—B	X = B (20)	295
		"	$ClB(C_2H_5)_2$	X = $B(C_2H_5)_2$ (48)	295
		EtLi/ether, TMEDA/ 34°/1 hr	C_6H_5I/CuBr	X = C_6H_5 (41)	68
		2.5 BuLi/hexane, TMEDA/25°/0.5 hr	$CO_2(CH_2N_2)$	Dimethyl 1-methylpyrrole-2,4-dicarboxylate (I) (100)	40, 581
		2.5 BuLi/hexane, TMEDA/reflux/ 0.5 hr	"	I (41), Dimethyl 1-methylpyrrole-2,4-dicarboxylate (II) (7), dimethyl 1-methylpyrrole-2,5-dicarboxylate (III) (50)	40
		4.5 BuLi/hexane, TMEDA/reflux/ 120 hr	"	I (4), II (40), III (2)	40
		BuLi/hexane/ reflux/0.5 hr	DMF	1-Methylpyrrole-2,4-dicarboxaldehyde (—)	581

TABLE IV. PYRROLES (ALPHA) (*Continued*)

Formula	Compound Lithiated	Conditions	Substrate	Product and Yield (%)	Refs.
$C_{10}H_9N$	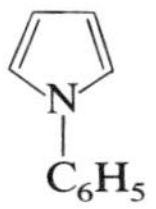 N	BuLi/ether/ 25°/8 hr	CO_2	1-Phenylpyrrole-2-carboxylic acid (14)	81
	C_6H_5				
		2 BuLi/ether/ reflux/14 hr	”	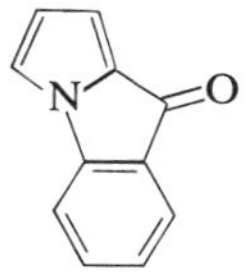 N, O (5)	81
$C_{13}H_{11}NS$	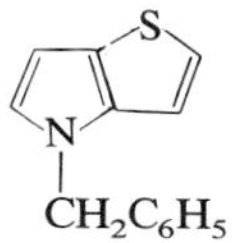 S, N	BuLi/ether, pentane, heptane/25°/ 15 hr	$CO_2(CH_2N_2)$	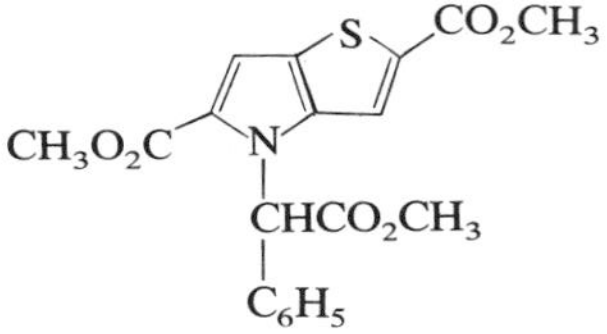 S, CO_2CH_3, CH_3O_2C, N, $CHCO_2CH_3$, C_6H_5 (85)	582
	$CH_2C_6H_5$				

Note: References 360–607 are on pp. 355–360.

TABLE V. Indoles (Alpha)

Formula	Compound Lithiated	Conditions	Substrate	Product and Yield (%)	Refs.
C_9H_9N	N-CH$_3$ (1-methylindole)	BuLi/ether/ reflux/8 hr	CO_2	1-methylindole-2-X (N-CH$_3$); $X = CO_2H$ (78)	83
		"	$(CO_2C_2H_5)_2$ (OH^-)	$X = COCO_2H$ (59)	583, 584
		"	C_6H_5NCO	$X = CONHC_6H_5$ (42)	83
		"	p-ClC_6H_4CHO	$X = CH(OH)C_6H_4Cl$-p (50)	83
		"	o-$CH_3C_6H_4NCO$	$X = CONHC_6H_4CH_3$-o (63)	83
		"	p-$CH_3OC_6H_4NCO$	$X = CONHC_6H_4OCH_3$-p (40)	83
		"	p-$CH_3C_6H_4SO_2OCH_3$	$X = CH_3$ (45)	83
		"	quinoline (N) $(C_6H_5NO_2)$	X = 2-quinolyl (N) (54)	83
		"	1-Naphthyliso- cyanate	X = CONH-1-naphthyl (52)	83
		"	$(C_6H_5)_2CO$	$X = C(C_6H_5)_2OH$ (53)	83

TABLE V. INDOLES (ALPHA) (*Continued*)

Formula	Compound Lithiated	Conditions	Substrate	Product and Yield (%)	Refs.
$C_{10}H_{11}NO$	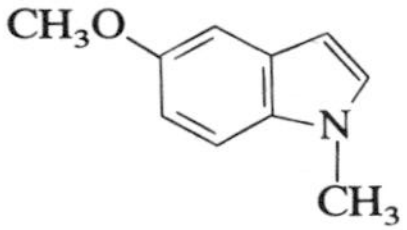 5-methoxy-1-methylindole (CH_3O, N–CH_3)	BuLi/ether/ reflux/13 hr	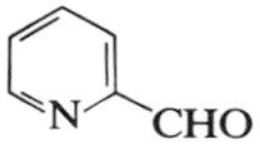 2-pyridinecarboxaldehyde (N, CHO)	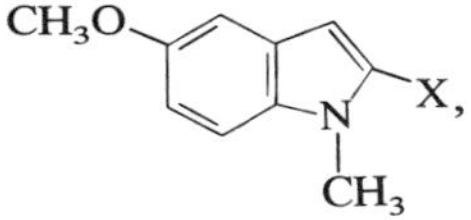 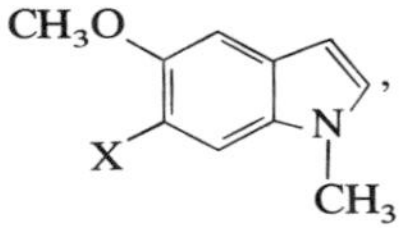 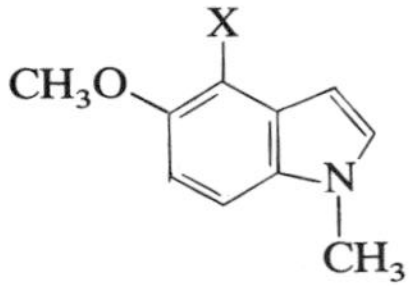 (I) 2-X-5-CH_3O-1-CH_3-indole, (II) 6-X-5-CH_3O-1-CH_3-indole, (III) 4-X-5-CH_3O-1-CH_3-indole (I + II + III, 74) (I:II:III, 4:5:1)	84
		t-BuLi/THF/0°	"	I (39) X = CH(OH)-2-pyridyl 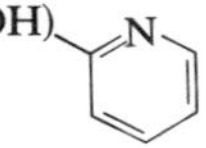	84
	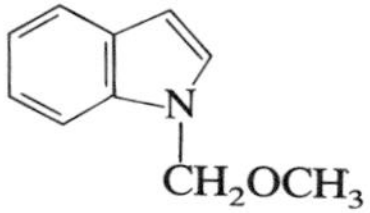 1-(CH_2OCH_3)indole	*t*-BuLi/ether/25°/ 1 hr	CO_2	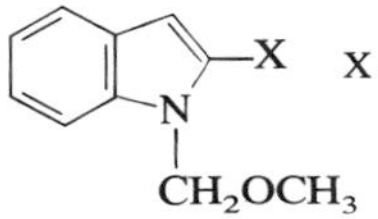 2-X-1-(CH_2OCH_3)indole X = CO_2H (80)	85
		"	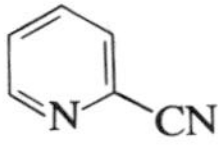 2-cyanopyridine (N, CN) (H_3O^+)	X = CO-2-pyridyl (56) 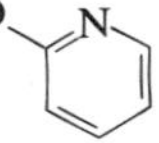	85

		”	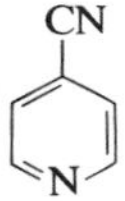 4-cyanopyridine (H_3O^+)	X 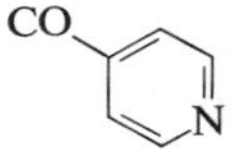 = CO-(4-pyridyl) (56)	85
		”	C_6H_5CN (H_3O^+)	$X = COC_6H_5$ (84)	85
		”	C_6H_5CHO	$X = CH(OH)C_6H_5$ (40)	85
		”	$C_6H_5N(CH_3)CHO$	$X = CHO$ (46)	85
		”	p-$CH_3OC_6H_4CN$ (H_3O^+)	$X = COC_6H_4OCH_3$-p (70)	85
$C_{13}H_{18}N_2$	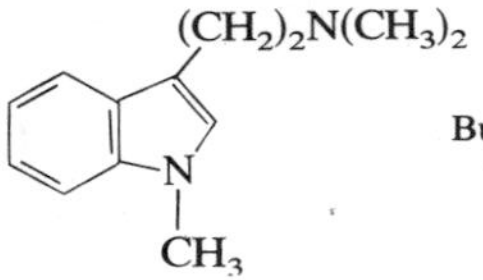 3-$(CH_2)_2N(CH_3)_2$-1-CH_3-indole	BuLi/THF/0°/ 75 min	D_2O	3-[2-(Dimethylamino)ethyl]-2-*d*-1-methylindole (74)[a]	153
$C_{14}H_{11}N$	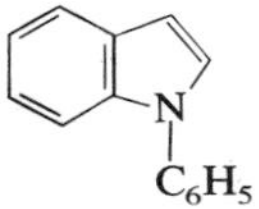 1-C_6H_5-indole	BuLi/ether/reflux/ 12 hr	CO_2	1-(*o*-Carboxyphenyl)indole-2-carboxylic acid (15), 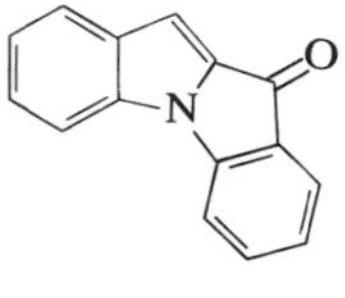 indolo[1,2-a]indol-10-one (42)	83
$C_{14}H_{11}NO_2S$	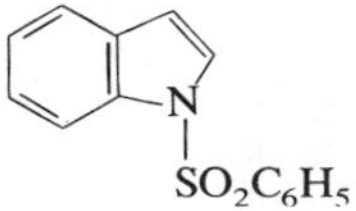 1-$SO_2C_6H_5$-indole	*t*-BuLi/THF/−12°–25°/20 min	CO_2	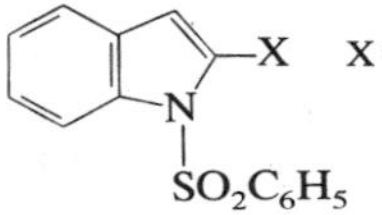 2-X-1-$SO_2C_6H_5$-indole, $X = CO_2H$ (63)	85

TABLE V. INDOLES (ALPHA) (*Continued*)

Formula	Compound Lithiated	Conditions	Substrate	Product and Yield (%)	Refs.
$C_{14}H_{11}NO_2S$ (*Contd.*)	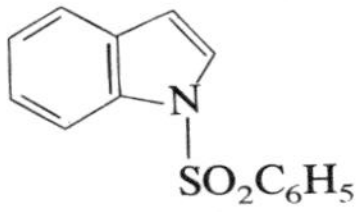 N–$SO_2C_6H_5$	*t*-BuLi/THF/ – 12°– 25°/20 min	$ClCO_2C_2H_5$	$X = CO_2C_2H_5$ (75)	85
		"	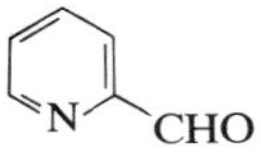 N, CHO	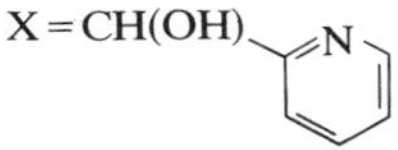 $X = CH(OH)$–N (32)	85
		"	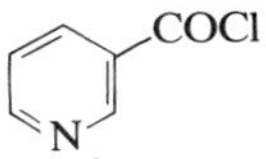 COCl, N	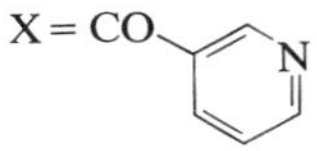 $X = CO$–N (60)	85
		"	C_6H_5CHO	$X = CH(OH)C_6H_5$ (55)	85
		"	C_6H_5COCl	$X = COC_6H_5$ (65)	85
		"	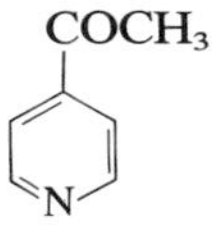 $COCH_3$, N	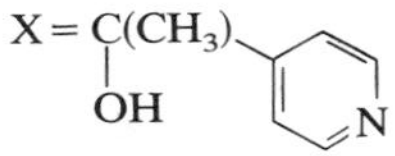 $X = C(CH_3)(OH)$–N (35)	85
			$C_6H_5COCH_3$	$X = C(CH_3)(OH)C_6H_5$ (64)	85
		"	*p*-$CH_3OC_6H_4CHO$	$X = CH(OH)C_6H_4OCH_3$-*p* (65)	85
		"	*p*-$CH_3OC_6H_4COCH_3$	$X = C(CH_3)(OH)C_6H_4OCH_3$-*p* (35)	85

		"	2-cyanopyridine (H_3O^+)	indole, 2-X, N–H	X = CO–(2-pyridyl)	(36)	85
		"	4-cyanopyridine (H_3O^+)		X = CO–(4-pyridyl)	(26)	85
		"	$C_6H_5CN(H_3O^+)$		$X = COC_6H_5$	(30)	85
		"	3-pyridyl–$CO_2C_2H_5$		X = CO–(3-pyridyl)	(22)	85
		"	4-pyridyl–$CO_2C_2H_5$		X = CO–(4-pyridyl)	(31)	85
		"	$C_6H_5CO_2C_2H_5$		$X = COC_6H_5$	(26)	85
$C_{15}H_{13}NO_3S$	5-CH_3O-1-$SO_2C_6H_5$-indole	t-BuLi/THF/0°–25°/45 min	2-pyridyl–CHO	5-CH_3O-1-$SO_2C_6H_5$-indole, 2-X	X = CH(OH)–(2-pyridyl)	(59)	84

TABLE V. INDOLES (ALPHA) (*Continued*)

Formula	Compound Lithiated	Conditions	Substrate	Product and Yield (%)	Refs.
$C_{15}H_{13}NO_3S$ (*Contd.*)	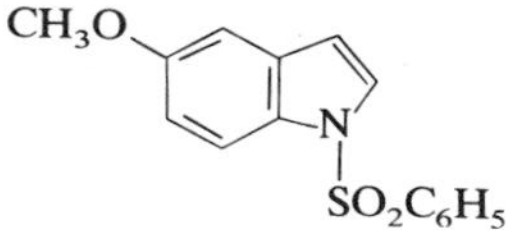 CH_3O, N, $SO_2C_6H_5$	"	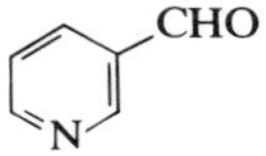 CHO, N	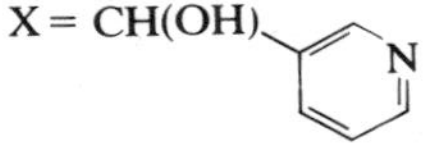 X = CH(OH), N (84)	84
	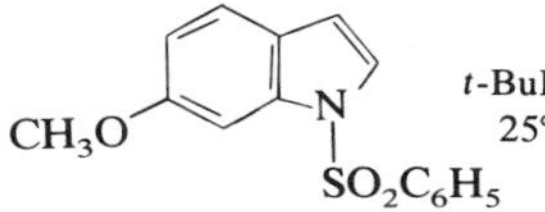 CH_3O, N, $SO_2C_6H_5$	*t*-BuLi/THF/0°–25°/45 min	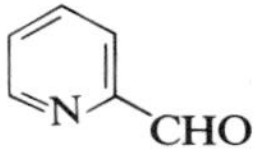 N, CHO	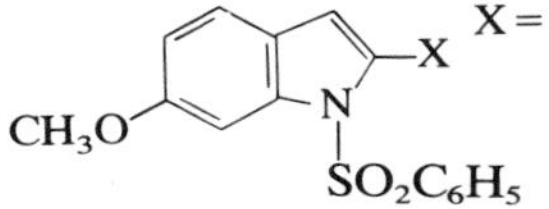 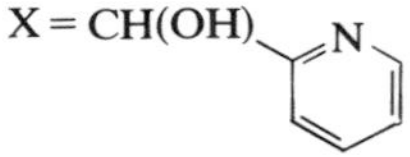 CH_3O, N, $SO_2C_6H_5$, X; X = CH(OH), N (62)	84
		"	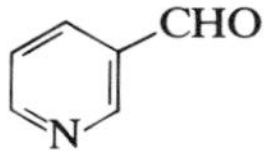 CHO, N	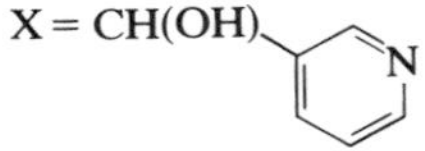 X = CH(OH), N (70)	84
		"	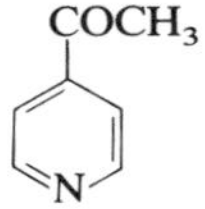 $COCH_3$, N	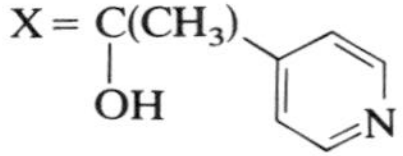 X = $C(CH_3)$, OH, N (58)	84

[a] The isolated yield was 74%; the deuterium incorporation was 75%.

Note: References 360–607 are on pp. 355–360.

TABLE VI. PYRAZOLES (ALPHA)

Formula	Compound Lithiated	Conditions	Substrate	Product and Yield (%)	Refs.
$C_3H_3BrN_2$	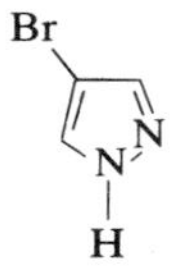 4-Br, N–H pyrazole	C_6H_5Li/ether/25°/ 2 hr	CO_2	4-Bromopyrazole-5-carboxylic acid (35)	87
$C_3H_4N_2$	 N–H pyrazole	BuLi/ether/ −30°/ 2 hr	CO_2	Pyrazole-5-carboxylic acid (9)	87
$C_4H_6N_2$	 N–CH_3 pyrazole	BuLi/ether/ −20°	CO_2	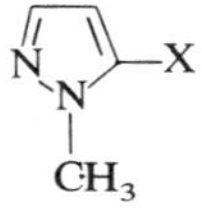 1-CH_3-pyrazole-5-X (I) $X = CO_2H$ (66)	87, 88
		BuLi/ether/ −30°– 20°/2 hr	$(CH_3O)_2SO_2$	$X = CH_3$ (75)	87
		BuLi/ether/ −30°– 20°/4 hr	$(C_6H_5)_2CO$	$X = C(C_6H_5)_2OH$ (87)	87
		BuLi/ether/25°/ 1.5 hr	C_6H_5CHO	I, 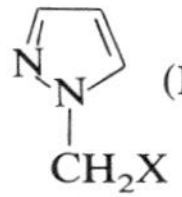 1-CH_2X-pyrazole (II) $X = CH(OH)C_6H_5$ (I + II, 88) (I : II, 66 : 34)	86

TABLE VI. PYRAZOLES (ALPHA) (*Continued*)

Formula	Compound Lithiated	Conditions	Substrate	Product and Yield (%)	Refs.
$C_5H_8N_2$	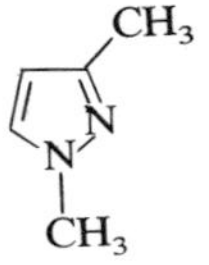 1,3-dimethylpyrazole (CH_3, N–CH_3)	BuLi/ether/ reflux/0.5 hr	C_6H_5CHO	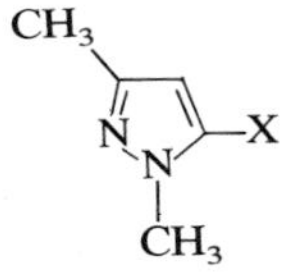 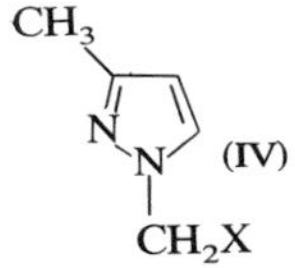 3-CH_3-1-CH_3-pyrazole-5-X (III), 3-CH_3-1-CH_2X-pyrazole (IV) X = $CH(OH)C_6H_5$ (III + IV, 90) (III : IV, 2 : 1)	86
$C_6H_{10}N_2$	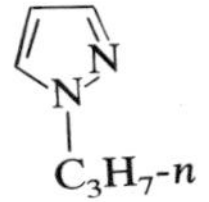 1-(C_3H_7-*n*)pyrazole	BuLi/ether/25°/ 1 hr	C_6H_5CHO	α-Phenyl-1-propylpyrazole-5-methanol (81)	86
$C_7H_{12}N_2$	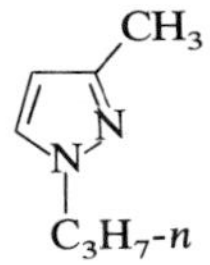 3-CH_3-1-(C_3H_7-*n*)pyrazole	BuLi/ether/25°/ 0.5 hr	C_6H_5CHO	3-Methyl-α-phenyl-1-propylpyrazole-5-methanol (95)	86
$C_9H_8N_2$	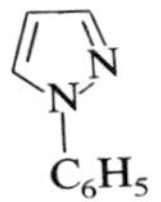 1-(C_6H_5)pyrazole	BuLi/ether/0°– 25°/2 hr	CO_2	1-Phenylpyrazole-5-carboxylic acid (39), 1-(*o*-Carboxyphenyl)pyrazole (10)	88
		2 BuLi/ether/25° 7 hr	"	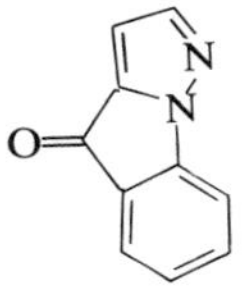 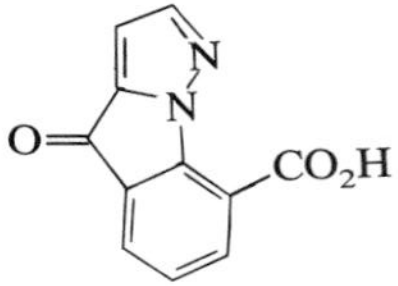 pyrazolo-indenone (O=) (8), pyrazolo-indenone with CO_2H (26)	88

$C_{10}H_{10}N_2$	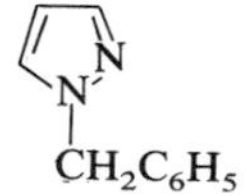 1-benzylpyrazole ($CH_2C_6H_5$ on N)	C_6H_5Li/ether/ 25°/3 hr	CO_2	1-Benzylpyrazole-5-carboxylic acid (57)	87
	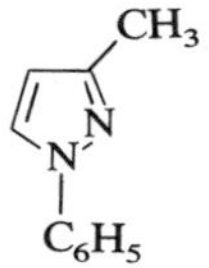 3-methyl-1-phenylpyrazole (CH_3, C_6H_5)	BuLi/ether/25°/ 2 hr	CO_2	3-Methyl-1-phenylpyrazole-5-carboxylic acid (—)	585
$C_{14}H_{15}ClN_2O$	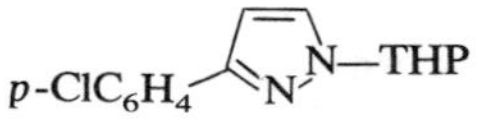 *p*-ClC_6H_4-pyrazole-N—THP	BuLi/ether/0°/ 0.5 hr	CH_2O	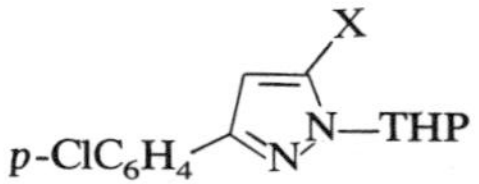 *p*-ClC_6H_4-pyrazole(X)-N—THP X = CH_2OH (73)	23
		,,	*t*-C_4H_9NCO	X = $CONHC_4H_9$-*t* (63)	23
		,,	$(C_6H_5S)_2$	X = SC_6H_5 (80)	23
$C_{18}H_{14}N_4$	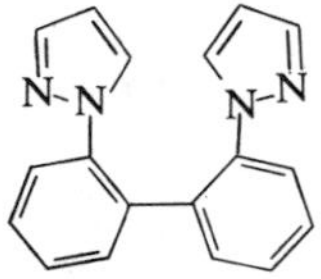 2,2'-bis(pyrazol-1-yl)biphenyl	BuLi/THF/20°	$CuI(O_2)$	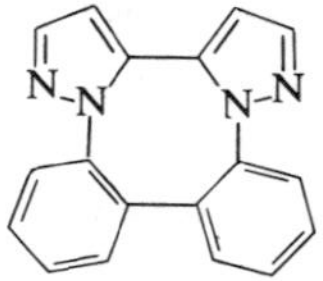 (cyclized bipyrazole-biphenyl) (11)	282

TABLE VII. IMIDAZOLES (ALPHA)

Formula	Compound Lithiated	Conditions	Substrate	Product and Yield (%)	Refs.
$C_4H_5BrN_2$	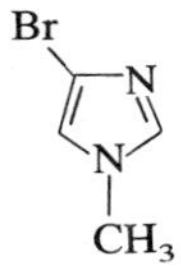 4-bromo-1-methylimidazole (structure)	BuLi/ether/−80°	CH_3CHO	4-Bromo-α,1-dimethylimidazole-2-methanol (50)	92
$C_4H_5ClN_2$	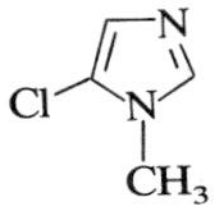 5-chloro-1-methylimidazole (structure)	BuLi/ether/−80°	CH_3CHO	5-Chloro-α,1-dimethylimidazole-2-methanol (64)	92
$C_4H_6N_2$	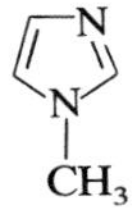 1-methylimidazole (structure)	BuLi/ether/−60°/ 2 hr, 25°/3 hr	CO_2	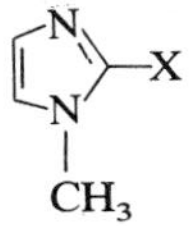 1-methylimidazole-2-X (structure): $X = CO_2H$ (32)	586
		BuLi/ether/−78°/ 1 hr	CH_3CHO	$X = CH(OH)CH_3$ (35)	92
		BuLi/ether/ reflux/1 hr	$(CH_3)_3SiCl$	$X = Si(CH_3)_3$ (56)	93
		BuLi/ether/25°/ 1 hr	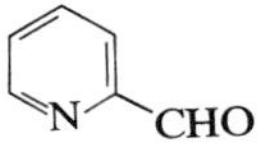 pyridine-2-CHO (structure)	$X = CH(OH)$-2-pyridyl (structure) (48) 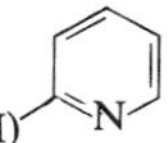	587
		”	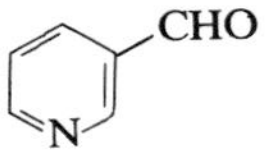 pyridine-3-CHO (structure)	$X = CH(OH)$-3-pyridyl (structure) (33) 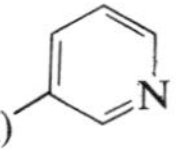	587
		—	$C_6H_5N_3$	$X = NH_2$ (—)	287

Formula	Substrate	Conditions	Electrophile	Product(s) and Yield(s) (%)	Refs.
		BuLi/ether/25°/ 1 hr	Cyclohexanone	X = 1-hydroxycyclohexyl (56)	587
		”	n-$C_6H_{13}CHO$	$X = CH(OH)C_6H_{13}$-n (41)	587
		”	$C_6H_5CH_2CHO$	$X = CH(OH)CH_2C_6H_5$ (44)	587
		”	3,4-$(CH_3O)_2C_6H_3CHO$	$X = CH(OH)C_6H_3(OCH_3)_2$-3,4 (23)	587
		”	p-$(CH_3)_2NC_6H_4CHO$	$X = CH(OH)C_6H_4N(CH_3)_2$-$p$ (48)	587
		”	1-Naphthyl-NCO	X = CONH-1-naphthyl (66)	586
		”	$(C_6H_5)_2CO$	$X = C(C_6H_5)_2OH$ (86)	586
		2 BuLi/ether/ −70°/2.5 hr	$(CH_3)_3SiCl$	1-Methyl-2,5-bis(trimethylsilyl)imidazole (32)	93
$C_5H_8N_2$	1,2-Dimethylimidazole	BuLi/ether/−10°/ 15 min	DMF	1,2-dimethyl-5-X-imidazole (I) X = CHO (20)	90
		”	$(C_2H_5)_2NBr$	X = Br (26)	90
		BuLi/ether/0°/ 1 hr	C_6H_5CHO	(I), 1-methyl-2-(XCH_2)imidazole (II) (I+II, 70) (I:II, 25:75) $X = CH(OH)C_6H_5$	588, 90

TABLE VII. IMIDAZOLES (ALPHA) (*Continued*)

Formula	Compound Lithiated	Conditions	Substrate	Product and Yield (%)		Refs.
$C_5H_8N_2$ (*Contd.*)	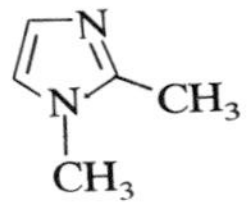 N, N, CH_3, CH_3	C_6H_5Li/ether, benzene/0°/ 0.5 hr	C_6H_5CHO	I, II (I+II, 60) (I:II, 1:1)		589
		BuLi/ether/−15°/ 5-20 min	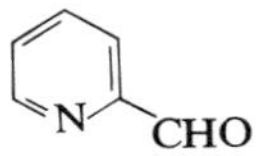 N, CHO	II (66)	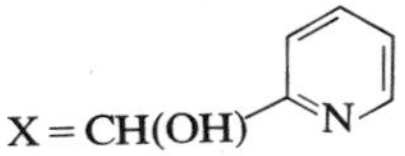 X = CH(OH)–, N	91
		BuLi/ether/−10°/ 15 min	$C_6H_5C\equiv CI$	I	X = I (48)	90
		"	$(C_6H_5)_2CO$	I	X = $C(C_6H_5)_2OH$ (73)	90
	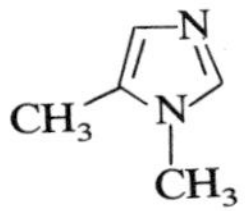 N, N, CH_3, CH_3	BuLi/ether/−80°	CH_3CHO	α,1,5-Trimethylimidazole-2-methanol (30)		92
$C_5H_8N_2O$	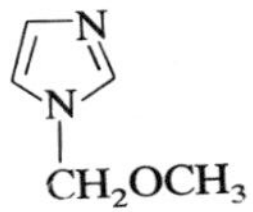 N, N, CH_2OCH_3	BuLi/ether	C_6H_5CHO	1-(Methoxymethyl)-α-phenylimidazole-2-methanol (45)		587
$C_6H_{10}N_2$	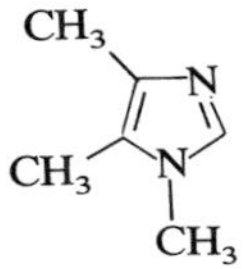 N, N, CH_3, CH_3, CH_3	BuLi/ether/−80°	CH_3CHO	α,1,4,5-Tetramethylimidazole-2-methanol (24)		92

$C_9H_8N_2$	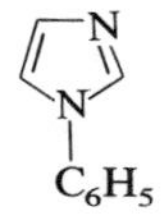 C_6H_5	BuLi/ether/25°/ 8 hr	CO_2	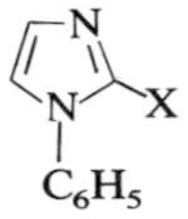 X, C_6H_5	$X = CO_2H$ (46)	586
		—	$C_6H_5N_3$		$X = NH_2$ (—)	287
		BuLi/ether/25°/ 8 hr	C_6H_5NCO		$X = CONHC_6H_5$ (39)	586
		,,	$(C_6H_5)_2CO$		$X = C(C_6H_5)_2OH$ (76)	586
		3 BuLi/ether/ reflux/12 hr	CO_2	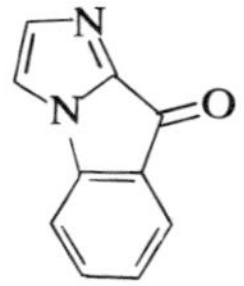 (5)		586
$C_9H_8N_2O_2S$	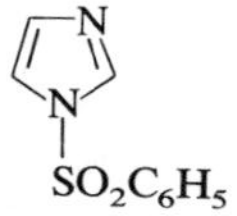 $SO_2C_6H_5$	*t*-BuLi/THF/ −20°/10 min	D_2O	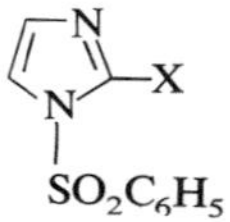 X, $SO_2C_6H_5$	$X = D$ (100)	573
		n-BuLi/THF/ − 10°/ 15 min	I_2		$X = I$ (7)	573
		t-BuLi/THF/ − 20°/0.5 hr	CH_2O		$X = CH_2OH$ (10)	573
		n-BuLi/THF/0°/ 5 min	Cyclohexanone		$X = HO$ 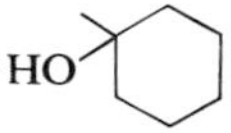 (15)	573
		t-BuLi/THF/0°/ 10 min	C_6H_5CHO		$X = CH(OH)C_6H_5$ (18)	573
$C_{10}H_{10}N_2$	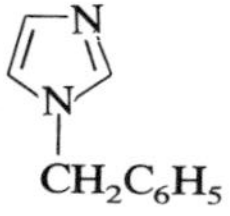 $CH_2C_6H_5$	BuLi/ether/ − 60°/ 2 hr, 25°/2 hr	CO_2	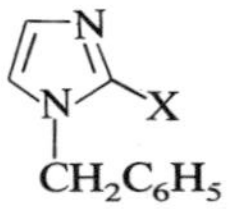 X, $CH_2C_6H_5$ (III)	$X = CO_2H$ (67)	586

TABLE VII. IMIDAZOLES (ALPHA) (*Continued*)

Formula	Compound Lithiated	Conditions	Substrate	Product and Yield (%)	Refs.
$C_{10}H_{10}N_2$ (*Contd.*)	1-benzylimidazole ($CH_2C_6H_5$)	BuLi/ether/0°/ 2 hr	CH_3CHO	$X = CH(OH)CH_3$ (22)	587
		”	$HCO_2C_2H_5$	$X = CH(OH)$–(1-benzylimidazol-2-yl, $CH_2C_6H_5$) (22)	587
		—	$C_6H_5N_3$	$X = NH_2$ (–)	287
		BuLi/ether/0°/ 2 hr	C_6H_5CHO	$X = CH(OH)C_6H_5$ (48)	587
		BuLi/ether/–60°/ 2 hr, 25°/2 hr	quinoline ($C_6H_5NO_2$)	X = 2-quinolyl (29)	586
		BuLi/ether/–70° –25°/2 hr	di-O-isopropylidene lactone [(CH₃)₂ acetonides], III	III (30), X–(1-benzylimidazol-5-yl, $CH_2C_6H_5$) (12); X = HO-substituted di-O-isopropylidene furanose [$(CH_3)_2$, $(CH_3)_2$]	590

Note: References 360–607 are on pp. 355–360.

TABLE VIII. Benzimidazoles (Alpha)

Formula	Compound Lithiated	Conditions	Substrate	Product and Yield (%)	Refs.
$C_7H_6N_2$	N, N, H	BuLi/ether/−70°	$(CH_3)_3SiCl$	2-(Trimethylsilyl)benzimidazole (17)	93
$C_8H_8N_2$	N, N, CH_3	BuLi/ether/25°	—	1,1′-Dimethyl-2,2′-bibenzimidazole (53)	591
		BuLi/ether/−60°/ 3 hr	CO_2	1-Methylbenzimidazole-2-carboxylic acid (45)	591
		BuLi/ether/−70°	$(CH_3)_3SiCl$	1-Methyl-2-(trimethylsilyl)benzimidazole (91)	93
$C_{12}H_{10}N_2$	N, N, CH_3	BuLi/−78°	CO_2	CO_2H, N, N, CH_3 (80)	607
	CH_3, N, N	BuLi/−78°	CO_2	CH_3, N, CO_2H, N (72)	607
$C_{13}H_{10}N_2$	N, N, C_6H_5	C_6H_5Li	CO_2	1-Phenylbenzimidazole-2-carboxylic acid (25)	592

TABLE VIII. BENZIMIDAZOLES (ALPHA) (*Continued*)

Formula	Compound Lithiated	Conditions	Substrate	Product and Yield (%)	Refs.
$C_{14}H_{12}N_2$	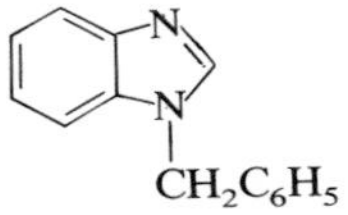 1-($CH_2C_6H_5$)benzimidazole	BuLi/ether/−70°–25°/2 hr	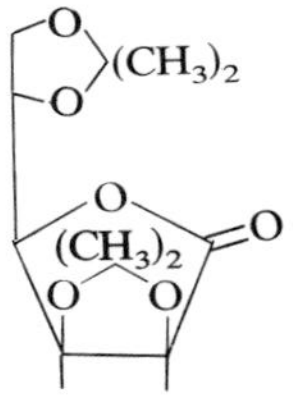 [structure: di-O-isopropylidene lactone, $(CH_3)_2$]	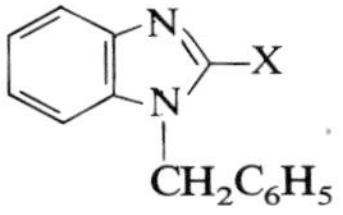 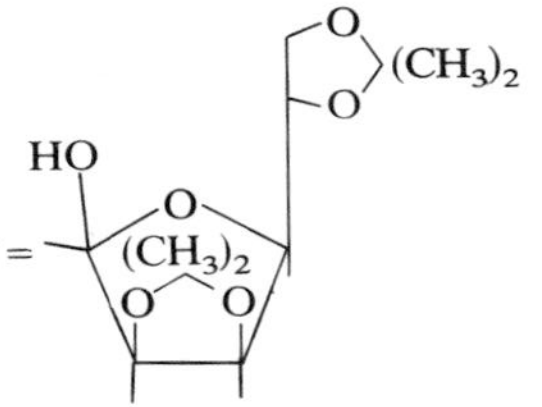 2-X-1-($CH_2C_6H_5$)benzimidazole, X = [structure: HO, $(CH_3)_2$, $(CH_3)_2$] (40)	590
		”	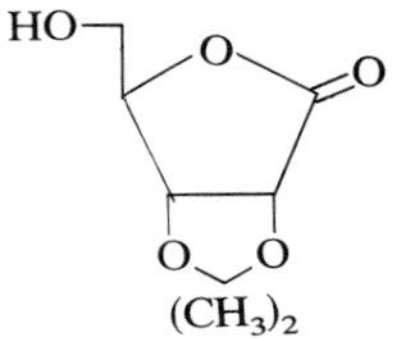 [structure: HO–, O-isopropylidene lactone, $(CH_3)_2$]	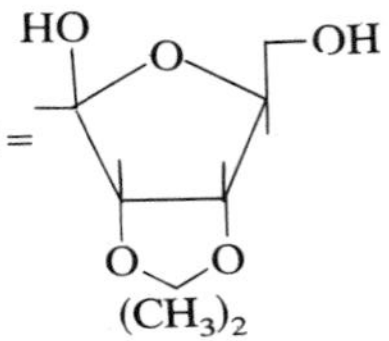 X = [structure: HO, –OH, $(CH_3)_2$] (25)	590

Note: References 360–607 are on pp. 355–360.

TABLE IX. IMIDAZO[1,2-*a*]PYRIDINES (ALPHA)

Formula	Compound Lithiated	Conditions	Substrate	Product and Yield (%)	Refs.
$C_7H_6N_2$	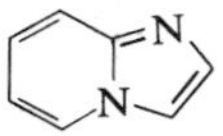	C_6H_5Li/ether/10 min	DMF	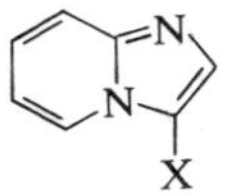 X = CHO (24)	94
		”	Cyclohexanone	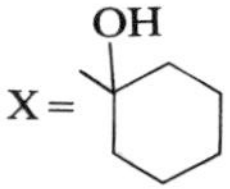 X = 1-hydroxycyclohexyl (35)	94
		”	C_6H_5NCO	X = $CONHC_6H_5$ (15)	94
$C_8H_8N_2$	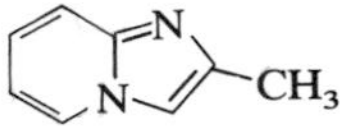	C_6H_5Li/ether/10 min	Cyclohexanone	1-(2-Methylimidazo[1,2-*a*]pyridin-3-yl) cyclohexanol (14)	94
	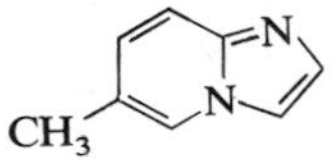	C_6H_5Li/ether/10 min	Cyclohexanone	1-(6-Methylimidazo[1,2-*a*]pyridin-3-yl) cyclohexanol (40)	94
	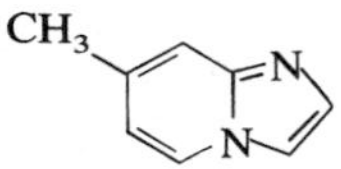	C_6H_5Li/ether/10 min	Cyclohexanone	1-(7-Methylimidazo[1,2-*a*]pyridin-3-yl) cyclohexanol (48)	94
	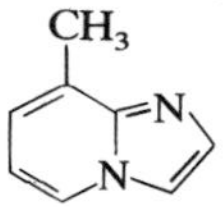	C_6H_5Li/ether/10 min	Cyclohexanone	1-(8-Methylimidazo[1,2-*a*]pyridin-3-yl) cyclohexanol (42)	94

TABLE X. TRIAZOLES (ALPHA)

Formula	Compound Lithiated	Conditions	Substrate	Product and Yield (%)	Refs.
$C_8H_7N_3$	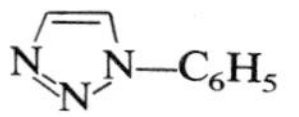 N N—C_6H_5 N	BuLi/THF/−20° to −60°/0.5–1 hr	CH_3I	5-Methyl-1-phenyl-1H-1,2,3-triazole (94)	95
$C_9H_9N_3$	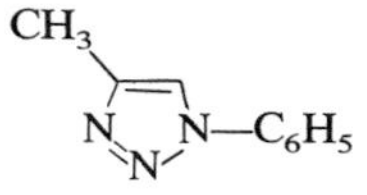 CH_3 N N—C_6H_5 N	BuLi/THF/−20° to −60°/0.5–1 hr	CH_3I	4,5-Dimethyl-1-phenyl-1H-1,2,3-triazole (81)	95
$C_{14}H_{11}N_3$	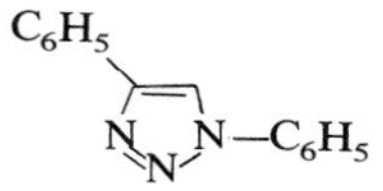 C_6H_5 N N—C_6H_5 N	BuLi/THF/−20° to −60°/0.5–1 hr	CO_2	1,4-Diphenyl-1H-1,2,3-triazole-5-carboxylic acid (62)	95
		"	CH_3I	5-Methyl-1,4-diphenyl-1H-1,2,3-triazole (78)	95
$C_{15}H_{13}N_3$	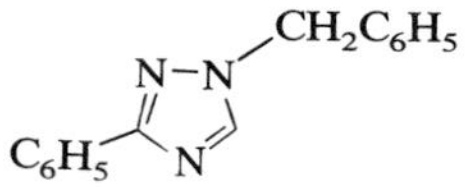 $CH_2C_6H_5$ N−N C_6H_5 N	BuLi/THF/−78°	CH_2O	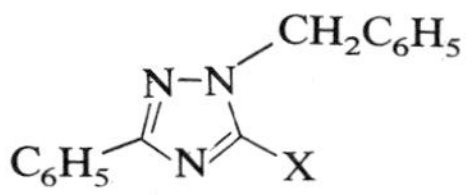 $CH_2C_6H_5$ N−N C_6H_5 N X; X = CH_2OH (78)	96
		"	$(C_6H_5)_2CO$	X = $C(C_6H_5)_2OH$ (92)	96
		"	$(p\text{-}CH_3OC_6H_4)_2CO$	X = $C(C_6H_4OCH_3\text{-}p)_2OH$ (94)	96
		"	$[p\text{-}(CH_3)_2NC_6H_4]_2CO$	X = $C[C_6H_4N(CH_3)_2\text{-}p]_2OH$ (91)	96

TABLE XI. TETRAZOLES (ALPHA)

Formula	Compound Lithiated	Conditions	Substrate	Product and Yield (%)	Refs.
$C_2H_4N_4$		BuLi/THF/ −50°/1 hr	$[(C_2H_5)_3P]_2NiCl_2$	Ni (80)	301
		BuLi/THF/ −65°	Br_2	X = Br (41)	98
		”	I_2	X = I (36)	98
		”	“S”	X = SH (67)	98
		”	C_6H_5CHO	$X = CH(OH)C_6H_5$ (63)	98
		”	$C_6H_5COCH_3$	$X = C(CH_3)(OH)C_6H_5$ (67)	98
		”	$C_6H_5CO_2CH_3$	$X = COC_6H_5$ (41)	98
		”	$(C_6H_5)_2CO$	$X = C(C_6H_5)_2OH$ (75)	98
		”	CH_3COCl	OCOX; $X = CH_3$ (35)	98
		”	C_6H_5COCl	$X = C_6H_5$ (68)	98
$C_7H_6N_4$		BuLi/THF/ −50°/ 1 hr	—	[1-Phenyl-2(1H)tetrazolyl]lithium (—)	301
$C_7H_{12}N_4$		BuLi/THF/ −50°/ 1 hr	$[(C_2H_5)_3P]_2NiCl_2$	Ni (−)	301

TABLE XII. PYRIDINES AND CONDENSED PYRIDINES (ALPHA)

Formula	Compound Lithiated	Conditions	Substrate	Product and Yield (%)	Refs.
C_5H_5N	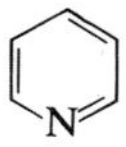	LDA/ether, HMPA/ −70°/1 hr	—	2,2′-Bipyridine (50)	99
C_9H_7N	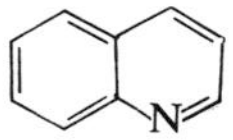 N	LDA/ether, HMPA/ −70°/1 hr	—	2,2′-Biquinoline (74)	99
	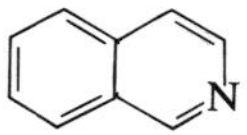 N	LDA/ether, HMPA/ −70°/1 hr	—	1,1′-Biisoquinoline (55)	99

TABLE XIII. Pyridine-N-Oxides (Alpha)

Formula	Compound Lithiated	Conditions	Substrate	Product and Yield (%)	Refs.
C_5H_4ClNO	4-Cl-pyridine N^+–O^-	—	CO_2	4-Chloropicolinic acid 1-oxide (49)	593, 101
		BuLi/ether/−65°/ 15 min	Cyclohexanone	4-Cl-2-X-pyridine N^+–O^- (36), 4-Cl-2,6-X$_2$-pyridine N^+–O^- (21); X = 1-hydroxycyclohexyl (OH)	594
C_5H_5NO	pyridine N^+–O^-	BuLi/THF, ether/ −65°/1 hr	Br_2	2-Br-6-X-pyridine N^+–O^-: X = Br (3), X = 6-bromopyridin-2-yl N^+–O^- (8), X = 6-bromopyridin-2-yl (6)	595, 596
		”	Cl_2	2,6-Dichloropyridine 1-oxide (5)	595, 596
		BuLi/ether/−65°–25°	“S”	1-Hydroxy-2(1H)-pyridinethione (10)	102

TABLE XIII. PYRIDINE-N-OXIDES (ALPHA) (*Continued*)

Formula	Compound Lithiated	Conditions	Substrate	Product and Yield (%)			Refs.
C_5H_5NO (*Contd.*)	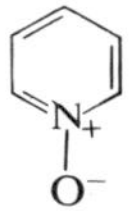 [structure: pyridine N-oxide]	BuLi/THF/−65°/ 15 min	CH_3CHO	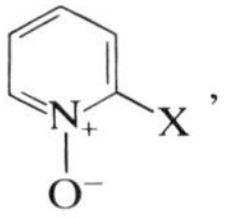 [structure: 2-X-pyridine N-oxide], (36)	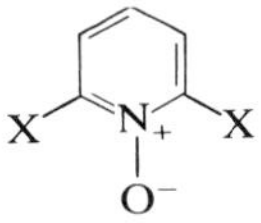 [structure: 2,6-X₂-pyridine N-oxide] (30)	X = CH(OH)CH_3	594
		”	CH_3COCH_3	(18)	(—)	X = C$(CH_3)_2$OH	594
		BuLi/ether/25°	Cyclohexanone	(13)	(36)	X 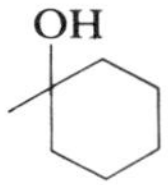 = 1-hydroxycyclohexyl [structure]	594
		—	”	(15)	(—)	X 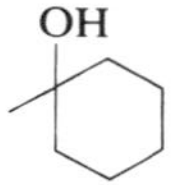 = 1-hydroxycyclohexyl [structure]	593, 594
C_6H_6ClNO	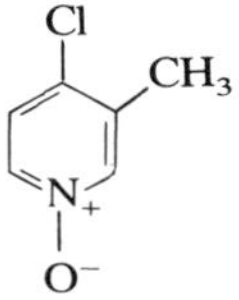 [structure: 4-chloro-3-methylpyridine N-oxide]	BuLi/THF/−65°/ 1 hr	“S”	4-Chloro-1-hydroxy-5-methyl-2(1H)-pyridinethione (12)			102, 596
		BuLi/ether/−65°	CO_2	4-Chloro-5-methylpicolinic acid 1-oxide (24)			593, 101
		BuLi/ether/−65°/ 15 min	Cyclohexanone	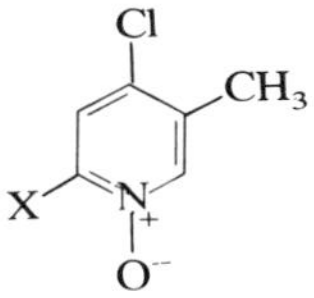 [structure: 2-X-4-chloro-5-methylpyridine N-oxide], (44)	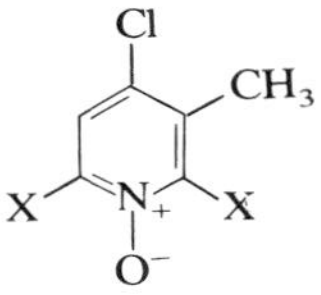 [structure: 2,6-X₂-4-chloro-3-methylpyridine N-oxide] (5)	X = 1-hydroxycyclohexyl [structure] 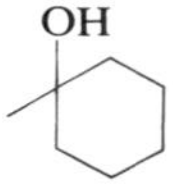	594

Formula	Substrate	Conditions	Electrophile	Product(s) and yield(s) (%)		Refs.
		BuLi/ether/−65°	”	(38)	(—)	593
		BuLi/ether/−65°/ 15 min	C_6H_5CHO	(—)	(9) $X = CH(OH)C_6H_5$	594
		BuLi/ether	C_6H_5CN	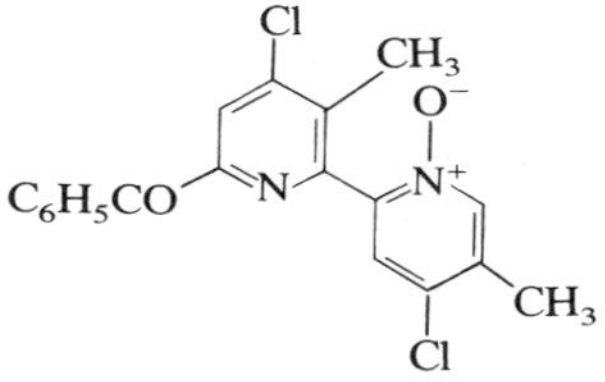 (12)		101
C_6H_7NO	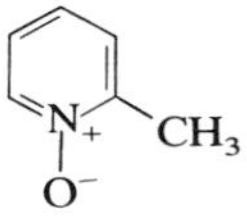	BuLi	CO_2	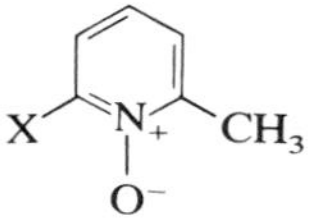 (14), 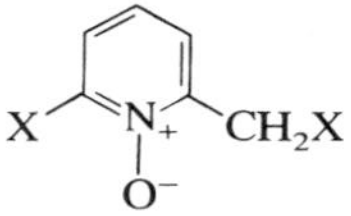	(10) $X = CO_2H$	101
		BuLi/THF/−65°/ 15 min	Cyclohexanone	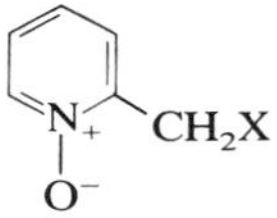 (4), 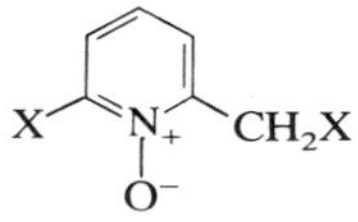	(20) X 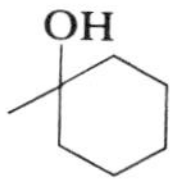 =	594
	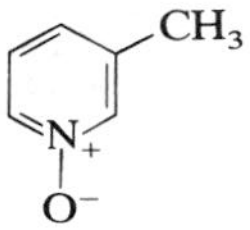	BuLi/THF, ether/ −65°/15 min	Cyclohexanone	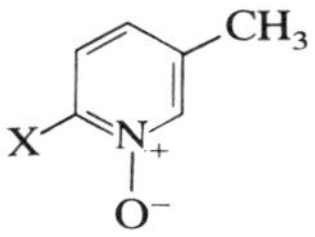 (25), 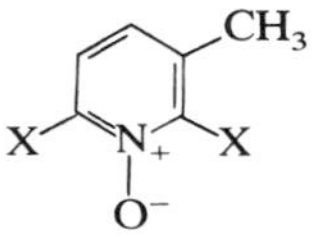	(8) X 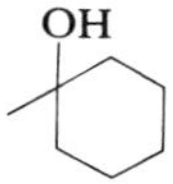 =	594

TABLE XIII. PYRIDINE-N-OXIDES (ALPHA) (*Continued*)

Formula	Compound Lithiated	Conditions	Substrate	Product and Yield (%)	Refs.
C_6H_7NO (*Contd.*)	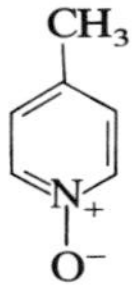 4-methylpyridine N-oxide [structure]	BuLi/THF/−65°/ 1 hr	Br_2	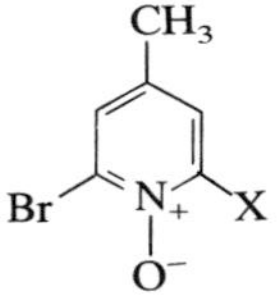 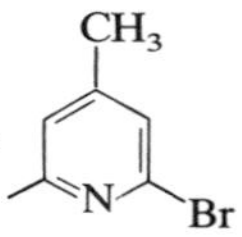 2-Br-6-X-4-methylpyridine N-oxide [structure]: X = Br (5), X 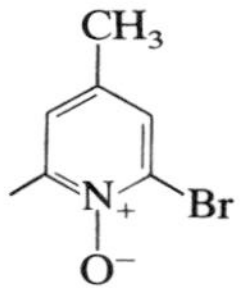 = 6-methyl-4-methyl-2-bromopyridine N-oxide residue [structure] (13), X = 6-methyl-4-methyl-2-bromopyridine residue [structure] (18)	595, 596
		"	O_2	1-Hydroxy-4-methyl-2(1H)-pyridinone (13)	102, 596
		"	"S"	1-Hydroxy-4-methyl-2(1H)-pyridinethione (39)	102, 596
		BuLi/THF/−65°	CO_2	4-Methylpyridine-2,6-dicarboxylic acid 1-oxide (48)	101, 593
		BuLi/THF/−65°/ 15 min	Cyclohexanone	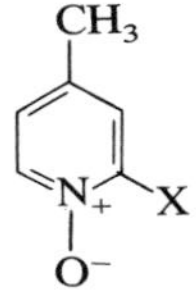 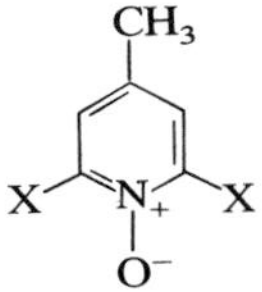 2-X-4-methylpyridine N-oxide [structure] (21), 2,6-X₂-4-methylpyridine N-oxide [structure] (27); X = 1-hydroxycyclohexyl [structure] 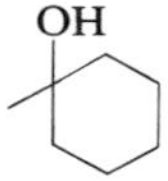	594, 593

C_7H_9NO		BuLi	Br_2	$X = Br$ (13–23), $X =$ (2–4)	595, 596
		”	Cl_2	2,6-Dichloro-3,4-dimethylpyridine 1-oxide (9)	595, 596
		BuLi/THF/−65°/ 1 hr	O_2	1-Hydroxy-3,4-dimethyl-2(1H)-pyridinone (10), 1-hydroxy-4,5-dimethyl-2(1H)-pyridinone (14)	102, 596
		”	“S”	1-Hydroxy-3,4-dimethyl-2(1H)-pyridinethione (13), 1-hydroxy-4,5-dimethyl-2(1H)-pyridinethione (24) (37)	102, 596
		BuLi/THF/−65°	CO_2	4,5-Dimethylpicolinic acid 1-oxide (18)	101

TABLE XIII. PYRIDINE-N-OXIDES (ALPHA) (*Continued*)

Formula	Compound Lithiated	Conditions	Substrate	Product and Yield (%)			Refs.
C_7H_9NO (*Contd.*)	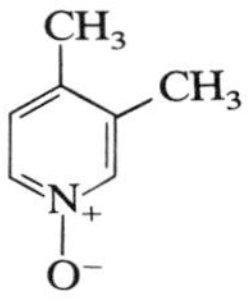 CH_3, CH_3, N^+–O^-	BuLi/THF	$CH_3CO_2C_2H_5$	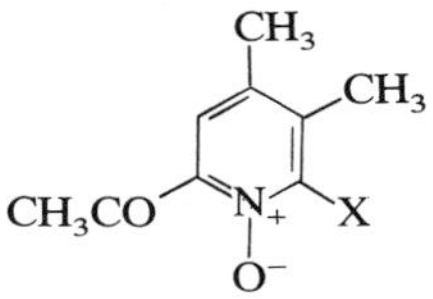 CH_3CO, CH_3, CH_3, X, N^+–O^-		X = H (65)	101
		BuLi/THF/−65°	DMA	”		X 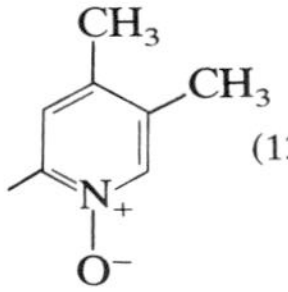 = CH_3, CH_3, N^+–O^- (13)	101
		BuLi/THF/−65°/ 15 min	n-C_3H_7CHO	4,5-Dimethyl-α-propyl-2-pyridinemethanol 1-oxide (15)			594
		BuLi/ether/−65°/ 15 min	Cyclohexanone	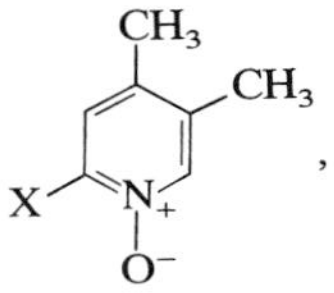 X, CH_3, CH_3, N^+–O^- ,	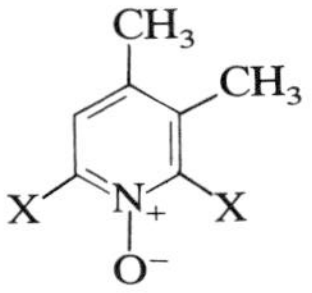 X, X, CH_3, CH_3, N^+–O^-	X = OH (cyclohexyl) 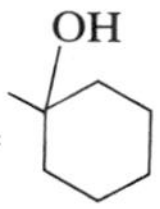	
				(56)	(16)		594
		—	”	(84)	(—)		593
		BuLi/THF/−65°/ 1 hr	$C_6H_5N{=}CHC_6H_5$	(12)	(62)	X = $CH(C_6H_5)NHC_6H_5$	102

$C_7H_9NO_2$	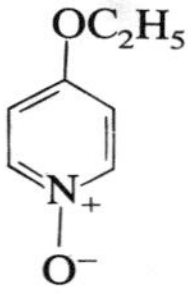 4-ethoxypyridine N-oxide (OC_2H_5, N^+–O^-)	BuLi/THF/−65°/ 15 min	n-$C_3H_7CO_2C_2H_5$	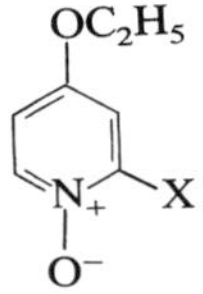 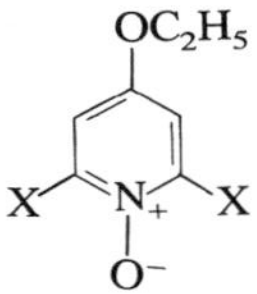 (19), (17)	$X = COC_3H_7$-n	101
		"	Cyclohexanone	(20), (21)	X 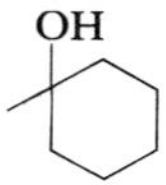 = 1-hydroxycyclohexyl (OH)	594, 593

Note: References 360–607 are on pp. 355–360.

TABLE XIV. PYRIMIDINES (ALPHA)

Formula	Compound Lithiated	Conditions	Substrate	Product and Yield (%)	Refs.
$C_5H_6N_2$	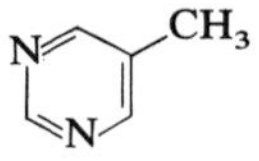 CH$_3$, N, N	LDA/ether/ 0°–35°/1 hr	$(C_6H_5)_2CO$	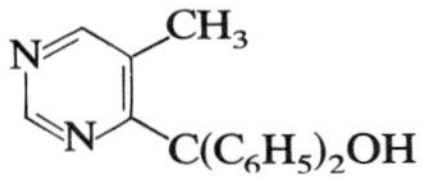 CH$_3$, N, N, $C(C_6H_5)_2OH$ (30)	99
$C_8H_6N_4$	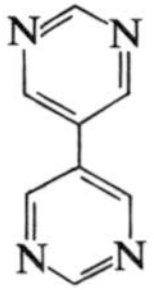 N, N, N, N	LDA/THF/ −70°–0°	—	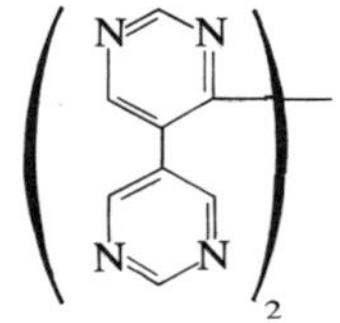 (N, N, N, N)$_2$ (38),	597
				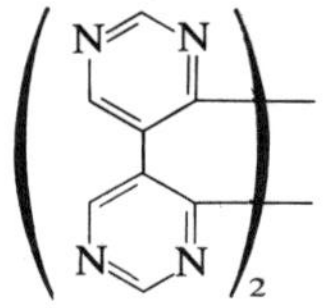 (N, N, N, N)$_2$ (I) (12)	597
$C_{16}H_{10}N_2$	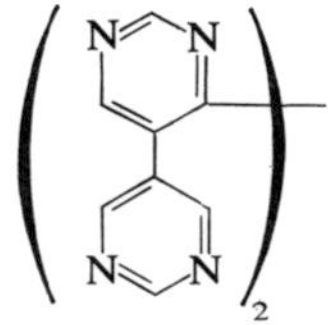 (N, N, N, N)$_2$	LDA/THF/ −70°–20°	—	I (16)	597

Note: References 360–607 are on pp. 355–360.

TABLE XV. Alkyl Vinyl Ethers and Allenic Ethers (Alpha)

Formula	Compound Lithiated	Conditions	Substrate	Product and Yield (%)		Refs.
C_3H_6O	$CH_2{=}CHOCH_3$	t-BuLi*	CuI	$[CH_2{=}C(OCH_3)]_2CuLi$ (—)		105
		t-BuLi/THF/ −65°–0°	$CH_3CH{=}CHCHO$	$CH_2{=}C(OCH_3)X$	$X = CH(OH)CH{=}CHCH_3$ (74)	104
		”	Cyclopentanone		X = 1-hydroxycyclopentyl (88)	104
		”	$(CH_3)_2C{=}CHCOCH_3$		$X = C(OH)(CH_3)CH{=}C(CH_3)_2$ (75)	104
		”	C_6H_5CHO		$X = CH(OH)C_6H_5$ (78)	104
		”	C_6H_5CN		$X = COC_6H_5$ (70)	104
		”	$C_6H_5CO_2H$		$X = COC_6H_5$ (62)	104
		”	$C_6H_5CH_2COCH_3$		$X = C(OH)(CH_3)CH_2C_6H_5$ (90)	104
		”	n-C_3H_7CHO (H_3O^+)	CH_3COX	$X = CH(OH)C_3H_7$-n (63)	104
		”	$(CH_3)_2C{=}CHCH_2Br$ (H_3O^+)		$X = CH_2CH{=}C(CH_3)_2$ (74)	104
		”	Cyclohexanone (H_3O^+)		X = 1-hydroxycyclohexyl (90)	104
		”	n-$C_8H_{17}I$ (H_3O^+)		$X = C_8H_{17}$-n (80)	104
		”	n-$C_4H_9CO_2CH_3$	$CH_3O(CH_2{=})C{-}C(R)(OH){-}C({=}CH_2)OCH_3$	$R = n$-C_4H_9 (82)	104, 563
		”	$C_2H_5CH(CH_3)CO_2CH_3$		$R = CH(CH_3)C_2H_5$ (93)	563
		”	t-$C_4H_9CO_2CH_3$		$R = C_4H_9$-t (95)	563
		”	$C_6H_5CO_2CH_3$		$R = C_6H_5$ (75)	563, 104

TABLE XV. ALKYL VINYL ETHERS AND ALLENIC ETHERS (ALPHA) (*Continued*)

Formula	Compound Lithiated	Conditions	Substrate	Product and Yield (%)	Refs.
C_3H_6O (*Contd.*)	$CH_2{=}CHOCH_3$	*t*-BuLi/THF/ −65°–0°	$CH_3CH{=}CHCO_2CH_3$ (H_3O^+)	$(CH_3CO)_2C(OH)CH{=}CHCH_3$ (46)	104
		"	$C_6H_5COCH_2Br$ (H_3O^+)	3,4-Dihydroxy-3-phenyl-2-butanone (58)	104
		t-BuLi/THF/ −80°	$(i\text{-}C_4H_9)_3B$ (H_3O^+, H_2O_2, OH^-)	$CH_3C(OH)R_2$ R = C_4H_9-*i* (77)	598
		"	$(s\text{-}C_4H_9)_3B$ (H_3O^+, H_2O_2, OH^-)	R = C_4H_9-*s* (92)	598
		"	(cyclopentyl)$_3$B (H_3O^+, H_2O_2, OH^-)	R = cyclopentyl (87)	598
		"	(cyclohexyl)$_3$B (H_3O^+, H_2O_2, OH^-)	R = cyclohexyl (94)	598
		"	$(n\text{-}C_6H_{13})_3B$ (H_3O^+, H_2O_2, OH^-)	R = C_6H_{13}-*n* (100)	598
		t-BuLi/THF/ −65°–0°	Estrone methyl ether	3-O-Methyl-17α(α-methoxyvinyl)estra-3,17β-diol (83)	365, 104
C_4H_5ClO	3-chloro-2,5-dihydrofuran (4-chloro-2,3-dihydrofuran structure)	BuLi/THF/ −78°	CH_3I	4-chloro-5-X-2,3-dihydrofuran; X = CH_3 (74)	266
		"	CH_3COCl	X = C(CH$_3$)(OAc)-(4-chloro-2,3-dihydrofuran-5-yl) (78)	266
		"	$n\text{-}C_4H_9I$	X = C_4H_9-*n* (62)	266

C_4H_6O	$CH_2{=}C{=}CHOCH_3$	BuLi/THF, ether/ −40°/45 min	CH_3COCH_2Cl	$CH_2{=}C{=}C(OCH_3)X$	X 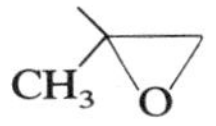 = CH_3 O (50–75)	599
		BuLi/THF, ether/ −20°/10 min	n-C_3H_7Br		X = C_3H_7-n (47)	108a
		"	n-C_3H_7I		X = C_3H_7-n (51)	108a
		BuLi/THF, ether/ −40°/45 min	$CH_3COCH(CH_3)Cl$		X 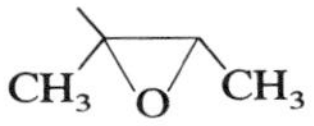 = CH_3 O CH_3 (50–75)	599
		BuLi/THF, ether/ −20°/10 min	n-C_4H_9Br		X = C_4H_9-n (67)	108a
		BuLi/THF, ether/ −40°/45 min	$CH_3COC(CH_3)_2Cl$		X 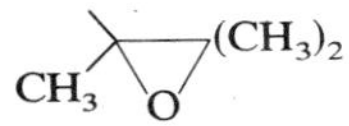 = CH_3 O $(CH_3)_2$ (50–75)	599
		"	t-$C_4H_9COCH_2Cl$		X 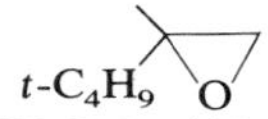 = t-C_4H_9 O (50–75)	599
		BuLi/THF, ether/ −20°/10 min	$C_6H_5CH_2Br$		X = $CH_2C_6H_5$ (64)	108a
		BuLi/THF, ether/ −40°/45 min	$C_6H_5COCH_2Cl$		X 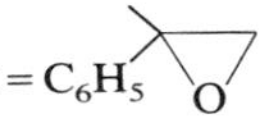 = C_6H_5 O (50–75)	599
	 O	t-BuLi/ −78°–5°/ 0.5 hr	$(CH_3)_2C{=}CHCH_2Br$	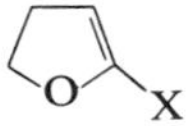 O X	X = $CH_2CH{=}C(CH_3)_2$ (67)	576
		"	n-$C_6H_{13}I$		X = C_6H_{13}-n (64)	576
$C_4H_7ClO_2$	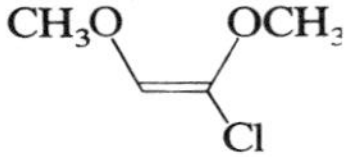 CH_3O OCH_3 Cl	s-BuLi/THF/ −100°/0.5 hr	$HgCl_2$	Bis[(Z)-2-chloro-1,2-dimethoxyvinyl]mercury (—)		600
		"	CO_2	(E)-3-Chloro-2,3-dimethoxyacrylic acid (45)		600

TABLE XV. Alkyl Vinyl Ethers and Allenic Ethers (Alpha) (*Continued*)

Formula	Compound Lithiated	Conditions	Substrate	Product and Yield (%)		Refs.
C_4H_8O	$CH_2{=}CHOC_2H_5$	t-BuLi/THF/ −65°–0°*	$(CH_3)_2SiHCl$	$CH_2{=}C(OC_2H_5)X$	$X = SiH(CH_3)_2$ (44)	601
		”	$(t\text{-}C_4H_9)_2SiF_2$		$X = SiF(C_4H_9\text{-}t)_2$ (50)	601
		t-BuLi/ TMEDA/ −30°/40 min	C_6H_5CHO		$X = CH(OH)C_6H_5$ (43)	103
		t-BuLi/THF/ −65°–0°	$(CH_3)_3SiCl$	Acetyltrimethylsilane (31)		601
	$CH_3CH{=}CHOCH_3$	t-BuLi/TMEDA	C_6H_5CHO	1-Hydroxy-1-phenyl-2-butanone (55)		104
C_5H_7ClO	[5-chloro-3,4-dihydro-2H-pyran]	BuLi/THF/25°/ 2 hr	CH_3I	[5-chloro-6-X-3,4-dihydro-2H-pyran]	$X = CH_3$ (>65)	562
		”	C_2H_5I		$X = C_2H_5$ (>65)	562
		”	$n\text{-}C_4H_9I$		$X = C_4H_9\text{-}n$ (>65)	562
C_5H_8O	[3,4-dihydro-2H-pyran]	BuLi/THF/50°/ 1 hr	CH_3I	[6-X-3,4-dihydro-2H-pyran]	$X = CH_3$ (35)	562
		”	$n\text{-}C_4H_9I$		$X = C_4H_9\text{-}n$ (35–75)	562
		t-BuLi/ pentane/ −78°–5°/0.5 hr	[3-bromocyclohexene]		X = [cyclohex-2-enyl] (60)	576
		”	$CH_3CO(CH_2)_2CO_2CH_3$		X = [2-methyl-5-oxotetrahydrofuran-2-yl] (68)	576
		BuLi/THF/50°/ 1 hr	$n\text{-}C_6H_{13}I$		$X = C_6H_{13}\text{-}n$ (75)	562, 576
C_5H_8O	$CH_2{=}CH{-}CH{=}CHOCH_3$	t-BuLi/THF/ −65°–0°	C_6H_5CHO	1-Hydroxy-1-phenyl-3-buten-2-one (30)		104

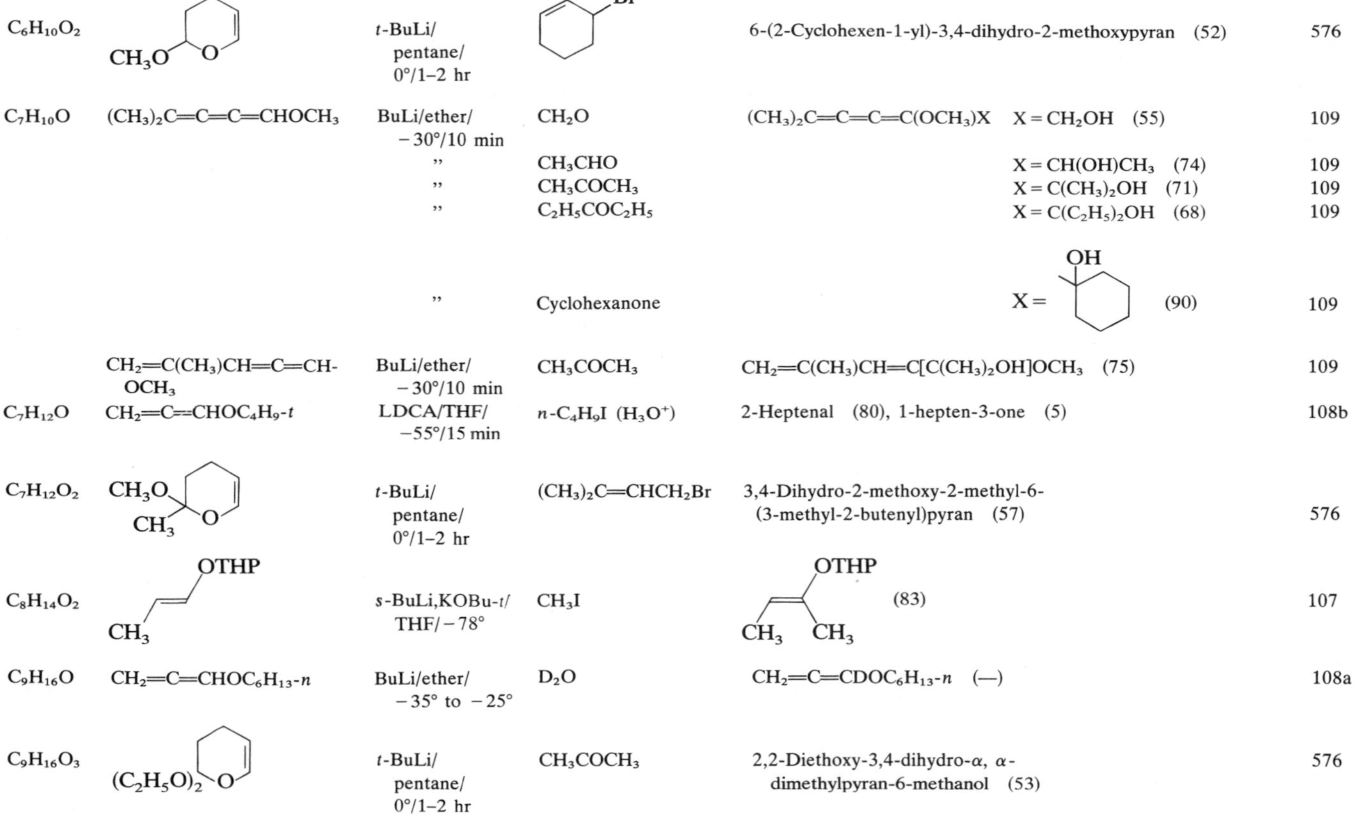

Formula	Substrate	Conditions	Reactant	Product(s) (Yield, %)	Refs.
$C_6H_{10}O_2$	(2-methoxy-3,4-dihydro-2H-pyran; CH_3O, O)	t-BuLi/ pentane/ 0°/1–2 hr	(3-bromocyclohexene; Br)	6-(2-Cyclohexen-1-yl)-3,4-dihydro-2-methoxypyran (52)	576
$C_7H_{10}O$	$(CH_3)_2C{=}C{=}C{=}CHOCH_3$	BuLi/ether/ −30°/10 min	CH_2O	$(CH_3)_2C{=}C{=}C{=}C(OCH_3)X$ $X = CH_2OH$ (55)	109
		"	CH_3CHO	$X = CH(OH)CH_3$ (74)	109
		"	CH_3COCH_3	$X = C(CH_3)_2OH$ (71)	109
		"	$C_2H_5COC_2H_5$	$X = C(C_2H_5)_2OH$ (68)	109
		"	Cyclohexanone	X = (1-hydroxycyclohexyl; OH) (90)	109
	$CH_2{=}C(CH_3)CH{=}C{=}CH$-$OCH_3$	BuLi/ether/ −30°/10 min	CH_3COCH_3	$CH_2{=}C(CH_3)CH{=}C[C(CH_3)_2OH]OCH_3$ (75)	109
$C_7H_{12}O$	$CH_2{=}C{=}CHOC_4H_9$-t	LDCA/THF/ −55°/15 min	n-C_4H_9I (H_3O^+)	2-Heptenal (80), 1-hepten-3-one (5)	108b
$C_7H_{12}O_2$	(2-methoxy-2-methyl-3,4-dihydropyran; CH_3O, CH_3, O)	t-BuLi/ pentane/ 0°/1–2 hr	$(CH_3)_2C{=}CHCH_2Br$	3,4-Dihydro-2-methoxy-2-methyl-6-(3-methyl-2-butenyl)pyran (57)	576
$C_8H_{14}O_2$	(CH_3CH=CH-OTHP; OTHP, CH_3)	s-BuLi,KOBu-t/ THF/ −78°	CH_3I	(OTHP, CH_3, CH_3) (83)	107
$C_9H_{16}O$	$CH_2{=}C{=}CHOC_6H_{13}$-n	BuLi/ether/ −35° to −25°	D_2O	$CH_2{=}C{=}CDOC_6H_{13}$-n (—)	108a
$C_9H_{16}O_3$	(2,2-diethoxy-3,4-dihydropyran; $(C_2H_5O)_2$, O)	t-BuLi/ pentane/ 0°/1–2 hr	CH_3COCH_3	2,2-Diethoxy-3,4-dihydro-α, α-dimethylpyran-6-methanol (53)	576

TABLE XV. Alkyl Vinyl Ethers and Allenic Ethers (Alpha) (*Continued*)

Formula	Compound Lithiated	Conditions	Substrate	Product and Yield (%)		Refs.
$C_{10}H_9LiO$	$C_6H_5(Li)C{=}C{=}C(H)OCH_3$	BuLi/ether/ −75°/15 min	$CH_3I, (CH_3)_3SiCl$	$C_6H_5(X)C{=}C{=}C(Y)OCH_3$		
				X	Y	
				CH_3	$Si(CH_3)_3$ (78)	602
		BuLi/ether/ −75°	C_2H_5Br, —	C_2H_5	H (87)	603
		BuLi/ether/ −75°/15 min	C_2H_5Br, CH_3I	$C_6H_5(X)C{=}C{=}C(Y)OCH_3$		
				X	Y	
				C_2H_5	CH_3 (76)	602, 603
		"	$C_2H_5Br, (CH_3)_3SiCl$	C_2H_5	$Si(CH_3)_3$ (82)	602, 603
		"	$(CH_3O)_2SO_2, (CH_3O)_2SO_2$	CH_3	CH_3 (70)	602, 603
		"	$(CH_3)_3SiCl, CO_2$	$Si(CH_3)_3$	CO_2H (90)	602
		"	$(CH_3)_3SiCl, CH_3I$	$Si(CH_3)_3$	CH_3 (70)	602, 603
		"	$(CH_3)_3SiCl, CH_3COCH_3$	$Si(CH_3)_3$	$C(CH_3)_2OH$ (80)	602
		"	$(CH_3)_3SiCl, (CH_3)_3SiCl$	$Si(CH_3)_3$	$Si(CH_3)_3$ (80)	602, 603
		BuLi/ether/ −75°	n-C_4H_9Br,—	n-C_4H_9	H (86)	603

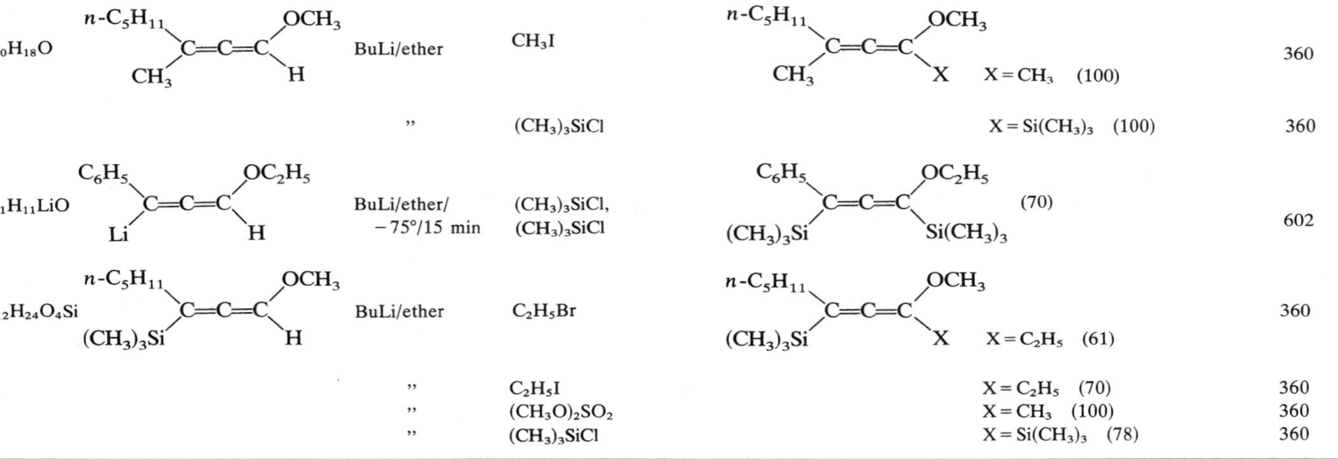

Formula	Allene	Base/Conditions	Electrophile	Product (Yield, %)	Refs.
$C_{10}H_{18}O$	$n\text{-}C_5H_{11}(CH_3)C{=}C{=}C(H)OCH_3$	BuLi/ether	CH_3I	$n\text{-}C_5H_{11}(CH_3)C{=}C{=}C(X)OCH_3$ $X = CH_3$ (100)	360
		”	$(CH_3)_3SiCl$	$X = Si(CH_3)_3$ (100)	360
$C_{11}H_{11}LiO$	$C_6H_5(Li)C{=}C{=}C(H)OC_2H_5$	BuLi/ether/ $-75°$/15 min	$(CH_3)_3SiCl$, $(CH_3)_3SiCl$	$C_6H_5[(CH_3)_3Si]C{=}C{=}C[Si(CH_3)_3]OC_2H_5$ (70)	602
$C_{12}H_{24}O_4Si$	$n\text{-}C_5H_{11}[(CH_3)_3Si]C{=}C{=}C(H)OCH_3$	BuLi/ether	C_2H_5Br	$n\text{-}C_5H_{11}[(CH_3)_3Si]C{=}C{=}C(X)OCH_3$ $X = C_2H_5$ (61)	360
		”	C_2H_5I	$X = C_2H_5$ (70)	360
		”	$(CH_3O)_2SO_2$	$X = CH_3$ (100)	360
		”	$(CH_3)_3SiCl$	$X = Si(CH_3)_3$ (78)	360

Note: References 360–607 are on pp. 355–360.

TABLE XVI. FURANS (ALPHA)

Formula	Compound Lithiated	Conditions	Substrate	Product and Yield (%)	Refs.
$C_4H_2Br_2O$	[2,3-dibromofuran structure]	LDA/THF or ether/−70°	$(CH_3)_3SiCl$	(4,5-Dibromo-2-furyl)trimethylsilane (—)	246
C_4H_3BrO	[3-bromofuran structure]	LDA/THF/−80°/ 2.5 hr	CH_2O	[3-bromo-2-X-furan structure] X = CH_2OH (52)	566
		"	CH_3OCH_2Cl	X = CH_2OCH_3 (49)	566
		LDA/THF or ether/−70°	$(CH_3)_3SiCl$	X = $Si(CH_3)_3$ (—)	246
		LDA/THF/−80°/ 2.5 hr	$(CH_3)_2C{=}CHCH_2Br$	X = $CH_2CH{=}C(CH_3)_2$ (66)	566
C_4H_3ClO	[3-chlorofuran structure]	LDA/THF/−80°/ 2.5 hr	$(CH_3)_2C{=}CHCH_2Br$	3-Chloro-2-(3-methyl-2-butenyl)furan (41)	566
C_4H_4O	[furan structure]	C_6H_5Li/ether	"S" + CH_3I	[2-X-furan structure] X = SCH_3 (35)	292
		BuLi/ether/−35°–reflux/4 hr	"S" + C_2H_5I	X = SC_2H_5 (53)	361
		C_6H_5Li/ether	"S" + $CH_2{=}CHCH_2Br$	X = $SCH_2CH{=}CH_2$ (70)	292
		"	"S" + Ac_2O	X = SAc (45)	292
		"	"S" + $C_6H_5CH_2Cl$	X = $SCH_2C_6H_5$ (60)	292
		"	"Se" + CH_3I	X = $SeCH_3$ (49)	292
		BuLi/ether/−20°*	"Se" + C_2H_5I	X = SeC_2H_5 (51)	362
		C_6H_5Li/ether	"Se" + $CH_2{=}CHCH_2Br$	X = $SeCH_2CH{=}CH_2$ (61)	292
		"	"Se" + Ac_2O	X = SeAc (64)	292
		"	"Se" + $C_6H_5CH_2Cl$	X = $SeCH_2C_6H_5$ (58)	292

Conditions	Electrophile	Product (Yield, %)	Refs.
BuLi/ether/25°–reflux/6 hr	SO_2	X = SO_2Li (55)	363
BuLi/ether/−20°–reflux/4 hr	CO_2	X = CO_2H (77)	112
BuLi/THF/−15°/4 hr	CH_3I	X = CH_3 (89)	566
LDA/THF/−15°/4 hr	"	X = CH_3 (36)	566
—	$CF_2{=}CCl_2$	X = $CF{=}CCl_2$ (53)	364
EtLi/ether/25°	CCl_3CCl_3	X = Cl (48)	365
BuLi/ether/20°/3 hr	Ethylene oxide	X = $(CH_2)_2OH$ (57)	366
BuLi/THF/−15°/6.5 hr	Propylene oxide	X = $CH_2CH(OH)CH_3$ (98)	367, 368, 561
BuLi/ether/reflux/3 hr	$(CH_3)_3SiCl$	X = $Si(CH_3)_3$ (52)	369
—	Pyrimidine ($KMnO_4$)	X = 4-pyrimidinyl (−)	370
BuLi/ether/reflux/4 hr	Ac_2O	X = $COCH_3$ (26)	371
BuLi/THF/−15°/6.5 hr	1,2-Epoxybutane (ethyloxirane, C_2H_5)	X = $CH_2CH(OH)C_2H_5$ (85)	367
BuLi	$ClB[N(CH_3)_2]_2$	X = $B[N(CH_3)_2]_2$ (15)	295
BuLi/ether/−20°–reflux/4 hr	$(CH_3)_2CHCHO$	X = $CH(OH)CH(CH_3)_2$ (93)	112
"	$CH_3COC_2H_5$	X = $C(OH)(CH_3)C_2H_5$ (88)	112

TABLE XVI. FURANS (ALPHA) (*Continued*)

Formula	Compound Lithiated	Conditions	Substrate	Product and Yield (%)	Refs.
C_4H_4O (*Contd.*)	furan	”	$CH_3CO_2C_2H_5$	2-X-furan; X = $C(OH)(CH_3)$(2-furyl) (94)	112
		BuLi/THF/−25° to −15°/4hr	n-C_4H_9Br	X = C_4H_9-n (77)	112
		BuLi/THF/−15°/ 4 hr	$(CH_3)_2C{=}CHCHO$	X = $CH(OH)CH{=}C(CH_3)_2$ (79)	566
		”	$(CH_3)_2C{=}CHCH_2Br$	X = $CH_2CH{=}C(CH_3)_2$ (76)	566
		BuLi/ether/−20°–reflux/4 hr	Cyclohexanone	X = 1-hydroxycyclohexyl (95)	112
		BuLi/THF/−15°/ 6 hr	$BrCH_2CH(OC_2H_5)_2$	X = $CH_2CH(OC_2H_5)_2$ (70)	367
		BuLi/ether/−20°–reflux/4 hr	C_6H_5CN (H_3O^+)	X = COC_6H_5 (89)	112
		BuLi/ether/−20°–reflux/4 hr	C_6H_5CHO	X = $CH(OH)C_6H_5$ (98)	112
		”	$C_6H_5COCH_3$	X = $C(OH)(CH_3)C_6H_5$ (96)	112
		”	$C_6H_5CO_2C_2H_5$	X = $C(OH)(C_6H_5)$(2-furyl) (44)	112
		EtLi/ether/−40°–reflux/0.5 hr	$(n$-$C_4H_9O)_3B$ (H_3O^+)	X = $B(OH)_2$ (55)	115

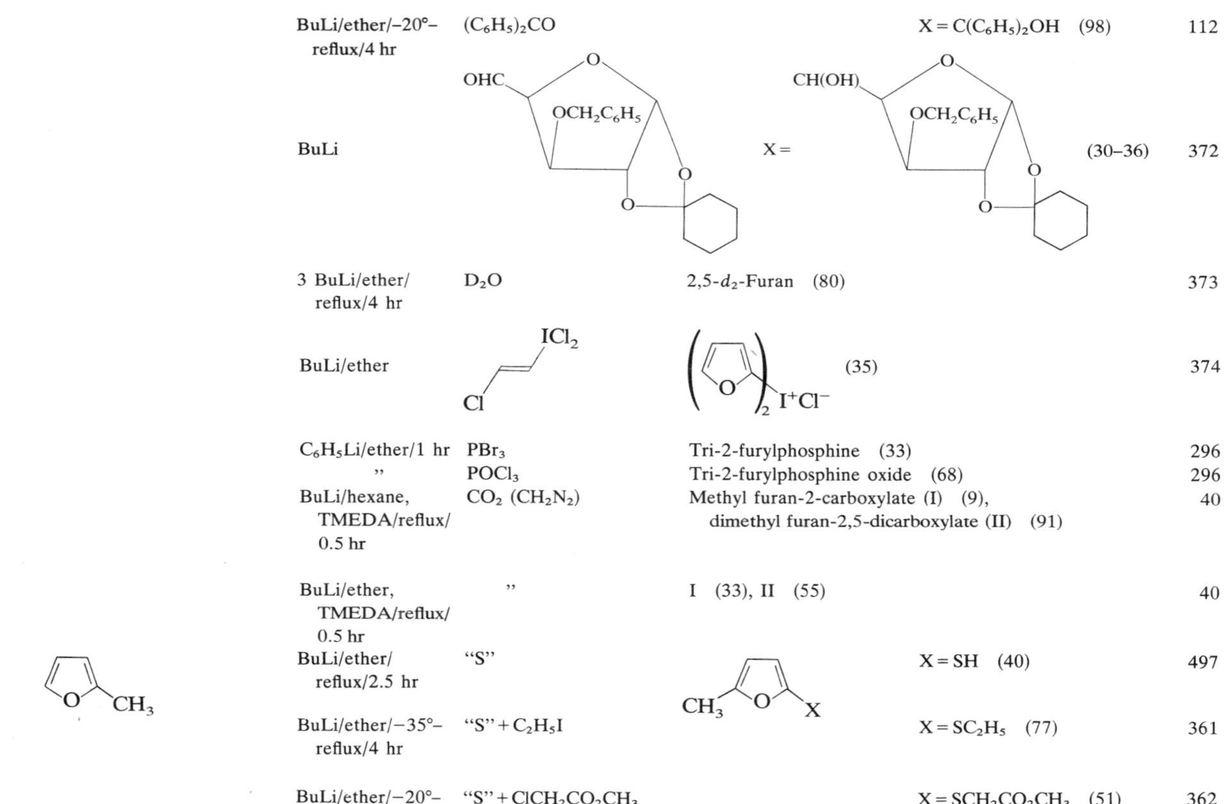

		BuLi/ether/−20°–reflux/4 hr	$(C_6H_5)_2CO$	$X = C(C_6H_5)_2OH$ (98)	112
		BuLi	OHC–(furanose: $OCH_2C_6H_5$, O-cyclohexylidene)	X = CH(OH)–(furanose: $OCH_2C_6H_5$, O-cyclohexylidene) (30–36)	372
		3 BuLi/ether/reflux/4 hr	D_2O	2,5-d_2-Furan (80)	373
		BuLi/ether	Cl–CH=CH–ICl_2	(2-furyl)$_2I^+Cl^-$ (35)	374
		C_6H_5Li/ether/1 hr	PBr_3	Tri-2-furylphosphine (33)	296
		”	$POCl_3$	Tri-2-furylphosphine oxide (68)	296
		BuLi/hexane, TMEDA/reflux/0.5 hr	CO_2 (CH_2N_2)	Methyl furan-2-carboxylate (I) (9), dimethyl furan-2,5-dicarboxylate (II) (91)	40
		BuLi/ether, TMEDA/reflux/0.5 hr	”	I (33), II (55)	40
C_5H_6O	2-methylfuran	BuLi/ether/reflux/2.5 hr	“S”	5-methyl-2-X-furan, X = SH (40)	497
		BuLi/ether/−35°–reflux/4 hr	“S” + C_2H_5I	$X = SC_2H_5$ (77)	361
		BuLi/ether/−20°–reflux	“S” + $ClCH_2CO_2CH_3$	$X = SCH_2CO_2CH_3$ (51)	362

Formula	Compound Lithiated	Conditions	Substrate	Product and Yield (%)		Refs.
C_5H_6O (*Contd.*)	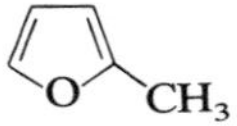 O CH_3	BuLi/ether/−35°–reflux/4 hr	"S" + n-C_4H_9I	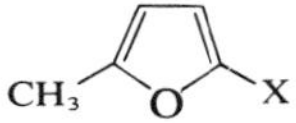 CH_3 O X	$X = C_4H_9$-n (66)	361
		"	"S" + i-C_4H_9I		$X = C_4H_9$-i (59)	361
		BuLi/ether/reflux/4 hr	CuBr		X 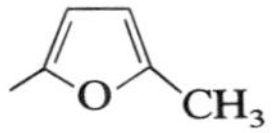 = O CH_3 (25)	367
		BuLi/ether/−20°–reflux	"Se" + C_2H_5I		$X = SeC_2H_5$ (41)	362
		"	"Se" + $ClCH_2CO_2CH_3$		$X = SeCH_2CO_2CH_3$ (43)	362
		BuLi/hexane, TMEDA/reflux/0.5 hr	CO_2 (CH_2N_2)		$X = CO_2CH_3$ (62)	40
		BuLi/THF/−15°/6 hr	$BrCH_2CH(OC_2H_5)_2$		$X = CH_2CH(OC_2H_5)_2$ (45)	367
		—	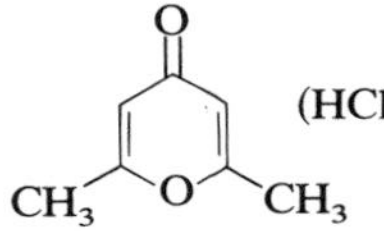 O, CH_3 O CH_3 ($HClO_4$)		X 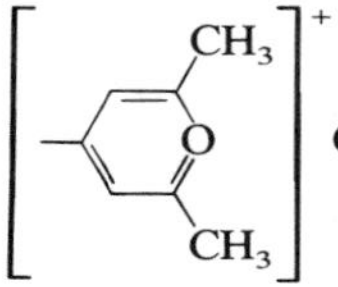 = [CH_3 O CH_3]$^+$ ClO_4^- (−)	375
		BuLi/ether/0°/1.5 hr	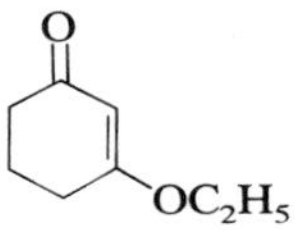 O, OC_2H_5		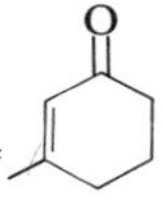 X = O (58)	284
		BuLi/THF/−25°/1 hr, −15°/4 hr	$CH_2{=}CHCH(CH_3)Br$	(III + IV, 59) (III : IV, 1 : 2)	$X = CH_2CH{=}CHCH_3$ (III), $X = CH(CH_3)CH{=}CH_2$ (IV)	367

Formula	Substrate	Conditions	Electrophile	Product (yield, %)	Ref.
		BuLi	$BrCH(CH_3)CH(OC_2H_5)_2$	$X = CH(CH_3)CH(OC_2H_5)_2$ (V), 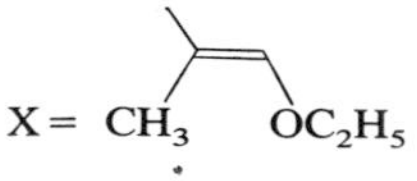 $X = C(CH_3)=CH(OC_2H_5)$ (VI) (V + VI, 50) (V : VI, 2 : 1)	367
	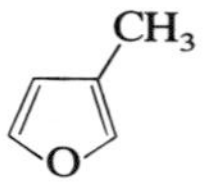 CH_3	BuLi/hexane, TMEDA/reflux/ 0.5 hr	CO_2 (CH_2N_2)	Dimethyl 3-methylfuran-2,5-dicarboxylate (66), methyl 3-methylfuran-2-carboxylate (18), methyl 4-methylfuran-2-carboxylate (4)	40
C_6H_8O	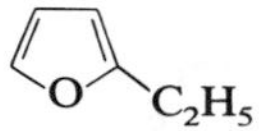 C_2H_5	BuLi/ether/−35°–reflux/4 hr	"S" + C_2H_5I	2-Ethyl-5-(ethylthio)furan (76)	361
C_6H_8OS	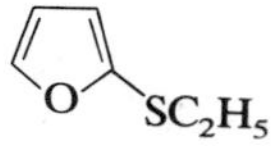 SC_2H_5	BuLi/ether/−35°–reflux/4 hr	"S" + C_2H_5I	2,5-Di(ethylthio)furan (77)	361
$C_7H_7BrO_3$	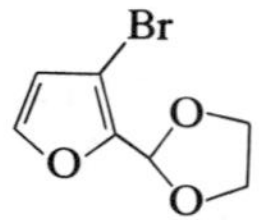 Br	LDA/THF or ether/−70°	D_2O	2-(3-Bromo-2-furyl-5-*d*)-1,3-dioxolane (—)	246
$C_7H_8O_3$	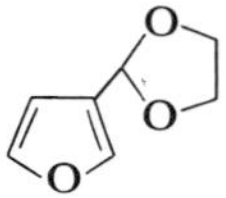	BuLi/ether/−10°–reflux/15 min	CO_2 (H_3O^+)	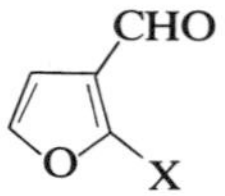 CHO, X; $X = CO_2H$ (50)	376
		"	DMA (H_3O^+)	$X = COC_2H_5$ (40)	376
		BuLi/ether/0°/ 0.5 hr	 $CON(CH_3)_2$ (H_3O^+)	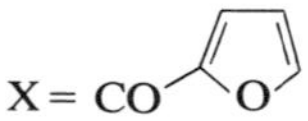 $X = CO$ (41)	113

TABLE XVI. FURANS (ALPHA) (*Continued*)

Formula	Compound Lithiated	Conditions	Substrate	Product and Yield (%)	Refs.
$C_7H_8O_3$ (*Contd.*)	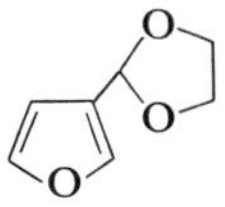 O, O, O	”	 S, $CON(CH_3)_2$ (H_3O^+)	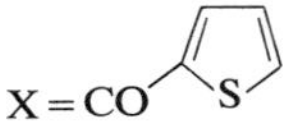 $X = CO$–S (30)	113
		”	$C_6H_5CON(CH_3)_2$ (H_3O^+)	$X = COC_6H_5$ (61)	113
		BuLi/ether/ −30°–25°/0.5 hr	(n-$C_4H_9O)_3B$ (H_3O^+)	$X = B(OH)_2$ (36)	115, 114
$C_7H_{10}LiNO_2$	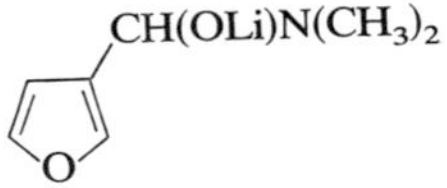 $CH(OLi)N(CH_3)_2$, O	EtLi/ether/ reflux/2 hr	(n-$C_4H_9O)_3B$ (H_3O^+)	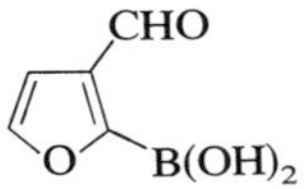 CHO, O, $B(OH)_2$ (32)	114, 115
$C_7H_{10}O_2$	 O, $CH_2CH(CH_3)OH$	BuLi/THF/−15°/ 6 hr	Ethylene oxide	5-[2-Hydroxyethyl]-α-methyl-2-furanethanol (80)	367
$C_7H_{12}OSi$	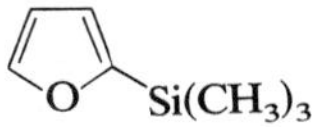 O, $Si(CH_3)_3$	BuLi/ether/ reflux/4 hr	CO_2	5-(Trimethylsilyl)-2-furoic acid (62)	369
$C_9H_{11}NO_2$	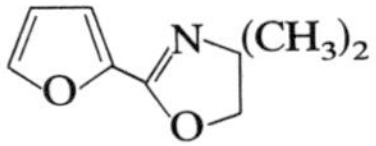 O, N, $(CH_3)_2$, O	BuLi/THF/−70°/ 1 hr	C_6H_5CHO	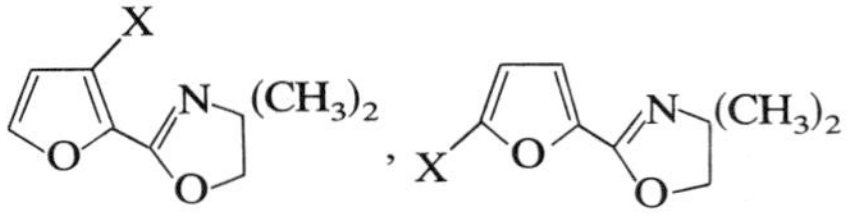 X, O, N, $(CH_3)_2$, O (49), X, O, N, $(CH_3)_2$, O (36); $X = CH(OH)C_6H_5$	377
$C_9H_{14}O_3$	 O, $CH(OC_2H_5)_2$	BuLi/ether/−10°– 25°/4 hr	D_2O	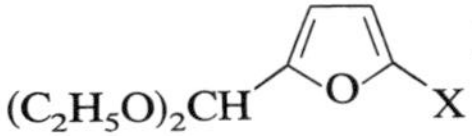 $(C_2H_5O)_2CH$, O, X; $X = D$ (93)	297
		”	$(CH_3)_3SiCl$	$X = Si(CH_3)_3$ (80)	297
		”	$(CH_3)_2C_2H_5SiCl$	$X = SiC_2H_5(CH_3)_2$ (78)	297
		”	$(CH_3)_3SiCH_2Cl$	$X = CH_2Si(CH_3)_3$ (33)	297

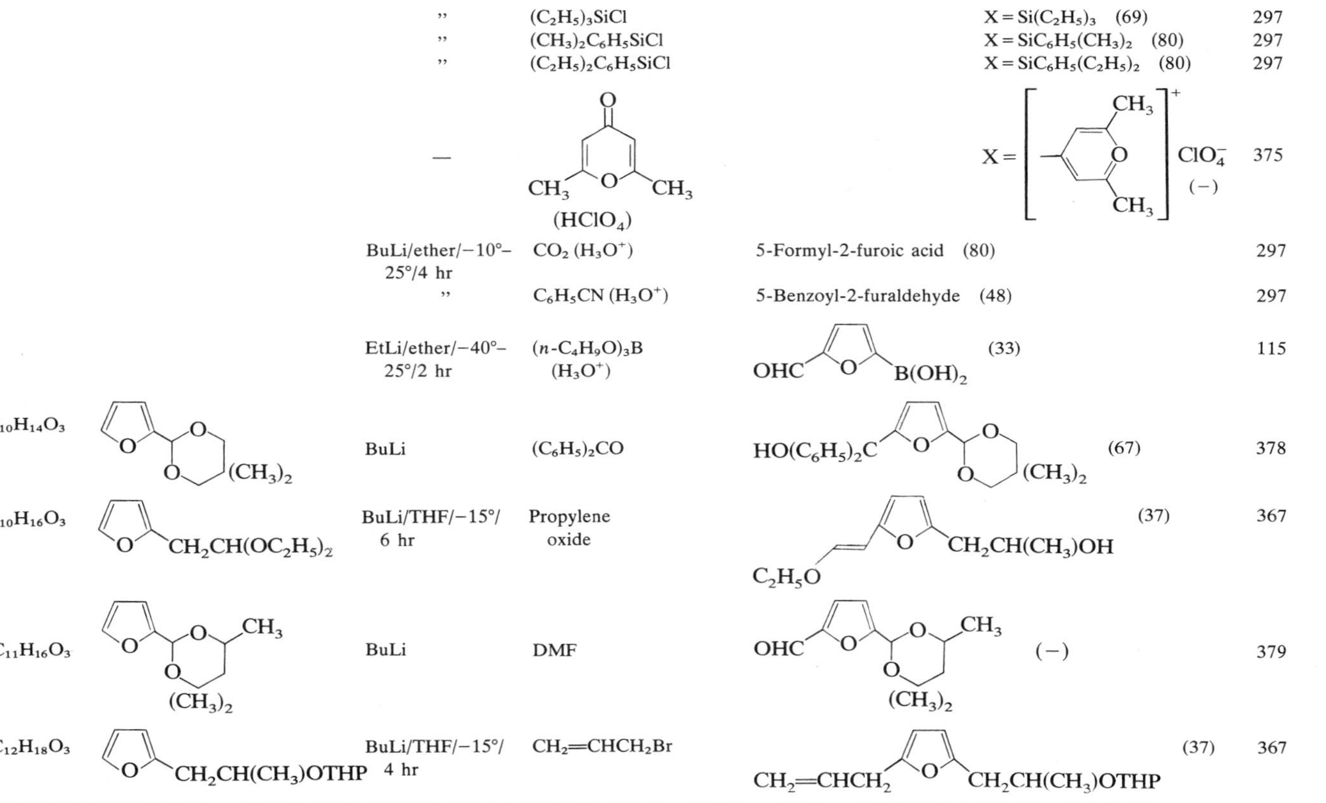

Formula	Reactant	Metalating agent/conditions	Electrophile	Product (% yield)	Refs.
		”	$(C_2H_5)_3SiCl$	$X = Si(C_2H_5)_3$ (69)	297
		”	$(CH_3)_2C_6H_5SiCl$	$X = SiC_6H_5(CH_3)_2$ (80)	297
		”	$(C_2H_5)_2C_6H_5SiCl$	$X = SiC_6H_5(C_2H_5)_2$ (80)	297
		—	2,6-dimethyl-4-pyrone ($HClO_4$)	X = 2,6-dimethylpyrylium ClO_4^- (–)	375
		BuLi/ether/−10°–25°/4 hr	CO_2 (H_3O^+)	5-Formyl-2-furoic acid (80)	297
		”	C_6H_5CN (H_3O^+)	5-Benzoyl-2-furaldehyde (48)	297
		EtLi/ether/−40°–25°/2 hr	$(n\text{-}C_4H_9O)_3B$ (H_3O^+)	OHC–furan–$B(OH)_2$ (33)	115
$C_{10}H_{14}O_3$	2-(2-furyl)-dioxane $(CH_3)_2$	BuLi	$(C_6H_5)_2CO$	$HO(C_6H_5)_2C$–furan–dioxane $(CH_3)_2$ (67)	378
$C_{10}H_{16}O_3$	furan–$CH_2CH(OC_2H_5)_2$	BuLi/THF/−15°/6 hr	Propylene oxide	C_2H_5O–CH=CH–furan–$CH_2CH(CH_3)OH$ (37)	367
$C_{11}H_{16}O_3$	2-(2-furyl)-dioxane CH_3, $(CH_3)_2$	BuLi	DMF	OHC–furan–dioxane CH_3, $(CH_3)_2$ (–)	379
$C_{12}H_{18}O_3$	furan–$CH_2CH(CH_3)OTHP$	BuLi/THF/−15°/4 hr	$CH_2{=}CHCH_2Br$	$CH_2{=}CHCH_2$–furan–$CH_2CH(CH_3)OTHP$ (37)	367

Note: References 360–607 are on pp. 355–360.

TABLE XVII. Condensed Furans (Alpha)

Formula	Compound Lithiated	Conditions	Substrate	Product and Yield (%)	Refs.
C_8H_5ClO	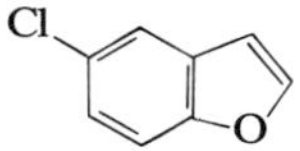	BuLi/ether/ reflux/2 hr	D_2O	5-Chloro-2-*d*-benzofuran (—)	380
		BuLi	Ethylene oxide	5-Chloro-2-(2-hydroxyethyl)-benzofuran (45–62)	116
C_8H_6O	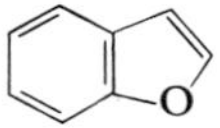	BuLi/ether/ 25°/40 min	D_2O	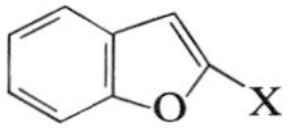 X = D (70)	381, 380
		BuLi/ether/ −10°/1 hr	CO_2	X = CO_2H (70)	479
		"	CH_3I	X = CH_3 (63)	479
		BuLi/ether/ 22°/0.5 hr	Ethylene oxide	X = $(CH_2)_2OH$ (45)	382, 116
		BuLi/ether/ −10°/1 hr	DMF	X = CHO (70)	479
C_9H_8O	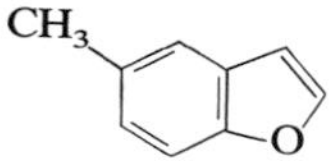	BuLi	Ethylene oxide	2-(2-Hydroxyethyl)-5-methyl-benzofuran (45–62)	116
$C_9H_8O_2$	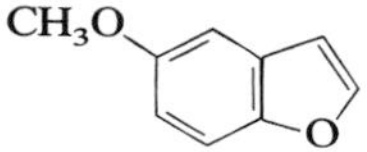	BuLi	Ethylene oxide	2-(2-Hydroxyethyl)-5-methoxy-benzofuran (45–62)	116
$C_{12}H_8O$	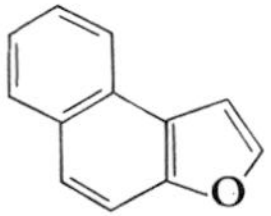	BuLi/ether/ reflux/10 min	DMF	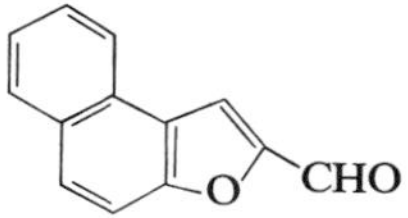 CHO (45)	278

Note: References 360–607 are on pp. 355–360.

TABLE XVIII. OXAZOLES AND OXAZOLINES (ALPHA)

Formula	Compound lithiated	Conditions	Substrate	Product and Yield (%)	Refs.
C_5H_9NO	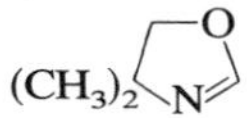 $(CH_3)_2$ [oxazoline ring: O, N]	BuLi/THF	D_2O	2-*d*-4,4-Dimethyl-2-oxazoline (99)	119
		BuLi/THF/−60°	D_2O	2-*d*-5-Phenyloxazole (96)	118
C_9H_7NO	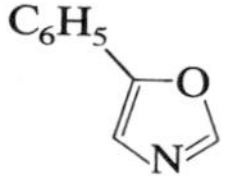 C_6H_5 [oxazole ring: O, N]	BuLi	C_6H_5CHO	α,5-Diphenyl-2-oxazolemethanol (—)	383
C_9H_9NO	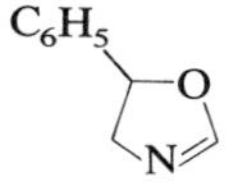 C_6H_5 [oxazoline ring: O, N]	BuLi/THF/−70°/ 10 min	CH_3OD	2-*d*-5-Phenyl-2-oxazoline (—)	120
$C_{15}H_{11}NO$	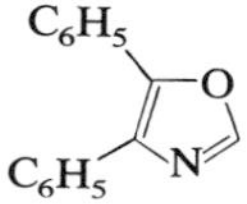 C_6H_5, C_6H_5 [oxazole ring: O, N]	BuLi/THF/−60°	D_2O	2-*d*-4,5-Diphenyloxazole (95)	118
		"	C_6H_5CHO	α,4,5-Triphenyl-2-oxazolemethanol (67)	118
$C_{15}H_{13}NO$	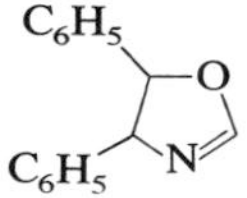 C_6H_5, C_6H_5 [oxazoline ring: O, N]	BuLi/THF/−70°/ 10 min	C_6H_5CHO	α,4,5-Triphenyl-2-oxazoline-2-methanol (58)	120

Note: References 360–607 are on pp. 355–360.

TABLE XIX. VINYL SULFIDES AND ALLENIC THIOETHERS (ALPHA)

Formula	Compound Lithiated	Conditions	Substrate	Product and Yield (%)	Refs.
C_4H_8S	$CH_2{=}CHSC_2H_5$	s-BuLi/THF, HMPA/ −78°/0.5 hr	$Br(CH_2)_3Br$ ($HgCl_2$)	3-Methyl-2-cyclohexen-1-one (52)	121
		"	$Br(CH_2)_4Br$ ($HgCl_2$)	CH_3COX $X = (CH_2)_4COCH_3$ (60)	121
		"	C_6H_5CHO ($HgCl_2$)	$X = CH(OH)C_6H_5$ (64)	121
		"	Styrene oxide ($HgCl_2$)	$X = CH{=}CHC_6H_5$ (68)	121
		"	n-$C_8H_{17}Br$ ($HgCl_2$)	$X = C_8H_{17}$-n (90)	121
		"	n-$C_8H_{17}CHO$ ($HgCl_2$)	$X = CH(OH)C_8H_{17}$-n (58)	121
$C_6H_4S_4$	(tetrathiafulvalene structure)	BuLi	CO_2	(1,3-Dithiol-2-ylidene)-1,3-dithiole-4-carboxylic acid (56)	162
$C_6H_{10}S$	$CH_3CH{=}C{=}CHSC_2H_5$	CH_3Li/ether	CH_2O	$CH_3CH(CH_2OH)C{\equiv}CSC_2H_5$ (I), $CH_3CH{=}C{=}C(SC_2H_5)CH_2OH$ (II) (I+II, 75) (I:II, 3:2)	130
$C_6H_{10}S_2$	(1,4-bis(methylthio)-1,3-pentadiene structure: SCH_3, SCH_3)	BuLi/THF, TMEDA/−40°	CH_3I	1,4-Bis (methylthio)-1,3-pentadiene (96)	605
		BuLi/THF, TMEDA, HMPA/25°	H_2O	2-(Methylthio)thiophene (75–80)	605
		"	CH_3I	2-Methyl-5-(methylthio)thiophene (75–80)	605
$C_6H_{12}S_2$	$C_2H_5SCH{=}CHSC_2H_5$ (cis)	LDA/THF/−80°/10 min	CH_3OD	$C_2H_5SCH{=}C(X)SC_2H_5$ $X = D$ (85)	128
		"	CH_3SO_3F	$X = CH_3$ (68)	128
		t-BuLi/THF/ −80°/10 min	$(CH_3S)_2$	$X = SCH_3$ (100)	128
		"	C_6H_5CHO	$X = CH(OH)C_6H_5$ (87)	128

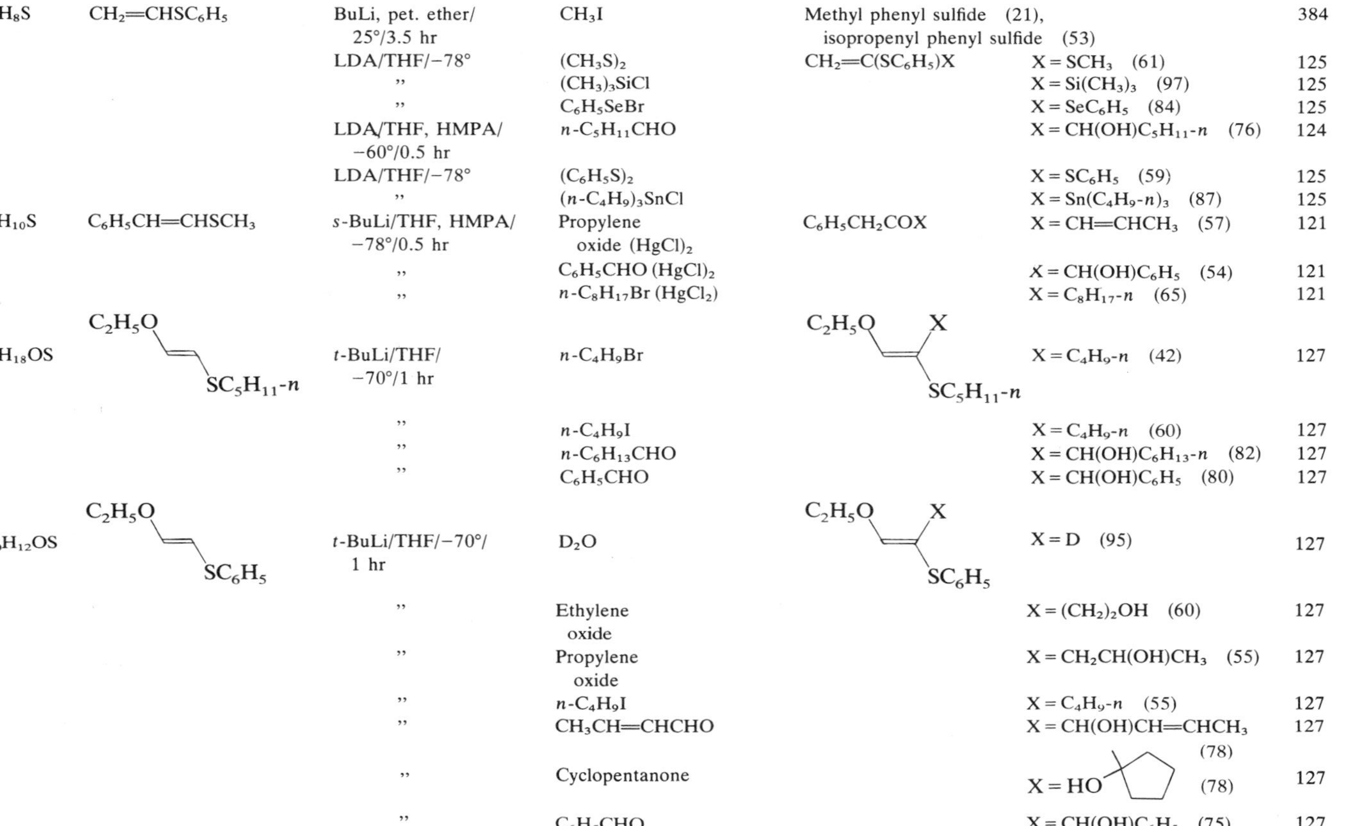

C_8H_8S	$CH_2{=}CHSC_6H_5$	BuLi, pet. ether/ 25°/3.5 hr	CH_3I	Methyl phenyl sulfide (21), isopropenyl phenyl sulfide (53)	384
		LDA/THF/−78°	$(CH_3S)_2$	$CH_2{=}C(SC_6H_5)X$ $X = SCH_3$ (61)	125
		”	$(CH_3)_3SiCl$	$X = Si(CH_3)_3$ (97)	125
		”	C_6H_5SeBr	$X = SeC_6H_5$ (84)	125
		LDA/THF, HMPA/ −60°/0.5 hr	n-$C_5H_{11}CHO$	$X = CH(OH)C_5H_{11}$-n (76)	124
		LDA/THF/−78°	$(C_6H_5S)_2$	$X = SC_6H_5$ (59)	125
		”	$(n$-$C_4H_9)_3SnCl$	$X = Sn(C_4H_9$-$n)_3$ (87)	125
$C_9H_{10}S$	$C_6H_5CH{=}CHSCH_3$	s-BuLi/THF, HMPA/ −78°/0.5 hr	Propylene oxide $(HgCl)_2$	$C_6H_5CH_2COX$ $X = CH{=}CHCH_3$ (57)	121
		”	C_6H_5CHO $(HgCl)_2$	$X = CH(OH)C_6H_5$ (54)	121
		”	n-$C_8H_{17}Br$ $(HgCl_2)$	$X = C_8H_{17}$-n (65)	121
$C_9H_{18}OS$	C_2H_5O–CH=CH–SC_5H_{11}-n	t-BuLi/THF/ −70°/1 hr	n-C_4H_9Br	C_2H_5O–CH=C(X)–SC_5H_{11}-n $X = C_4H_9$-n (42)	127
		”	n-C_4H_9I	$X = C_4H_9$-n (60)	127
		”	n-$C_6H_{13}CHO$	$X = CH(OH)C_6H_{13}$-n (82)	127
		”	C_6H_5CHO	$X = CH(OH)C_6H_5$ (80)	127
$C_{10}H_{12}OS$	C_2H_5O–CH=CH–SC_6H_5	t-BuLi/THF/−70°/ 1 hr	D_2O	C_2H_5O–CH=C(X)–SC_6H_5 $X = D$ (95)	127
		”	Ethylene oxide	$X = (CH_2)_2OH$ (60)	127
		”	Propylene oxide	$X = CH_2CH(OH)CH_3$ (55)	127
		”	n-C_4H_9I	$X = C_4H_9$-n (55)	127
		”	$CH_3CH{=}CHCHO$	$X = CH(OH)CH{=}CHCH_3$ (78)	127
		”	Cyclopentanone	X = HO–(1-cyclopentyl) (78)	127
		”	C_6H_5CHO	$X = CH(OH)C_6H_5$ (75)	127

TABLE XIX. VINYL SULFIDES AND ALLENIC THIOETHERS (ALPHA) (*Continued*)

Formula	Compound Lithiated	Conditions	Substrate	Product and Yield (%)	Refs.
$C_{10}H_{12}OS$ (*Contd.*)	C_2H_5O–CH=CH–SC_6H_5	"	n-$C_6H_{13}CHO$	X = $CH(OH)C_6H_{13}$-n (84)	127
$C_{11}H_{15}NS$	$(CH_3)_2NCH_2$–CH=CH–SC_6H_5	BuLi/ether/0°/ 1 hr	D_2O	$(CH_3)_2NCH_2$–CH=C(X)–SC_6H_5 X = D (>90%)	23
		"	$(CH_3S)_2$	X = SCH_3 (>90%)	23
$C_{11}H_{22}S$	n-$C_8H_{17}CH{=}CHSCH_3$	s-BuLi/THF, HMPA/−78°/ 0.5 hr	C_6H_5CHO ($HgCl_2$)	n-$C_9H_{19}COX$ X = $CH(OH)C_6H_5$ (51)	121
		"	n-$C_8H_{17}Br$ ($HgCl_2$)	X = C_8H_{17}-n (82)	121
$C_{11}H_{12}S_2$	C_6H_5S–CH=CH–CH=CH–SCH_3	BuLi/THF, TMEDA/−40°	CH_3I	1-(Methylthio)-4-(phenylthio)-1,3-butadiene (96)	605
		BuLi/THF, TMEDA, HMPA/25°	H_2O	2-(Methylthio)thiophene (75–80)	605
		"	CH_3I	2-Methyl-5-(methylthio)thiophene (75–80)	605
$C_{14}H_{12}S_2$	C_6H_5S–CH=CH–SC_6H_5 (cis)	BuLi/ether/−10°	$(CH_3O)_2SO_2$	$C_6H_5C{\equiv}CX$ X = H (26), X = CH_3 (45)	129
	C_6H_5S–CH=CH–SC_6H_5 (trans)	BuLi/ether/—10°	$(CH_3O)_2SO_2$	$C_6H_5C{\equiv}CX$ X = H (17), X = CH_3 (53)	129

Note: References 360–607 are on pp. 355–360.

TABLE XX. Vinyl Sulfoxides (Alpha)

Formula	Compound Lithiated	Conditions	Substrate	Product and Yield (%)	Refs.
C_8H_9NOS	2-pyridyl–S(→O)CH=CHCH$_3$	LDA/THF/−100°/ 5 min	CH_3I	(E)-1-Methylpropenyl 2-pyridyl sulfoxide (96)	132
		"	C_6H_5CHO	(E)-1-Phenyl-2-(2-pyridylsulfinyl)-2-buten-1-ol (91)	132
$C_9H_{10}OS$	$C_6H_5S(\rightarrow O)CH{=}CHCH_3$	LDA/THF/−100°/ 10 min	CH_3I	$C_6H_5S(\rightarrow O)C(X){=}CHCH_3$ X = CH_3 (80)	132
		"	Propylene oxide	X = $CH_2CH(OH)CH_3$ (45)	132
		"	C_6H_5CHO	X = $CH(OH)C_6H_5$ (93)	132
		"	Styrene oxide	X = $CH(C_6H_5)CH_2OH$ (23)	132
$C_{10}H_{12}O_2S$	$C_6H_5S(\rightarrow O)CH{=}CHCH_2OCH_3$	LDA/THF/−100°/ 5 min	D_2O	$C_6H_5S(\rightarrow O)C(X){=}CHCH_2OCH_3$ X = D (71)	132

TABLE XX. VINYL SULFOXIDES (ALPHA) (*Continued*)

Formula	Compound Lithiated	Conditions	Substrate	Product and Yield (%)	Refs.
$C_{10}H_{12}O_2S$ (*Contd.*)	$C_6H_5S(\rightarrow O)CH{=}CHCH_2OCH_3$	"	CH_3I	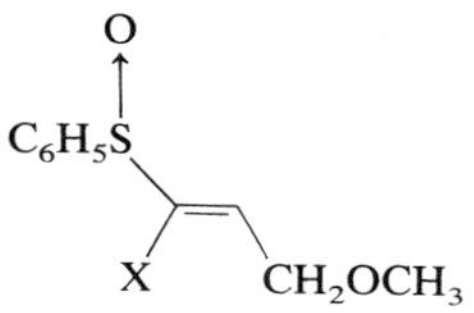 $C_6H_5S(\rightarrow O)C(X){=}CHCH_2OCH_3$, $X = CH_3$ (73)	132
		"	C_6H_5CHO	$X = CH(OH)C_6H_5$ (80)	132
		"	$C_6H_5COCH_3$	$X = C(OH)(CH_3)C_6H_5$ (74)	132
$C_{15}H_{15}NOS$	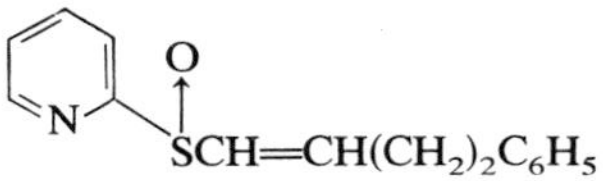 2-Pyridyl-$S(\rightarrow O)CH{=}CH(CH_2)_2C_6H_5$	LDA/THF/−100°/ 5 min	CH_3I	(E)-1-Methyl-4-phenyl-1-butenyl 2-pyridyl sulfoxide (99)	132
$C_{16}H_{16}OS$	$C_6H_5S(\rightarrow O)CH{=}CH(CH_2)_2C_6H_5$	LDA/THF/−100°/ 10 min	CH_3I	(E)-1-Methyl-4-phenyl-1-butenyl phenyl sulfoxide (89)	132

TABLE XXI. THIOPHENES (ALPHA)

Formula	Compound Lithiated	Conditions	Substrate	Product and Yield (%)	Refs.
C_4H_3BrS	2-bromothiophene	LDA/THF or ether/−70°	$(CH_3)_3SiCl$	(5-Bromo-2-thienyl)trimethylsilane (—)	246
	3-bromothiophene	C_6H_5Li/ether/ overnight	CO_2	3-Bromo-2-thiophenecarboxylic acid (72)	48
C_4H_3FS	3-fluorothiophene	BuLi/ether/ reflux/15 min	CO_2	3-Fluoro-2-thiophenecarboxylic acid (75)	385
C_4H_3IS	2-iodothiophene	C_6H_5Li/ether, TMEDA	CO_2	5-Iodo-2-thiophenecarboxylic acid (30)	68
	3-iodothiophene	1-methyl-2-pyrrolyl-Li/ether, TMEDA/ −10°/0.5 hr	CO_2	3-Iodo-2-thiophenecarboxylic acid (80), 4-iodo-2-thiophenecarboxylic acid (20)	68
$C_4H_3NO_2S$	3-nitrothiophene	LDA/THF or ether/−70°	$(CH_3)_3SiCl$	Trimethyl(3-nitro-2-thienyl)silane (—)	246
C_4H_3ST	2-tritiothiophene	BuLi/ether/ reflux/0.5 hr	CO_2	5-*t*-2-Thiophenecarboxylic acid (I), 2-thiophenecarboxylic acid (II) (I+II, 87) (I:II, 89:11)	386, 37

TABLE XXI. THIOPHENES (ALPHA) (*Continued*)

Formula	Compound Lithiated	Conditions	Substrate	Product and Yield (%)	Refs.
C_4H_4S	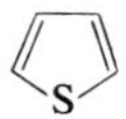 S	BuLi/ether/0°/ 1 hr	$FClO_3$	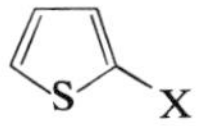 S X; X = F (49)	149
		BuLi/THF/−30° to −20°/1 hr	"S"	X = SH (65–70)	357
		BuLi/ether/ reflux/2 hr	"S" + $ClCH_2CO_2H$	X = SCH_2CO_2H (80)	387
		BuLi/ether	"S" + $ClCH_2CO_2CH_3$	X = $SCH_2CO_2CH_3$ (85)	388
		BuLi/ether/ reflux/6 hr	SO_2	X = SO_2Li (57)	363
		BuLi/ether/−20°	"Se" + C_2H_5I	X = SeC_2H_5 (73)	362
		BuLi/ether/ reflux/0.5 hr	$CF_2{=}CCl_2$	X = $CF{=}CCl_2$ (81)	364
		BuLi/ether	Ethylene oxide	X = $(CH_2)_2OH$ (78)	389
		BuLi/ether/25°/ 1 hr	C_2H_5Br	X = C_2H_5 (61)	20
		BuLi/ether/−35°	$ClCH_2OCH_3$	X = CH_2OCH_3 (76)	390
		BuLi/ether/ reflux/3 hr	$(CH_3)_3SiCl$	X = $Si(CH_3)_3$ (75)	369
		BuLi/ether/25°/ 1 hr	$(CH_3O)_2SO_2$	X = CH_3 (65)	20
		BuLi/ether 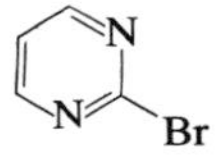 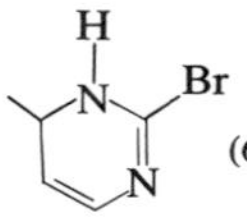	N, N, Br	X = H, N, Br, N (64)	391
		"	Br, N, N	X = H, N, N, Br (90)	391

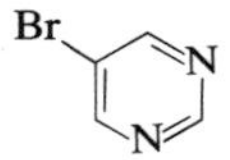

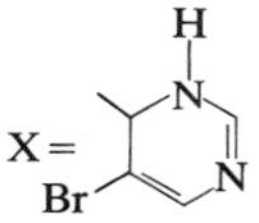

—	 N, N (ring) (KMnO$_4$)	X 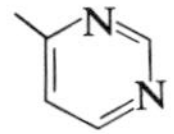 = (N, N ring) (−)	370
BuLi/ether/ reflux/3 hr	 ClB (ring)	 X = B (ring) (35)	295
BuLi/ether/25°/ 1 hr	n-C_4H_9Br	X = C_4H_9-n (47)	20
BuLi/ether/ reflux/3 hr	$ClB[N(CH_3)_2]_2$	X = $B[N(CH_3)_2]_2$ (17)	295
BuLi/ether/0°	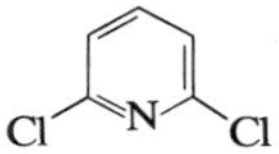 Cl, N, Cl (ring)	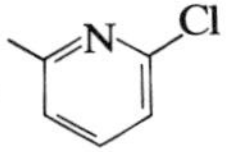 X = (N, Cl ring) (36)	137
,,	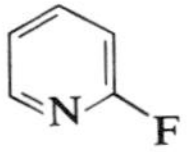 N, F (ring)	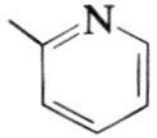 X = (N ring) (48)	137
BuLi/THF/−15°/ 2 hr	C_6H_5Br	X = C_6H_5 (31)	20
BuLi/THF/25°/ 6 hr	$BrCH_2CH(OC_2H_5)_2$	X = $CH_2CH(OC_2H_5)_2$ (40)	392
BuLi/ether	C_6H_5SCN	X = CN (19), X = SC_6H_5 (59)	393
BuLi/ether/0°/ 2 hr	o-FC_6H_4CHO	X = $CH(OH)C_6H_4F$-o (90)	23
BuLi/ether/25°/ 1 hr	$C_6H_5CH_2Br$	X = $CH_2C_6H_5$ (62)	20
,,	n-$C_8H_{17}Br$	X = C_8H_{17}-n (46)	20
LDA/ether/0°/ 4.5 hr	$(C_6H_5)_2CO$	X = $C(C_6H_5)_2OH$ (31)	23

TABLE XXI. THIOPHENES (ALPHA) (*Continued*)

Formula	Compound Lithiated	Conditions	Substrate	Product and Yield (%)	Refs.
C_4H_4S (*Contd.*)	 (thiophene, S)	BuLi/hexane, TMEDA/ reflux/0.5 hr	$CO_2(CH_2N_2)$	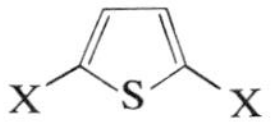 X–(thiophene, S)–X $X = CO_2CH_3$ (III) (100)	40
		BuLi/ether/ reflux/3 hr	$ClB(CH_3)N(CH_3)_2$	$X = B(CH_3)N(CH_3)_2$ (22)	295
		BuLi/ether, TMEDA/ reflux/0.5 hr	$CO_2(CH_2N_2)$	III (46), methylthiophene-2-carboxylate (39)	40
		BuLi/ether	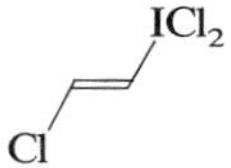 $ClCH{=}CHICl_2$	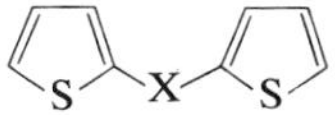 (thienyl, S)–X–(thienyl, S) $X = I^+Cl^-$ (69)	394
		BuLi/ether/ reflux/3 hr	$Cl_2BN(CH_3)_2$	$X = BN(CH_3)_2$ (17)	295
		BuLi/ether/0°	½ 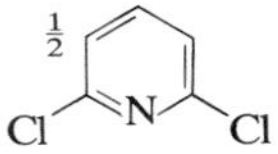 2,6-dichloropyridine (Cl, N, Cl)	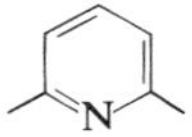 X = 2,6-pyridinediyl (N) (18)	137
		BuLi/reflux/ 20 min	$(n\text{-}C_4H_9O)_3B$ (H_2O_2)	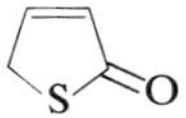 (S, O) (28)	289
		C_6H_5Li/ether/ reflux/2 hr	$RCH_2CO_2C_2H_5$	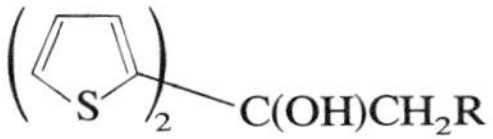 (thienyl, S)$_2$C(OH)CH_2R	
		"	$R = CH(CH_3)NH_2$	$R = CH(CH_3)NH_2$ (34)	395
		"	$R = CH_2N(CH_3)_2$	$R = CH_2N(CH_3)_2$ (72)	395
		"	$R = CH(CH_3)N(CH_3)_2$	$R = CH(CH_3)N(CH_3)_2$ (71)	395

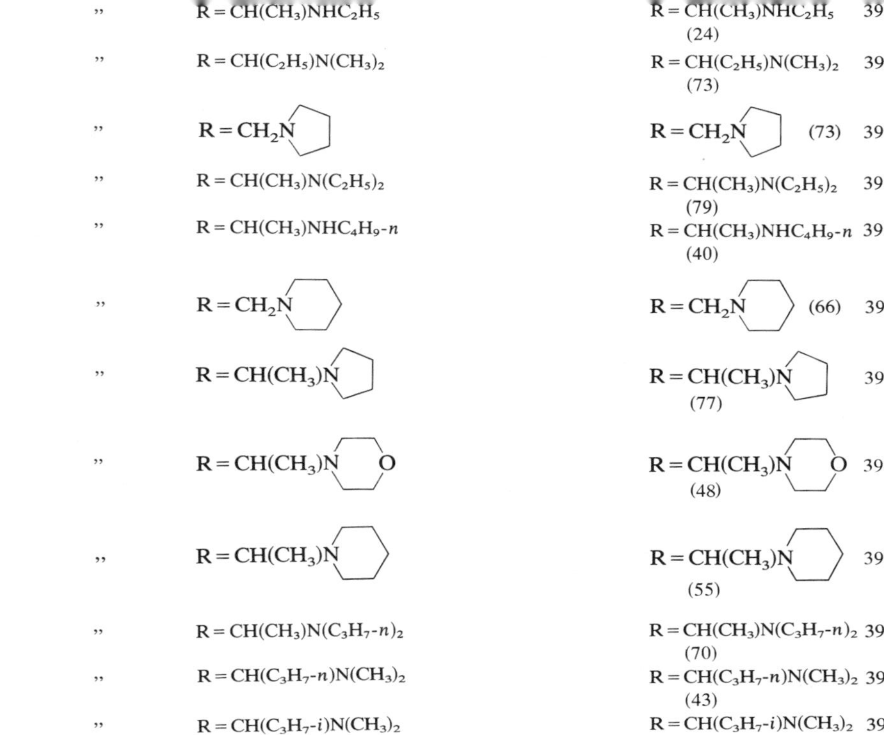

"	$R = CH(CH_3)NHC_2H_5$	$R = CH(CH_3)NHC_2H_5$ (24)	395
"	$R = CH(C_2H_5)N(CH_3)_2$	$R = CH(C_2H_5)N(CH_3)_2$ (73)	395
"	$R = CH_2N$	$R = CH_2N$ (73)	395
"	$R = CH(CH_3)N(C_2H_5)_2$	$R = CH(CH_3)N(C_2H_5)_2$ (79)	395
"	$R = CH(CH_3)NHC_4H_9\text{-}n$	$R = CH(CH_3)NHC_4H_9\text{-}n$ (40)	395
"	$R = CH_2N$	$R = CH_2N$ (66)	395
"	$R = CH(CH_3)N$	$R = CH(CH_3)N$ (77)	395
"	$R = CH(CH_3)N$ O	$R = CH(CH_3)N$ O (48)	395
"	$R = CH(CH_3)N$	$R = CH(CH_3)N$ (55)	395
"	$R = CH(CH_3)N(C_3H_7\text{-}n)_2$	$R = CH(CH_3)N(C_3H_7\text{-}n)_2$ (70)	395
"	$R = CH(C_3H_7\text{-}n)N(CH_3)_2$	$R = CH(C_3H_7\text{-}n)N(CH_3)_2$ (43)	395
"	$R = CH(C_3H_7\text{-}i)N(CH_3)_2$	$R = CH(C_3H_7\text{-}i)N(CH_3)_2$ (71)	395

TABLE XXI. THIOPHENES (ALPHA) (*Continued*)

Formula	Compound Lithiated	Conditions	Substrate	Product and Yield (%)	Refs.
C_4H_4S (*Contd.*)	thiophene		$RCH_2CO_2H_5$	(2-thienyl)$_2$C(OH)CH$_2$R	
		,,	$R = CH(CH_3)N(CH_3)CH_2C_6H_5$	$R = CH(CH_3)N(CH_3)$-$CH_2C_6H_5$ (56)	395
C_5H_3NS	3-cyanothiophene	BuLi/ether/ −70°/1 hr	CO_2	3-Cyano-2-thiophenecarboxylic acid (68)	217, 396
$C_5H_4O_2S$	3-thiophenecarboxylic acid	LDA/THF or ether, HMPA/−70°	$(CH_3)_3SiCl$	2-(Trimethylsilyl)-3-thiophenecarboxylic acid (—)	246
C_5H_5BrS	3-bromo-5-methylthiophene	LDA/ether/25°	CO_2	3-Bromo-5-methyl-2-thiophenecarboxylic acid (56)	397
C_5H_6OS	2-methoxythiophene	EtLi/ether/ reflux/0.5 hr	$FClO_3$	5-X-2-methoxythiophene, X = F (34)	385
		C_6H_5Li/ether/ 25°/1 hr	CO_2	$X = CO_2H$ (61)	231
		,,	$(CH_3O)_2SO_2$	$X = CH_3$ (40)	231
		,,	$CH_2{=}CHCH_2Cl$	$X = CH_2CH{=}CH_2$ (77)	231
		,,	DMF	$X = CHO$ (67)	231

Reactant	Conditions	Electrophile	Product (Yield, %)	Refs.
	C_6H_5Li/ether/ 25°/45 min	p-$CH_3OC_6H_4CH_2$–$COC_2H_5(H_3O^+)$	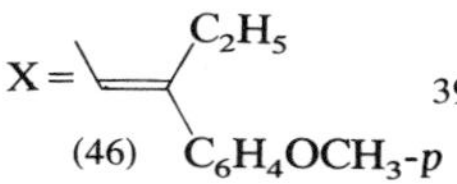 X = C_2H_5 / $C_6H_4OCH_3$-p (46)	398
	BuLi/ether/ reflux/0.5 hr	$CH_3OSO_2C_6H_4CH_3$-p	X = CH_3 (78)	399
	BuLi/reflux/ 20 min	t-$C_4H_9OCO_2C_6H_5$	X = OC_4H_9-t (72)	289
	C_6H_5Li/ether/25°/ 45 min	C_6H_5CHO (H_3O^+)	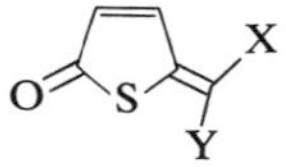 X = H, Y = C_6H_5 (17)	400
	"	$C_6H_5COCH_3$ (H_3O^+)	X = CH_3, Y = C_6H_5 (11)	400
	"		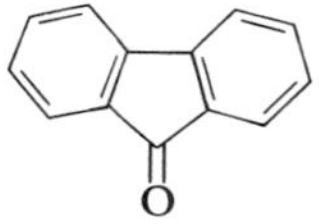 X, Y 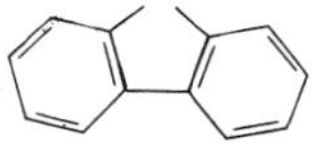 = (58)	400
	"	$(C_6H_5)_2CO$	X = Y = C_6H_5 (72)	400
	"	p-$CH_3OC_6H_4CH(C_2H_5)$–COC_2H_5 (H_3O^+)	X = $CH(C_2H_5)C_6H_4$-OCH_3-p, Y = C_2H_5 (30)	398
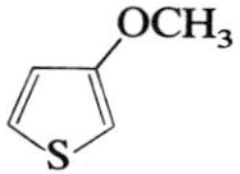 3-OCH_3 thiophene	BuLi/ether/ 25°–reflux/ 2 hr	I_2	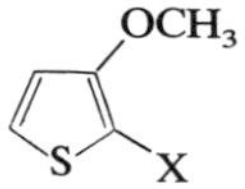 X = I (42)	567, 401
	BuLi/ether/ reflux/0.5 hr	CO_2	X = CO_2H (86)	402

TABLE XXI. THIOPHENES (ALPHA) (*Continued*)

Formula	Compound Lithiated	Conditions	Substrate	Product and Yield (%)	Refs.
C_5H_6OS (*Contd.*)	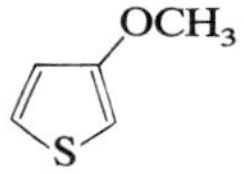	BuLi	DMF	X = CHO (83)	165
		"	DMA	X = $COCH_3$ (32)	165
		BuLi/ether/ reflux/15 min	$(CH_3O)_2SO_2$	3-Methoxy-2-methylthiophene (IV), 4-methoxy-2-methylthiophene (V) (IV + V, 75) (IV : V, 93 : 7)	145
		BuLi/reflux/ 20 min	$(n\text{-}C_4H_9O)_3B$ (H_2O_2)	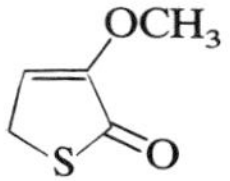 (24)	289
C_5H_6S	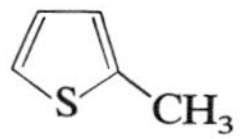	EtLi/ether/ reflux/0.5 hr	$FClO_3$	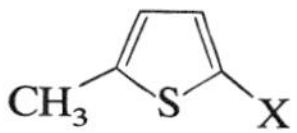 X = F (53)	385, 149
		BuLi	"S"	X = SH (57)	290, 497
		BuLi/ether/ reflux/2 hr	"S" + CH_3I	X = SCH_3 (79)	403
		BuLi/ether/ reflux/1 hr	"S" + $BrCH_2CO_2C_2H_5$ (OH^-)	X = SCH_2CO_2H (—)	404
		BuLi	CO_2	X = CO_2H (84)	405
		BuLi/hexane, TMEDA/reflux/ 0.5 hr	CO_2 (CH_2N_2)	X = CO_2CH_3 (95)	40
		BuLi/ether/ reflux/0.5 hr	$(CH_3S)_2$	X = SCH_3 (75)	399
		BuLi/ether/ reflux/4 hr	Ac_2O	X = $COCH_3$ (33)	371
		BuLi/ether/ reflux/3 hr	$ClB(C_2H_5)_2$	X = $B(C_2H_5)_2$ (28)	295

Formula	Compound	Conditions	Electrophile	Product(s) (Yield %)	Refs.
		BuLi/reflux/ 20 min	$(n\text{-}C_4H_9O)_3B$ (H_2O_2)	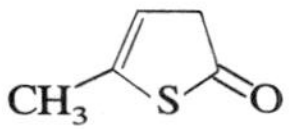 CH_3, S, O (47)	289
	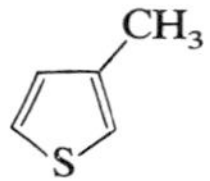 CH_3, S	C_6H_5Li/ether/ reflux/2 hr	CO_2	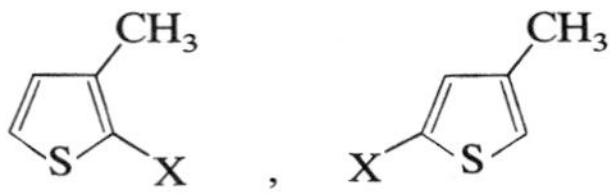 CH_3, S, X , X, S, CH_3 (VI) (19) (VII) (68) $X = CO_2H$	20, 48
		BuLi/ether, TMEDA/reflux/ 15 min	$(CH_3O)_2SO_2$	(VI + VII, 79) (VI : VII, 7 : 93) $X = CH_3$	145, 406
		BuLi/ether/ reflux/0.5 hr	DMF	(VI + VII, 59) (VI : VII, 17 : 83) $X = CHO$	399, 407
		BuLi/hexane, TMEDA/reflux/ 0.5 hr	CO_2 (CH_2N_2)	Dimethyl 3-methylthiophene-2,5-dicarboxylate (100)	40
		BuLi/reflux/ 20 min	$(n\text{-}C_4H_9O)_3B$ (H_2O_2)	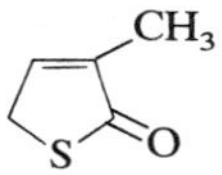 CH_3, S, O (50)	289
C_5H_6SSe	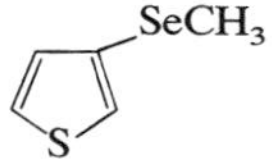 $SeCH_3$, S	BuLi/2 hr	CO_2	3-(Methylselenyl)-2-thiophenecarboxylic acid (VIII), 4-(methylselenyl)-2-thiophenecarboxylic acid (IX) (VIII + IX, 75) (VIII : IX, 56 : 44)	140

TABLE XXI. THIOPHENES (ALPHA) (*Continued*)

Formula	Compound Lithiated	Conditions	Substrate	Product and Yield (%)	Refs.
$C_5H_6S_2$	2-(methylthio)thiophene (SCH_3)	BuLi	"S" + CH_3I	2,5-Bis(methylthio)thiophene (57)	404
		BuLi/ether/0° 15 min	CO_2	5-(Methylthio)-2-thiophenecarboxylic acid (87)	404
	3-(methylthio)thiophene (SCH_3)	BuLi/ether/ reflux/0.5 hr	CO_2	3-(Methylthio)-2-thiophenecarboxylic acid (70)	248
C_6H_7NOS	3-thiophene-$CONHCH_3$	BuLi/ether/25°–35°/1 hr	D_2O	2-*d*-N-Methyl-3-thiophenecarboxamide (55)	134
		BuLi/ether/25°–35°/3 hr	$(C_6H_5)_2CO$	2-(Hydroxymethyl)-N-methyl-α,α-diphenyl-3-thiophenecarboxamide (23)	134
C_6H_8OS	3-thiophene-CH_2OCH_3	BuLi/ether/25°/ 10 hr	CO_2	3-(CH_2OCH_3)-2-X-thiophene, $X = CO_2H$ (89)	142
		BuLi/ether/25°–35°/1 – 1.5 hr	$(CH_3S)_2$	$X = SCH_3$ (61)	134, 43
		"	DMF	$X = CHO$ (72)	43, 134
		"	$(C_6H_5)_2CO$	$X = C(C_6H_5)_2OH$ (46)	134
C_6H_8S	2-thiophene-C_2H_5	EtLi/ether/ reflux/0.5 hr	$FClO_3$	C_2H_5-2-thiophene-5-X, $X = F$ (55)	385
		BuLi/ether/ reflux/1 hr	"S" + C_2H_5I	$X = SC_2H_5$ (78)	408, 403
		—	"Se" + C_2H_5I	$X = SeC_2H_5$ (65)	362
		—	"Se" + $ClCH_2CO_2CH_3$	$X = SeCH_2CO_2CH_3$ (46)	362

		BuLi/ether/−35°	$ClCH_2OCH_3$	$X = CH_2OCH_3$ (65)	390
	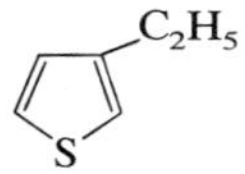	BuLi/ether, TMEDA reflux/15 min	$(CH_3O)_2SO_2$	3-Ethyl-2-methylthiophene (X), 4-ethyl-2-methylthiophene (XI) (X+XI, 79) (X:XI, 3:97)	145
	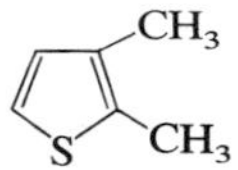	BuLi/ether/ reflux/ 2 hr	$CuCl_2$	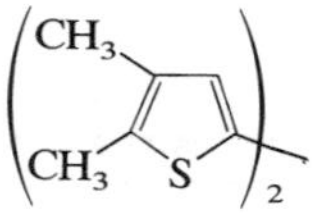 (68)	409
$C_6H_8S_2$	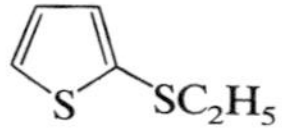	—	CO_2	5-(Ethylthio)-2-thiophenecarboxylic acid (77)	404
$C_6H_9NO_2S_2$	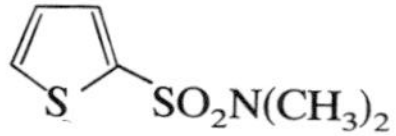	BuLi/THF/25°/ 20 min	$(CH_3)_3SiCl$	N,N-Dimethyl-5-(trimethylsilyl)-2-thiophenesulfonamide (41)	135
C_7H_6S	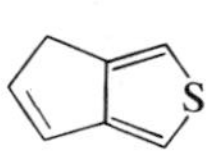	BuLi/ether/ −10°–0°/ 0.5 hr	CO_2 (CH_2N_2)	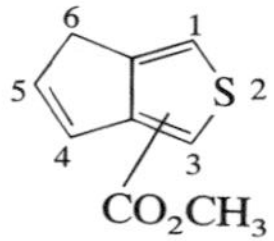 1 (10), 3 (21), 4 (39), 6 (29)	410
$C_7H_8O_2S$	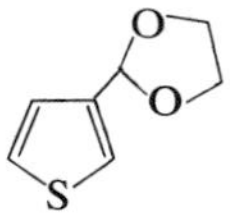	BuLi/ether/ reflux/20 min	"S" + $ClCH_2CO_2CH_3$ (NaOEt)	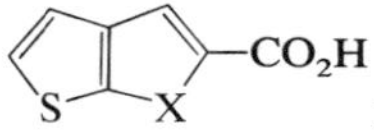 X = S (94)	143
		BuLi/ether/ 25°–reflux/ 15 min	"Se" + $ClCH_2CO_2CH_3$ (NaOEt)	X = Se (48)	411
		BuLi/ether/ reflux/15 min	CO_2 (H_3O^+)	3-Formyl-2-thiophenecarboxylic acid (78)	412
		BuLi/ether/−70°	C_6H_5CN (H_3O^+)	2-Benzoyl-3-thiophenecarboxaldehyde (16)	413

TABLE XXI. THIOPHENES (ALPHA) (*Continued*)

Formula	Compound Lithiated	Conditions	Substrate	Product and Yield (%)	Refs.
C_7H_9NOS	3-$CON(CH_3)_2$-thiophene (S)	BuLi/ether/25°/ 20 min	—	(thiophene-fused quinone: O, S, S, O) (41)	134, 43
$C_7H_{10}LiNOS$	3-$CH(OLi)N(CH_3)_2$-thiophene (S)	BuLi/reflux/ 135 min	CO_2	3-Formyl-2-thiophenecarboxylic acid (44)	414
$C_7H_{10}S$	2-C_3H_7-*n*-thiophene (S)	EtLi/ether/ reflux/0.5 hr	$FClO_3$	2-Fluoro-5-propylthiophene (55)	385
	3-C_3H_7-*i*-thiophene (S)	BuLi/ether, TMEDA/ reflux/15 min	$(CH_3O)_2SO_2$	3-Isopropyl-2-methylthiophene (XII), 4-isopropyl-2-methylthiophene (XIII) (XII + XIII, 66) (XII : XIII, 1 : 99)	145
$C_7H_{10}S_2$	2-$CH_2SC_2H_5$-thiophene (S)	Buli/ether	CO_2	5-[(Ethylthio)methyl]-2-thiophene-carboxylic acid (53)	415
$C_7H_{11}NS$	2-$CH_2N(CH_3)_2$-thiophene (S)	BuLi/THF	D_2O	2-$(CH_3)_2NCH_2$-5-X-thiophene (S), X = D (88)	134
		"	DMF	X = CHO (40)	134
		"	$(CH_3)_3SiCl$	X = $Si(CH_3)_3$ (71)	134
		"	C_6H_5CHO	X = $CH(OH)C_6H_5$ (50)	416
		"	$(C_6H_5)_2CO$	X = $C(C_6H_5)_2OH$ (47)	416

Formula	Substrate	Conditions	Reagent	Product (Yield %)	Refs.
	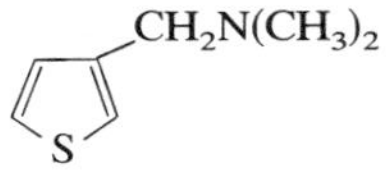 3-thienyl–$CH_2N(CH_3)_2$	BuLi/ether/25°–35°/4 hr	D_2O	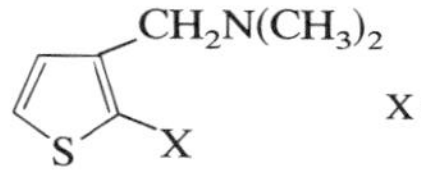 3-$CH_2N(CH_3)_2$, 2-X thiophene: X = D (80)	43, 134
		BuLi/ether/25°–35°/1 hr	DMF	X = CHO (75)	134, 43
		BuLi/ether/25°–35°/1 hr	C_6H_5CN (H_3O^+)	X = COC_6H_5 (54)	134, 43
		,,	$(C_6H_5)_2CO$	X = $C(C_6H_5)_2OH$ (67)	134, 43
		,,	p-$CH_3C_6H_4SO_2Br$	X = Br (32)	134
		,,	p-$CH_3C_6H_4SO_2Cl$	X = Cl (25)	134
$C_7H_{12}SSi$	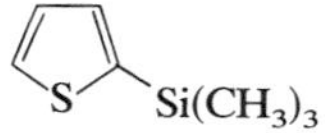 2-thienyl–$Si(CH_3)_3$	BuLi/ether/reflux/4 hr	CO_2	5-(Trimethylsilyl)-2-thiophenecarboxylic acid (62)	369
$C_8H_4Br_2S_2$	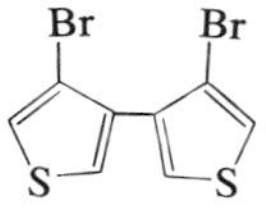 4,4′-Br, Br-3,3′-bithiophene	EtLi/THF/−70°/2 min	CO_2	4,4′-Dibromo-[3,3′-bithiophene]-5-carboxylic acid (51)	417
$C_8H_6S_2$	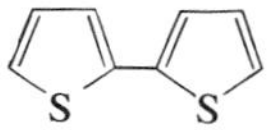 2,2′-bithiophene	C_6H_5Li/ether/reflux/15 min	CO_2	[2,2′-Bithiophene]-5-carboxylic acid (74)	418
		BuLi/reflux/45 min	,,	[2,2′-Bithiophene]-5,5′-dicarboxylic acid (72)	419
$C_8H_6S_2$	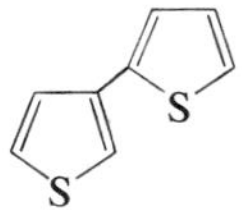 2,3′-bithiophene	C_6H_5Li/reflux/15 min	CO_2	[2,3′-Bithiophene]-2′-carboxylic acid (XIV), [2,3′-bithiophene]-5′-carboxylic acid (XV) (XIV + XV, 75) (XIV:XV, 38:52)	418

TABLE XXI. THIOPHENES (ALPHA) (*Continued*)

Formula	Compound Lithiated	Conditions	Substrate	Product and Yield (%)	Refs.
$C_8H_6S_3$	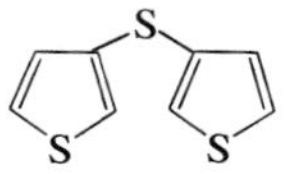	BuLi/ether/ reflux/1 hr	$CuCl_2$	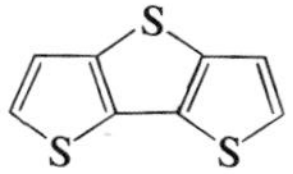 (52)	144
$C_8H_8S_2$	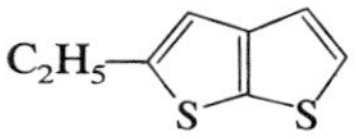 C_2H_5	—	"S" + C_2H_5I	5-Ethyl-2-(ethylthio)thieno[2,3-*b*]-thiophene (79)	420
	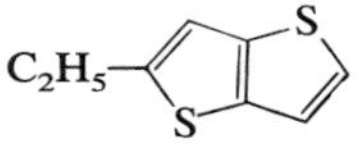 C_2H_5	BuLi/ether	"S" + C_2H_5I	5-Ethyl-2-(ethylthio)thieno[3,2-*b*]-thiophene (69)	420
$C_8H_{10}O_2S$	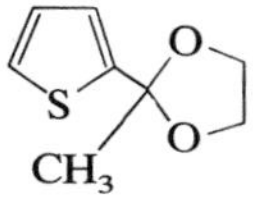 CH_3	BuLi/ether/25°/ 1 hr	$(CH_3)_3SiCl$	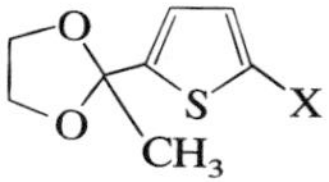 CH_3; X = $Si(CH_3)_3$ (96)	421, 298
		BuLi/ether/25°/ 1.5 hr	$(C_2H_5)_3SiCl$	X = $Si(C_2H_5)_3$ (74)	298
		BuLi/ether/25°/ 1 hr	CO_2 (H_3O^+)	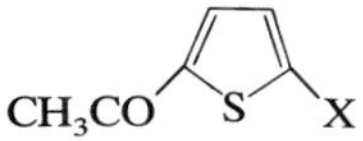 CH_3CO; X = CO_2H (70)	421
		BuLi/ether/25°/ 1.5 hr	$(CH_3)_3SiCl$ (H_3O^+)	X = $Si(CH_3)_3$ (93)	298
		"	$C_2H_5(CH_3)_2SiCl$ (H_3O^+)	X = $Si(CH_3)_2C_2H_5$ (85)	298
		BuLi/ether/25°/ 1 hr	C_6H_5CN (H_3O^+)	X = COC_6H_5 (66)	421
		BuLi/ether/25°/ 1.5 hr	$CH_3(C_6H_5)_2SiCl$ (H_3O^+)	X = $Si(C_6H_5)_2CH_3$ (67)	298
		"	$(C_6H_5)_3SiCl$ (H_3O^+)	X = $Si(C_6H_5)_3$ (73)	298

Formula	Reactant	Conditions	Electrophile	Product (Yield, %)	Refs.
		BuLi/ether/25°/ 1 hr	C_6H_5CHO	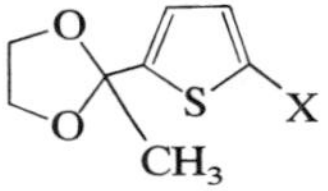 $X = CH(OH)C_6H_5$ (60)	421
		”	C_6H_5NCO	$X = CONHC_6H_5$ (65)	421
	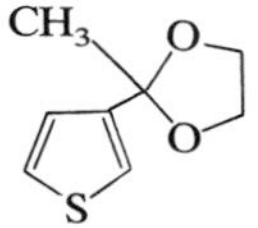 CH_3	BuLi/ether/25°/ 45 min, reflux/ 15 min	C_6H_5CN (H_3O^+)	3-Acetyl-2-benzoylthiophene (27)	413
$C_8H_{12}LiNOS$	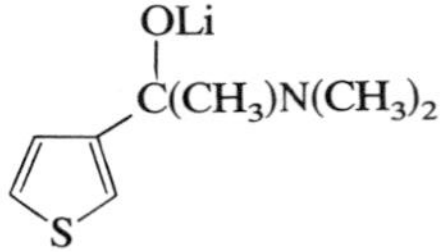 $C(CH_3)N(CH_3)_2$ (OLi)	BuLi	"S" + $BrCH_2CO_2CH_3$	Methyl [(3-acetyl-2-thienyl)thio]acetate (—)	422
$C_8H_{12}OS$	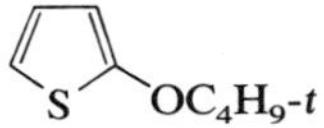 OC_4H_9-t	BuLi/ether/ reflux/2 hr	"S"	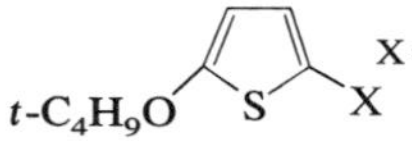 t-C_4H_9O … $X = SH$ (73)	275
		”	CO_2	$X = CO_2H$ (75)	275
		”	$(CH_3S)_2$	$X = SCH_3$ (87)	275
		”	$(CH_3O)_2SO_2$	$X = CH_3$ (87)	275
		”	$ClCO_2C_2H_5$	$X = CO_2C_2H_5$ (46)	275
		”	$CH_2{=}CHCH_2Br$	$X = CH_2CH{=}CH_2$ (86)	275
		”	$C_6H_5CH_2Cl$	$X = CH_2C_6H_5$ (73)	275
		”	t-$C_4H_9OCO_2C_6H_5$ ($MgBr_2$)	$X = OC_4H_9$-t (78)	275
		”	Cyclopentanone (H_3O^+)	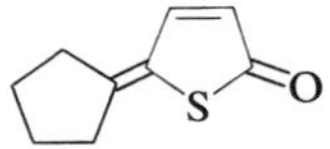 (80)	275

TABLE XXI. THIOPHENES (ALPHA) (*Continued*)

Formula	Compound Lithiated	Conditions	Substrate	Product and Yield (%)	Refs.
$C_8H_{12}OS$	3-(OC_4H_9-*t*)thiophene	BuLi/ether/ reflux/0.5 hr	CO_2	3-(OC_4H_9-*t*)-2-X-thiophene: $X = CO_2H$ (62)	423
		BuLi/ether/ reflux/2 hr	$(CH_3O)_2SO_2$	$X = CH_3$ (87)	141
		"	$ClCO_2C_2H_5$	$X = CO_2C_2H_5$ (75)	141
		BuLi/ether/−30°– reflux/1 hr	DMF	$X = CH(OLi)N(CH_3)_2$ (—)	148
		BuLi/ether/ reflux/2 hr	Ac_2O ($MgBr_2$)	$X = COCH_3$ (75)	141
		"	*t*-$C_4H_9OCO_2C_6H_5$ ($MgBr_2$)	$X = OC_4H_9$-*t* (70)	141
$C_8H_{12}O_2S_2$	2-($SO_2C_4H_9$-*t*)thiophene	BuLi/THF/−20°/ 9.5 hr	CO_2	5-(*t*-Butylsulfonyl)-2,4-thiophene-dicarboxylic acid (68)	136
		"	DMF	5-(*t*-Butylsulfonyl)-2,4-thiophene-dicarboxaldehyde (59)	136
	3-($SO_2C_4H_9$-*t*)thiophene	BuLi	CO_2	3-(*t*-Butylsulfonyl)-2,4-thiophenedicarboxylic acid (42), 3-(*t*-butylsulfonyl)-2-thiophenecarboxylic acid (4), 3-(*t*-butylsulfonyl)-2,5-thiophenedicarboxylic acid (8)	565
$C_8H_{12}S$	2-(C_4H_9-*n*)thiophene	EtLi/ether/ reflux/0.5 hr	$FClO_3$	2-Butyl-5-fluorothiophene (55)	385

	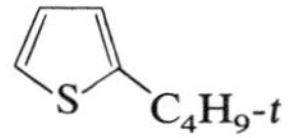 2-t-butylthiophene (structure: C_4H_9-t at 2-position)	EtLi/ether/ reflux/0.5 hr	$FClO_3$	2-t-Butyl-5-fluorothiophene (49)	385
		BuLi/ether/ reflux/45 min	CO_2	5-t-Butyl-2-thiophenecarboxylic acid (55)	399
	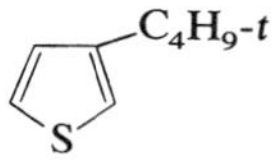 3-t-butylthiophene (structure: C_4H_9-t at 3-position)	BuLi/ether, TMEDA/ reflux/15 min	$(CH_3O)_2SO_2$	3-t-Butyl-5-methylthiophene (72)	145
$C_8H_{12}SSe$	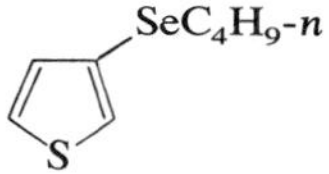 3-(n-butylseleno)thiophene (structure: SeC_4H_9-n at 3-position)	BuLi/ether/−30°	"S" + CH_3I	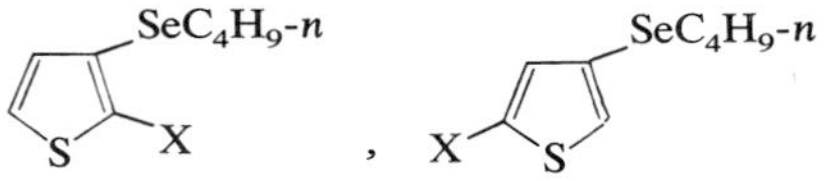 (XVI): 3-SeC_4H_9-n, 2-X thiophene , (XVII): 4-SeC_4H_9-n, 2-X thiophene	
				(XVI:XVII, 70:30) X = SCH_3	140
		"	"Se" + CH_3I	(XVI + XVII, 67) X = $SeCH_3$	140
		BuLi/2 hr	CO_2	(XVI + XVII, 79) (XVI:XVII, 60:40) X = CO_2H	140
$C_8H_{12}S_2$	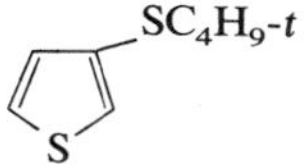 3-(t-butylthio)thiophene (structure: SC_4H_9-t at 3-position)	BuLi	CO_2	3-(t-Butylthio)-2-thiophenecarboxylic acid (—), 4-(t-butylthio)-2-thiophenecarboxylic acid (—), 3-(t-butylthio)-2,5-thiophenecarboxylic acid (—)	424
		"	—	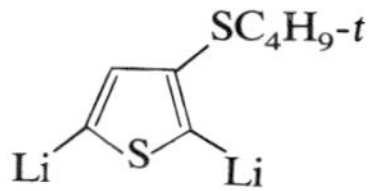 2,5-dilithio-3-(SC_4H_9-t)thiophene (structure) (—)	565
$C_8H_{13}NO_2S_2$	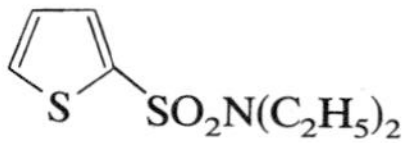 N,N-diethyl-2-thiophenesulfonamide (structure: $SO_2N(C_2H_5)_2$ at 2-position)	BuLi/ether/25°/ 1 hr	CO_2	5-(Diethylsulfamoyl)-2-thiophenecarboxylic acid (82)	135
		BuLi/ether/25°/ 0.5 hr	$(CH_3)_3SiCl$	N,N-Diethyl-5-(trimethylsilyl)-2-thiophenesulfonamide (79)	135

TABLE XXI. THIOPHENES (ALPHA) (*Continued*)

Formula	Compound Lithiated	Conditions	Substrate	Product and Yield (%)			Refs.
C_9H_6ClNS	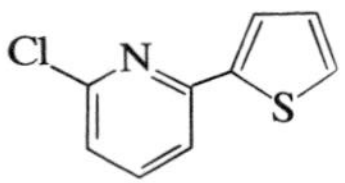			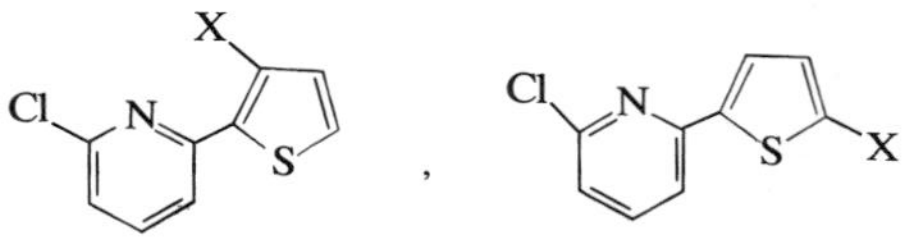			
		BuLi/THF/0°/ 0.5 hr	$(CH_3)_3SiCl$	(0.4)	(91)	$X = Si(CH_3)_3$	138
		BuLi/ether/0°/ 0.5 hr	$(CH_3)_3SiCl$	(36)	(43)	$X = Si(CH_3)_3$	138
		t-BuLi/THF/−60°	$(C_6H_5)_2CO$	(—)	(83)	$X = C(C_6H_5)_2OH$	137
C_9H_7NS	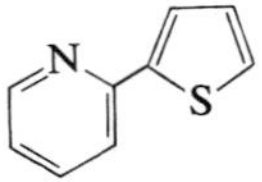	*t*-BuLi/THF/−60°	$CuCl_2(O_2)$	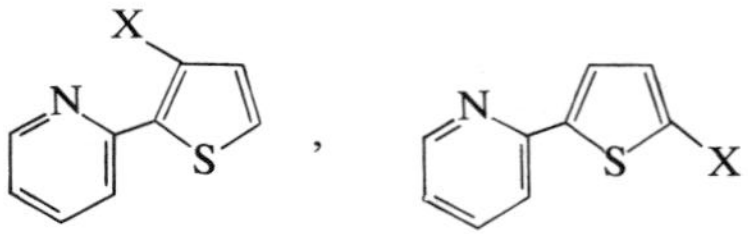			
				(—)	(63)	X = 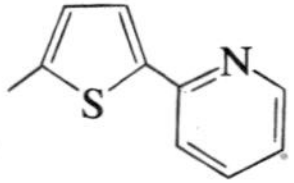	137
		BuLi/ether/0°/ 0.5 hr	$(CH_3)_3SiCl$	(62)	(13)	$X = Si(CH_3)_3$	138
		BuLi/THF/0°/ 0.5 hr	"	(4)	(93)	$X = Si(CH_3)_3$	138
		t-BuLi/THF/−60°	$(C_6H_5)_2CO$	(—)	(87)	$X = C(C_6H_5)_2OH$	137

Formula	Compound	Metalation conditions	Electrophile	Product(s) and Yield(s) (%)	Refs.
	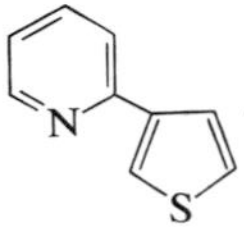 (2-(3-thienyl)pyridine; N, S)	BuLi/ether/0°/ 15 min	$CuCl_2$	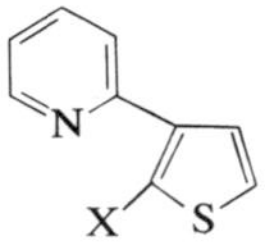 (X 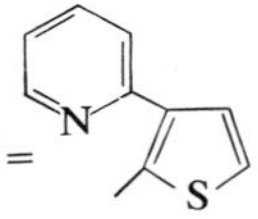 = 2-(3-pyridyl... structure; N, S) (45)	280
		,,	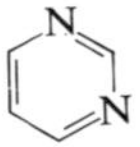 pyrimidine (N, N) (ox.)	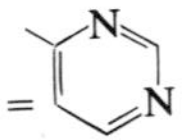 X = pyrimidinyl (N, N) (55)	280
C_9H_8OS	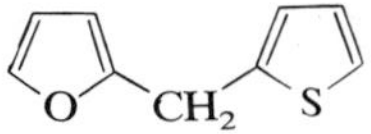 (O) CH_2 (S)	BuLi/ether/0°/ 2 hr	Ac_2O	5-Furfuryl-2-thienyl methyl ketone (16)	371
$C_9H_8S_2$	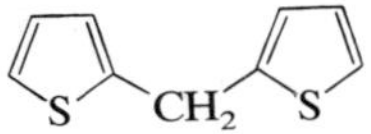 (S) CH_2 (S)	—	CO_2	5-(2-Thienyl)-2-thiophenecarboxylic acid (—)	389
		BuLi/ether/ 10 min	Ethylene oxide	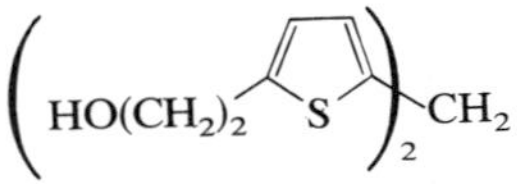 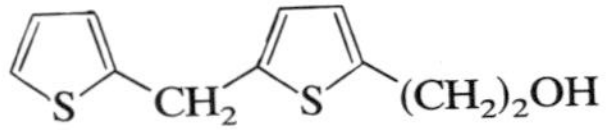 $(HO(CH_2)_2$-thienyl$)_2CH_2$ (13); thienyl-CH_2-thienyl-$(CH_2)_2OH$ (74)	389
		2 BuLi	,,	(77); (13)	389
$C_9H_9OPS_2$	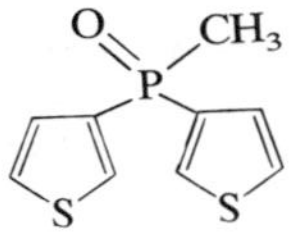 O=P(CH_3)(3-thienyl)$_2$	BuLi	CO_2	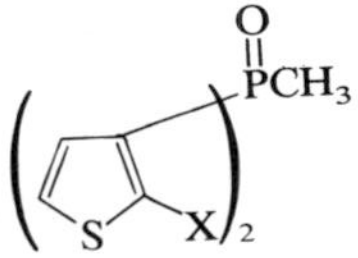 $(\text{S}, \text{X})_2P(O)CH_3$, X = CO_2H (33)	270

TABLE XXI. THIOPHENES (ALPHA) (*Continued*)

Formula	Compound Lithiated	Conditions	Substrate	Product and Yield (%)	Refs.
$C_9H_9OPS_2$ (*Contd.*)	O, CH_3, P, S, S	"	CH_3COCH_3	O, PCH_3, S, X, 2; X = $C(CH_3)_2OH$ (40)	270
$C_9H_{11}NOS$	S, N, O, $(CH_3)_2$	BuLi/THF/−70°/ 1 hr	C_6H_5CHO	X, S, N, O, $(CH_3)_2$, X, S, N, O, $(CH_3)_2$; (36) (55) X = $CH(OH)C_6H_5$	139
		BuLi/ether/−70°–25°	"	(91) (4) X = $CH(OH)C_6H_5$	139
$C_9H_{14}OS$	OC_4H_9-*t*, S, CH_3	BuLi/ether/ reflux/2 hr	$ClCO_2C_2H_5$	Ethyl 4-*t*-butoxy-5-methyl-2-thiophenecarboxylate (66)	141
	CH_3, OC_4H_9-*t*, S	BuLi/ether/ reflux/0.5 hr	*t*-$C_4H_9OCO_2C_6H_5$	2,3-Di-*t*-butoxy-4-methylthiophene (79)	147
$C_9H_{14}O_2S$	S, $CH(OC_2H_5)_2$	BuLi/2.5 hr	$ClCH_2OCH_3$	$(C_2H_5O)_2HC$, S, X; X = CH_2OCH_3 (56)	390
		BuLi/ether/ −30°–25°/ 2.5 hr	DMF	X = CHO (—)	425

		BuLi/ether/25°/ 1 hr	$(CH_3)_3SiCl$ (H_3O^+)	$X = Si(CH_3)_3$ (94)	298
		"	$(C_2H_5)_3SiCl$ (H_3O^+)	$X = Si(C_2H_5)_3$ (87)	298
	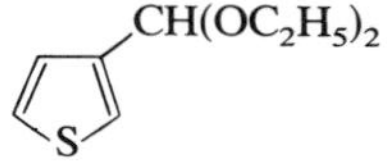	BuLi/ether/ −10°–25°/ 2.5 hr	CO_2 (H_3O^+)	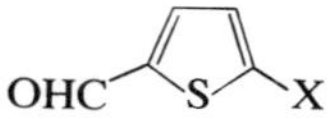 OHC, S, X; $X = CO_2H$ (60)	426
	$CH(OC_2H_5)_2$, S	BuLi/ether/25°/ 1 hr	$C_2H_5(CH_3)_2SiCl$ (H_3O^+)	$X = Si(CH_3)_2C_2H_5$ (83)	298
		"	$CH_3(C_6H_5)_2SiCl$ (H_3O^+)	$X = Si(C_6H_5)_2CH_3$ (49)	298
		"	$(C_6H_5)_3SiCl$ (H_3O^+)	$X = Si(C_6H_5)_3$ (72)	298
		BuLi/ether/−30°– reflux/20 min	"S" + $ClCH_2CO_2CH_3$ (H_3O^+)	Methyl [3-formyl-2-thienyl)thio]acetate (77)	327
		BuLi	DMF (H_3O^+)	2,3-Thiophenedicarboxaldehyde (85)	428
$C_{10}H_7BrS$	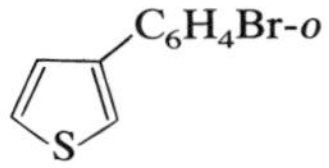 C_6H_4Br-*o*, S	BuLi/−35°– 25°	CO_2 (CH_2N_2)	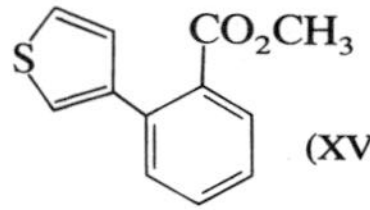 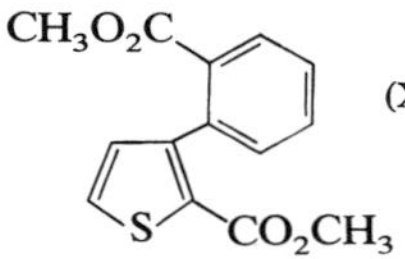 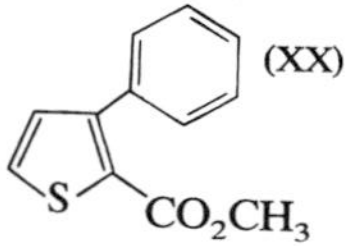 S, CO_2CH_3 (XVIII), CH_3O_2C, S, CO_2CH_3 (XIX), S, CO_2CH_3 (XX) (XVIII + XIX + XX, 44) (XVIII : XIX : XX, 1 : 2 : 3)	429

TABLE XXI. THIOPHENES (ALPHA) (*Continued*)

Formula	Compound Lithiated	Conditions	Substrate	Product and Yield (%)	Refs.
$C_{10}H_7BrS$	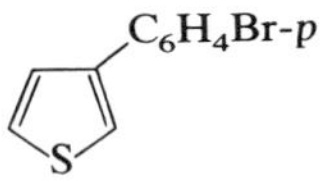 3-(C_6H_4Br-*p*)thiophene	BuLi/ether/ reflux	$CO_2(CH_2N_2)$	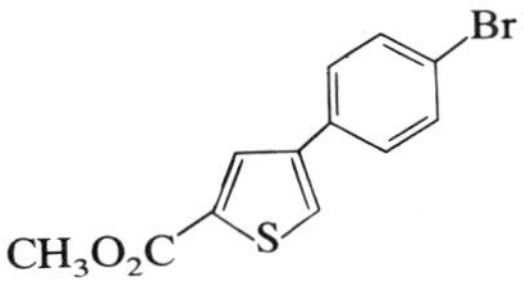 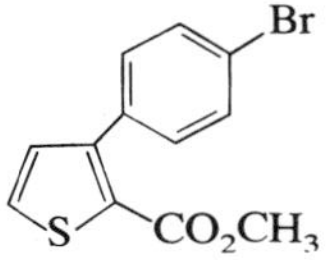 4-(p-bromophenyl)-2-(CH_3O_2C)thiophene (XXI), 3-(p-bromophenyl)-2-(CO_2CH_3)thiophene (XXII) (XXI + XXII, 49) (XXI : XXII, 2 : 3)	429
$C_{10}H_8Br_2S_2$	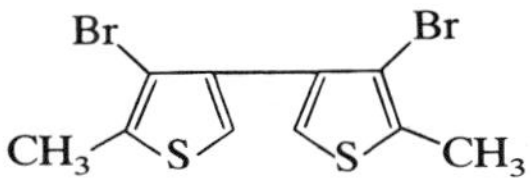 4,4'-Br, 5,5'-CH_3 bithiophene	BuLi/ether/−70°	CO_2	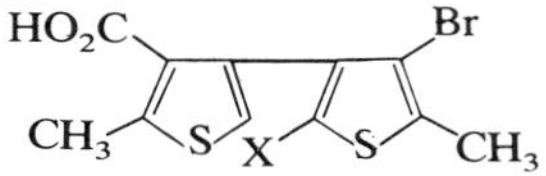 HO_2C, Br, CH_3, X bithiophene: X = H (XXIII); X = CO_2H (XXIV) (XXIII + XXIV, 56) (XXIII : XXIV, 1 : 1)	417
$C_{10}H_8O_2S_2$	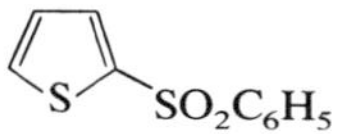 2-($SO_2C_6H_5$)thiophene	BuLi/ether/0°	CO_2	5-(Phenylsulfonyl)-2-thiophenecarboxylic acid (42)	69
$C_{10}H_8S$	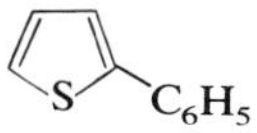 2-(C_6H_5)thiophene	EtLi/ether/ reflux/0.5 hr	$FClO_3$	2-Fluoro-5-phenylthiophene (55)	385

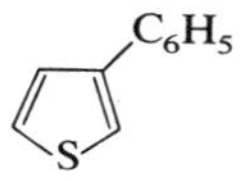

Formula	Substrate	Conditions	Reagent	Product(s) (yield %)	Ref.
	C_6H_5 S	BuLi/ether/25°/ 0.5 hr	CO_2	C_6H_5 S X (XXV) , C_6H_5 X S (XXVI) $X = CO_2H$ (XXV + XXVI, 72) (XXV : XXVI, 45 : 55)	146
		BuLi/ether/0°	CO_2 (CH_2N_2)	(XXV + XXVI, 59) (XXV : XXVI, 55 : 45) $X = CO_2CH_3$	429
$C_{10}H_{10}S_2$	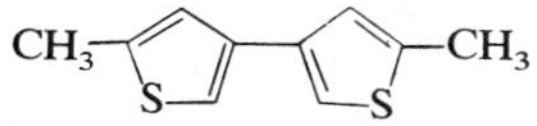 CH_3 S S CH_3	BuLi/ether/ reflux/2 hr	CO_2	5,5′-Dimethyl-[3,3′-bithiophene]-2,2′-dicarboxylic acid (45)	430, 431
	CH_3 CH_3 S S	BuLi/ether/ reflux/1.5 hr	CO_2 ($SOCl_2$, MeOH)	Methyl 4,4′-dimethyl-[3,3′-bithiophene]-2,2′-dicarboxylate XXVIII, methyl 4,4′-dimethyl-[3,3′-bithiophene]-5,5′-dicarboxylate XXIX (XXVIII + XXIX, 82) (XXVIII : XXIX, 2 : 3)	432
	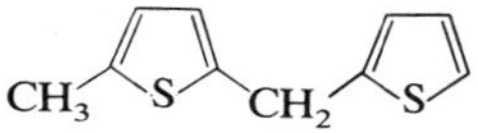 CH_3 S CH_2 S	BuLi/ether/−5°	CO_2	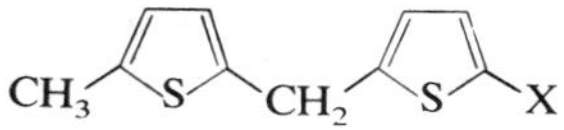 CH_3 S CH_2 S X $X = CO_2H$ (31)	433
		”	Ethylene oxide	$X = (CH_2)_2OH$ (59)	433
		”	DMF	$X = CHO$ (37)	433

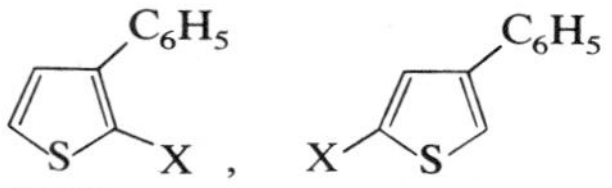

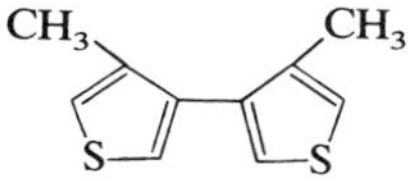

TABLE XXI. THIOPHENES (ALPHA) (*Continued*)

Formula	Compound Lithiated	Conditions	Substrate	Product and Yield (%)	Refs.
$C_{10}H_{14}S$	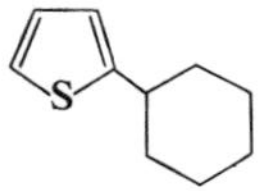 S	BuLi	$ClCH_2OCH_3$	2-Cyclohexyl-5-(methoxymethyl)thiophene (78)	390
$C_{10}H_{15}NS$	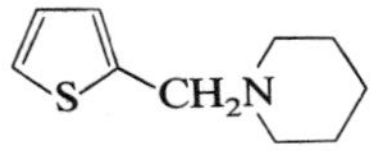 S, CH_2N	BuLi	$ClCH_2OCH_3$	1-[5-(Methoxymethyl)-2-thienyl]piperidine (50)	390
$C_{10}H_{16}O_2S$	 S, $CH_2CH(OC_2H_5)_2$	BuLi/ether/25°/ 5 min	$(CH_3)_3SiCl$	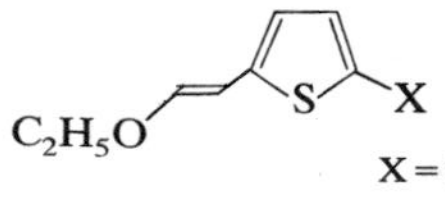 C_2H_5O, S, X; $X = Si(CH_3)_3$ (87)	392
		BuLi/ether/25°/ 10 min	C_6H_5CHO	$X = CH(OH)C_6H_5$ (66)	392
	 S, $C(OC_2H_5)_2CH_3$	BuLi/2.5 hr	$ClCH_2OCH_3$	5-(Methoxymethyl)-2-thienyl methyl ketone, diethyl acetal (50)	390
		BuLi	DMF	5-(1,1-Diethoxyethyl)thiophene-2-carboxaldehyde (48)	434
		BuLi/−10°/ 0.5 hr	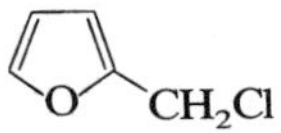 O, CH_2Cl	5-Furfuryl-2-thienyl methyl ketone, diethyl acetal (22)	371
$C_{11}H_8LiNS$	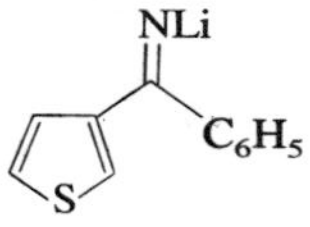 NLi, S, C_6H_5	BuLi/ether/ reflux/2 hr	CO_2 (H_3O^+)	3-Benzoylthiophene-2-carboxylic acid (70)	570
		,,	C_6H_5CN (H_3O^+)	2,3-Dibenzoylthiophene (47)	570

Formula	Conditions	Electrophile	Product(s) (yield %)	Ref.
$C_{11}H_8S$	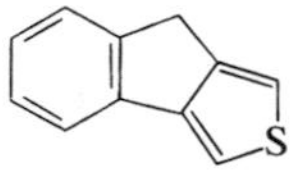 BuLi/ether/ reflux	CO_2	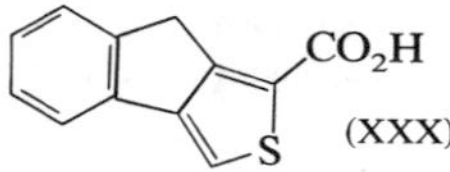 (XXX), 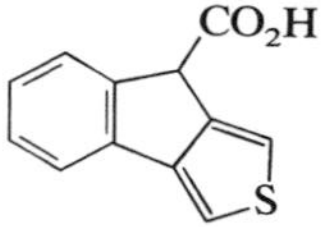 (XXXI), 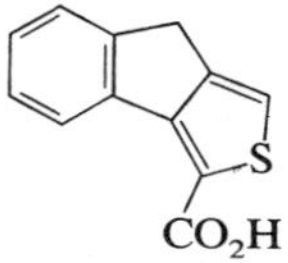 (XXXII) (XXX + XXXI + XXXXII, 45) (XXX : XXXI : XXXII, 7 : 19 : 24)	435
$C_{11}H_{10}O_2S_2$	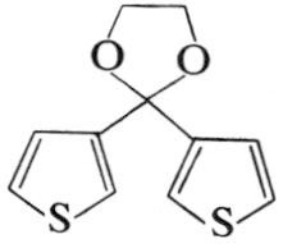 BuLi/ether/−35° to −20°/20 min	"S" (H_3O^+)	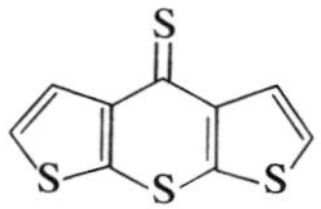 (35), 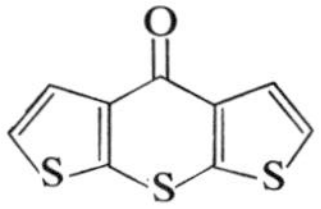 (9), 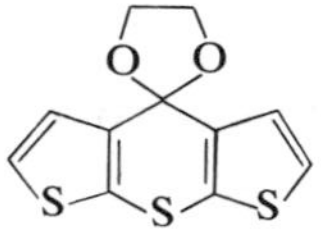 (6)	436
	BuLi/ether/−35°	I_2	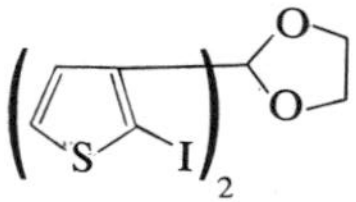 (56)	294
	BuLi/ether/ reflux/15 min	CO_2 (H_3O^+)	3,3′-Carbonyldi-2-thiophenecarboxylic acid (93)	437

TABLE XXI. THIOPHENES (ALPHA) (*Continued*)

Formula	Compound Lithiated	Conditions	Substrate	Product and Yield (%)	Refs.
$C_{11}H_{10}S$	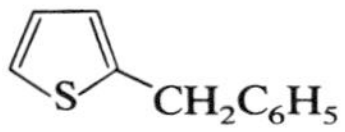 $CH_2C_6H_5$	BuLi/ether/ reflux/0.5 hr	CO_2	5-Benzyl-2-thiophenecarboxylic acid (77)	438
	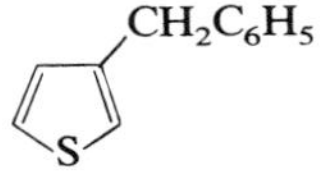 $CH_2C_6H_5$	BuLi/ether/ reflux	CO_2	3-Benzyl-2-thiophenecarboxylic acid (XXXII), 4-benzyl-2-thiophenecarboxylic acid (XXXIV) (XXXIII + XXXIV, 80) (XXXIII : XXXIV, 3 : 22)	439
$C_{11}H_{18}LiNO_2S$	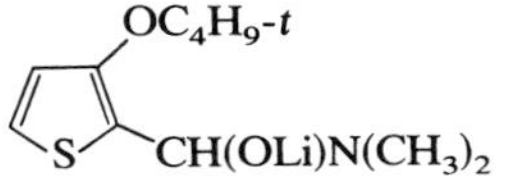 OC_4H_9-*t*, $CH(OLi)N(CH_3)_2$	BuLi/ether/−30°– reflux	DMF	3-*t*-Butoxy-2,5-thiophenedicarboxaldehyde (60)	148
	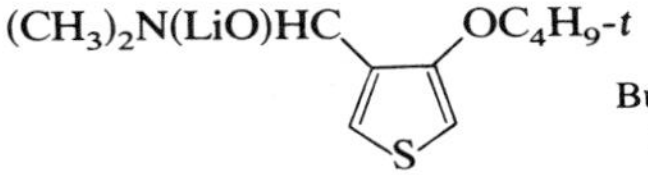 $(CH_3)_2N(LiO)HC$, OC_4H_9-*t*	BuLi/ether/25°/ 3 hr	DMF	4-*t*-Butoxy-2,3-thiophenedicarboxaldehyde (60)	148
$C_{12}H_8N_2S$	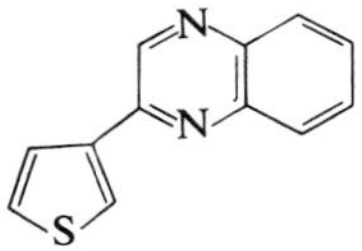	LDA/ether	CuI_2	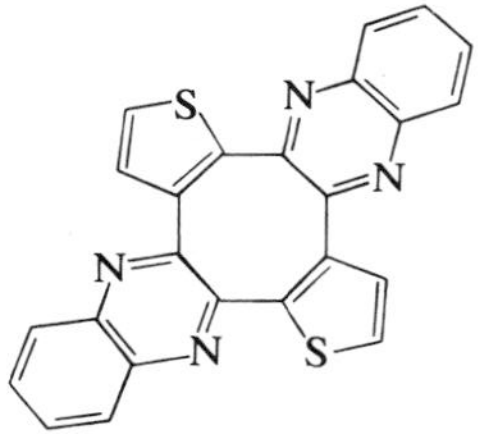 (9)	440
$C_{12}H_{14}S_2$	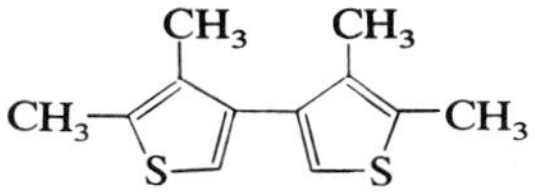 CH_3, CH_3, CH_3, CH_3	BuLi/ether/ reflux/2 hr	CO_2	4,4′,5,5′-Tetramethyl-[3,3′-bithiophene]-2,2′-dicarboxylic acid (84)	409

Formula	Structure	Conditions	Electrophile	Product(s) (Yield %)	Refs.
	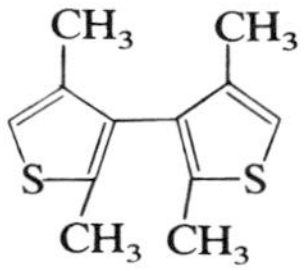 CH$_3$ CH$_3$ S S CH$_3$ CH$_3$	EtLi/ether/ reflux/3 hr	CO_2	2,2′,4,4′-Tetramethyl-[3,3′-bithiophene]-5,5′-dicarboxylic acid (77)	406
		EtLi/ether/ reflux/2 hr	$(CH_3O)_2SO_2$	2,2′,4,4′,5,5′-Hexamethyl-3,3′-bithiophene (84)	406
$C_{12}H_{20}O_2S$	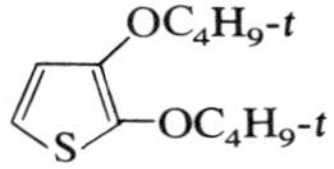 OC$_4$H$_9$-*t* OC$_4$H$_9$-*t* S	BuLi/ether/20°/ 0.5 hr	$(CH_3O)_2SO_2$	2,3-Di-*t*-butoxy-5-methylthiophene (69)	147
		BuLi/ether/ reflux/1.5 hr	$ClCO_2C_2H_5$	Ethyl 4,5-di-*t*-butoxy-2-thiophenecarboxylate (51)	141
	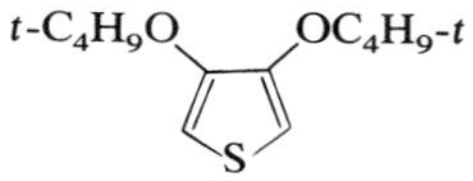 *t*-C$_4$H$_9$O OC$_4$H$_9$-*t* S	BuLi/ether/20°/ 0.5 hr	$(CH_3O)_2SO_2$	3,4-Di-*t*-butoxy-2-methylthiophene (91)	147
		BuLi/ether/ −20°– reflux/1 hr	DMF	3,4-Di-*t*-butoxy-2-thiophenecarboxaldehyde (50)	148
$C_{13}H_9NS$	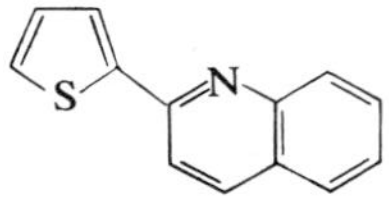 S N	BuLi/ether/0°/ 0.5 hr	$(CH_3)_3SiCl$	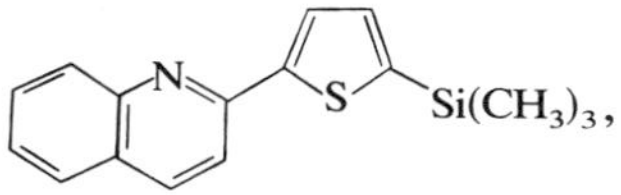 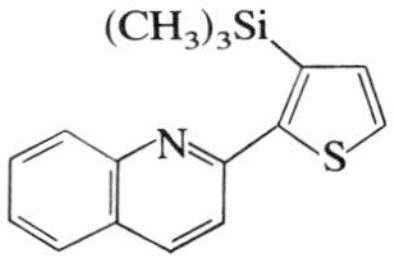 N S Si(CH$_3$)$_3$, (CH$_3$)$_3$Si N S (8) (45)	138
		BuLi/THF/0°/ 0.5 hr	”	(85) (0)	138

TABLE XXI. Thiophenes (Alpha) (*Continued*)

Formula	Compound Lithiated	Conditions	Substrate	Product and Yield (%)	Refs.
$C_{13}H_{11}FO_2S$	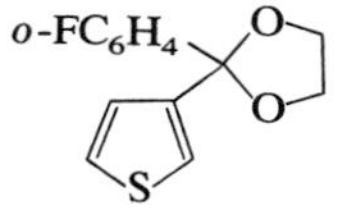 3-[2-(o-FC$_6$H$_4$)-1,3-dioxolan-2-yl]thiophene	BuLi/ether/0°/ 2 hr	CH_3CHO	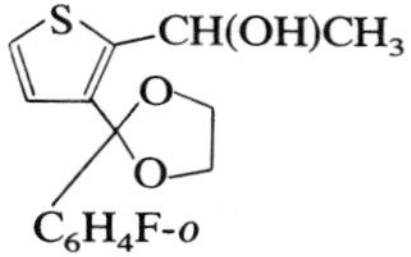 CH(OH)CH$_3$; C$_6$H$_4$F-o (60)	23
$C_{13}H_{12}O_2S$	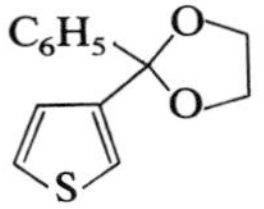 3-[2-(C$_6$H$_5$)-1,3-dioxolan-2-yl]thiophene	BuLi/ether/ reflux/2 hr	CO_2 (H_3O^+)	3-Benzoyl-2-thiophenecarboxylic acid (43)	413
$C_{13}H_{22}O_2S$	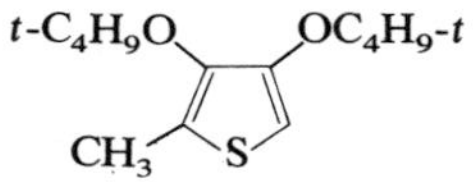 t-C$_4$H$_9$O, OC$_4$H$_9$-t, CH$_3$	BuLi/ether/20°/ 0.5 hr	$(CH_3O)_2SO_2$	3,4-Di-*t*-butoxy-2,5-dimethylthiophene (81)	147
$C_{14}H_{11}OPS_2$	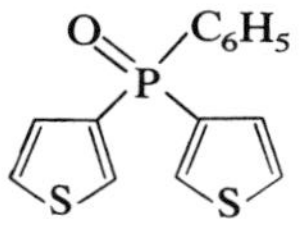 O=P(C$_6$H$_5$)(3-thienyl)$_2$	BuLi/THF/25°/ 1 hr	Br_2	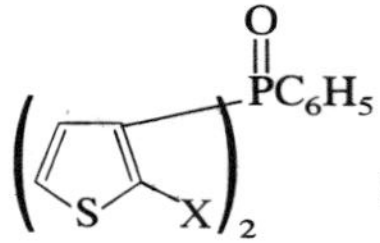 O=PC$_6$H$_5$ (2-X-3-thienyl)$_2$, X = Br (65)	270
		"	CO_2	X = CO$_2$H (75)	270
		BuLi/THF/25°/ 1 hr	CH_3COCH_3	X = C(CH$_3$)$_2$OH (45)	270

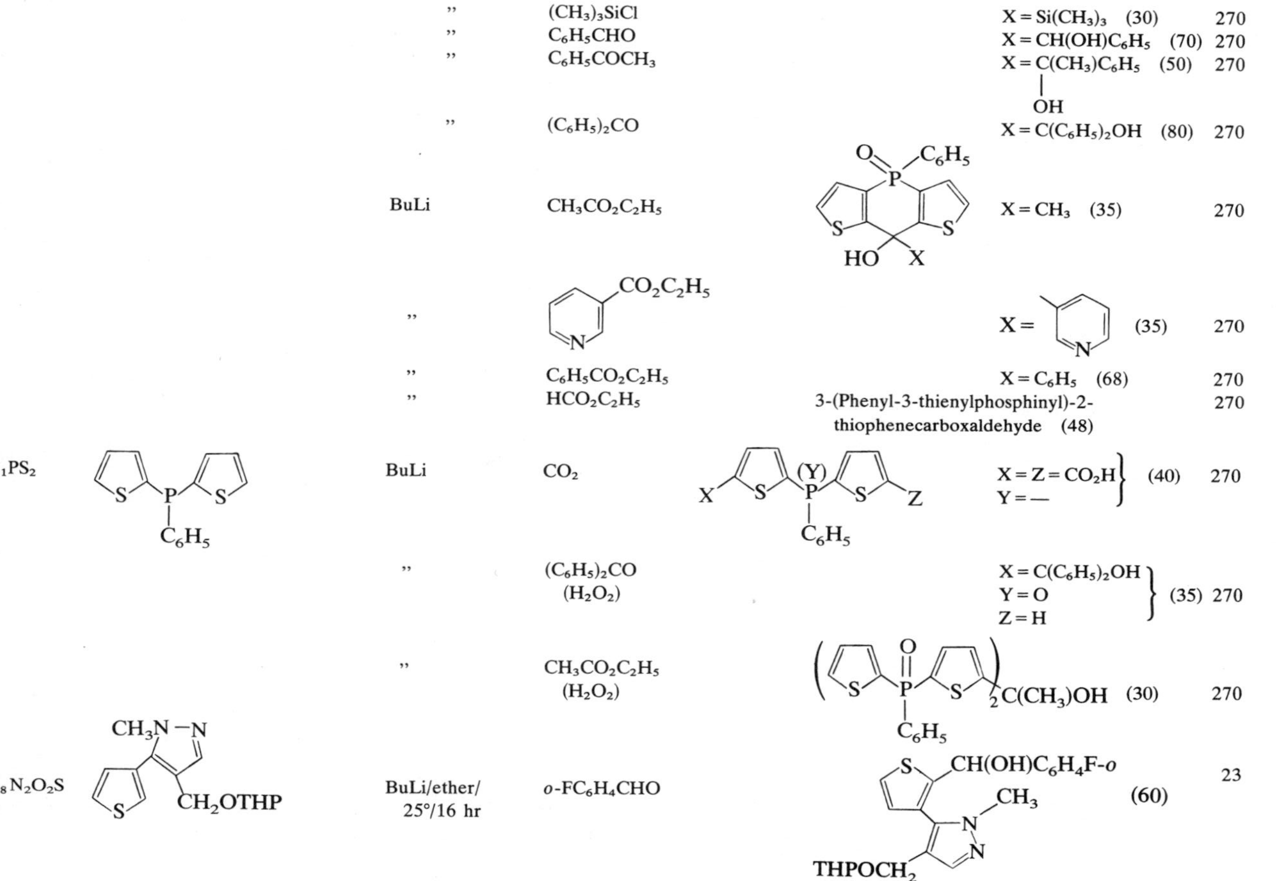

Formula	Reactant	Conditions	Electrophile	Product (Yield %)	Ref.
		”	$(CH_3)_3SiCl$	$X = Si(CH_3)_3$ (30)	270
		”	C_6H_5CHO	$X = CH(OH)C_6H_5$ (70)	270
		”	$C_6H_5COCH_3$	$X = C(CH_3)C_6H_5$ (with OH) (50)	270
		”	$(C_6H_5)_2CO$	$X = C(C_6H_5)_2OH$ (80)	270
		BuLi	$CH_3CO_2C_2H_5$	$X = CH_3$ (35)	270
		”	3-pyridyl-$CO_2C_2H_5$	X = 3-pyridyl (35)	270
		”	$C_6H_5CO_2C_2H_5$	$X = C_6H_5$ (68)	270
		”	$HCO_2C_2H_5$	3-(Phenyl-3-thienylphosphinyl)-2-thiophenecarboxaldehyde (48)	270
$C_{14}H_{11}PS_2$		BuLi	CO_2	$X = Z = CO_2H$, $Y = —$ (40)	270
		”	$(C_6H_5)_2CO$ (H_2O_2)	$X = C(C_6H_5)_2OH$, $Y = O$, $Z = H$ (35)	270
		”	$CH_3CO_2C_2H_5$ (H_2O_2)	$C(CH_3)OH$ (30)	270
$C_{14}H_{18}N_2O_2S$	CH_2OTHP	BuLi/ether/ 25°/16 hr	o-FC_6H_4CHO	$CH(OH)C_6H_4F$-o, CH_3, $THPOCH_2$ (60)	23

TABLE XXI. THIOPHENES (ALPHA) (*Continued*)

Formula	Compound Lithiated	Conditions	Substrate	Product and Yield (%)	Refs.
$C_{14}H_{24}O_4S$	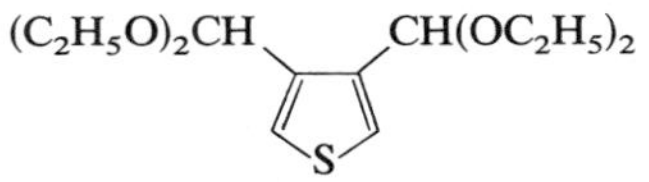	BuLi/ether/ −40°/1 hr	DMF	2,3,4-Thiophenetricarboxaldehyde, 3,4-bis(diethyl acetal) (30)	283
$C_{15}H_{26}LiNO_3S$	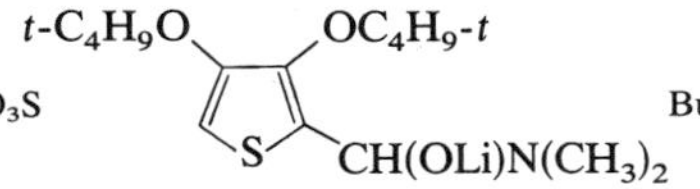	BuLi/reflux	DMF	3,4-Di-*t*-butoxy-2,5-thiophenedicarboxaldehyde (30)	148
$C_{16}H_{13}OPS$	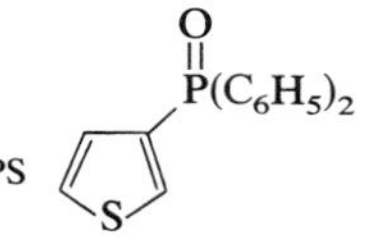	BuLi/THF/25°/ 1 hr	CO_2	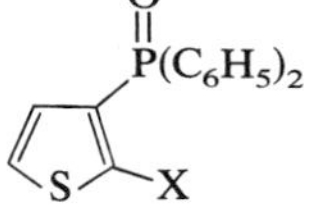 X = CO_2H (30)	270
		"	$(C_6H_5)_2CO$	X = $C(C_6H_5)_2OH$ (50)	270
$C_{19}H_{34}O_6S$	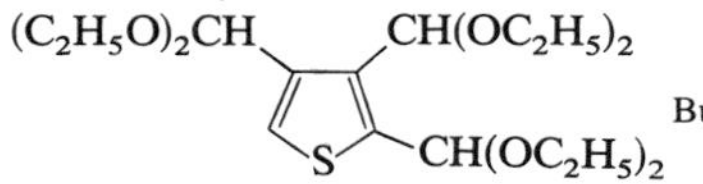	BuLi/"long time"	DMF	2,3,4,5-Thiophenetetracarboxaldehyde, 2,3,4-tris(diethyl acetal) ("good")	441
$C_{22}H_{14}S$	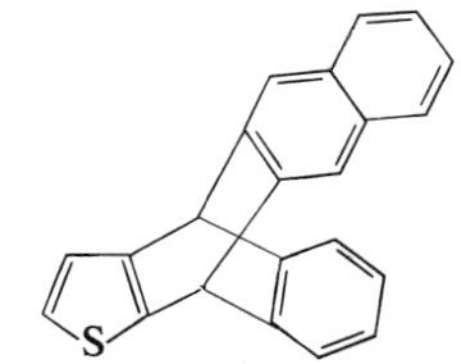	BuLi/ether/ reflux/0.5 hr	CO_2	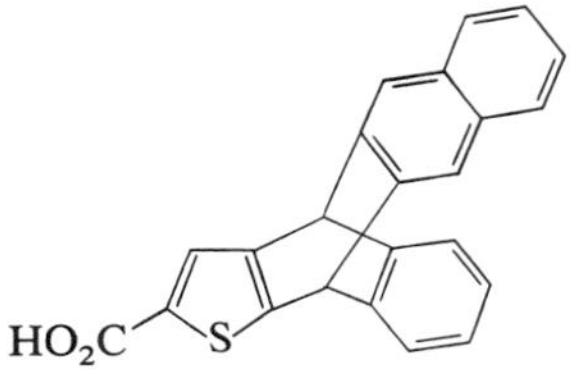 (63)	442

Note: References 360–607 are on pp. 355–360.

TABLE XXII. CONDENSED THIOPHENES (ALPHA)

Formula	Compound Lithiated	Conditions	Substrate	Product and Yield (%)	Refs.
C_7H_5NS	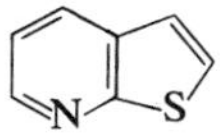 thieno[2,3-b]pyridine	CH_3Li/ether/ −25°/4 hr	D_2O	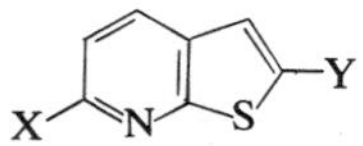 6-X-2-Y-thieno[2,3-b]pyridine: X = CH_3, Y = H (11),	443
				X = H, Y = D (40)	443
		BuLi/hexane, TMEDA/−70°	DMF	X = H, Y = CHO (66)	444
C_8H_5ClS	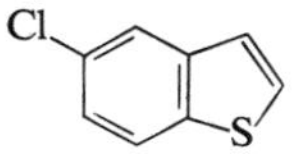 5-chlorobenzo[b]thiophene	BuLi/ether/0°/* 1 hr	CO_2	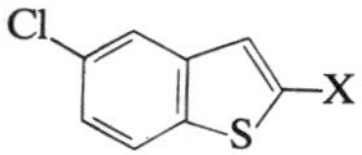 5-Cl-2-X-benzo[b]thiophene: X = CO_2H (83)	445
		"	$(CH_3S)_2$	X = SCH_3 (82)	445
		"	$(CH_3O)_2SO_2$	X = CH_3 (96)	445
C_8H_6S	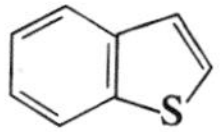 benzo[b]thiophene	BuLi	Br_2	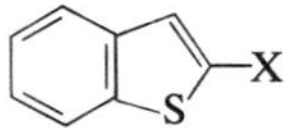 2-X-benzo[b]thiophene: X = Br (39)	279
		BuLi/ether/0°	$FClO_3$	X = F (70)	149
		BuLi/ether	"S" + C_2H_5I	X = SC_2H_5 (69)	420
		BuLi/ether/25°/ "overnight"	CO_2	X = CO_2H (82)	48, 279, 446, 281
		BuLi/hexane, TMEDA/reflux/ 0.5 hr	CO_2 (CH_2N_2)	X = CO_2CH_3 (55), Dimethyl benzo[b]thiophene-2,7-dicarboxylate (12)	40
		BuLi	C_2H_5I	X = C_2H_5 (30)	446
		BuLi	CH_3CHO	X = $CH(OH)CH_3$ (72)	279
		EtLi/ether/ reflux/45 min	CCl_3CHO	X = $CH(OH)CCl_3$ (86)	447
		BuLi/ether/0°/* 1 hr	$(CH_3S)_2$	X = SCH_3 (87)	445

TABLE XXII. Condensed Thiophenes (Alpha) (*Continued*)

Formula	Compound Lithiated	Conditions	Substrate	Product and Yield (%)	Refs.
C_8H_6S (*Contd.*)	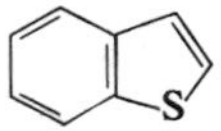 (S)	BuLi	$(CH_3O)_2SO_2$	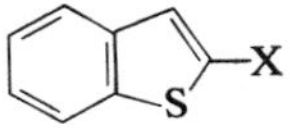 (S)–X; X = CH_3 (91)	274, 448
		BuLi/THF/−10°–25°/2 hr	$(CH_3)_3SiCl$	X = $Si(CH_3)_3$ (25)	251
		BuLi	DMA	X = $COCH_3$ (79)	446
		—	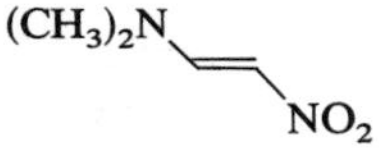 $(CH_3)_2N$–CH=CH–NO_2	X = CH=CHNO$_2$ (26)	285
		BuLi	$(C_2H_5O)_2SO_2$	X = C_2H_5 (81)	274, 448
		”	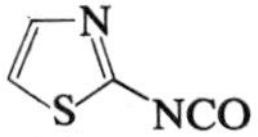 (N, S) NCO	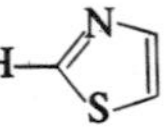 X = CONH–(N, S) (71)	279
		”	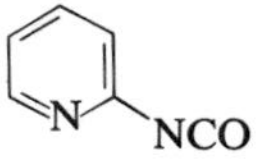 (N) NCO	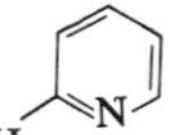 X = CONH–(N) (54)	279
		”	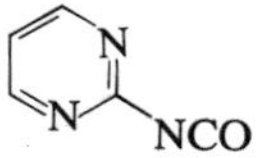 (N, N) NCO	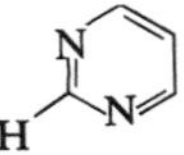 X = CONH–(N, N) (74)	279
		BuLi/ether/25°/24 hr	C_6H_5F	X = C_6H_5 (55)	449
		BuLi	o-BrC_6H_4Cl	X = C_6H_5 (7)	450
		”	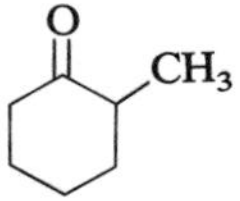 (O, CH_3)	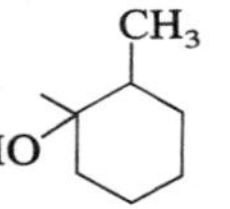 X = HO (CH$_3$) (41)	450
		”	C_6H_5CHO	X = CH(OH)C_6H_5 (70)	279

Formula	Compound	Metalation conditions	Reagent	Product (Yield %)	Ref.
		"	p-ClC_6H_4CHO	$X = CH(OH)C_6H_4Cl$-p (68)	279
		"	p-$CH_3C_6H_4CHO$	$X = CH(OH)C_6H_4CH_3$-p (—)	279
		"	p-$(CH_3)_2NC_6H_4CHO$	$X = CH(OH)C_6H_4N(CH_3)_2$-$p$ (47)	279
		"	C_6H_5NCO	$X = CONHC_6H_5$ (81)	279
		"	o-$CH_3C_6H_4NCO$	$X = CONHC_6H_4CH_3$-o (41)	279
		BuLi/ether/−10°	$C_6H_5COCH_3$	$X = C(OH)(CH_3)C_6H_5$ (77)	451
		BuLi/ether/−20°/ 1.5 hr	$(C_6H_5)_3SiCl$	$X = Si(C_6H_5)_3$ (71)	452
		BuLi	3-HO-4-(HO_2C)C_6H_3NCO	$X = CONH$-[3-OH-4-CO_2H-C_6H_3] (23)	279
		"	$C_6H_5N(CH_3)CHO$	$X = CHO$ (42)	278
		"	$CH_3OSO_2C_6H_5CH_3$-p	$X = CH_3$ (43)	279
		"	quinoline ($C_6H_5NO_2$)	$X =$ 2-quinolyl (49)	281
		BuLi/ether/ 25°/1 hr	$(n$-$C_4H_9O)_3B$ (H_3O^+, H_2O_2)	benzo[b]thiophen-2(3H)-one (76)	151, 150
		BuLi	$(4$-OCN-$C_6H_4)_2SO_2$	$(\text{benzo[b]thiophen-2-yl-CONH-}C_6H_4)_2SO_2$ (17)	279
C_9H_8S	3-methylbenzo[b]thiophene	BuLi/ether/ reflux	CO_2	3-Methylbenzo[b]thiophene-2-carboxylic acid (65)	18

TABLE XXII. CONDENSED THIOPHENES (ALPHA) (*Continued*)

Formula	Compound Lithiated	Conditions	Substrate	Product and Yield (%)	Refs.
C_9H_8S (*Contd.*)	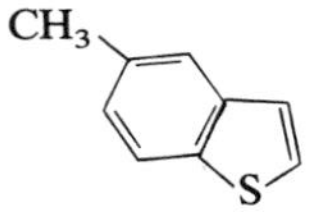 CH_3, S	BuLi/ether/* 0°/1 hr	$(CH_3S)_2$	5-Methyl-2-(methylthio)benzo[*b*]thiophene (91)	445
		”	$(CH_3O)_2SO_2$	2,5-Dimethylbenzo[*b*]thiophene (85)	445
$C_{10}H_6S_2$	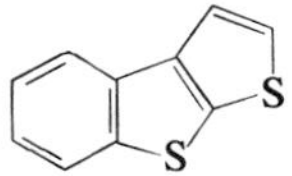 S, S	BuLi/ether/−40°–25°/2 hr	DMF	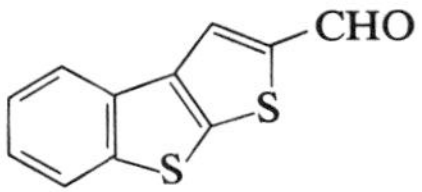 CHO, S, S (70)	453
	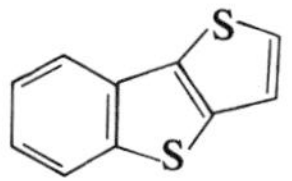 S, S	BuLi/ether/−40°–25°/2 hr	CO_2	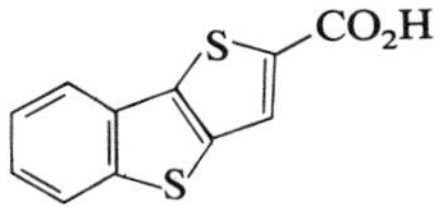 CO_2H, S, S (63)	453
	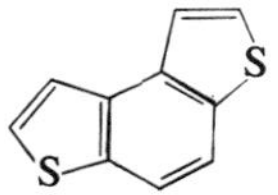 S, S	BuLi/ether/25°/15 min	$C_6H_5N(CH_3)CHO$	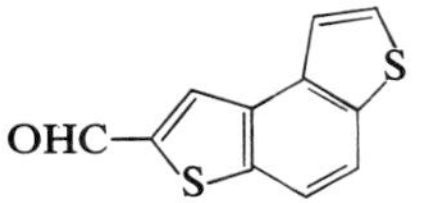 OHC, S, S (54)	278
		4 BuLi/ether/reflux/0.5 hr	”	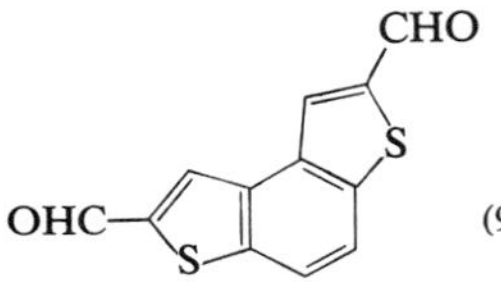 OHC, CHO, S, S (99)	278
$C_{10}H_8O_2S$	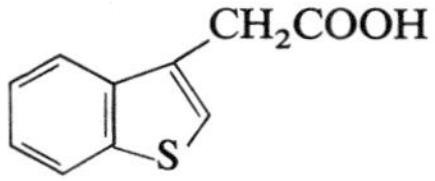 CH_2COOH, S	BuLi/THF/−78°/0.5 hr	D_2O	2-*d*-Benzo[*b*]thiophene-3-acetic acid (81)[a]	153

$C_{10}H_{11}NS$	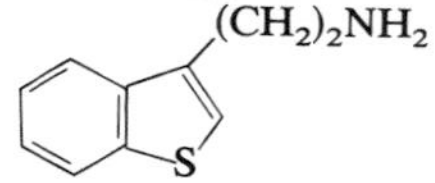	BuLi/THF/−78°/ 0.5 hr	D_2O	2-*d*-Benzo[*b*]thiophene-3-ethylamine (83)[b]	153
$C_{11}H_{12}LiNOS$	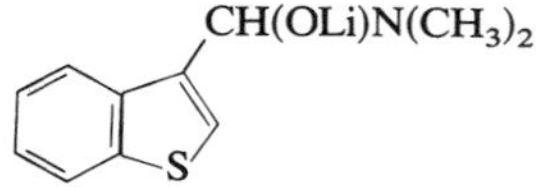	BuLi/−70° to −30°	"S"	2-Mercaptobenzo[*b*]thiophene-3-carboxaldehyde (—)	454
$C_{12}H_8S$	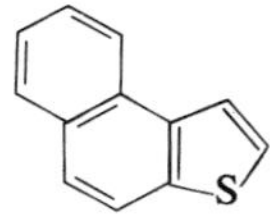	BuLi/ether/25°/ 2 hr, reflux/ 2 hr	CO_2	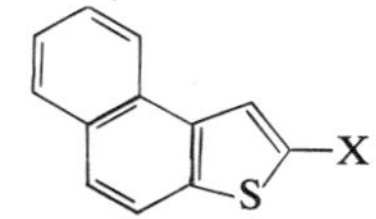 $X = CO_2H$ (75)	455
			$C_6H_5N(CH_3)CHO$	$X = CHO$ (68)	455, 278
$C_{12}H_{15}NS$	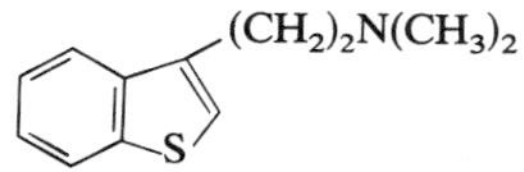	BuLi/THF/−78°/ 0.5 hr	D_2O	N,N-Dimethyl-2-*d*-benzo[*b*]thiophene-3-ethylamine (83)	153
$C_{14}H_{10}S$	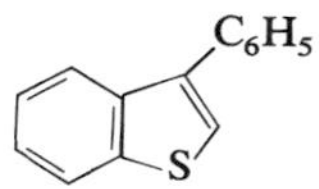	BuLi/ether/0°–25°/1 hr	CO_2	3-Phenylbenzo[*b*]thiophene-2-carboxylic acid (53)	456
$C_{16}H_{10}N_2S$	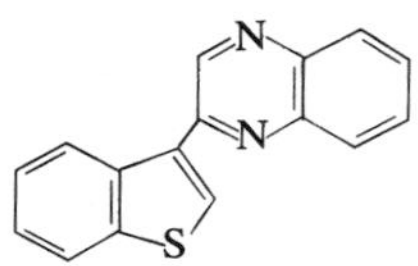	LDA/ether	CuI_2	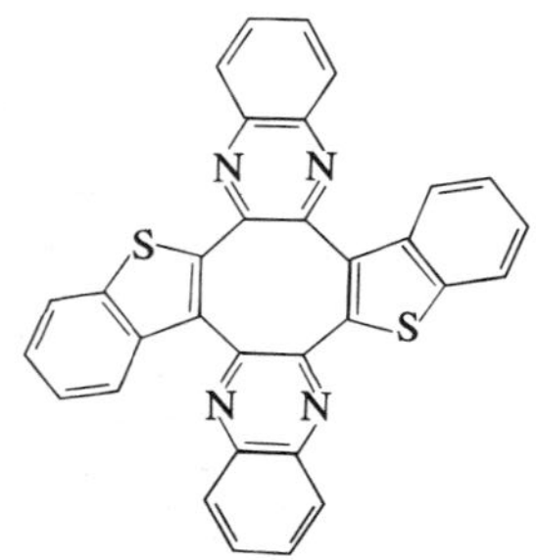 (61)	440

TABLE XXII. CONDENSED THIOPHENES (ALPHA) (*Continued*)

Formula	Compound Lithiated	Conditions	Substrate	Product and Yield (%)	Refs.
$C_{24}H_{12}S_3$	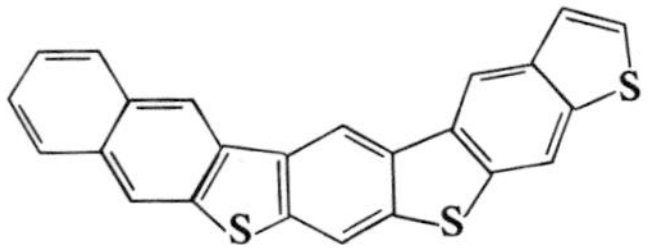	BuLi/ether/0°–25°/10 min	$C_6H_5N(CH_3)CHO$	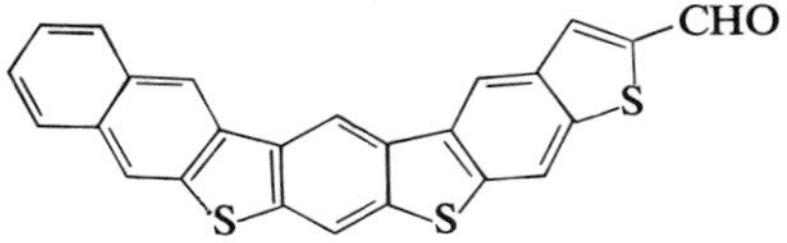 CHO (65)	278
$C_{24}H_{14}N_2S_2$	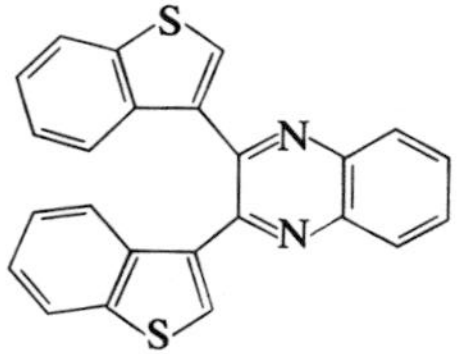	LDA/THF/−50°/1 hr	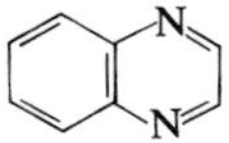	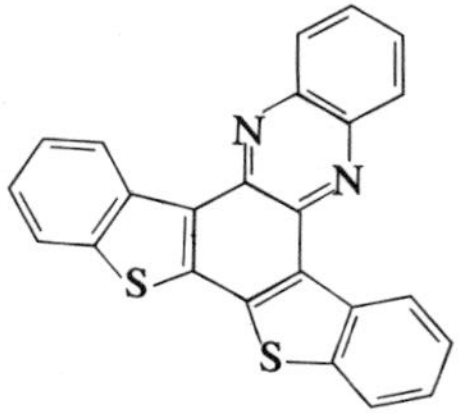 (69),	282
				(18)	

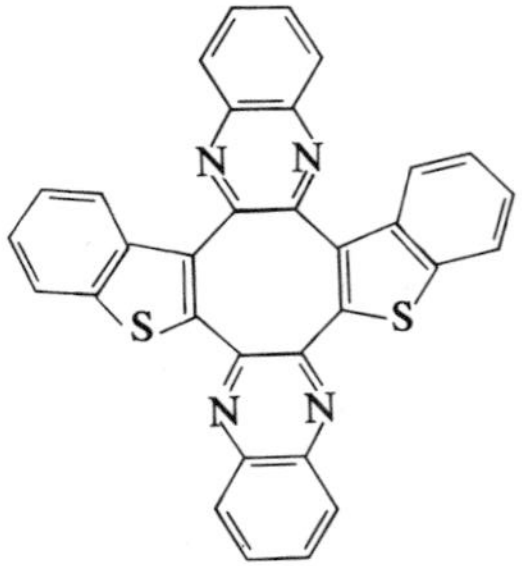

[a] The isolated yield was 81%; the deuterium incorporation was 70%.
[b] The isolated yield was 83%; the deuterium incorporation was 80%.

Note: References 360–607 are on pp. 355–360.

TABLE XXIII. ISOTHIAZOLES (ALPHA)

Formula	Compound Lithiated	Conditions	Substrate	Product and Yield (%)	Refs.
C_3H_2BrNS	4-Br-isothiazole	BuLi/THF/ −65°/15 min*	CO_2	4-Bromo-5-isothiazolecarboxylic acid (70)	154
		”	DMF	4-Bromo-5-isothiazolecarboxaldehyde (73)	154
C_3H_2ClNS	4-Cl-isothiazole	BuLi/THF/ −65°/15 min*	CO_2	4-Chloro-5-isothiazolecarboxylic acid (68)	154
		”	DMF	4-Chloro-5-isothiazolecarboxaldehyde (65)	154
C_3H_2INS	4-I-isothiazole	BuLi/THF/ −65°/15 min*	DMF	4-Iodo-5-isothiazolecarboxaldehyde (33)	154
C_3H_3NS	isothiazole	BuLi/THF/ −65°/15 min	Br_2	5-X-isothiazole, X = Br (34)	154
		”	CO_2	X = CO_2H (48)	154
		”	CH_3I	X = CH_3 (40)	154
		”	DMF	X = CHO (75)	154
$C_4H_3NO_2S$	4-CO_2H-isothiazole	BuLi/THF/ −65°/15 min*	CO_2	4,5-Isothiazoledicarboxylic acid (15)	154
C_4H_4BrNS	3-CH_3-4-Br-isothiazole	BuLi/THF/ −65°/15 min*	CO_2	3-CH_3-4-Br-5-X-isothiazole, X = CO_2H (56)	154

TABLE XXIII. Isothiazoles (Alpha) (*Continued*)

Formula	Compound Lithiated	Conditions	Substrate	Product and Yield (%)	Refs.
C_4H_4BrNS (*Contd.*)	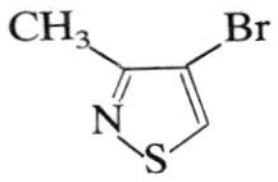 CH_3 Br N S	,,	CH_3I	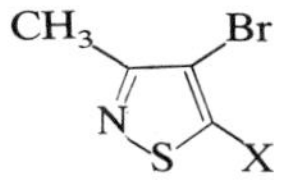 CH_3 Br N S X; $X = CH_3$ (40)	154
		,,	DMF	$X = CHO$ (51)	154
		,,	C_2H_5I*	$X = C_2H_5$ (34)	154
		,,	n-C_3H_7I*	$X = C_3H_7$-n (28)	154
		,,	$C_6H_5CH_2Br$	$X = CH_2C_6H_5$ (13)	154
C_4H_4ClNS	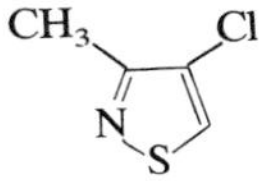 CH_3 Cl N S	BuLi/THF/ −65°/15 min*	CO_2	4-Chloro-3-methyl-5-isothiazole-carboxylic acid (75)	154
		,,	DMF	4-Chloro-3-methyl-5-isothiazole-carboxaldehyde (47)	154
C_4H_4INS	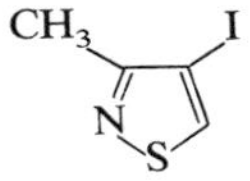 CH_3 I N S	BuLi/THF/ −65°/15 min*	CO_2	4-Iodo-3-methyl-5-isothiazolecarboxylic acid (58)	154
		,,	DMF	4-Iodo-3-methyl-5-isothiazolecarboxaldehyde (68)	154
C_4H_5NS	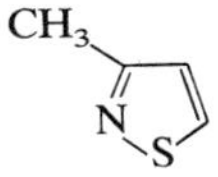 CH_3 N S	BuLi/THF/−70°	"S" + $BrCH_2CO_2C_2H_5$ (OH^-)	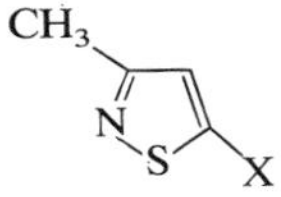 CH_3 N S X; $X = SCH_2CO_2H$ (36)	457
		BuLi/THF/ −65°/15 min	CO_2	$X = CO_2H$ (50)	154
		,,	DMF	$X = CHO$ (50)	154

	CH_3 N S	BuLi/THF/−70°	"S" + $BrCH_2CO_2C_2H_5$ (OH^-)	CH_3 N S X	$X = SCH_2CO_2H$ (51)	457
		BuLi/THF/ −65°/15 min	CO_2		$X = CO_2H$ (40)	154
		"	DMF		$X = CHO$ (55)	154
$C_5H_5NO_2S$	CH_3 CO_2H N S	BuLi/THF/ −70°/15 min	Br_2	CH_3 CO_2H N S X	$X = Br$ (52)	154
		BuLi/THF/ −65°/15 min*	CO_2		$X = CO_2H$ (29)	154
		"	DMF		$X = CHO$ (25)	154

Note: References 360–607 are on pp. 355–360.

TABLE XXIV. THIAZOLES (ALPHA)

Formula	Compound Lithiated	Conditions	Substrate	Product and Yield (%)	Refs.
C_3H_2ClNS	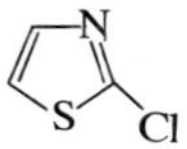 N S Cl	BuLi/ether/−80°	CH_3CHO	2-Chloro-α-methyl-5-thiazolemethanol (88)	157
C_3H_3NS	 N S	C_6H_5Li/−60°	CO_2	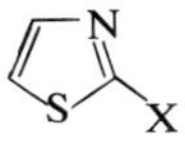 N S X $X = CO_2H$ (40)	156
		BuLi/ether/ −60°/0.5 hr	Ethylene oxide	$X = (CH_2)_2OH$ (30)	564
		"	CH_3CHO	$X = CH(OH)CH_3$ (30)	564
		"	C_2H_5CHO	$X = CH(OH)C_2H_5$ (50)	564
		"	n-C_3H_7CHO	$X = CH(OH)C_3H_7$-n (90)	564
		"	$(CH_3)_2CHCHO$	$X = CH(OH)CH(CH_3)_2$ (85)	564
		"	n-$C_6H_{13}CHO$	$X = CH(OH)C_6H_{13}$-n (90)	564
		"	$(C_6H_5)_2CO$	$X = C(C_6H_5)_2OH$ (22)	564
C_4H_5NS	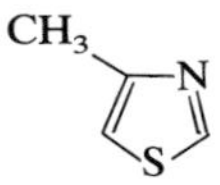 CH_3 N S	BuLi/ether/ −60°/0.5 hr	CH_3I	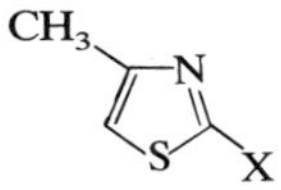 CH_3 N S X $X = CH_3$ (26)	564
		"	Ethylene oxide	$X = (CH_2)_2OH$ (42)	564
		BuLi/ether/ −75°/10 min	CH_3CHO	$X = CH(OH)CH_3$ (48)	564
		BuLi/ether/ −60°/0.5 hr	Propylene oxide	$X = CH_2CH(OH)CH_3$ (51)	564
		"	n-C_3H_7CHO	$X = CH(OH)C_3H_7$-n (93)	564
	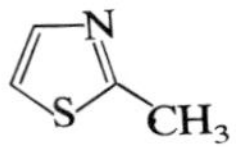 N S CH_3	BuLi/ether/−70°	CO_2	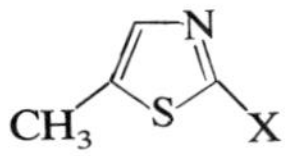 CH_3 N S X $X = CO_2H$ (42)	158

Formula	Substrate	Conditions	Electrophile	Product(s) (Yield, %)	Refs.
		BuLi/ether/$-25°$	CH_3I	X = CH_3 (70)	158
		BuLi/ether/$-70°$	C_6H_5CN (H_3O^+)	X = COC_6H_5 (50)	158
		”	n-C_3H_7CHO	X = $CH(OH)C_3H_7$-n (74), 2-Methyl-α-propyl-5-thiazolemethanol (26)	158
C_5H_7NS	(thiazole: CH_3, N, S, CH_3)	BuLi/THF/ $-78°$/0.5–1 hr	CH_3I	(thiazole: CH_3, N, S, X, CH_3) , (thiazole: CH_3, N, S, CH_2X)	
				(12) X = CH_3 (88)	159, 460
		BuLi/ether/ $-60°$/0.5 hr	CH_3CHO	(—) X = $CH(OH)CH_3$ (57)	564
		”	n-C_3H_7CHO	(—) X = $CH(OH)C_3H_7$-n (72)	564
		BuLi/THF/ $-78°$/0.5–1 hr	$C_6H_5CH_2Cl$	(10) X = $CH_2C_6H_5$ (90)	159
	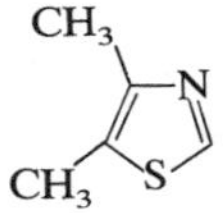 (thiazole: CH_3, N, CH_3, S)	C_6H_5Li/ether/ 25°/1 hr	CH_2O	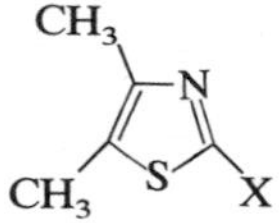 (thiazole: CH_3, N, CH_3, S, X) X = CH_2OH (58)	459
		”	$ClCH_2OCH_3$	X = CH_2OCH_3 (42)	459
		”	C_6H_5CHO	X = $CH(OH)C_6H_5$ (55)	459
C_6H_9NS	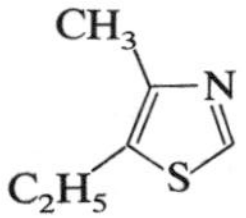 (thiazole: CH_3, N, C_2H_5, S)	BuLi/ether/ $-60°$/0.5 hr	Propylene oxide	5-Ethyl-α,4-dimethylthiazole-2-ethanol (76)	564
$C_{10}H_8ClNS$	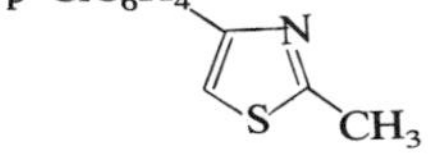 (thiazole: p-ClC_6H_4, N, S, CH_3)	BuLi/THF/$-78°$/ 0.5–1 hr	CH_3I	4-(p-Chlorophenyl)-2,5-dimethylthiazole (93), 4-(p-chlorophenyl)-2-ethylthiazole (3)	159

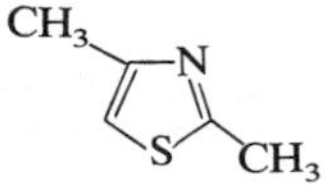

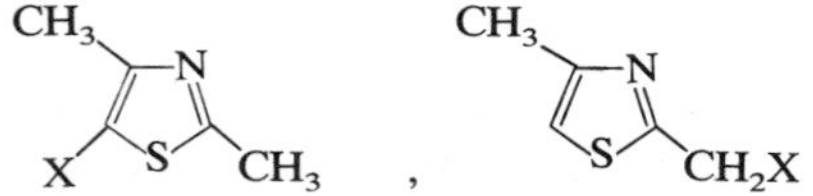

TABLE XXIV. THIAZOLES (ALPHA) (*Continued*)

Formula	Compounded Lithiated	Conditions	Substrate	Product and Yield (%)	Refs.
$C_{10}H_9NS$	4-C_6H_5-2-CH_3-thiazole	BuLi/THF/−78°/ 0.5–1 hr	CH_3I	4-C_6H_5-2-CH_2X-thiazole , 4-C_6H_5-5-X-2-CH_3-thiazole	
				(5) $X = CH_3$ (95)	460, 159
		"	C_2H_5I	(7) $X = C_2H_5$ (86)	159
		"	$(CH_3)_3SiCl$	(4) $X = Si(CH_3)_3$ (96)	159
		"	C_6H_5CHO	(—) $X = CH(OH)C_6H_5$ (97)	159
$C_{11}H_{11}NOS$	4-(p-$CH_3OC_6H_4$)-2-CH_3-thiazole	BuLi/THF/−78°/ 0.5–1 hr	CH_3I	4-(*p*-Methoxyphenyl)-2,5-dimethylthiazole (86), 2-ethyl-4-(*p*-methoxyphenyl)thiazole (6)	159

Note: References 360–670 are on pp. 355–360.

TABLE XXV. Benzothiazoles (Alpha)

Formula	Compound Lithiated	Conditions	Substrate	Product and Yield (%)	Refs.
C_7H_5NS	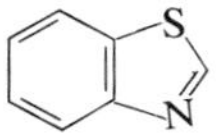 [benzothiazole structure]	BuLi/ether/−70°	$CF_2{=}CCl_2$	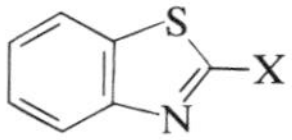 [2-X-benzothiazole structure] X = $CF{=}CCl_2$ (59)	364
		BuLi/ether/−75°/ 15 min	$(CH_3)_3SiCl$	X = $Si(CH_3)_3$ (77)	160, 461
		BuLi/THF/−78°	$C_2H_5(CH_3)_2SiCl$	X = $Si(CH_3)_2C_2H_5$ (27)	462
		"	$(C_2H_5)_3SiCl$	X = $Si(C_2H_5)_3$ (30)	462
		—	 [pyrimidine structure] ($KMnO_4$)	X 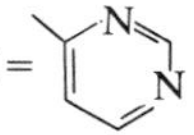 = [pyrimidinyl structure] (−)	370
		—	$C_6H_5N_3$	X = NH_2 (—)	287
		BuLi/ether/−70°– 25°/2 hr 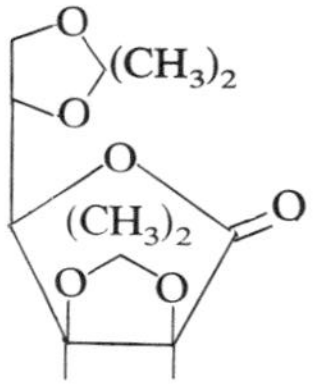 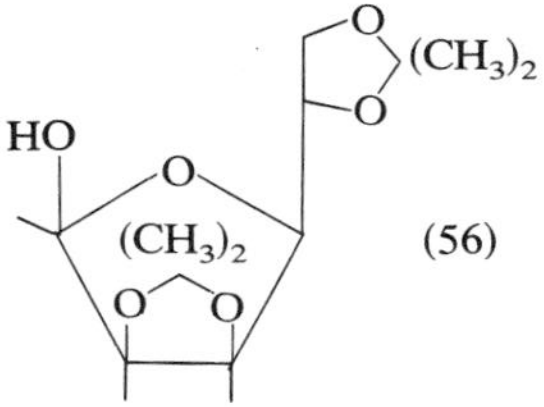	[lactone structure with two $(CH_3)_2$ acetonide groups]	X = [HO-substituted furanose structure with two $(CH_3)_2$ acetonide groups] (56)	590
		"	[lactone structure with $HOCH_2$ and $(CH_3)_2$ acetonide]	X = [HO-substituted furanose structure with CH_2OH and $(CH_3)_2$ acetonide] (23)	590

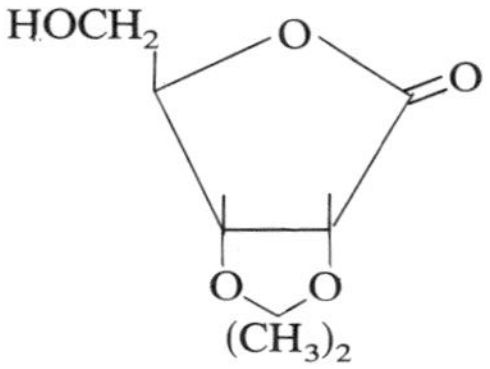

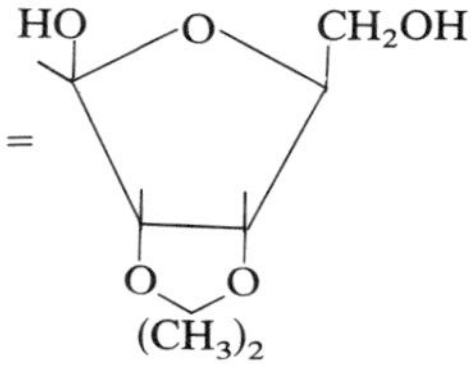

TABLE XXV. BENZOTHIAZOLES (ALPHA) (*Continued*)

Formula	Compound Lithiated	Conditions	Substrate	Product and Yield (%)	Refs.
C_7H_5NS (*Contd.*)	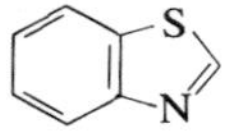 [benzothiazole structure]	BuLi/ether/−65°/ 5 min	$(C_6H_5)_3SiBr$	$X = Si(C_6H_5)_3$ (19)	452
		BuLi	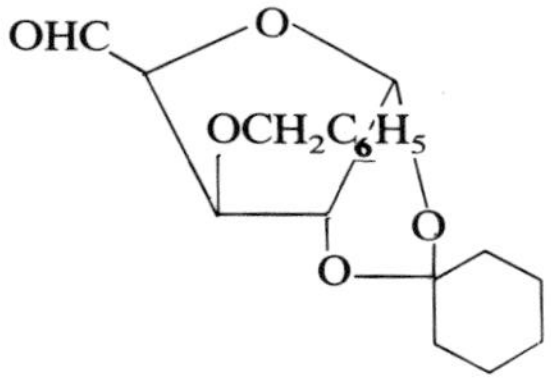 OHC–[furanose: O, $OCH_2C_6H_5$, O–cyclohexylidene–O]	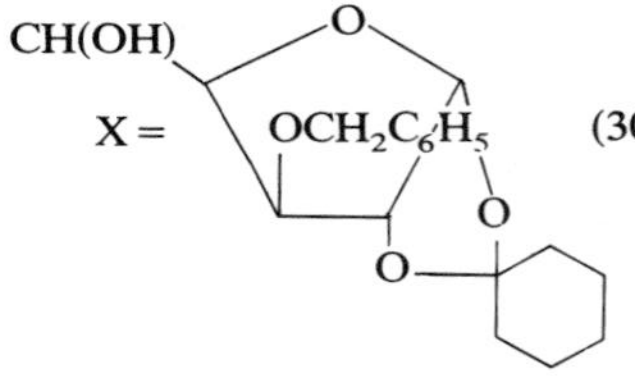 CH(OH)– ; X = [furanose: O, $OCH_2C_6H_5$, O–cyclohexylidene–O] (30–36)	372
		BuLi/THF/−78°/ 5 hr	$SiCl_4$	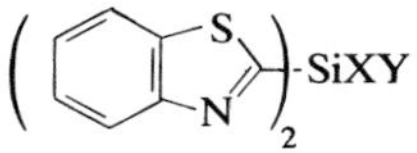 (benzothiazol-2-yl)$_2$SiXY, $X = Y = Cl$ (33)	461
		”	CH_3SiCl_3	$X = CH_3$, $Y = Cl$ (39)	461
		”	$(CH_3)_2SiCl_2$	$X = Y = CH_3$ (51)	461

Note: References 360–607 are on pp. 355–360.

TABLE XXVI. VINYL SELENIDES (ALPHA)

Formula	Compound Lithiated	Conditions	Substrate	Product and Yield (%)	Refs.
C_8H_8Se	$CH_2{=}CHSeC_6H_5$	LDA/THF/−78°/1 hr	D_2O	$CH_2=C(SeC_6H_5)X$ X = D (80)	161
		"	n-$C_5H_{11}CHO$	X = $CH(OH)C_5H_{11}$-n (40)	161
		LDA/THF, HMPA	n-$C_{10}H_{21}Br$	X = $C_{10}H_{21}$-n (40)	161

TABLE XXVII. SELENOPHENES (ALPHA)

Formula	Compound Lithiated	Conditions	Substrate	Product and Yield (%)	Refs.
C_4H_4Se	selenophene	BuLi/ether/1 hr	SO_2	2-X-selenophene, X = SO_2Li (57)	163
		BuLi/ether/ reflux/0.5 hr	$CuCl_2$	X = 2-selenienyl (47)	575
		BuLi/ether/ reflux/15 min	"Se" + 2-bromoselenophene	X = Se-(2-selenienyl) (48)	575
		"	CO_2	X = CO_2H (56)	163
		BuLi/ether/25°	CCl_3CCl_3	X = Cl (45)	365
		BuLi/ether/ reflux/15 min	$(C_2H_5Se)_2$	X = SeC_2H_5 (66)	575
		BuLi/ether	Cl–CH=CH–ICl_2	(2-selenienyl)$_2I^+Cl^-$ (71)	374
C_5H_3NSe	2-cyanoselenophene	BuLi/ether/0.5 hr	CO_2	5-Cyano-2-selenophenecarboxylic acid (—)	396
		"	DMF	5-Formyl-2-selenophenecarboxamide (—)	396
	3-cyanoselenophene	BuLi/ether/0.5 hr	CO_2	3-Cyano-2-selenophenecarboxylic acid (—)	396
		"	DMF	2-Formyl-3-selenophenecarboxamide (—)	396
C_5H_6Se	2-methylselenophene	BuLi/ether/ reflux/2.5 hr	"S"	5-X-2-methylselenophene, X = SH (35)	497
		BuLi/ether/ 1 hr	SO_2	X = SO_2Li (62)	163
		"	CO_2	X = CO_2H (43)	163

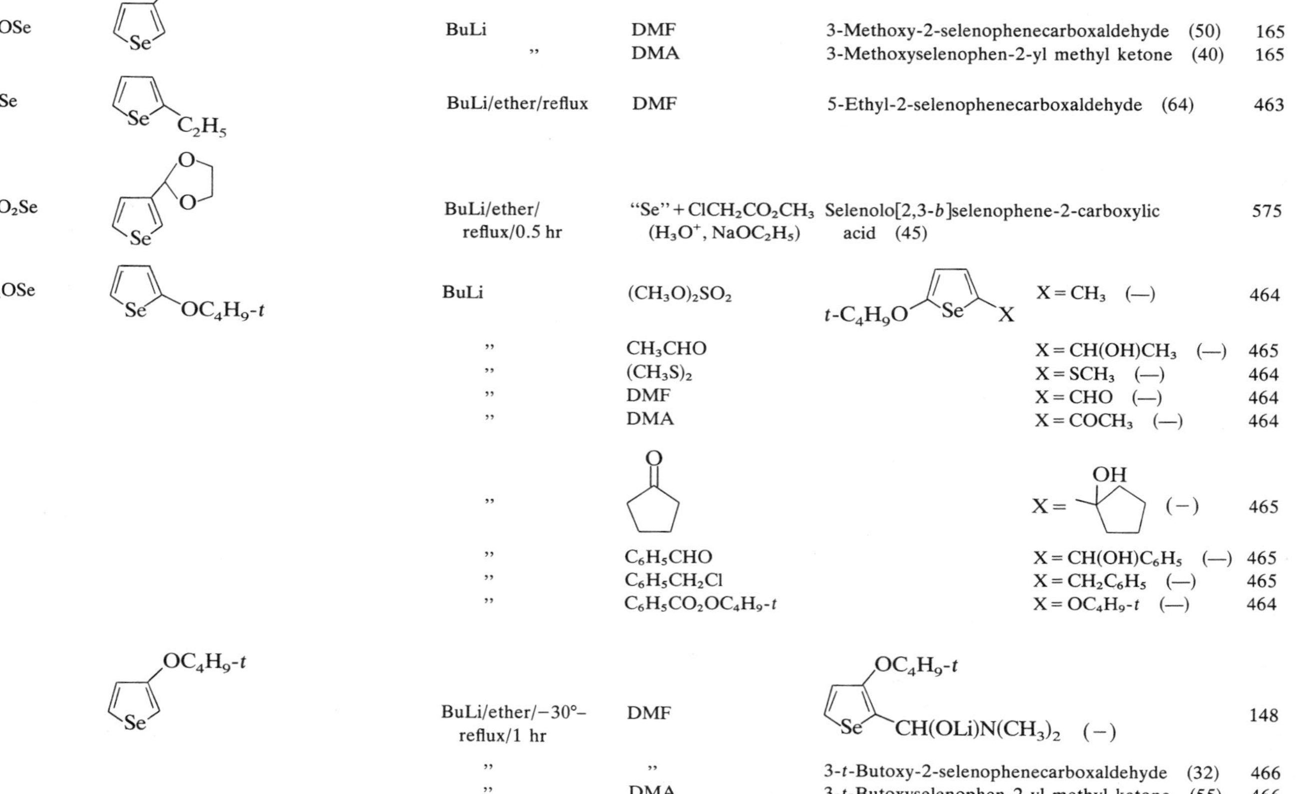

Formula	Substrate	Metalation conditions	Electrophile	Product (yield, %)	Refs.
C_5H_6OSe	3-methoxyselenophene	BuLi	DMF	3-Methoxy-2-selenophenecarboxaldehyde (50)	165
		”	DMA	3-Methoxyselenophen-2-yl methyl ketone (40)	165
C_6H_8Se	2-ethylselenophene	BuLi/ether/reflux	DMF	5-Ethyl-2-selenophenecarboxaldehyde (64)	463
$C_7H_8O_2Se$	3-(1,3-dioxolan-2-yl)selenophene	BuLi/ether/ reflux/0.5 hr	“Se” + $ClCH_2CO_2CH_3$ (H_3O^+, $NaOC_2H_5$)	Selenolo[2,3-*b*]selenophene-2-carboxylic acid (45)	575
$C_8H_{12}OSe$	2-OC_4H_9-*t*-selenophene	BuLi	$(CH_3O)_2SO_2$	*t*-C_4H_9O-selenophene-X: X = CH_3 (—)	464
		”	CH_3CHO	X = $CH(OH)CH_3$ (—)	465
		”	$(CH_3S)_2$	X = SCH_3 (—)	464
		”	DMF	X = CHO (—)	464
		”	DMA	X = $COCH_3$ (—)	464
		”	cyclopentanone	X = 1-hydroxycyclopentyl (–)	465
		”	C_6H_5CHO	X = $CH(OH)C_6H_5$ (—)	465
		”	$C_6H_5CH_2Cl$	X = $CH_2C_6H_5$ (—)	465
		”	$C_6H_5CO_2OC_4H_9$-*t*	X = OC_4H_9-*t* (—)	464
	3-OC_4H_9-*t*-selenophene	BuLi/ether/–30°– reflux/1 hr	DMF	3-OC_4H_9-*t*-2-$CH(OLi)N(CH_3)_2$-selenophene (–)	148
		”	”	3-*t*-Butoxy-2-selenophenecarboxaldehyde (32)	466
		”	DMA	3-*t*-Butoxyselenophen-2-yl methyl ketone (55)	466

TABLE XXVII. SELENOPHENES (ALPHA) (*Continued*)

Formula	Compound Lithiated	Conditions	Substrate	Product and Yield (%)	Refs.
$C_9H_{14}O_2Se$	2-selenienyl-$CH(OC_2H_5)_2$ (ring: Se)	BuLi/ether/reflux	I_2	2-(Diethoxymethyl)-5-iodoselenophene (—)	396
		BuLi/ether/ −30° to −10°	DMF	2,5-Selenophenedicarboxaldehyde, 5-(diethyl acetal) (40)	467
	3-selenienyl-$CH(OC_2H_5)_2$ (ring: Se)	BuLi/ether/reflux	I_2	3-(Diethoxymethyl)-2-iodoselenophene (—)	396
		BuLi/ether/−40°/ 1 hr	DMF	2,3-Selenophenedicarboxaldehyde, 3-(diethyl acetal) (80)	283
$C_{11}H_{18}LiNO_2Se$	3-OC_4H_9-*t*, 2-$CH(OLi)N(CH_3)_2$ selenophene (ring: Se)	BuLi/ether/−30°–reflux/1 hr	DMF	3-*t*-Butoxy-2-selenophenecarboxaldehyde (90), 3-*t*-butoxy-2,5-selenophenedicarboxaldehyde (10)	148
	4-*t*-C_4H_9O, 3-$CH(OLi)N(CH_3)_2$ selenophene (ring: Se)	BuLi/ether/reflux 1 hr	DMF	4-*t*-Butoxy-2,3-selenophenedicarboxaldehyde (50)	148
$C_{12}H_{20}O_2Se$	3,4-di-(OC_4H_9-*t*) selenophene (*t*-C_4H_9O, OC_4H_9-*t*; ring: Se)	BuLi/ether/−20°–reflux/1 hr	DMF	3,4-Di-*t*-butoxy-2-selenophenecarboxaldehyde (5)	148

$C_{14}H_{24}O_4Se$	3-$(C_2H_5O)_2CH$-2-$CH(OC_2H_5)_2$-selenophene	BuLi/ether/ −50°/5 hr	DMF	2,3,5-Selenophenetricarboxaldehyde, 3,5-bis(diethyl acetal) (40)	283
	3-$(C_2H_5O)_2CH$-4-$CH(OC_2H_5)_2$-selenophene	BuLi/ether/−40°/ 1 hr	DMF	2,3,4-Selenophenetricarboxaldehyde, 3,4-bis(diethyl acetal) (40)	283
$C_{19}H_{34}O_6Se$	3-$(C_2H_5O)_2CH$-4-$CH(OC_2H_5)_2$-2-$CH(OC_2H_5)_2$-selenophene	BuLi/"long time"	DMF	2,3,4,5-Selenophenetetracarboxaldehyde, 2,3,4-tris(diethyl acetal) ("good")	441

Note: References 360–607 are on pp. 355–360.

TABLE XXVIII. TELLUROPHENES (ALPHA)

Formula	Compound Lithiated	Conditions	Substrate	Product and Yield (%)	Refs.
C_4H_4Te	 [tellurophene structure]	BuLi/ether/25°/ 45 min	CO_2	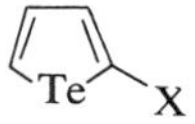 [2-X-tellurophene structure] $X = CO_2H$ (37)	164
		BuLi/ether/25°/ 0.5 hr	Cl_3CCCl_3	$X = Cl$ (53)	568
		”	Br_3CCBr_3	$X = Br$ (44)	568
		BuLi/ether/25°/ 45 min	$(CH_3O)_2SO_2$	$X = CH_3$ (75)	164
		”	CH_3CHO	$X = CH(OH)CH_3$ (60)	164
		”	$C_6H_5N(CH_3)CHO$	$X = CHO$ (24)	164
		BuLi/ether/25°/ 0.5 hr	“Te”	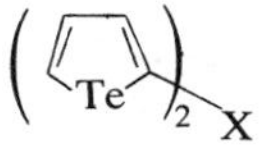 [(2-tellurophenyl)$_2$X structure] $X = Te$ (—)	568
		”	$ClCH{=}CHICl_2$ [structure]	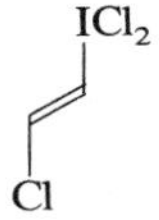 $X = I^+Cl^-$ (—)	568
C_5H_6Te	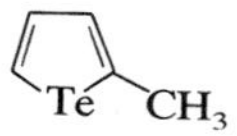 [2-methyltellurophene structure]	BuLi/ether/25°/ 45 min	CO_2	5-Methyl-2-tellurophenecarboxylic acid (35)	164

Note: References 360–607 are on pp. 355–360.

TABLE XXIX. Fluoro-, Chloro-, and Bromoalkenes (Alpha)

Formula	Compound Lithiated	Conditions	Substrate	Product and Yield (%)	Refs.
C_2HClF_2	$CF_2{=}CClH$	BuLi/ether/−100° to −78°/0.5–1 hr	CH_3COCH_3	2-Chloro-3-methylcrotonic acid (15)	172
		"	CF_3COCF_3	$CF_2{=}CClX$ $X = C(CF_3)_2OH$ (56)	172
		"	CH_3COCF_3	$X = C(CF_3)(OH)CH_3$ (61)	172
		"	$(C_2H_5)_3SiCl$	$X = Si(C_2H_5)_3$ (10)	173
C_2HCl_2F	$CFCl{=}CClH$	BuLi/ether/−100° to −78°/0.5–1 hr	C_6H_5CHO	(Z)-2-Chlorocinnamic acid (I) (44)	172
		"	CF_3COCF_3	$CFCl{=}CClX$ $X = C(CF_3)_2OH$ (66)	172
		"	CH_3COCH_3	$X = C(CH_3)_2OH$ (60)	172
		"	$(C_2H_5)_3SiCl$	$X = Si(C_2H_5)_3$ (55)	173
C_2HCl_3	$CCl_2{=}CClH$	BuLi/THF, ether, pet. ether/ −110°/50 min	CO_2	Trichloroacrylic acid (81)	167, 168
C_2HF_3	$CF_2{=}CFH$	BuLi/ether/−100° to −78°/0.5–1 hr	CH_3COCH_3	2-Fluoro-3-methylcrotonic acid (30)	172
		"	CO_2	$CF_2{=}CFX$ $X = CO_2H$ (57)	172
		"	CF_3COCF_3	$X = C(CF_3)_2OH$ (63)	172
		"	$(C_2H_5)_3SiCl$	$X = Si(C_2H_5)_3$ (79)	173
$C_2H_2Cl_2$	(Z)-ClCH=CHCl	BuLi/THF, ether, pet. ether/ −100°/20 min	CO_2	Chloropropiolic acid (100)	167, 168
	(E)-ClCH=CHCl	BuLi/THF, ether, pet. ether/ −110°/40 min	Br_2	(Z)-1-Bromo-1,2-dichloroethylene (26)	167
		"	CO_2	(E)-2,3-Dichloroacrylic acid (99)	167, 168

TABLE XXIX. FLUORO-, CHLORO-, AND BROMOALKENES (ALPHA) (*Continued*)

Formula	Compound Lithiated	Conditions	Substrate	Product and Yield (%)	Refs.
C_2H_3Cl	$CHCl{=}CH_2$	BuLi/THF, ether, pet. ether/−110°/ 70 min	CO_2	2-Chloroacrylic acid (100)	167, 168
C_4H_7BrO	(Z)-BrCH=CHOC$_2$H$_5$	BuLi/THF, hexane/ −100°	CO_2	(Z)-2-Bromo-3-ethoxyacrylic acid (22), (Z)-3-ethoxyacrylic acid (10)	174a
	(E)-BrCH=CHOC$_2$H$_5$	BuLi/ether/ −78°–50°/5 hr	Cyclopentanone (H_3O^+)	α-Bromo-$\Delta^{1,\alpha}$-cyclopentaneacetaldehyde (30)	174b
		"	t-C_4H_9CHO	(E)-2-Bromo-1-ethoxy-4,4-dimethyl-1-penten-3-ol (58)	174b
C_4H_7ClO	(Z)-ClCH=CHOC$_2$H$_5$	BuLi/THF, hexane/ −100°	CO_2	(Z)-2-Chloro-3-ethoxyacrylic acid (40)	174a
		"	Cyclohexanone	α-Chloro-$\Delta^{1,\alpha}$-cyclohexeneacetaldehyde (40)	174a
		2 BuLi/THF, hexane/ −100°	"	Ethyl $\Delta^{1,\alpha}$-cyclohexeneacetate (20)	174a
	(E)-ClCH=CHOC$_2$H$_5$	BuLi/THF, hexane/ −100°	CO_2	(E)-2-Chloro-3-ethoxyacrylic acid (100)	174a
C_8H_7Cl	(Z)-C$_6$H$_5$CH=CHCl	BuLi/−115°/2 min	CO_2	Phenylacetylene (8), I (13)	169
	(E)-C$_6$H$_5$CH=CHCl	BuLi/THF, ether, hexane/−80°/ 15 min	CO_2	(E)-2-Chlorocinnamic acid (90)	169

Formula	Substrate	Conditions	Electrophile	Product (Yield, %)	Refs.
		”	C_2H_5OD	(E)-β-*d*-β-Chlorostyrene (80)	169
$C_{14}H_9Cl_3$	$(o\text{-}ClC_6H_4)_2C{=}CHCl$	BuLi/THF, ether, pet. ether/−108°/45 min	CO_2	2-Chloro-3,3-bis(*o*-chlorophenyl)acrylic acid (76)	171
	$(p\text{-}ClC_6H_4)_2C{=}CHCl$	BuLi/THF, ether, pet. ether/−41°/20 min	CO_2	2-Chloro-3,3-bis(*p*-chlorophenyl)acrylic acid (86)	171, 170
$C_{14}H_{10}Cl$	$(p\text{-}ClC_6H_4)(C_6H_5)C{=}CHCl$ (Cl *cis* to *p*-ClC_6H_4)	BuLi/THF/−108°/40 min	CO_2	(Z)-2-Chloro-3-(*p*-chlorophenyl)-3-phenylacrylic acid (84)	170
	$(C_6H_5)(p\text{-}ClC_6H_4)C{=}CHCl$ (Cl *cis* to C_6H_5)	BuLi/THF/−108°/40 min	CO_2	(E)-2-Chloro-3-(*p*-chlorophenyl)-3-phenylacrylic acid (87)	170
$C_{14}H_{11}Cl$	$(C_6H_5)_2C{=}CHCl$	BuLi/THF, ether, pet. ether/−93°/17 min	Br_2	$(C_6H_5)_2C{=}C(Cl)X$, X = Br (94)	358
		BuLi/THF, ether, pet. ether/−93°/10 min	I_2	X = I (98)	358

TABLE XXIX. Fluoro-, Chloro-, and Bromoalkenes (Alpha) (*Continued*)

Formula	Compound Lithiated	Conditions	Substrate	Product and Yield (%)	Refs.
$C_{14}H_{11}Cl$ (*Contd.*)	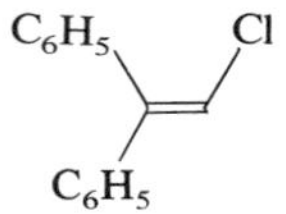 C_6H_5 Cl C_6H_5	BuLi/THF, ether/ −93°/2 hr	$HgCl_2$	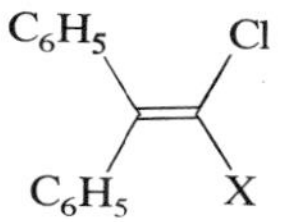 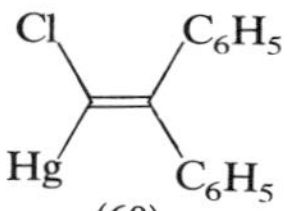 C_6H_5 Cl C_6H_5 X; X = Cl C_6H_5 Hg C_6H_5 (69)	358
		BuLi/THF/−71°/ 1 hr	CO_2	X = CO_2H (83)	171, 358, 170
		BuLi/THF, ether/ −93°/2 hr	CH_3I	X = CH_3 (86)	358
		BuLi/THF/−100°/ 30 min	CH_3OCH_2I	X = CH_2OCH_3 (30)	468
		”	$CH_2{=}CHCH_2I$	X = $CH_2CH{=}CH_2$ (66)	468
		BuLi/THF/−100°/ 30 min	C_6H_5COI	X = COC_6H_5 (62)	468
		”	$C_6H_5CH_2I$	X = $CH_2C_6H_5$ (68)	468
$C_{14}H_{21}Cl$	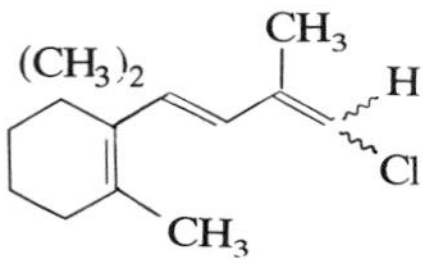 $(CH_3)_2$ CH_3 H Cl CH_3	BuLi/THF, ether, pet. ether/ −105°/15 hr	CO_2	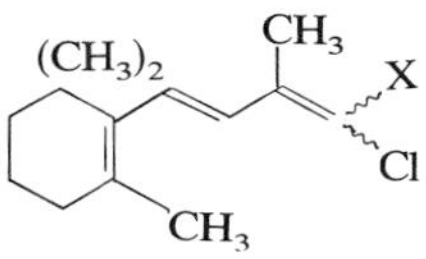 $(CH_3)_2$ CH_3 X Cl CH_3; X = CO_2H (82)	469

		”	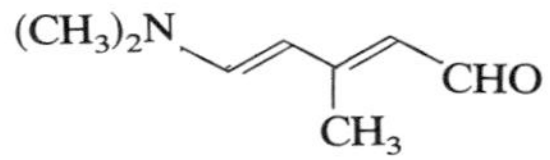 ($HClO_4$)	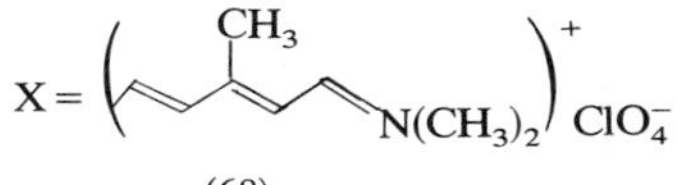 (68)	469
$C_{16}H_{15}Cl$	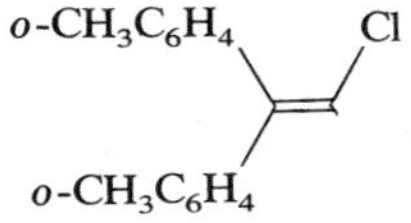	BuLi/THF, ether, pet. ether/ −108°/50 min	CO_2	2-Chloro-3,3-di-*o*-tolylacrylic acid (45)	171
	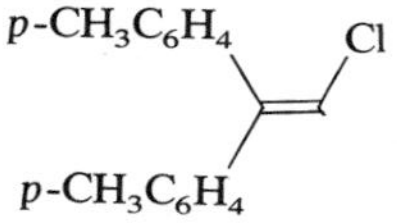	BuLi/THF/−108°/ 45 min	CO_2	2-Chloro-3,3-di-*p*-tolylacrylic acid (78)	170
	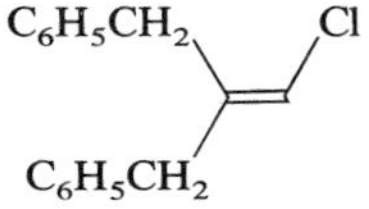	BuLi/−100°	I_2	2-Benzyl-1-chloro-1-iodo-3-phenylpropene (97)	470
		BuLi/THF, ether, pet. ether/ −100°	CO_2	3-Benzyl-2-chloro-4-phenylcrotonic acid (93)	471, 470
$C_{16}H_{15}ClO_2$	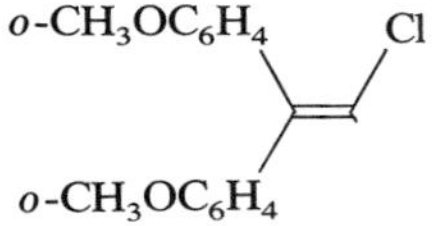	BuLi/THF/−120°/ 35 min	CO_2	2-Chloro-3,3-bis(*o*-methoxyphenyl)acrylic acid (61)	171, 170
	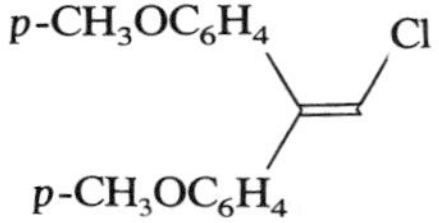	BuLi/THF, ether, pet. ether/ −108°/45 min	CO_2	2-Chloro-3,3-bis(*p*-methoxyphenyl)acrylic acid (67)	171, 170

Note: References 360–607 are on pp. 355–360.

TABLE XXX. MISCELLANEOUS MESOIONIC COMPOUNDS

Formula	Compound Lithiated	Conditions	Substrate	Product and Yield (%)	Refs.
$C_5H_4N_4$	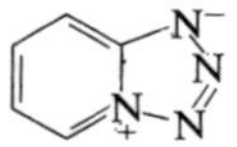	C_6H_5Li/ether/ 25°/3 hr*	CO_2	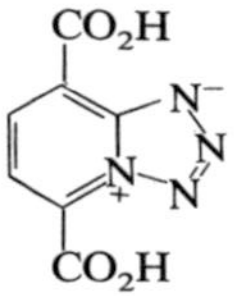 CO_2H, CO_2H (28)	472
$C_6H_6N_4$	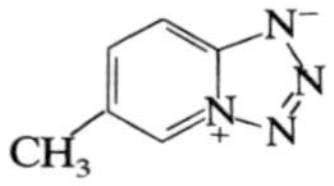 CH_3	C_6H_5Li/ether/ 25°/3 hr*	CO_2	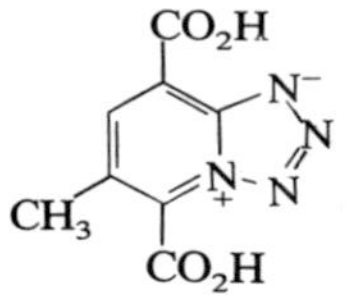 CO_2H, CH_3, CO_2H (11)	472
	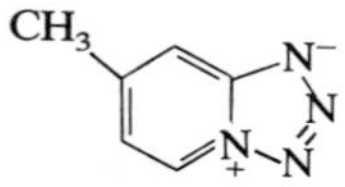 CH_3	C_6H_5Li/ether/ 25°/3 hr*	CO_2	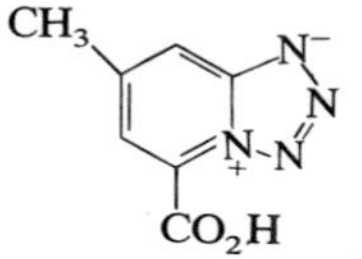 CH_3, CO_2H (5)*	472
	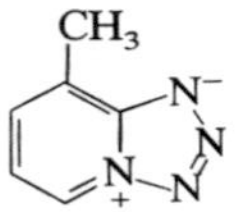 CH_3	C_6H_5Li/ether/ 25°/3 hr*	CO_2	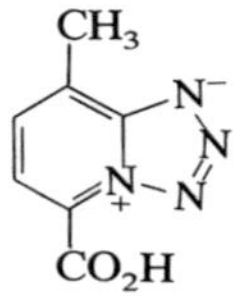 CH_3, CO_2H (18)	472
$C_7H_5N_3O_2$	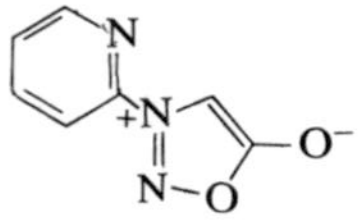	BuLi/ether, benzene/−20°/ 1 hr	CH_3COCH_3	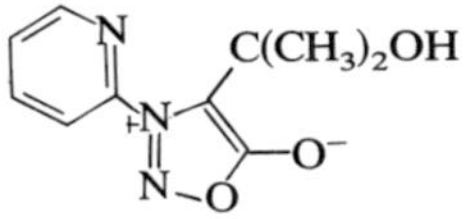 $C(CH_3)_2OH$ (64)	473

Formula	Reactant	Conditions	Electrophile	Product(s) and Yield(s) (%)		Refs.
$C_8H_6N_2O_2$	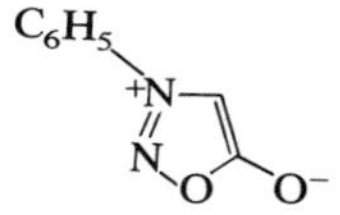 3-C_6H_5-sydnone (C_6H_5–N^+, N, O, O^-)	BuLi/ether, benzene/−20°/ 1 hr	CH_3CHO	3-C_6H_5-4-X-sydnone (C_6H_5–N^+, N, O, O^-, X)	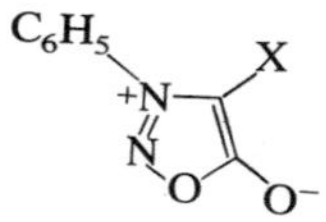 $X = CH(OH)CH_3$ (43)	473
		”	$CH_3COC_3H_7$-i		$X = C(OH)(CH_3)C_3H_7$-i (39)	473
		BuLi/THF/−20°/ 1 hr	$(C_6H_5)_2CO$		$X = C(C_6H_5)_2OH$ (84)	474
$C_8H_{12}N_2O_2$	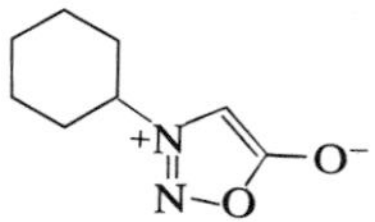 3-cyclohexylsydnone (N^+, N–O, O^-)	BuLi/ether, benzene/−20°/ 1 hr	CH_3COCH_3	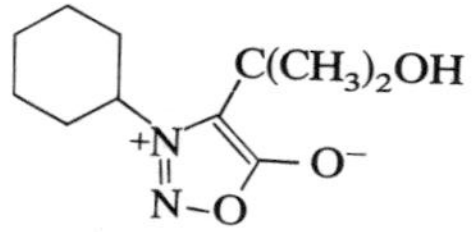 3-cyclohexyl-4-$C(CH_3)_2OH$-sydnone (N^+, N–O, O^-) (57)		473

Note: References 360–607 are on pp. 355–360.

TABLE XXXI. MONO-, DI-, AND TRIARYLAMINES (ORTHO)

Formula	Compound Lithiated	Conditions	Substrate	Product and Yield (%)	Refs.
$C_8H_{11}N$	$C_6H_5N(CH_3)_2$	BuLi/hexane/ reflux/16 hr	CF_3COCF_3	2-X-$C_6H_4N(CH_3)_2$; $X = C(CF_3)_2OH$ (49)	179
		"	$CH_3COC_2H_5$	$X = C(CH_3)(OH)C_2H_5$ (20)	179
		"	Cyclohexanone	X = 1-hydroxycyclohexyl (31)	179
		"	Cyclohexene oxide	X = cyclohexen-1-yl (14), 2-substituted cyclohexanol (7)	179
		"	$C_6H_5COCH_3$	$X = C(CH_3)(OH)C_6H_5$ (34)	179
		BuLi/hexane, TMEDA/25°/4 hr	$(C_6H_5)_2CO$	$X = C(C_6H_5)_2OH$ (71)	63, 179, 180

$C_9H_{13}N$	o-$CH_3C_6H_4N(CH_3)_2$	BuLi/ether/25°/ 40 hr	D_2O	2-CH_2D-$C_6H_4N(CH_3)_2$ (I), 2-CH_2D-6-D-$C_6H_3N(CH_3)_2$ (II) (I+II, 94) (I:II, 1.5:1)	63
	p-$CH_3C_6H_4N(CH_3)_2$	BuLi/hexane, TMEDA/25°/ 3 hr	D_2O	2-d-N,N-Dimethyl-p-toluidine (>90)	63
		BuLi/hexane, TMEDA/25°/4 hr	$(C_6H_5)_2CO$	[6-(Dimethylamino)-m-tolyl]diphenyl-methanol (80)	63, 359
$C_9H_{13}NO$	m-$CH_3OC_6H_4N(CH_3)_2$	BuLi/ether/35°/ 12 hr	$(C_6H_5)_2CO$	[2-(Dimethylamino)-6-methoxyphenyl]-diphenylmethanol (71)	19
$C_{10}H_9N$	1-Phenylpyrrole (C_6H_5–N pyrrole)	BuLi/ether/25°/ 8 hr	CO_2	1-Phenylpyrrole-2-carboxylic acid (14)	81
		2 BuLi/ether/ reflux/14 hr	”	9H-Pyrrolo[1,2-a]indol-9-one (5)	81
	1-Naphthylamine (NH_2)	BuLi/ether/ reflux/50 hr	CO_2	Benz[cd]indol-2(1H)-one (O, NH) (10)	486
$C_{11}H_{14}ClNO$	p-$ClC_6H_4NHCOC_4H_9$-t	BuLi/THF/0°/ 2 hr	CH_3I	4′-Chloro-o-pivalotoluidide (71)	181a
		”	$(CH_3S)_2$	4′-Chloro-2′-(methylthio)pivalanilide (79)	181a

TABLE XXXI. Mono-, Di-, and Triarylamines (Ortho) (*Continued*)

Formula	Compound Lithiated	Conditions	Substrate	Product and Yield (%)	Refs.
$C_{11}H_{14}ClNO$ (*Contd.*)	p-$ClC_6H_4NHCOC_4H_9$-t	,,	o-FC_6H_4CN	2-t-Butyl-6-chloro-4-(o-fluorophenyl)-quinazoline (57)	181a
$C_{11}H_{15}NO$	$C_6H_5NHCOC_4H_9$-t	BuLi/ether, THF/ 25°/18 hr	$(CH_3S)_2$	2′-(Methylthio)pivalanilide (78)	181a
		,,	DMF	2′-Formylpivalanilide (53)	181a
$C_{12}H_9NS$	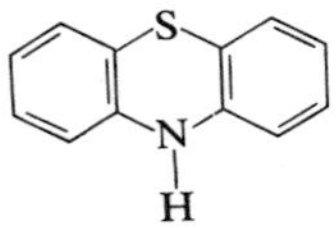	BuLi/ether/25°/ 30 hr	CO_2	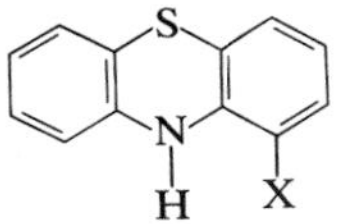 $X = CO_2H$ (53)	177
		,,	CH_3CO_2Li	$X = COCH_3$ (40)	177
		,,	$C_2H_5CO_2Li$	$X = COC_2H_5$ (33)	177
		,,	$C_6H_5CO_2Li$	$X = COC_6H_5$ (41)	177
		BuLi	$(C_6H_5)_2CO$	$X = C(C_6H_5)_2OH$ (70)	178
$C_{12}H_{17}NO_2$	m-$CH_3OC_6H_4NHCOC_4H_9$-t	BuLi/THF/0°/ 2 hr	$(CH_3S)_2$	3′-Methoxy-2′-(methylthio)pivalanilide (82)	181a
		,,	C_6H_5CHO	3′-Methoxy-2′-[(α-phenyl)hydroxymethyl]-pivalanilide (79)	181a
	p-$CH_3OC_6H_4NHCOC_4H_9$-t	BuLi/ether, THF/25°/18 hr	$(CH_3S)_2$	4′-Methoxy-2′-(methylthio)pivalanilide (53), 2′,5′-bis(methylthio)-4′-methoxypivalanilide (28), 4′-methoxy-3′-(methylthio)-pivalanilide (15)	181a
$C_{13}H_{11}NS$	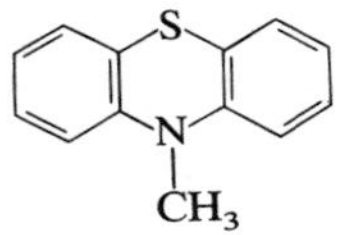	BuLi/ether/30 hr	CO_2	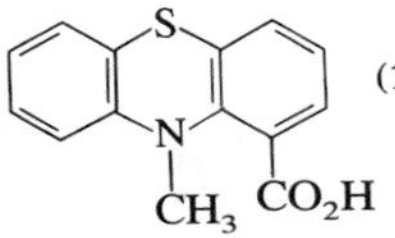 (14), 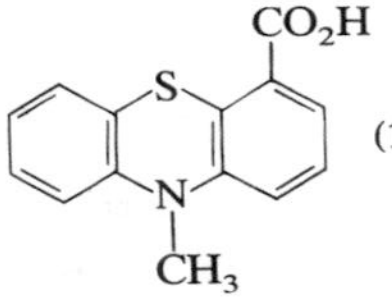 (16)	177, 183a

$C_{14}H_{11}N$	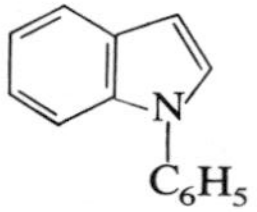 [structure: 1-phenylindole, N–C_6H_5]	BuLi/ether/ reflux/12 hr	CO_2	1-(*o*-Carboxyphenyl)indole-2-carboxylic acid (15), 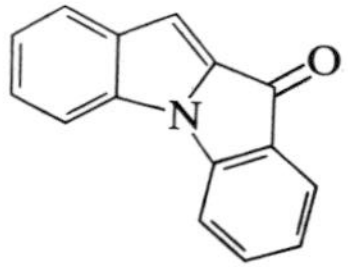 [structure] (42)	83
$C_{14}H_{13}N$	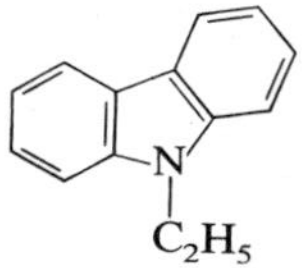 [structure: 9-ethylcarbazole, N–C_2H_5]	BuLi/THF, ether/ 25°/5 hr	CO_2	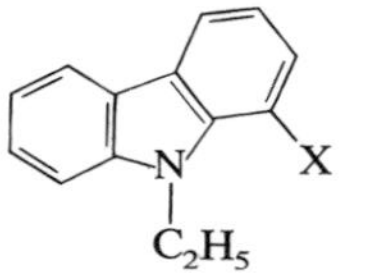 [structure: 9-ethylcarbazole with X, N–C_2H_5] $X = CO_2H$ (21)	560
		BuLi/ether/ reflux/44 hr	$(C_6H_5)_3SiBr$	$X = Si(C_6H_5)_3$ (12)	452
$C_{14}H_{13}NS$	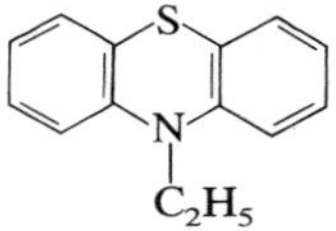 [structure: 10-ethylphenothiazine, N–C_2H_5]	BuLi/ether/ 30 hr	CO_2	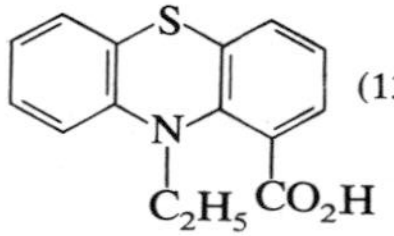 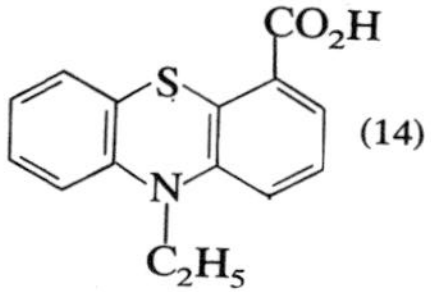 [structure: CO_2H, N–C_2H_5] (13), [structure: CO_2H, N–C_2H_5] (14)	177, 183a
$C_{16}H_{11}NS$	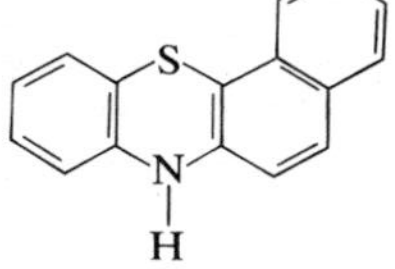 [structure: benzo-phenothiazine, N–H]	BuLi/ether/ reflux/24 hr	CO_2	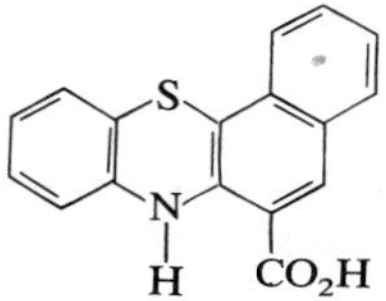 [structure: N–H, CO_2H] (41)	183b

TABLE XXXI. MONO-, DI-, AND TRIARYLAMINES (ORTHO) (*Continued*)

Formula	Compound Lithiated	Conditions	Substrate	Product and Yield (%)	Refs.
$C_{16}H_{11}NS$ (*Contd.*)	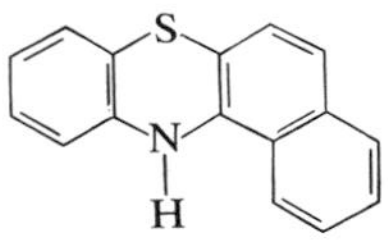 S, N, H	BuLi/ether/ 25°	CO_2	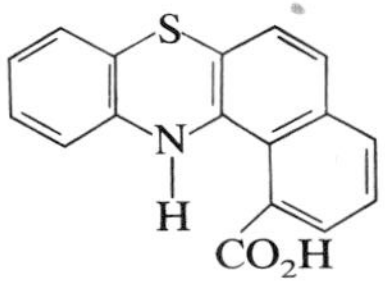 S, N, H, CO_2H (94)	176
$C_{16}H_{13}N$	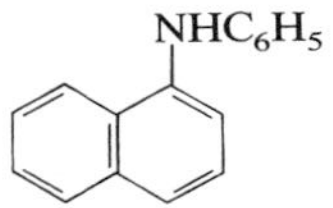 NHC_6H_5	BuLi/ether/25°/ 20 hr	CO_2	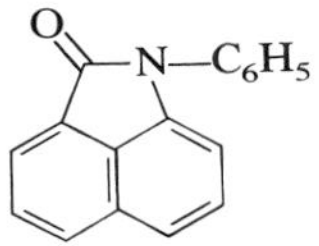 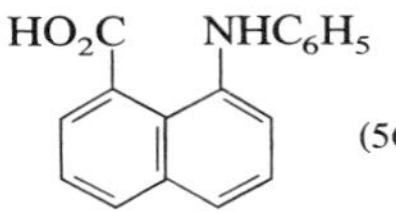 O, N—C_6H_5 (25), HO_2C NHC_6H_5 (56)	181b
	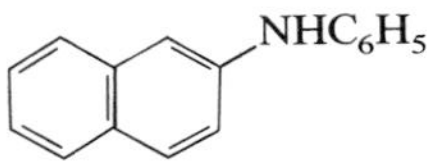 NHC_6H_5	BuLi/ether/25°/ 20 hr	CO_2	2-Anilino-3-naphthoic acid (53)	181b
$C_{20}H_{13}NS$	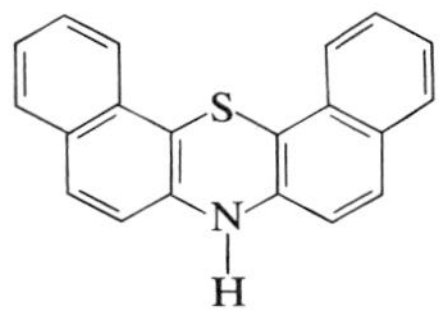 S, N, H	BuLi/ether/ reflux/4.5 hr	CO_2	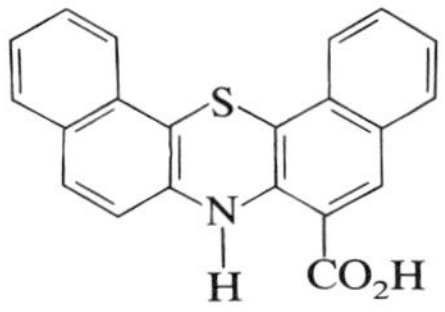 S, N, H, CO_2H (77)	182

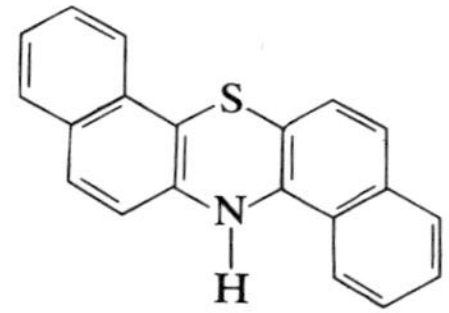	BuLi/ether/ reflux/5 hr	CO_2	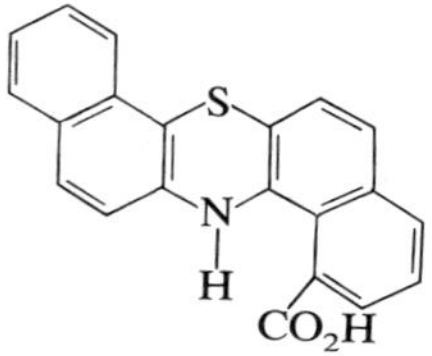 CO_2H (80)	182
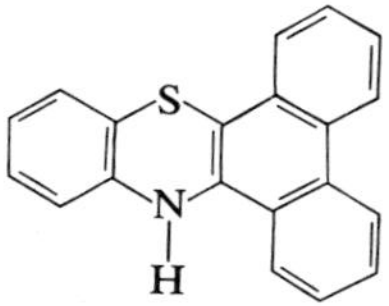	BuLi/ether/ 25°/23 hr	CO_2	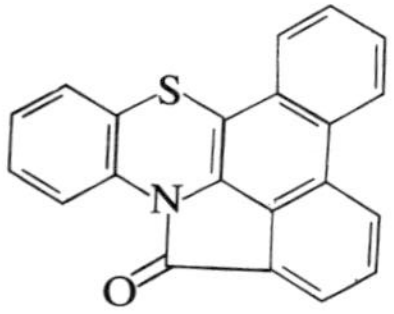 (85)	182

Note: References 360–607 are on pp. 355–360.

TABLE XXXII. α-LITHIO-(N-ALKYLIDENE)ARYLAMINES (ARYL ISOCYANIDES) (ORTHO)

Formula	Compound Lithiated	Conditions	Substrate	Product and Yield (%)		Ref.
$C_{11}H_{14}LiN$	$C_6H_5N{=}C(Li)C_4H_9$-*t*[a]	*t*-BuLi/ether, TMEDA/25°/4 hr	O_2	$N{=}C(X)C_4H_9$-*t*, X	X = OH (29)	184
		”	CH_3I		X = CH_3 (92)	184
		”	CO_2 (H_3O^+)	Anthranilic acid (54)		184
		”	SCl_2	N, X, C_4H_9-*t*	X = S (65)	184
		”	$(CH_3)_2SiCl_2$		X = $Si(CH_3)_2$ (53)	184
		”	$(CH_3)_2GeCl_2$		X = $Ge(CH_3)_2$ (68)	184
		”	$(CH_3)_2SnCl_2$		X = $Sn(CH_3)_2$ (41)	184
		”	$C_6H_5PCl_2$		X = PC_6H_5 (52)	184
		”	$(C_6H_5)_2SiCl_2$		X = $Si(C_6H_5)_2$ (63)	184

[a] This compound was formed by the reaction of phenyl isocyanide and *t*-BuLi.[184]

TABLE XXXIII. Aralkylamines and Allylamines (Ortho, Beta)

Formula	Compound Lithiated	Conditions	Substrate	Product and Yield (%)		Refs.
$C_7H_{11}NS$	3-thienyl-$CH_2N(CH_3)_2$	BuLi/ether/ 25°–35°/4 hr	D_2O	2-X-3-thienyl-$CH_2N(CH_3)_2$	X = D (80)	134, 43
		BuLi/ether/ 25°–35°/1 hr	DMF		X = CHO (75)	134, 43
		”	C_6H_5CN (H_3O^+)		X = COC_6H_5 (54)	134, 43
		”	$(C_6H_5)_2CO$		X = $C(C_6H_5)_2OH$ (67)	134, 43
		”	p-$CH_3C_6H_5SO_2Br$		X = Br (32)	134
		”	p-$CH_3C_6H_4SO_2Cl$		X = Cl (25)	134
$C_7H_{12}N_2O$	5-CH_3-3-isoxazolyl-$CH_2N(CH_3)_2$	BuLi/THF	CH_3I	4,5-di-CH_3-3-isoxazolyl-$CH_2N(CH_3)_2$ (I), (I + II, –)	5-C_2H_5-3-isoxazolyl-$CH_2N(CH_3)_2$ (II) (I : II, 1 : 1)	186
$C_8H_{11}N$	$C_6H_5CH_2NHCH_3$	BuLi/ether, TMEDA/ 5 hr	D_2O	2-X-$C_6H_4CH_2NHCH_3$	X = D (100)	188
		”	Cyclohexanone		X = 1-hydroxycyclohexyl (51)	188
		”	C_6H_5CHO		X = $CH(OH)C_6H_5$ (63)	188
		”	$C_6H_5COCH_3$		X = $C(CH_3)(OH)C_6H_5$ (40)	188
		”	$C_6H_5COC_2H_5$		X = $C(C_2H_5)(OH)C_6H_5$ (43)	188

TABLE XXXIII. ARALKYLAMINES AND ALLYLAMINES (ORTHO, BETA) (*Continued*)

Formula	Compound Lithiated	Conditions	Substrate	Product and Yield (%)	Refs.
$C_8H_{11}N$ (*Contd.*)	$C_6H_5CH_2NHCH_3$	"	$(C_6H_5)_2CO$	$X = C(C_6H_5)_2OH$ (48–52)	188
		BuLi/hexane, TMEDA/ 0.5 hr	C_6H_5CHO	2-[(Methylamino)methyl]-α,α′-diphenyl- *m*-xylene-α,α′-diol (65–75)	188
$C_8H_{13}NS$	CH_3 S $CH_2N(CH_3)_2$	BuLi/ether/ 25°–35°/4 hr	$(C_6H_5)_2CO$	2-[(Dimethylamino)methyl]-α,α-diphenyl-5-methyl-3-thiophenemethanol (65)	43, 134
$C_9H_{11}Cl_2N$	$2,4\text{-}Cl_2C_6H_3CH_2N(CH_3)_2$	BuLi/ether/ 25°/10 min	*o*-ClC_6H_4CN	2′-Chloro-3,5-dichloro-2-[(dimethylamino)methyl]benzophenone (70)	23
$C_9H_{12}ClN$	*o*-$ClC_6H_4CH_2N(CH_3)_2$	BuLi/ether/ 25°	Cu Br	Cl, $CH_2N(CH_3)_2$, X $X = Cu$ (25)	190
		BuLi/ether/ 25°/24 hr	$(C_6H_5)_2CO$	$X = C(C_6H_5)_2OH$ (81)	45
	p-$ClC_6H_4CH_2N(CH_3)_2$	BuLi/ether/ 25°	CuBr	$CH_2N(CH_3)_2$, Cl, X $X = Cu$ (41)	190
		BuLi/ether/ 0°/12 hr	$(CH_3S)_2$	$X = SCH_3$ (77)	23
		BuLi/ether/ 25°/3 hr	$ClCON(CH_3)_2$	$X = CON(CH_3)_2$ (85)	24
		"	*t*-C_4H_9NCO	$X = CONHC_4H_9\text{-}t$ (29)	24
		BuLi/ether/ 0°/12 hr	*o*-ClC_6H_4CN (H_3O^+)	$X = COC_6H_4Cl\text{-}o$ (73)	23
		"	*o*-FC_6H_4CN (H_3O^+)	$X = COC_6H_4F\text{-}o$ (78)	23

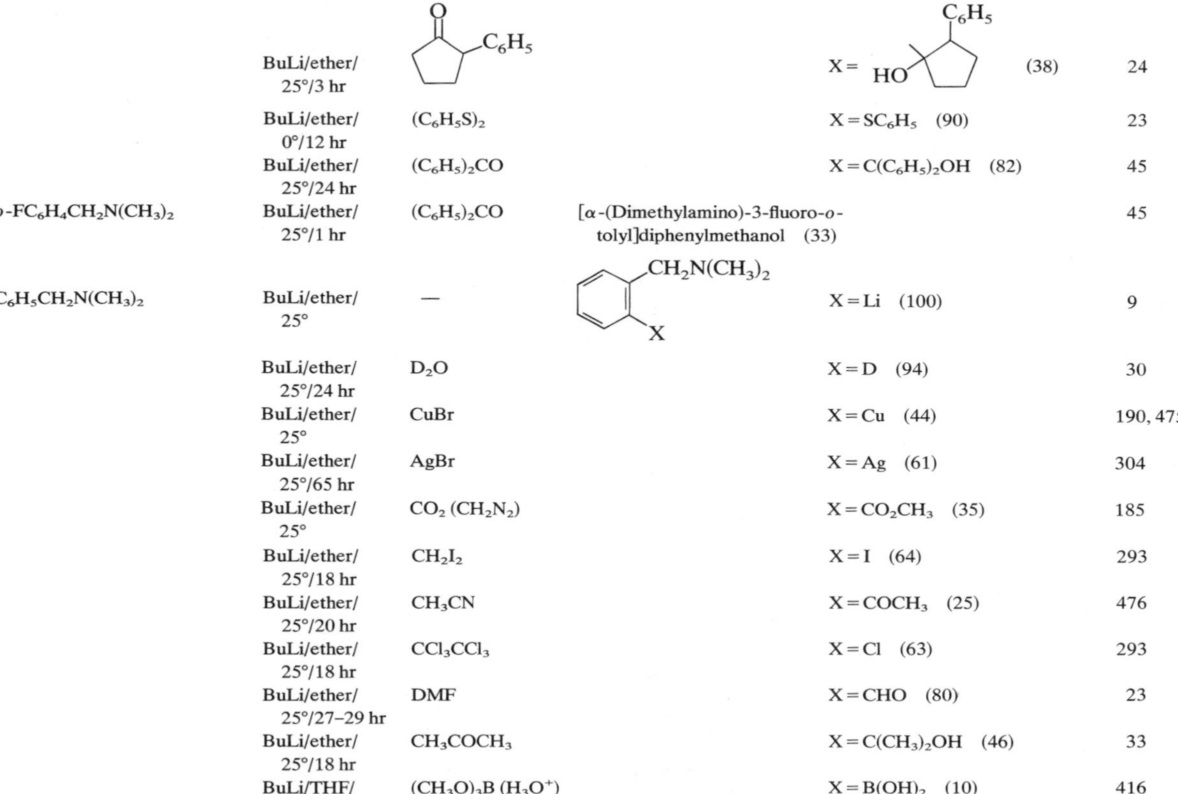

		BuLi/ether/ 25°/3 hr	2-phenylcyclopentanone (structure)	X = 1-hydroxy-2-phenylcyclopentyl (structure) (38)	24
		BuLi/ether/ 0°/12 hr	$(C_6H_5S)_2$	$X = SC_6H_5$ (90)	23
		BuLi/ether/ 25°/24 hr	$(C_6H_5)_2CO$	$X = C(C_6H_5)_2OH$ (82)	45
$C_9H_{12}FN$	o-$FC_6H_4CH_2N(CH_3)_2$	BuLi/ether/ 25°/1 hr	$(C_6H_5)_2CO$	[α-(Dimethylamino)-3-fluoro-*o*-tolyl]diphenylmethanol (33)	45
$C_9H_{13}N$	$C_6H_5CH_2N(CH_3)_2$	BuLi/ether/ 25°	—	o-$XC_6H_4CH_2N(CH_3)_2$ (structure), X = Li (100)	9
		BuLi/ether/ 25°/24 hr	D_2O	X = D (94)	30
		BuLi/ether/ 25°	CuBr	X = Cu (44)	190, 475
		BuLi/ether/ 25°/65 hr	AgBr	X = Ag (61)	304
		BuLi/ether/ 25°	CO_2 (CH_2N_2)	$X = CO_2CH_3$ (35)	185
		BuLi/ether/ 25°/18 hr	CH_2I_2	X = I (64)	293
		BuLi/ether/ 25°/20 hr	CH_3CN	$X = COCH_3$ (25)	476
		BuLi/ether/ 25°/18 hr	CCl_3CCl_3	X = Cl (63)	293
		BuLi/ether/ 25°/27–29 hr	DMF	X = CHO (80)	23
		BuLi/ether/ 25°/18 hr	CH_3COCH_3	$X = C(CH_3)_2OH$ (46)	33
		BuLi/THF/ 25°–45°/6 hr	$(CH_3O)_3B$ (H_3O^+)	$X = B(OH)_2$ (10)	416

TABLE XXXIII. ARALKYLAMINES AND ALLYLAMINES (ORTHO, BETA) (*Continued*)

Formula	Compound Lithiated	Conditions	Substrate	Product and Yield (%)	Refs.
$C_9H_{13}N$ (*Contd.*)	$C_6H_5CH_2N(CH_3)_2$	BuLi/ether/ 25°/18 hr	$(CH_3)_2CBrCBr(CH_3)_2$	2-X-$C_6H_4CH_2N(CH_3)_2$, X = Br (69)	293
		—	2-methyl-1,3-dioxolane-2-acetaldehyde (CH_3, CH_2CHO)	X = CH(OH)CH_2-(2-methyl-1,3-dioxolan-2-yl) (−)	477
		BuLi/ether/ 25°/18 hr	cyclohexene oxide	X = 2-hydroxycyclohexyl (36)	33
		"	o-ClC_6H_4CHO	X = CH(OH)C_6H_4Cl-o (80)	33
		BuLi/ether/ 25°/27–29 hr	m-ClC_6H_4CN (H_3O^+)	X = COC_6H_4Cl-m (50)	23
		"	o-ClC_6H_4CN (H_3O^+)	X = COC_6H_4Cl-o (90)	23
		"	p-ClC_6H_4CN (H_3O^+)	X = COC_6H_4Cl-p (85)	23
		BuLi/ether/ 25°/30 hr	C_6H_5CHO	X = CH(OH)C_6H_5 (78)	33
		BuLi/ether/ 25°/18 hr	C_6H_5CN (H_3O^+)	X = COC_6H_5 (63)	33, 344
		BuLi/ether/ 25°/30 hr	C_6H_5NCO	X = $CONHC_6H_5$ (65)	33
		BuLi/ether/ 25°/18 hr	p-$CH_3OC_6H_4CHO$	X = CH(OH)$C_6H_4OCH_3$-p (78)	33
		BuLi/ether/ 25°/ "overnight"	2-phenylcyclopentanone (C_6H_5)	X = 1-hydroxy-2-phenylcyclopentyl (HO, C_6H_5) (28)	24

BuLi/ether/ 25°/30 hr	$C_6H_4COC_4H_9$-*n*	$X = C(C_6H_5)(OH)C_4H_9$-*n* ("good")	33
BuLi/ether/ 25°/18 hr	3,4,5-$(CH_3O)_3C_6H_2CHO$	$X = CH(OH)C_6H_2(OCH_3)_3$-3,4,5 (56)	33
BuLi/ether/ 25°/27 hr	3,4,5-$(CH_3O)_3C_6H_2CN$ (H_3O^+)	$X = COC_6H_2(OCH_3)_3$-3,4,5 (90)	23
BuLi/ether/ 25°/30 hr	$(C_6H_5)_2CO$	$X = CO(C_6H_5)_2OH$ (77)	33, 45, 180
BuLi/ether/ 25°/30 hr	$C_6H_5CO_2C_2H_5$	Bis[(α-dimethylamino)-*o*-tolyl]phenylmethanol (59)	33
BuLi/hexane/ reflux/6 hr	$CoCl_2$	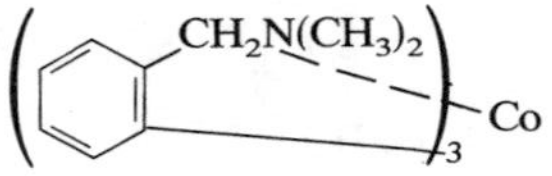 $CH_2N(CH_3)_2$ ... $)_3$Co (45)	300
BuLi/ether/ 25°/90 hr	½$AgBr_2$	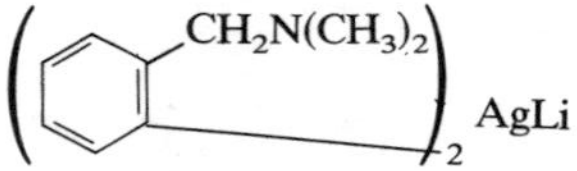 $CH_2N(CH_3)_2$ ... $)_2$AgLi (10)	305
BuLi/hexane/ reflux/16 hr	$Fe(CO)_5$	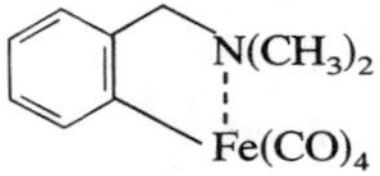 $N(CH_3)_2$, $Fe(CO)_4$ (−)	299
,,	$Ni(CO)_4$	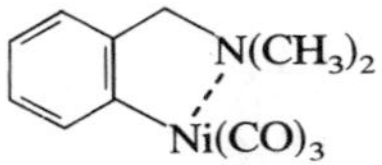 $N(CH_3)_2$, $Ni(CO)_3$ (−)	299
BuLi	$[(C_2H_5)_3P]_2NiCl_2$	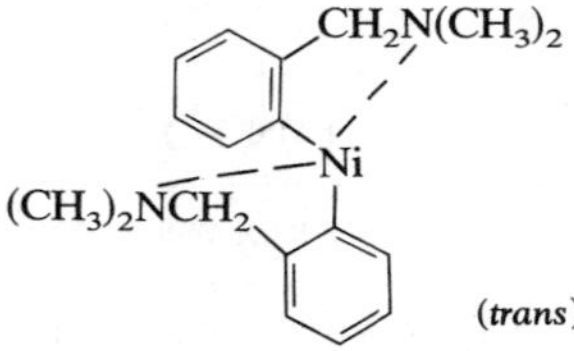 $CH_2N(CH_3)_2$, Ni, $(CH_3)_2NCH_2$ (*trans*) (60)	303

TABLE XXXIII. ARALKYLAMINES AND ALLYLAMINES (ORTHO, BETA) (*Continued*)

Formula	Compound Lithiated	Conditions	Substrate	Product and Yield (%)	Refs.
$C_9H_{13}N$ (*Contd.*)	$C_6H_5CH_2N(CH_3)_2$	BuLi/ether/ 25°	$(CH_3)_2SnBr_2$	2-[$CH_2N(CH_3)_2$]C_6H_4–$Sn(CH_3)RBr$ (N→Sn coordinated), R = CH_3 (50)	480
		,,	$(CH_3)C_6H_5SnBr_2$	R = C_6H_5 (90)	481
		BuLi/ether/ 25°	$(CH_3)_2SnBr_2$	(2-[$CH_2N(CH_3)_2$]C_6H_4)$_2Sn(CH_3)_2$ (–)	480
		BuLi	$[(C_2H_5)_2S]_2PdCl_2$	(2-[$CH_2N(CH_3)_2$]C_6H_4)$_2$Me, *cis*, Me = Pd (–)	303
		,,	$[(C_2H_5)_2S]_2PtCl_2$	Me = Pt (–)	303
		—	$(C_6H_5)_3PAuBr$	2-[$CH_2N(CH_3)_2$]$C_6H_4Au \cdot P(C_6H_5)_3$ (65)	307
$C_{10}H_{12}F_3N$	m-$CF_3C_6H_4CH_2N(CH_3)_2$	BuLi/ether/ 25°/1 hr	$(C_6H_5)_2CO$	[α^6-(Dimethylamino)-$\alpha^2,\alpha^2,\alpha^2$-trifluoro-2,6-xylyl]diphenylmethanol (72)	45
$C_{10}H_{13}NO_2$	3,4-(methylenedioxy)$C_6H_3CH_2N(CH_3)_2$	BuLi/ether/ 25°/1 hr	$(C_6H_5)_2CO$	[α-(Dimethylamino)-5,6-(methylenedioxy)-o-tolyl]diphenylmethanol (60)	483

$C_{10}H_{14}ClNS$	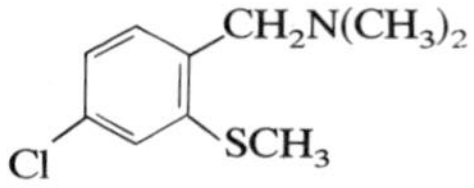 $CH_2N(CH_3)_2$, SCH_3, Cl	BuLi/ether/ 0°/2 hr	$(C_6H_5)_2CO$	[5-Chloro-α-(dimethylamino)-3-(methylthio)-*o*-tolyl]diphenylmethanol (35)	23
$C_{10}H_{15}N$	*p*-$CH_3C_6H_4CH_2N(CH_3)_2$	BuLi/ether/ 25°	CuBr	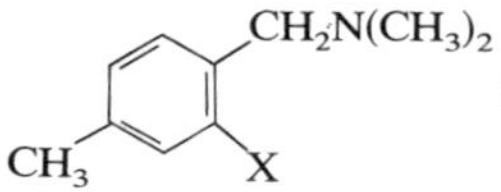 $CH_2N(CH_3)_2$, CH_3, X; X = Cu (66)	190, 39
		BuLi/ether/ 25°/30 hr	C_6H_5CN	X = COC_6H_5 (80)	23
		BuLi/ether/ 25°/"overnight"	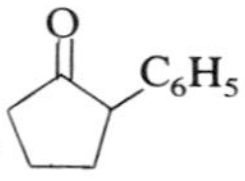 O, C_6H_5	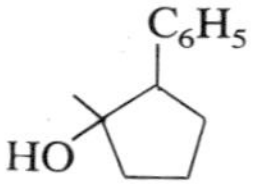 X = C_6H_5, HO (21)	24
		BuLi/ether/ 25°/24 hr	$(C_6H_5)_2CO$	X = $C(C_6H_5)_2OH$ (82)	45
$C_{10}H_{15}NO$	*m*-$CH_3OC_6H_4CH_2N(CH_3)_2$	BuLi/ether	Ethylene oxide	[2-(Dimethylamino)methyl]-6-methoxyphenethyl alcohol (—)	189
		BuLi/ether/ 27°/2 hr	$(C_6H_5)_2CO$	[α-(Dimethylamino)-6-methoxy-*o*-tolyl]diphenylmethanol (79)	19, 45, 189
	o-$CH_3OC_6H_4CH_2N(CH_3)_2$	BuLi/ether/ 27°/2 hr	$(C_6H_5)_2CO$	[α-(Dimethylamino)-2-methoxy-*m*-tolyl]diphenylmethanol (58), [α-(dimethylamino)-3-methoxy-*o*-tolyl]diphenylmethanol (<5)	19
	p-$CH_3OC_6H_4CH_2N(CH_3)_2$	BuLi/ether/ 27°/24 hr	D_2O	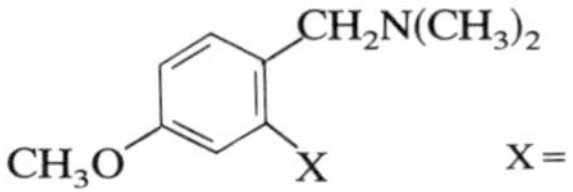 $CH_2N(CH_3)_2$, CH_3O, X (III); X = D (70)	19
		BuLi/ether/ 25°	CuBr	X = Cu (38)	190
		BuLi/ether	Ethylene oxide	X = $(CH_2)_2OH$ (—)	189

TABLE XXXIII. ARALKYLAMINES AND ALLYLAMINES (ORTHO, BETA) (*Continued*)

Formula	Compound Lithiated	Conditions	Substrate	Product and Yield (%)	Refs.
$C_{10}H_{15}NO$ (*Contd.*)	*p*-$CH_3OC_6H_4CH_2N(CH_3)_2$	BuLi/ether/ 25°/20 hr	Cyclopentanone	X 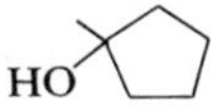 = HO (32)	24
		”	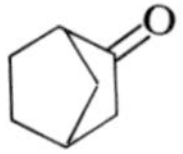 O	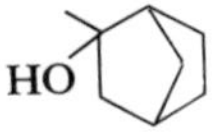 X = HO (39)	24
		”	$C_6H_5CH_2CHO$	X = $CH(OH)CH_2C_6H_5$ (10)	24
		”	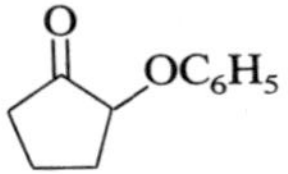 O, OC_6H_5	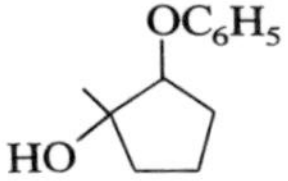 X = HO, OC_6H_5 (12)	24
		”	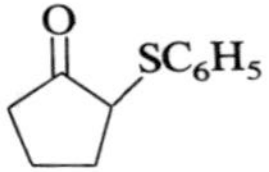 O, SC_6H_5	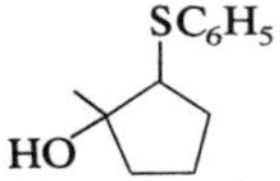 X = HO, SC_6H_5 (12)	24
		”	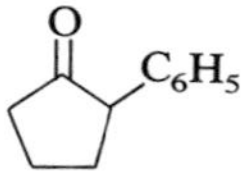 O, C_6H_5	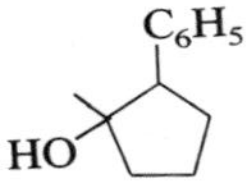 X = HO, C_6H_5 (20)	24
		”	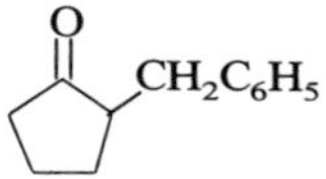 O, $CH_2C_6H_5$	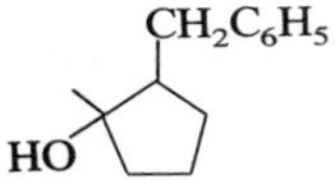 X = HO, $CH_2C_6H_5$ (31)	24
		”	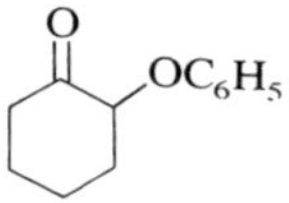 O, OC_6H_5	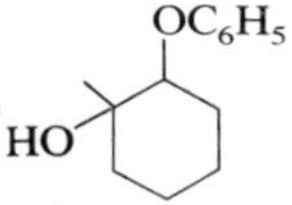 X = HO, OC_6H_5 (21)	24
		BuLi/ether/ 27°/24 hr	$(C_6H_5)_2CO$	X = $C(C_6H_5)_2OH$ (80)	19, 45, 180, 189

Formula	Substrate	Conditions	Electrophile	Product(s) (Yield, %)	Refs.
		BuLi/ether, TMEDA/ 27°/15 hr	D_2O	(III), CH_3O, X, $CH_2N(CH_3)_2$	
				(18) (48) X = D	19
		BuLi/ether, TMEDA/ 27°/2 hr	$(C_6H_5)_2CO$	(7) (55) $X = C(C_6H_5)_2OH$	19, 180
$C_{11}H_{14}ClNO$	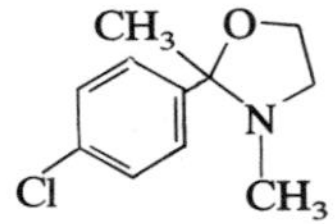 CH_3, O, N, Cl, CH_3	BuLi/ether/ 0°/4 hr	$(CH_3S)_2$	2-[4-Chloro-2-(methylthio)phenyl]-2,3-dimethyloxazolidine (80)	23
$C_{11}H_{15}NS$	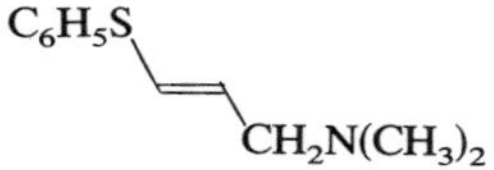 C_6H_5S, $CH_2N(CH_3)_2$	BuLi/ether/ 0°/1 hr	D_2O	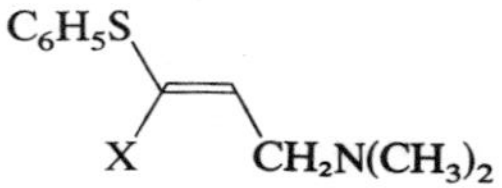 C_6H_5S, X, $CH_2N(CH_3)_2$ X = D (>90)	23
		"	$(CH_3S)_2$	$X = SCH_3$ (>90)	23
$C_{11}H_{15}NO$	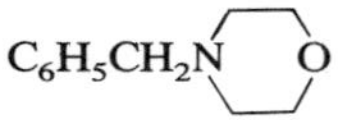 $C_6H_5CH_2N$, O	BuLi/ether/ 25°/20 hr	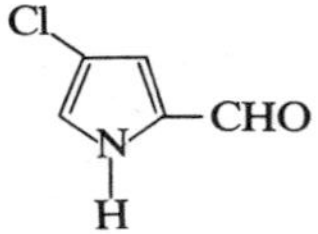 Cl, N, H, CHO	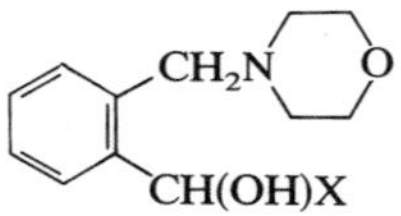 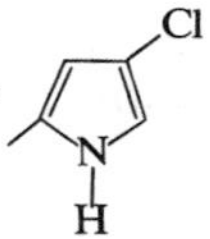 CH_2N, O, CH(OH)X X = Cl, N, H (−)	187
		"	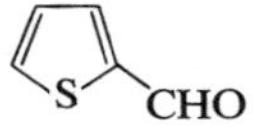 S, CHO	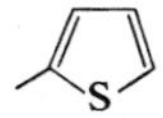 X = S (−)	187
		"	$3,4\text{-}Cl_2C_6H_3CHO$	$X = C_6H_3Cl_2\text{-}3,4$ (—)	187
$C_{11}H_{17}N$	$2,4\text{-}(CH_3)_2C_6H_3CH_2N(CH_3)_2$	BuLi/ether/ 25°/24 hr	$(C_6H_5)_2CO$	[2-[(Dimethylamino)methyl]-3,5-dimethylphenyl]diphenylmethanol (52)	45
	$C_6H_5C(CH_3)_2N(CH_3)_2$	BuLi/ether/ 18 hr	$(C_6H_5)_2CO$	[(α-(Dimethylamino)-*o*-cumenyl]-diphenylmethanol (57)	33

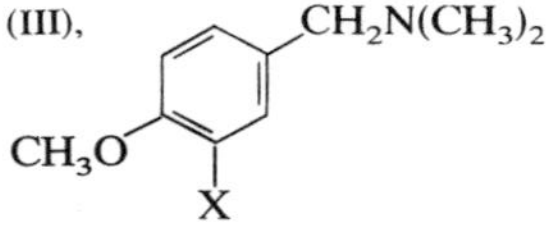

TABLE XXXIII. Aralkylamines and Allylamines (Ortho, Beta) (*Continued*)

Formula	Compound Lithiated	Conditions	Substrate	Product and Yield (%)	Refs.
$C_{11}H_{17}NO_2$	3,4-$(CH_3O)_2C_6H_3CH_2N(CH_3)_2$	BuLi/ether/ 0°/3 hr	I_2	2-X-3,4-$(CH_3O)_2C_6H_2CH_2N(CH_3)_2$: X = I (77)	23
		"	Ethylene oxide	X = $(CH_2)_2OH$ (60)	23
		"	CH_3CHO	X = $CH(OH)CH_3$ (65)	23
		"	CH_3OCH_2Cl	X = CH_2OCH_3 (66)	23
$C_{12}H_{17}N$	1-Methyl-2-phenylpiperidine (C_6H_5, N–CH_3 piperidine)	BuLi/ether/ 25°/20 hr	CH_3I	1-Methyl-2-(*o*-tolyl)piperidine (93)	24
$C_{12}H_{17}NO$	*p*-$CH_3C_6H_4CH_2N$(morpholino)	BuLi/ether/ 25°/20 hr	C_6H_5CHO	[5-Methyl-2-(4-morpholinomethyl)] benzhydrol (27)	187
$C_{12}H_{21}NSi$	*o*-$(CH_3)_3SiC_6H_4CH_2N(CH_3)_2$	BuLi/ether/ 25°	$(CH_3)_3SiCl$	N,N-Dimethyl-bis-2,6-(trimethylsilyl)benzylamine (—)	9
$C_{13}H_{13}N$	$C_6H_5CH_2NHC_6H_5$	BuLi/hexane, TMEDA/ 1.5–4 hr	CO_2	2-Phenylphthalimidine (8), *o*-X-$C_6H_4CH_2NHC_6H_5$: X = CO_2H (61)	188
		"	C_6H_5CHO	X = $CH(OH)C_6H_5$ (72–75)	188

Formula	Compound	Metalation Conditions	Reagent	Product(s) (Yield, %)	Refs.
		,,	$C_6H_5COCH_3$	$X = C(CH_3)(OH)C_6H_5$ (10)	188
		,,	fluorenone	X = 9-hydroxyfluoren-9-yl (OH) (93–98)	188
		,,	$(C_6H_5)_2CO$	$X = C(C_6H_5)_2OH$ (86)	188
$C_{13}H_{15}N$	1-naphthyl-$CH_2N(CH_3)_2$	BuLi/ether/ 25°/24 hr	D_2O	2-X-1-naphthyl-$CH_2N(CH_3)_2$ (IV), 8-X-1-naphthyl-$CH_2N(CH_3)_2$ (V)	
				(IV:V, 12:88) X = D	484
		,,	$(C_6H_5)_2CO$	(IV+V, 79) (IV:V, 8:92) $X = C(C_6H_5)_2OH$	191
		BuLi/ether/ 25°	$CO_2(O_2)$	N-Methyl-1,8-naphthalenedicarboximide (57)	185
	2-naphthyl-$CH_2N(CH_3)_2$	BuLi/ether/ 25°	D_2O	1-X-2-naphthyl-$CH_2N(CH_3)_2$ (VI), 3-X-2-naphthyl-$CH_2N(CH_3)_2$ (VII)	
				(VI:VII, 2:1) X = D	484
		,,	$CO_2(CH_2N_2)$	VI (29) VII (23) $X = CO_2CH_3$	185
		BuLi/ether/ 25°/48 hr	$(C_6H_5)_2CO$	(VI+VII, 79) (VI:VII, 43:58) $X = C(C_6H_5)_2OH$	191
$C_{13}H_{21}N$	p-t-$C_4H_9C_6H_4CH_2N(CH_3)_2$	BuLi/hexane/ reflux/6 hr	$CoCl_2$	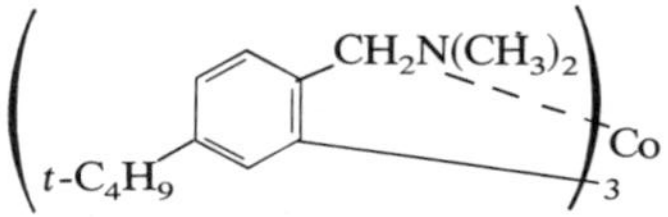 $(5\text{-}t\text{-}C_4H_9\text{-}2\text{-}[CH_2N(CH_3)_2]C_6H_3)_3Co$ (33)	300

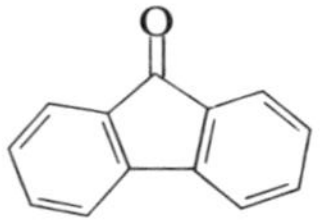

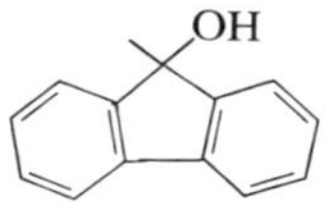

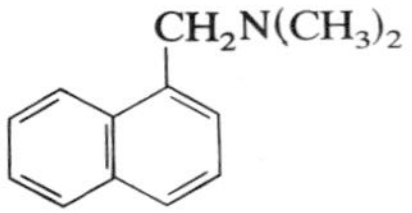

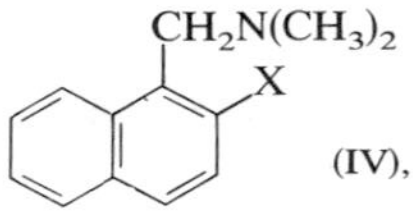

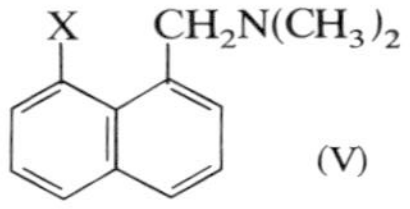

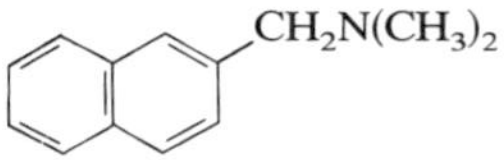

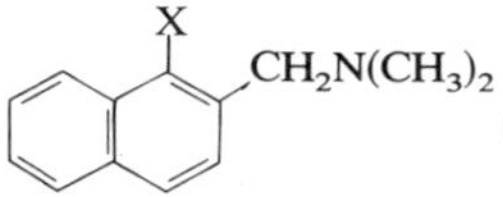

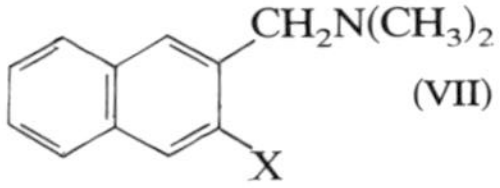

TABLE XXXIII. ARALKYLAMINES AND ALLYLAMINES (ORTHO, BETA) (*Continued*)

Formula	Compound Lithiated	Conditions	Substrate	Product and Yield (%)	Refs.
$C_{14}H_{17}NO$	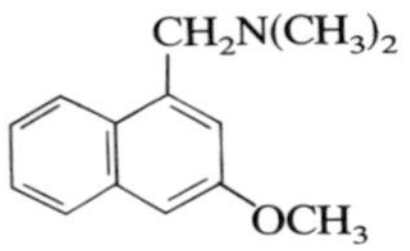 $CH_2N(CH_3)_2$, OCH_3	BuLi/ether/ 25°	$CO_2(CH_2N_2)$	Methyl [1-(dimethylamino)methyl]-3-methoxy-2-naphthoate (39)	185
	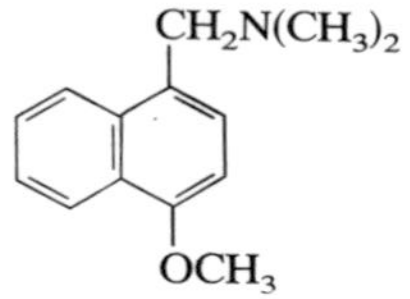 $CH_2N(CH_3)_2$, OCH_3	BuLi/25°/ 24 hr	CuBr	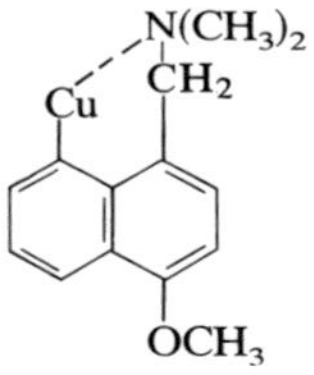 $N(CH_3)_2$, Cu, CH_2, OCH_3 (50)	190
	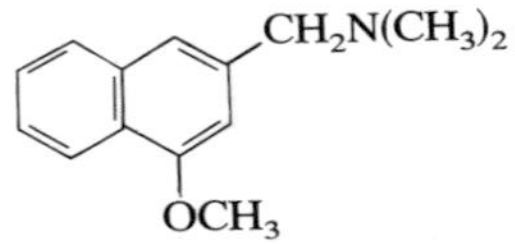 $CH_2N(CH_3)_2$, OCH_3	BuLi/ether/ 25°	$CO_2(CH_2N_2)$	Methyl [3-(dimethylamino)methyl]-1-methoxy-2-naphthoate (56)	185
	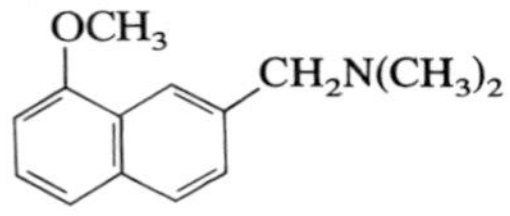 OCH_3, $CH_2N(CH_3)_2$	BuLi/ether/ 25°	$CO_2(CH_2N_2)$	Methyl [3-(dimethylamino)methyl]-5-methoxy-2-naphthoate (38)	185
$C_{15}H_{17}N$	$(C_6H_5CH_2)_2NCH_3$	BuLi/ether/ 25°/48 hr	D_2O	N-Methyl(dibenzyl)-2-*d*-amine (100)	30
		"	C_6H_5CHO	2-[(Benzylmethylamino)methyl]benzhydrol (56)	33

$C_{15}H_{19}NO_2$	OCH_3, $CH_2N(CH_3)_2$, OCH_3 (naphthalene structure)	BuLi/ether/ 25°	$CO_2(CH_2N_2)$	Methyl [3-(dimethylamino)methyl]-1,5-dimethoxy-2-naphthoate (52)	185
$C_{18}H_{24}N_2$	$[C_6H_5CH_2N(CH_3)CH_2]_2$	BuLi/25°/ 48 hr	D_2O	$C_6H_5CHDN(CH_3)(CH_2)_2N(CH_3)CH_2C_6H_5$ (30),	485
				$CH_2N(CH_3)(CH_2)_2N(CH_3)CH_2C_6H_5$, D (benzene structure) (65)	
$C_{23}H_{23}N$	$CH_2N(CH_3)_2$, C_6H_5, C_6H_5 (benzene structure)	BuLi/ether/ 25°/24 hr	$(C_6H_5)_2CO$	$C(C_6H_5)_2OH$, $CH_2N(CH_3)_2$, C_6H_5, C_6H_5 (benzene structure) (32)	45

Note: References 360–607 are on pp. 355–360.

TABLE XXXIV. 2-Arylethylamines (Ortho, Beta)

Formula	Compound Lithiated	Conditions	Substrate	Product and Yield (%)	Refs.
$C_9H_{13}NO$	$C_6H_5CH(OH)CH_2NHCH_3$	BuLi/ether/25°/ 24 hr	$(CH_3)_3SiCl$	α-[(Methylamino)methyl]-*o*-(trimethylsilyl)benzyl alcohol (21)	175
$C_{10}H_{11}NS$	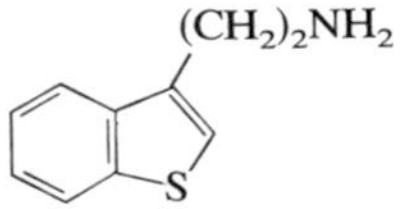 $(CH_2)_2NH_2$ (benzo[b]thiophene ring, S)	BuLi/THF/−78°/ 0.5 hr	D_2O	2-*d*-Benzo[b]thiophene-3-ethylamine (83)[a]	153
$C_{10}H_{15}N$	$C_6H_5(CH_2)_2N(CH_3)_2$	BuLi	C_6H_5CHO	2-[2-(Dimethylamino)ethyl]benzhydrol (—)	193
		BuLi/ether/25°/ 11 hr	$(C_6H_5)_2CO$	[*o*-2-(Dimethylamino)ethyl]phenyl]-diphenylmethanol (7)	194, 195, 193
$C_{10}H_{15}NO$	$C_6H_5CH(OH)CH_2N(CH_3)_2$	BuLi/ether/25°/ 24 hr	CH_3I	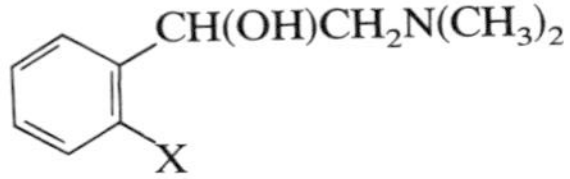 $CH(OH)CH_2N(CH_3)_2$, X (benzene ring) $X = CH_3$ (47)	175
		"	CH_2O	$X = CH_2OH$ (33)	175
		"	$(CH_3)_3SiCl$	$X = Si(CH_3)_3$ (61)	175
		"	$(C_6H_5)_2CO$	$X = C(C_6H_5)_2OH$ (48)	175
$C_{11}H_{17}NO_2$	*p*-$CH_3OC_6H_4CH(OH)CH_2N(CH_3)_2$	BuLi/ether/25°/ 24 hr	$(C_6H_5)_2CO$	α^1-[(Dimethylamino)methyl]-4-methoxy-α^2, α^2-diphenyl-*o*-xylene-α^1, α^2-diol (48)	175
$C_{12}H_{15}NS$	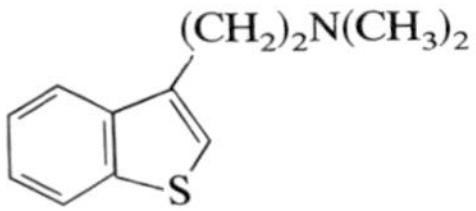 $(CH_2)_2N(CH_3)_2$ (benzo[b]thiophene ring, S)	BuLi/THF/−78°/ 0.5 hr	D_2O	N,N-Dimethyl-2-*d*-benzo[*b*]thiophene-3-ethylamine (83)	153

$C_{12}H_{19}N$	$C_6H_5C(CH_3)_2CH_2N(CH_3)_2$	BuLi/ether/25°/ 120 hr	$(C_6H_5)_2CO$	[*o*-[2-(Dimethylamino)-1,1-dimethylethyl]phenyl]-diphenylmethanol (17)	195
$C_{13}H_{18}N_2$	1-methyl-3-[$(CH_2)_2N(CH_3)_2$]indole (structure; N–CH_3)	BuLi/THF/0°/ 75 min	D_2O	3-[2-Dimethylamino)ethyl]-2-*d*-1-methylindole (74)[b]	153

[a] The isolated yield was 83%; the deuterium incorporation was 80%.
[b] The isolated yield was 74%; the deuterium incorporation was 74%.

TABLE XXXV. α-ALKOXIDOARALKYLAMINES (ORTHO, BETA)

Formula	Compound Lithiated	Conditions	Substrate	Product and Yield (%)	Refs.
$C_7H_{10}LiNO_2$	3-furyl-$CH(OLi)N(CH_3)_2$	Et/Li/ether/ reflux/2 hr	$(n\text{-}C_4H_9O)_3B$ (H_3O^+)	3-CHO-2-$B(OH)_2$-furan (32)	114, 115
$C_7H_{10}LiNOS$	3-thienyl-$CH(OLi)N(CH_3)_2$	BuLi/reflux/ 135 min	CO_2	3-Formyl-2-thiophenecarboxylic acid (44)	414
$C_8H_{12}LiNOS$	3-thienyl-$C(OLi)(CH_3)N(CH_3)_2$	BuLi	"S" + $BrCH_2CO_2H_3$	Methyl [(3-acetyl-2-thienyl)thio]acetate (—)	422
$C_{10}H_{13}ClLiNO$	$p\text{-}ClC_6H_4C(OLi)(CH_3)N(CH_3)_2$	BuLi/THF/0°–25°/16 hr	CH_3I	4-Cl-2-X-$C_6H_3COCH_3$: X = CH_3 (81)	196
		"	$(CH_3S)_2$	X = SCH_3 (45)	196
		"	DMF	X = CHO (56)	196
		"	$t\text{-}C_4H_9NCO$	X = $CONHC_4H_9\text{-}t$ (41)	196
		BuLi/THF/0°/ 8 hr	$o\text{-}ClC_6H_4CN$	X = $C(=NH)C_6H_4Cl\text{-}o$ (53)	196
		BuLi/THF/0°–25°/16 hr	$(C_6H_5)_2CO$	X = $C(C_6H_5)_2OH$ (47)	196
$C_{10}H_{14}LiNO$	$C_6H_5C(OLi)(CH_3)N(CH_3)_2$	BuLi/THF/0°–25°/16 hr	$(CH_3S)_2$	2′-(Methylthio)acetophenone (60)	196

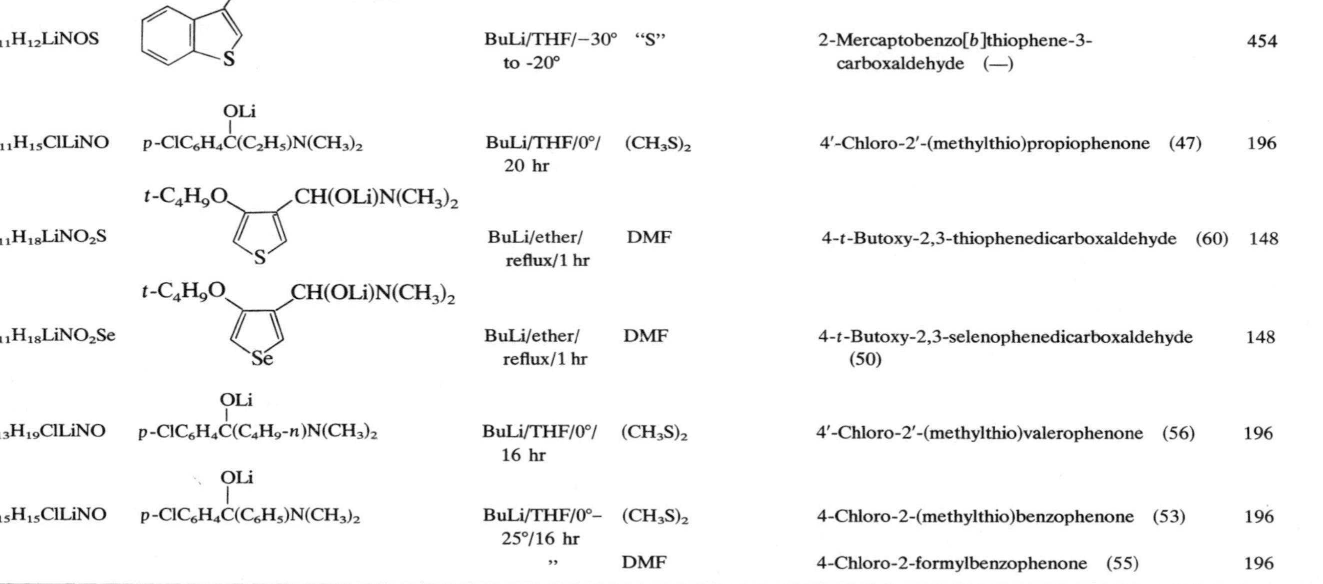

$C_{11}H_{12}LiNOS$	3-[$CH(OLi)N(CH_3)_2$]benzo[b]thiophene	BuLi/THF/−30° to -20°	"S"	2-Mercaptobenzo[*b*]thiophene-3-carboxaldehyde (—)	454
$C_{11}H_{15}ClLiNO$	p-$ClC_6H_4C(OLi)(C_2H_5)N(CH_3)_2$	BuLi/THF/0°/ 20 hr	$(CH_3S)_2$	4′-Chloro-2′-(methylthio)propiophenone (47)	196
$C_{11}H_{18}LiNO_2S$	t-C_4H_9O, $CH(OLi)N(CH_3)_2$ thiophene	BuLi/ether/ reflux/1 hr	DMF	4-*t*-Butoxy-2,3-thiophenedicarboxaldehyde (60)	148
$C_{11}H_{18}LiNO_2Se$	t-C_4H_9O, $CH(OLi)N(CH_3)_2$ selenophene	BuLi/ether/ reflux/1 hr	DMF	4-*t*-Butoxy-2,3-selenophenedicarboxaldehyde (50)	148
$C_{13}H_{19}ClLiNO$	p-$ClC_6H_4C(OLi)(C_4H_9$-$n)N(CH_3)_2$	BuLi/THF/0°/ 16 hr	$(CH_3S)_2$	4′-Chloro-2′-(methylthio)valerophenone (56)	196
$C_{15}H_{15}ClLiNO$	p-$ClC_6H_4C(OLi)(C_6H_5)N(CH_3)_2$	BuLi/THF/0°–25°/16 hr	$(CH_3S)_2$	4-Chloro-2-(methylthio)benzophenone (53)	196
		"	DMF	4-Chloro-2-formylbenzophenone (55)	196

Note: References 360–607 are on pp. 355–360.

TABLE XXXVI. ARYLCARBOXAMIDES (ORTHO, BETA)

Formula	Compound Lithiated	Conditions	Substrate	Product and Yield (%)	Refs.
C_6H_7NOS	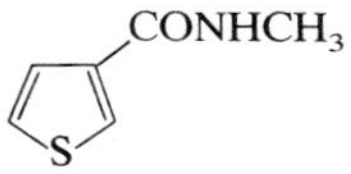 3-thienyl–$CONHCH_3$ (S)	BuLi/ether/ 25°–35°/ 1 hr	D_2O	2-*d*-N-Methyl-3-thiophenecarboxamide (55)	134
C_7H_9NOS	 3-thienyl–$CON(CH_3)_2$ (S)	BuLi/ether/ 25°/20 min	—	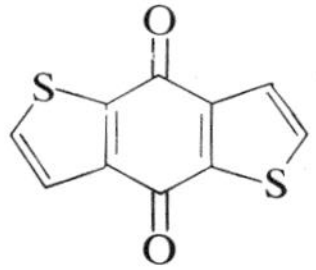 (S, O, O, S structure) (41)	134, 43
C_8H_8ClNO	*p*-$ClC_6H_4CONHCH_3$	BuLi/THF/ −30°–25°/ 2 hr	$I_2(CH_3I)$	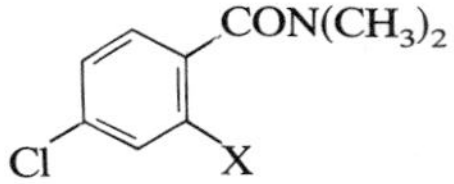 Cl–C_6H_3(X)–$CON(CH_3)_2$; X = I (30)	23
		”	*o*-ClC_6H_4CHO (CH_3I)	X = $CH(OCH_3)C_6H_4Cl$-*o* (90)	23
		”	*o*-FC_6H_4CHO (CH_3I)	X = $CH(OCH_3)C_6H_4F$-*o* (90)	23
		”	$(C_6H_5S)_2$ (CH_3I)	X = SC_6H_5 (62)	23
		”	Ethylene oxide	2-(2-Hydroxyethyl)-4-chloro-N-methylbenzamide (58)	23
		”	*o*-FC_6H_4CN	3-Amino-5-chloro-3-(*o*-fluorophenyl)-2-methylphthalimidine (26)	23
C_8H_9NO	$C_6H_5CONHCH_3$	BuLi/THF/0°/ 1 hr	Cyclohexanone	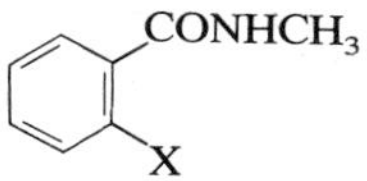 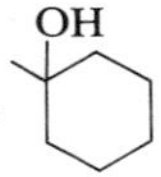 C_6H_4(X)–$CONHCH_3$; X = 1-hydroxycyclohexyl (OH) (40)	198

BuLi/THF/ reflux/15 min	C_6H_5CN	X = $C(=NH)C_6H_5$ (53)	54
BuLi/THF/5°/ 1 hr	2,4-$Cl_2C_6H_3CHO$	X = $CH(OH)C_6H_3Cl_2$-2,4 (—)	199
”	*p*-ClC_6H_4CHO	X = $CH(OH)C_6H_4Cl$-*p* (—)	199
BuLi/THF/0°/ 1 hr	C_6H_5CHO	X = $CH(OH)C_6H_5$ (28)[a] (—)	198, 199
BuLi/THF/5°/ 1 hr	*m*-$CF_3C_6H_4CHO$	X = $CH(OH)C_6H_4CF_3$-*m* (—)	199
”	*o*-$CH_3C_6H_4CHO$	X = $CH(OH)C_6H_4CH_3$-*o* (—)	199
”	*p*-$CH_3C_6H_4CHO$	X = $CH(OH)C_6H_4CH_3$-*p* (—)	199
”	*p*-$CH_3OC_6H_4CHO$	X = $CH(OH)C_6H_4OCH_3$-*p* (—)	199
BuLi/THF/ reflux/15 min	$C_6H_5COCH_3$	X = $C(OH)(CH_3)C_6H_5$ (43)	54, 198, 487
BuLi/THF/0°/ 1 hr	$C_6H_5COC_2H_5$	X = $C(OH)(C_2H_5)C_6H_5$ (70)	198
BuLi/THF/5°/ 1 hr	1-benzyl-4-piperidone (O=C ring, N–$CH_2C_6H_5$)	X = 4-hydroxy-1-benzyl-4-piperidyl (OH, $NCH_2C_6H_5$) (–)	488

TABLE XXXVI. ARYLCARBOXAMIDES (ORTHO, BETA) (*Continued*)

Formula	Compound Lithiated	Conditions	Substrate	Product and Yield (%)		Refs.
C_8H_9NO (*Contd.*)	$C_6H_5CONHCH_3$	BuLi/THF/0°/ 1 hr	O		X = OH (47)	198
		BuLi/THF/ reflux/ 15 min	p-$ClC_6H_4COC_6H_5$		$X = C(C_6H_5)C_6H_4Cl$-p (51) with OH	54
		,,	$(C_6H_5)_2CO$		$X = C(C_6H_5)_2OH$ (81)	54, 198, 487, 489
		BuLi/THF/5°/ 1 hr	p-$ClC_6H_4COC_6H_4CH_2N(CH_3)_2$-$o$		$X = C(C_6H_4Cl$-$p)C_6H_4$-$CH_2N(CH_3)_2$-o with OH (—)	490
		BuLi/THF/ reflux/ 15 min	Cyclohexanone (H_3O^+)	O, O, R_1, R_2	$R_1R_2 =$ 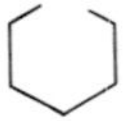 (27)	54
		,,	C_6H_5CHO (H_3O^+)		$R_1 = H$, $R_2 = C_6H_5$ (42)	54
		,,	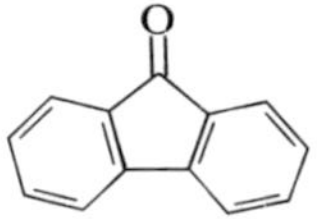 O (H_3O^+)		$R_1,R_2 =$ 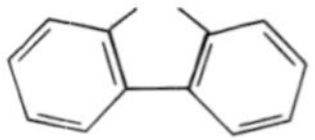 (58)	54

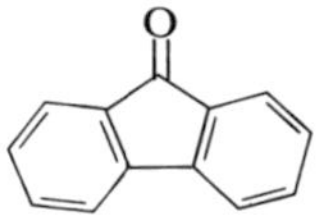

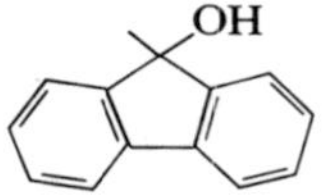

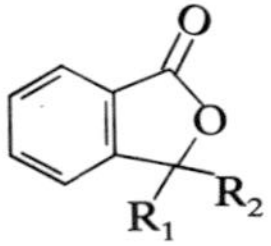

Formula	Substrate	Conditions	Reagent	Product(s) (Yield, %)	Refs.
		”	Cyclohexene oxide (H_3O^+)	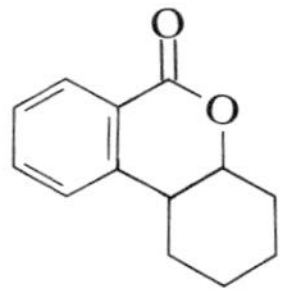 [structure] (11)	54
		BuLi/THF/0°/ 0.5 hr	C_6H_5CN	3-Amino-2-methyl-3-phenylphthalimidine (62)	201, 54
		BuLi/THF/ reflux/ 45 min	$C_6H_5N(CH_3)CHO$	3-Hydroxy-2-methylphthalimidine (53)	491
		BuLi/THF/5°/ 1 hr	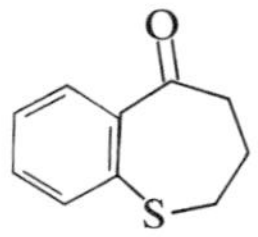 [structure]	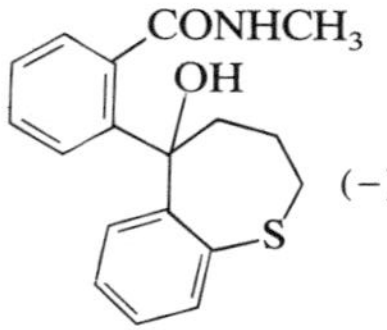 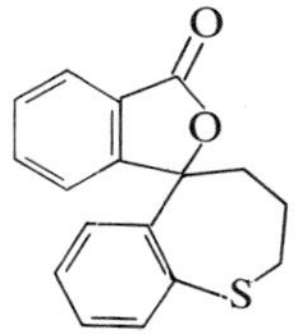 [structure] CONHCH$_3$, OH (−), [structure] (−)	492
		BuLi/THF/0°/ 0.5 hr	$(C_6H_5)_2SiCl_2$	 [structure] (CONHCH$_3$)$_2$ Si$(C_6H_5)_2$ (10)	571
$C_9H_{11}NO_2$	m-$CH_3OC_6H_4CONHCH_3$	BuLi/THF, ether/ reflux/1 hr	Ethylene oxide	3,4-Dihydro-5-methoxyisocoumarin (67)	493, 189
		BuLi/THF, ether/1 hr	$(C_6H_5)_2CO$	4-Methoxy-3,3-diphenylphthalide (91)	493, 189, 19
	o-$CH_3OC_6H_4CONHCH_3$	BuLi/THF, ether/ reflux/1 hr	Ethylene oxide	2-(2-Hydroxyethyl)-6-methoxy-N-methylbenzamide (50)	493, 494
		”	Propylene oxide	2-(2-Hydroxypropyl)-6-methoxy-N-methylbenzamide (60)	493, 494

TABLE XXXVI. ARYLCARBOXAMIDES (ORTHO, BETA) (*Continued*)

Formula	Compound Lithiated	Conditions	Substrate	Product and Yield (%)	Refs.
$C_9H_{11}NO_2$ (*Contd.*)	*o*-$CH_3OC_6H_4CONHCH_3$	BuLi/−70°–0°/ 1 hr	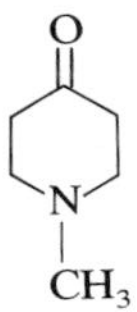 O, N, CH_3	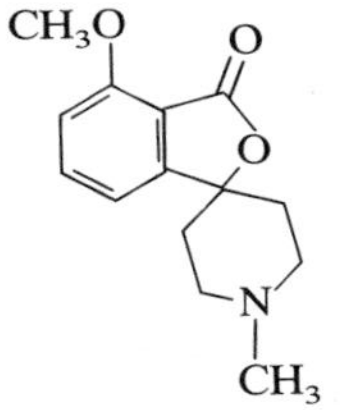 CH_3O, O, O, N, CH_3 (24)	200
		BuLi/THF, ether/ reflux/1 hr	$(C_6H_5)_2CO$	7-Methoxy-3,3-diphenylphthalide (60)	493, 189, 19
	p-$CH_3OC_6H_4CONHCH_3$	BuLi/ether, TMEDA/ 35°/5 hr	D_2O	2-*d*-N-Methyl-*p*-anisamide (50)	19
		BuLi/THF, ether/ reflux/1 hr	Ethylene oxide	3,4-Dihydro-6-methoxyisocoumarin (50)	493, 189
		BuLi/THF/ 65°/15 min	$(C_6H_5)_2CO$	α-Hydroxy-4-methoxy-N-methyl-α,α-diphenyl-*o*-toluamide (47)	19
$C_{11}H_{14}ClNO$	*m*-$ClC_6H_4CONHC_4H_9$-*t*	BuLi/THF/ −70°/3 hr	$(CH_3S)_2$	N-*t*-Butyl-3-chloro-2-(methylthio)benzamide (31)	24
$C_{11}H_{15}NO$	$C_6H_5CONHC_4H_9$-*t*	BuLi/THF/0°/ 1 hr	CO_2	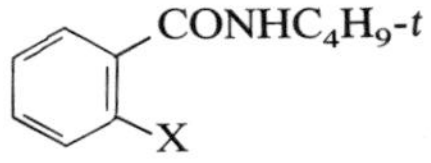 $CONHC_4H_9$-*t*, X; X = CO_2H (80)	24
		”	CH_3I	X = CH_3 (50)	24
		”	*t*-C_4H_9NCO	X = $CONHC_4H_9$-*t* (75)	24
		”	$C_6H_5CH_2Cl$	X = $CH_2C_6H_5$ (28)	24
		”	$(C_6H_5S)_2$	X = SC_6H_5 (62)	24
		”	DMF	2-*t*-Butyl-3-hydroxyphthalimidine (86)	24

	Reactant	Conditions	Electrophile	Product(s) (Yield, %)	Refs.
	$C_6H_5CON(C_2H_5)_2$	LTMP/THF/ 45 min	—	2-Benzoyl-N,N-diethylbenzamide (57)	55
$C_{11}H_{15}NO_4$	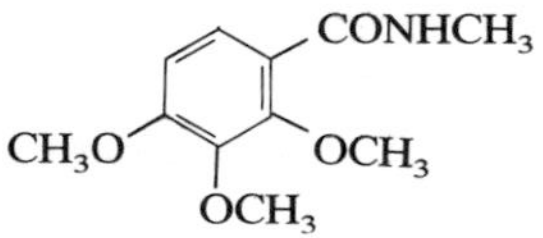 [structure: CONHCH₃, CH₃O, OCH₃, OCH₃]	BuLi/THF, ether/ reflux/1 hr	Propylene oxide	3,4-Dihydro-6,7,8-trimethoxy-3-methylisocoumarin (35)	202
$C_{11}H_{20}N_2O$	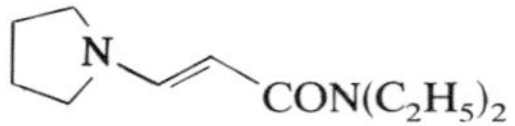 [structure: N, $CON(C_2H_5)_2$]	t-BuLi/THF/ −115°	CH_3OD	 [structure: N, X, $CON(C_2H_5)_2$] X = D (100)	57a
		"	CH_3I	X = CH_3 (95)	57a
		"	C_2H_5I	X = C_2H_5 (60)	57a
		"	$C_6H_5CO_2CH_3$	X = COC_6H_5 (95)	57a
		"	p-$CH_3C_6H_4CO_2CH_3$	X = $COC_6H_4CH_3$-p (60)	57a
$C_{12}H_{11}NO$	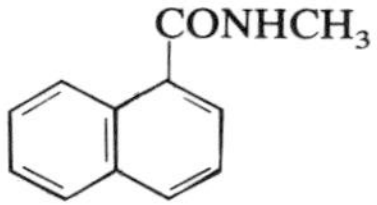 [structure: $CONHCH_3$]	BuLi/THF, ether/25°	Ethylene oxide (OH^-)	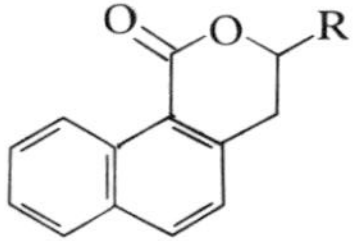 [structure: O, O, R] R = H (28)	495
		BuLi/THF, ether/ reflux/40 min	Propylene oxide	R = CH_3 (48)	482
	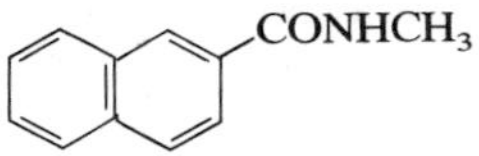 [structure: $CONHCH_3$]	BuLi/THF, ether/25°	Ethylene oxide (OH^-)	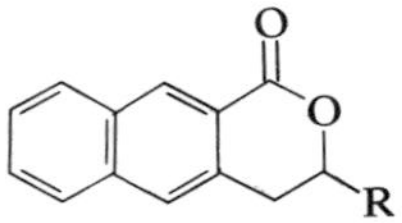 [structure: O, O, R] R = H (25)	495
		BuLi/THF, ether/ reflux/40 min	Propylene oxide	R = CH_3 (28)	482

TABLE XXXVI. ARYLCARBOXAMIDES (ORTHO, BETA) (*Continued*)

Formula	Compound Lithiated	Conditions	Substrate	Product and Yield (%)		Refs.
$C_{13}H_{10}ClNO$	p-$ClC_6H_4CONHC_6H_5$	BuLi/THF/ −10° −0°/0.5 hr	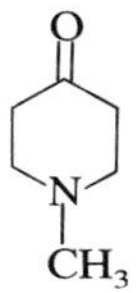 1-methyl-4-piperidone (O=, N–CH_3)	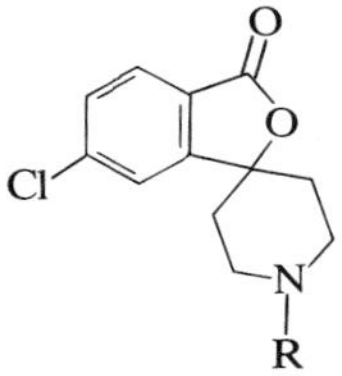 5-chloro spiro[isobenzofuran-1(3H),4'-piperidin]-3-one, N–R	$R = CH_3$ (25)	200
		"	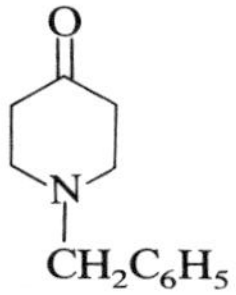 1-benzyl-4-piperidone (O=, N–$CH_2C_6H_5$)	"	$R = CH_2C_6H_5$ (17)	200
$C_{13}H_{10}FNO$	p-$FC_6H_4CONHC_6H_5$	BuLi/THF/ −70°– 0°/1 hr	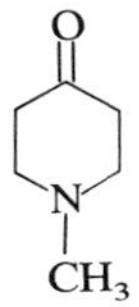 1-methyl-4-piperidone (O=, N–CH_3)	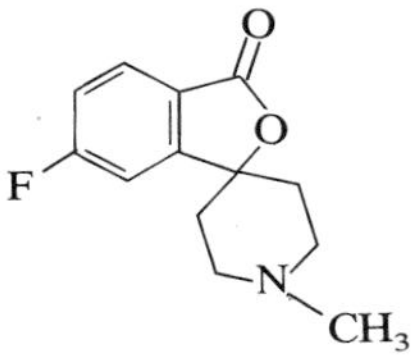 5-fluoro spiro[isobenzofuran-1(3H),4'-piperidin]-3-one, N–CH_3	(45)	200
$C_{13}H_{11}NO$	$C_6H_5CONHC_6H_5$	BuLi/THF/0°/ 1 hr	Cyclohexanone	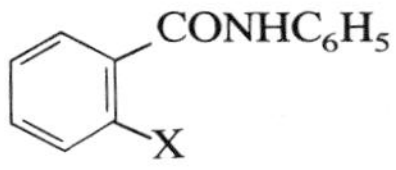 2-X-$C_6H_4CONHC_6H_5$	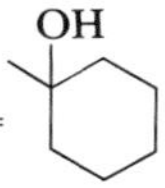 X = 1-hydroxycyclohexyl (OH) (56)	198
		"	$C_6H_5COC_2H_5$		$X = C(OH)(C_2H_5)C_6H_5$ (80)	198

"	9-fluorenone (O=)		X = 9-hydroxy-9-fluorenyl (OH) (55)	198
"	$(C_6H_5)_2CO$		$X = C(C_6H_5)_2OH$ (75)	198
BuLi/THF/ 0°/0.5 hr	$(CH_3)_2SiCl_2$	$\left(o\text{-}C_6H_4CONHC_6H_5\right)_m SiR_n$	$R = CH_3$, $m = n = 2$ (67)	571
"	$(CH_3)_3SiCl$		$R = CH_3$, $m = 1$, $n = 3$ (60)	571
"	$(C_6H_5)_2SiCl_2$		$R = C_6H_5$, $m = n = 2$ (70)	571
BuLi/THF/ −70°– 0°/1 hr	N-methyl-4-piperidone (O=, N–CH_3)	3,3-disubstituted phthalide (R_1, R_2, O, O)	R_1,R_2 = N-methylpiperidine-4,4-diyl (N–CH_3) (50)	200
"	$C_2H_5CO(CH_2)_2N(CH_3)_2$		$R_1 = C_2H_5$, $R_2 = (CH_2)_2N(CH_3)_2$ (55)	24
"	2-(dimethylamino)cyclohexanone (O=, $N(CH_3)_2$)		R_1,R_2 = 2-(dimethylamino)cyclohexane-1,1-diyl ($N(CH_3)_2$) (25)	24

TABLE XXXVI. ARYLCARBOXAMIDES (ORTHO, BETA) (*Continued*)

Formula	Compound Lithiated	Conditions	Substrate	Product and Yield (%)	Refs.
$C_{13}H_{11}NO$ (*Contd.*)	$C_6H_5CONHC_6H_5$	"	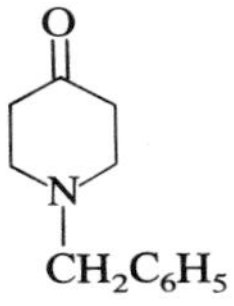 O, N, $CH_2C_6H_5$	R_1,R_2 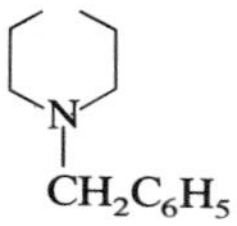 = N, $CH_2C_6H_5$ (42)	200
		BuLi/THF/0°/ 0.5 hr	C_6H_5CN	3-Amino-2,3-diphenylphthalimidine (65)	201
$C_{14}H_{13}NO$	$C_6H_5CONHCH_2C_6H_5$	BuLi/THF/0°/ 0.5 hr	C_6H_5NCO	N-Benzyl-N′-phenylphthalamide (20)	496
$C_{14}H_{13}NO_2$	*m*-$CH_3OC_6H_4CONHC_6H_5$	BuLi/THF/ −10°– 0°/0.5 hr	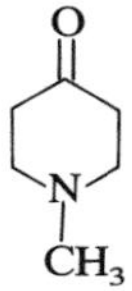 O, N, CH_3	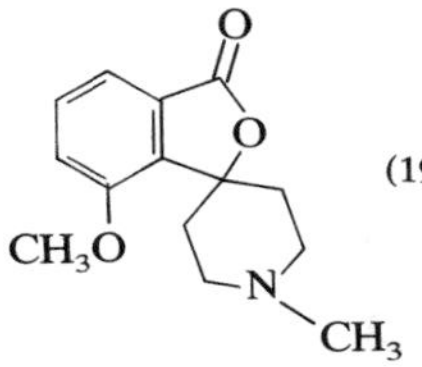 O, O, CH_3O, N, CH_3 (19)	200
	p-$CH_3OC_6H_4CONHC_6H_5$	BuLi/THF/ −10°– 0°/0.5 hr	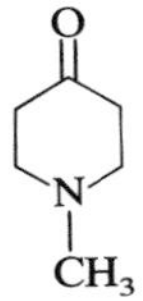 O, N, CH_3	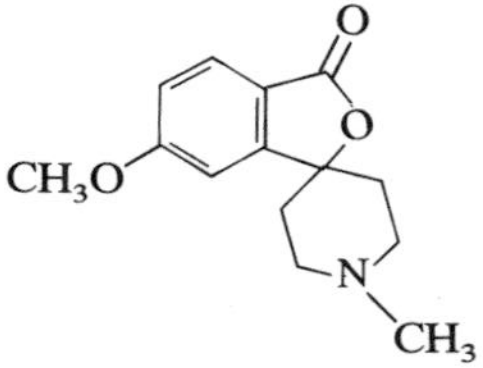 O, O, CH_3O, N, CH_3 (34)	200

$C_{14}H_{21}ClN_2O$	Cl-substituted benzene with $CH_2N(CH_3)_2$ and $CONHC_4H_9$-*t*	BuLi/THF/ −78°/1 hr	$(CH_3S)_2$	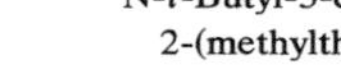 N-*t*-Butyl-3-chloro-6-[(dimethylamino)methyl]-2-(methylthio)benzamide (56)	24
$C_{16}H_{24}N_2O_2$	Benzene with two $CONHC_4H_9$-*n* groups (meta)	BuLi/THF, ether	Ethylene oxide	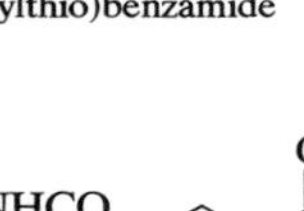 *n*-C_4H_9NHCO (lactone) (−)	51
	Benzene with two $CONHC_4H_9$-*t* groups (meta)	BuLi/THF/0°/ 6 hr	$(CH_3S)_2$	N,N′-Di-*t*-butyl-4-(methylthio)isophthalamide (51)	24

[a] The product was contaminated with the corresponding phthalide.

Note: References 360–607 are on pp. 355–360.

TABLE XXXVII. ARYLTHIOCARBOXAMIDES (ORTHO)

Formula	Compound Lithiated	Conditions	Substrate	Product and Yield (%)	Refs.
C_8H_8ClNS	p-$ClC_6H_4CSNHCH_3$	BuLi/THF/ −45°–10°	CH_3CHO (H_3O^+)	5-Chloro-3-methylphthalide (68)	65
		”	DMF	NCH_3, S, Cl, OH (77)	65
		”	t-C_4H_9NCO	$CSNHCH_3$, Cl, X; $X = CONHC_4H_9$-t (61)	65
		”	$(C_6H_5S)_2$	$X = SC_6H_5$ (91)	65
C_8H_9NS	$C_6H_5CSNHCH_3$	BuLi/THF/0°/ 4 hr	$(CH_3)_3SiCl$	N-Methyl-o-(trimethylsilyl)thiobenzamide (49)	65
$C_9H_{11}NOS$	p-$CH_3OC_6H_4CS$-$NHCH_3$	BuLi/THF/0°/ 25°/8.5 hr	$(CH_3S)_2$	4-Methoxy-N-methyl-2-(methylthio)thiobenzamide (78)	65

TABLE XXXVIII. 2-ARYLOXAZOLINES AND 2-ARYLOXAZINES (ORTHO, BETA)

Formula	Compound Lithiated	Conditions	Substrate	Product and Yield (%)	Refs.
$C_9H_{11}NOS$	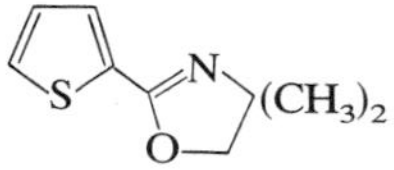 2-(2-thienyl)-4,4-(CH$_3$)$_2$-oxazoline	BuLi/THF/ −70°/1 hr	C_6H_5CHO	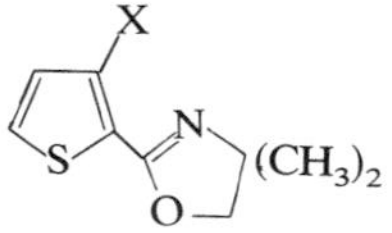 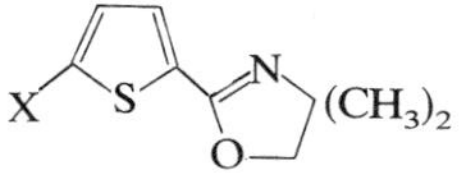 3-X-2-thienyl oxazoline (36), 5-X-2-thienyl oxazoline (55) X = CH(OH)C_6H_5	139
		BuLi/ether/ −70°–0°	"	(91), (4) X = CH(OH)C_6H_5	139
$C_9H_{11}NO_2$	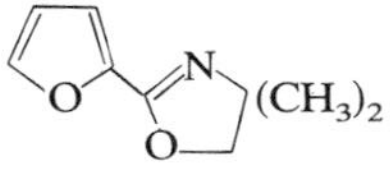 2-(2-furyl)-4,4-(CH$_3$)$_2$-oxazoline	BuLi/THF/ −70°/1 hr	C_6H_5CHO	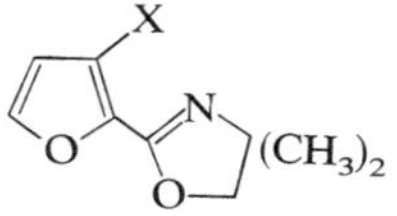 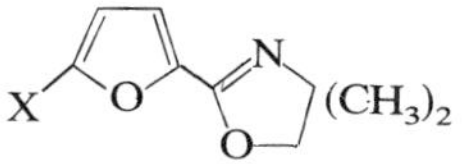 3-X-2-furyl oxazoline (49), 5-X-2-furyl oxazoline (36) X = CH(OH)C_6H_5	377
$C_{10}H_{12}N_2O$	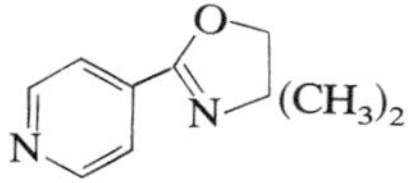 2-(4-pyridyl)-4,4-(CH$_3$)$_2$-oxazoline	CH_3Li/THF/ −78°–0°/ 1 hr	D_2O	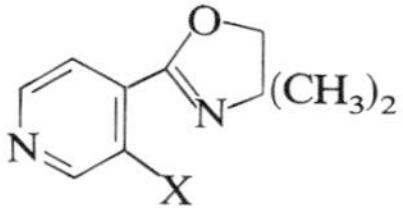 3-X-4-pyridyl oxazoline, X = D (80)	207
		"	CH_3I	X = CH_3 (63)	207
		"	C_2H_5I	X = C_2H_5 (56)	207
		"	$CH_2{=}CHCH_2Br$	X = $CH_2CH{=}CH_2$ (55)	207
		"	DMF	X = CHO (52)	207
		"	$C_2H_5COC_2H_5$	X = $C(C_2H_5)_2OH$ (76)	207
		"	C_6H_5CHO	X = CH(OH)C_6H_5 (83)	207
$C_{11}H_{11}D_2NO$	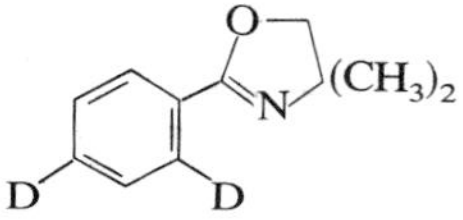 2-(2,4-dideuteriophenyl)-4,4-(CH$_3$)$_2$-oxazoline	BuLi/THF/ −45°/6 hr	D_2O	4,4-Dimethyl-2-(phenyl-2,4-6-d_3)-2-oxazoline (92)	204

TABLE XXXVIII. 2-ARYLOXAZOLINES AND 2-ARYLOXAZINES (ORTHO, BETA) (*Continued*)

Formula	Compound Lithiated	Conditions	Substrate	Product and Yield (%)	Refs.
$C_{11}H_{12}ClNO$	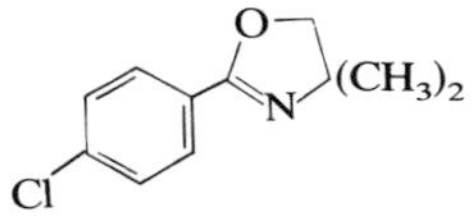 2-(4-chlorophenyl)-4,4-$(CH_3)_2$-2-oxazoline (structure)	BuLi/ether/ 0°/1 hr	I_2	2-(4-Chloro-2-iodophenyl)-4,4-dimethyl-2-oxazoline (66)	46
		"	CH_3I	2-(4-Chloro-*o*-tolyl)-4,4-dimethyl-2-oxazoline (71)	46
$C_{11}H_{12}DNO$	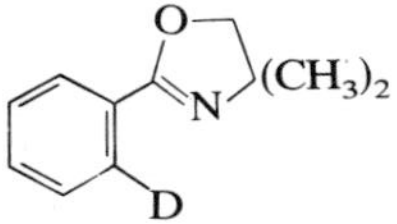 2-(2-D-phenyl)-4,4-$(CH_3)_2$-2-oxazoline (structure)	BuLi/THF/ −45°/6 hr	D_2O	4,4-Dimethyl-2-(phenyl-2,6-d_2)-2-oxazoline (90)	204
	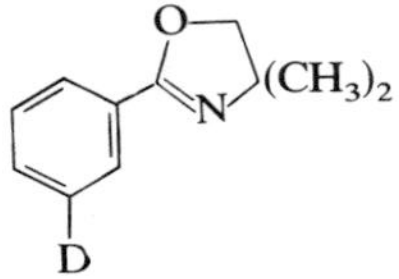 2-(3-D-phenyl)-4,4-$(CH_3)_2$-2-oxazoline (structure)	BuLi/THF/ −45°/7.5 hr,	D_2O	4,4-Dimethyl-2-(phenyl-2,3,6-d_3)-2-oxazoline (88)	204
	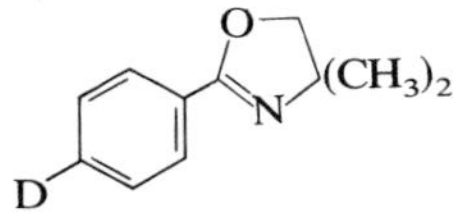 2-(4-D-phenyl)-4,4-$(CH_3)_2$-2-oxazoline (structure)	BuLi/THF/ −45°/1.5 hr	D_2O	4,4-Dimethyl-2-(phenyl-2,4-d_2)-2-oxazoline (90)	204
$C_{11}H_{13}NO$	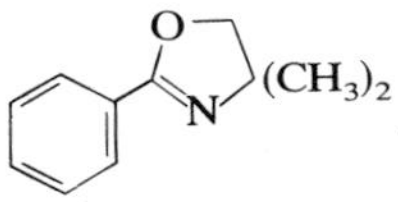 2-phenyl-4,4-$(CH_3)_2$-2-oxazoline (structure)	*s*-BuLi/ ether −70°–0°	D_2O	4,4-Dimethyl-2-(phenyl-2-*d*)-2-oxazoline (92)	46
		"	t-C_4H_9NCO	2-(2-*t*-Butylcarbamoylphenyl)-4,4-dimethyl-2-oxazoline (81)	46

$C_{12}H_{15}NO_2$	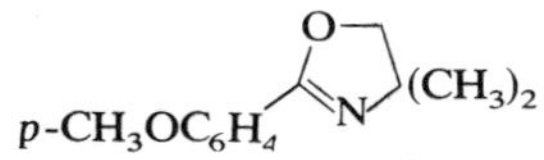 p-$CH_3OC_6H_4$ (oxazoline O, N, $(CH_3)_2$)	BuLi/ether/ 0°/4 hr	CH_3NCS	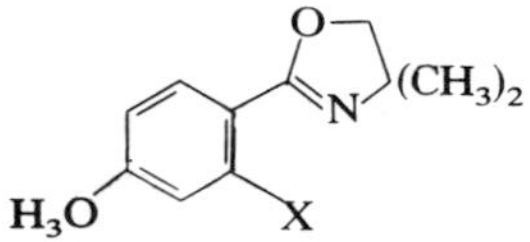 H_3O, X (oxazoline O, N, $(CH_3)_2$)	$X = CSNHCH_3$ (77)	46
		”	DMF		$X = CHO$ (70)	46
		”	$(C_6H_5S)_2$		$X = SC_6H_5$ (89)	46
$C_{13}H_{16}ClNO$	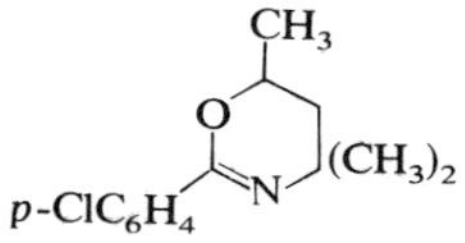 p-ClC_6H_4 (O, N, CH_3, $(CH_3)_2$)	BuLi/ether/ 0°/1 hr	o-FC_6H_4CHO	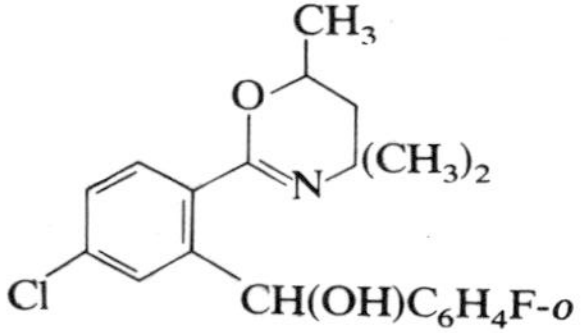 Cl, $CH(OH)C_6H_4F$-o (O, N, CH_3, $(CH_3)_2$)	(80) (2 isomers)	23
$C_{13}H_{17}NO_3$	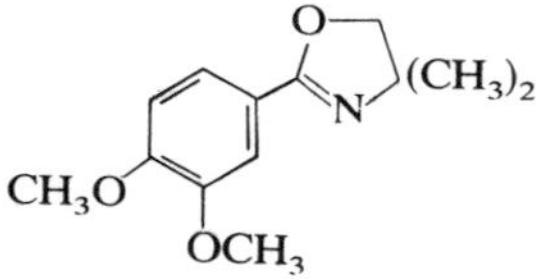 CH_3O, OCH_3 (oxazoline O, N, $(CH_3)_2$)	BuLi/THF/ −45°/1.5 hr	D_2O	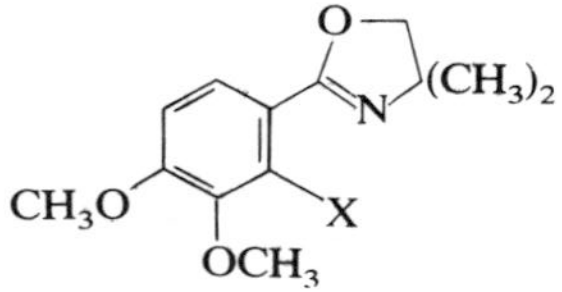 CH_3O, OCH_3, X (oxazoline O, N, $(CH_3)_2$)	$X = D$ (95)	204
		”	$(CH_3S)_2$		$X = SCH_3$ (92)	204
		”	N-Chloro-succinimide		$X = Cl$ (90)	204
$C_{16}H_{20}N_2O_2$	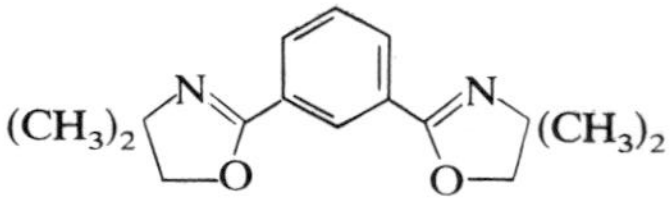 $(CH_3)_2$, N, O, N, O, $(CH_3)_2$	LDA/ benzene, TMEDA/ 25°/7 hr	CH_3I	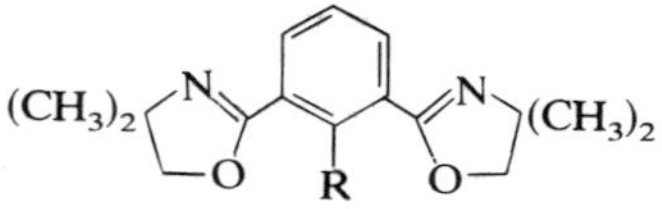 $(CH_3)_2$, N, O, R, N, O, $(CH_3)_2$	$R = CH_3$ (98)	208d
		”	$CH_2{=}CHCH_2Br$		$R = CH_2CH{=}CH_2$ (41)	208d

TABLE XXXIX. ARYLCARBIMINES (ORTHO, BETA)

Formula	Compound Lithiated	Conditions	Substrate	Product and Yield (%)		Refs.
C_9H_6ClNS	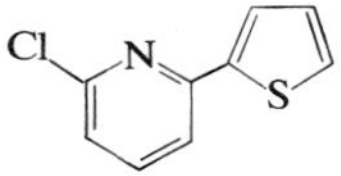	BuLi/ THF/0° 0.5 hr	$(CH_3)_3SiCl$	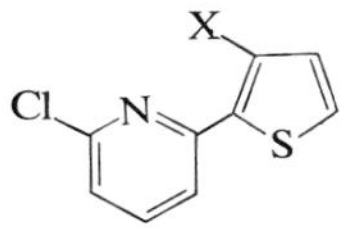 (0.4)	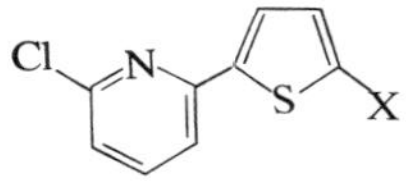 (91) $X = Si(CH_3)_3$	138
		BuLi/ ether/0° 0.5 hr	"	(36)	(43) $X = Si(CH_3)_3$	138
		t-BuLi/ ether/ −65°/0.5 hr	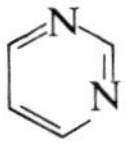 (ox.)	(27)	(—) X 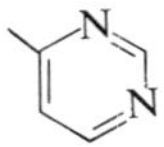 =	280
		t-BuLi/ THF/ −60°	$(C_6H_5)_2CO$	(—)	(83) $X = C(C_6H_5)_2OH$	137
C_9H_7NS	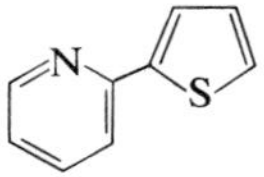	t-BuLi/ THF/ −60°	$CuCl_2(O_2)$	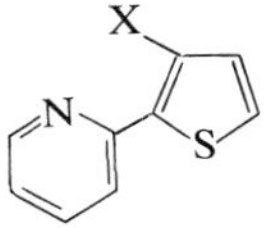 (−)	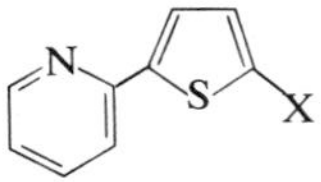 (63) X 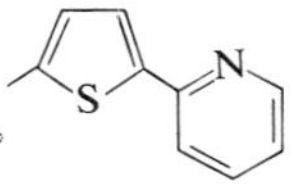 =	137
		BuLi/ ether/ 0°/0.5 hr	$(CH_3)_3SiCl$	(62)	(13) $X = Si(CH_3)_3$	138

Conditions	Electrophile	Yield (%)	Product (Yield %)	Ref.
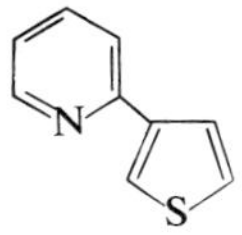 BuLi/ THF/ 0°/0.5 hr	”	(4)	(93) X = $Si(CH_3)_3$	138
BuLi/ ether/ 0°/0.5 hr	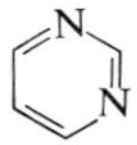	(87)	(—) X 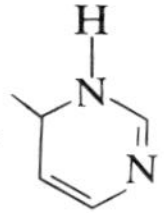 =	138
t-BuLi/ ether/ −65°/0.5 hr	” (ox.)	(79)	(—) X 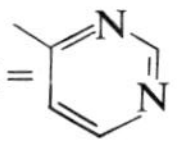 =	280
”	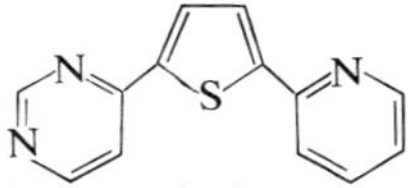 (ox.)	(28)	X 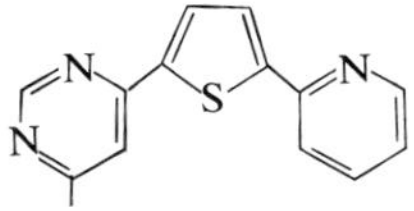 =	280
t-BuLi/ THF/ −60°	$(C_6H_5)_2CO$	(—)	(87) X = $C(C_6H_5)_2OH$	137
BuLi/ THF/ 0°/15 min	$CuCl_2$		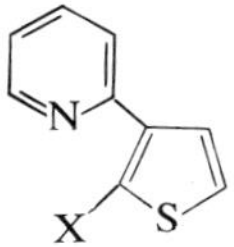 X 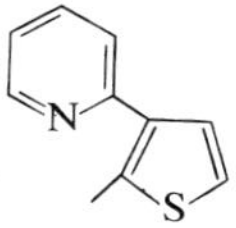 = (45)	280

TABLE XXXIX. Arylcarbimines (Ortho, Beta) (*Continued*)

Formula	Compound Lithiated	Conditions	Substrate	Product and Yield (%)	Refs.
C_9H_7NS (*Contd.*)	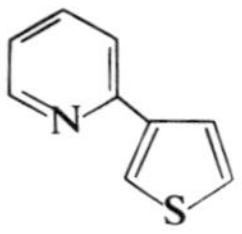 2-(3-thienyl)pyridine	BuLi/ ether/ 0°/15 min  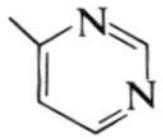	pyrimidine (ox.)	X = 4-pyrimidinyl (55)	280
$C_9H_9ClN_2$	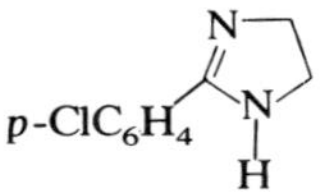 2-(p-ClC_6H_4)-2-imidazoline (N–H)	BuLi/ THF/ 50°/3 hr	p-$ClC_6H_4CO_2CH_3$	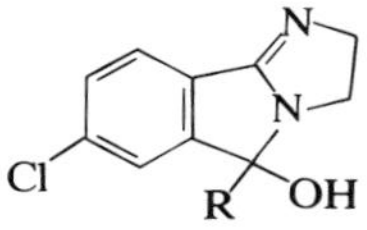 R = C_6H_4Cl-p (—)	308
		"	p-$FC_6H_4CO_2C_2H_5$	R = C_6H_4F-p (—)	308
$C_9H_{10}N_2$	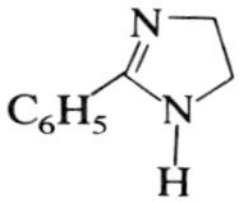 2-C_6H_5-2-imidazoline (N–H)	BuLi/ THF/ reflux/ 1 hr	p-ClC_6H_4CHO	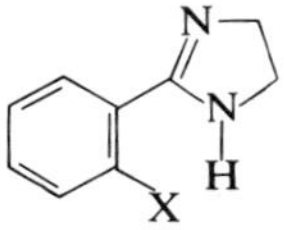 X = CH(OH)C_6H_4Cl-p (—)	216
		"	p-$ClC_6H_4COC_6H_5$	X = C(OH)C_6H_5(C_6H_4Cl-p) (—)	216
		"	$(C_6H_5)_2CO$	X = C$(C_6H_5)_2$OH (—)	216
$C_{10}H_{12}N_2$	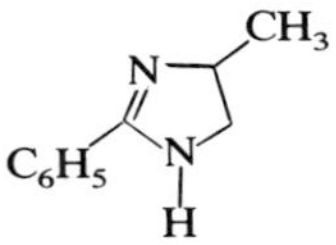 2-C_6H_5-4-CH_3-2-imidazoline (N–H)	BuLi/ THF/ 25°/24 hr	p-ClC_6H_4CN 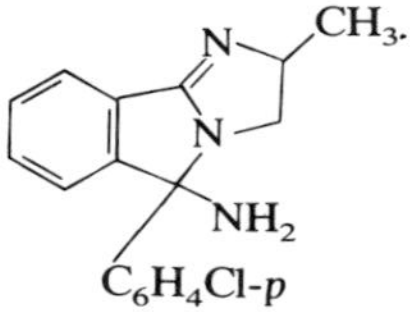	2-CH_3-5-NH_2-5-(C_6H_4Cl-p)-imidazo-isoindoline (–)	308

Formula	Substrate	Conditions	Electrophile	Product (yield, %)	Ref.
$C_{10}H_{13}ClN_2$	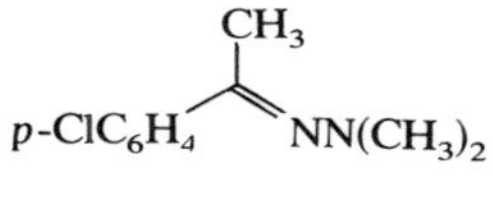 p-$ClC_6H_4C(CH_3)=NN(CH_3)_2$	BuLi/ THF/ −70°–0°/ 3 hr	$(CH_3S)_2$	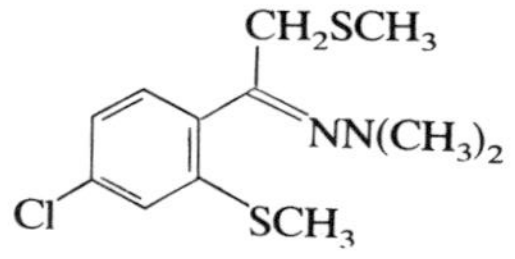 CH_2SCH_3, $NN(CH_3)_2$, Cl, SCH_3 [structure] (60)	23
$C_{11}H_8LiNS$	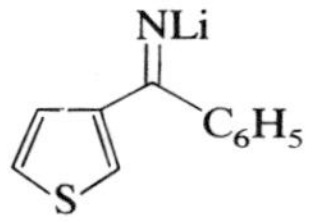 NLi, C_6H_5, S [structure]	BuLi/ ether/ reflux/ 2 hr	CO_2 (H_3O^+)	3-Benzoylthiophene-2-carboxylic acid (70)	570
		,,	C_6H_5CN (H_3O^+)	2,3-Dibenzoylthiophene (47)	570
$C_{12}H_8N_2S$	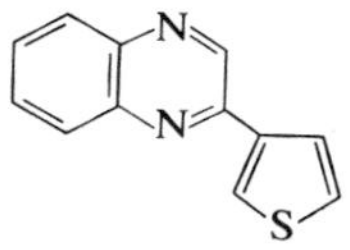 [structure]	LDA/ether	CuI_2	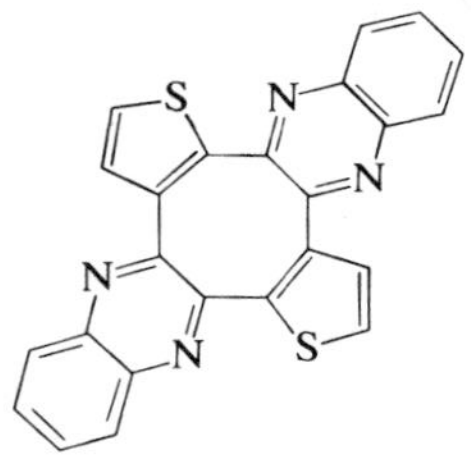 [structure] (9)	440
$C_{13}H_9NS$	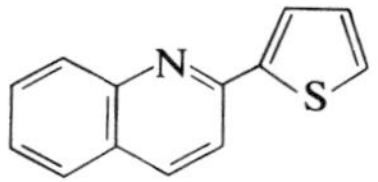 [structure]	BuLi/ ether/ 0°/0.5 hr	$(CH_3)_3SiCl$	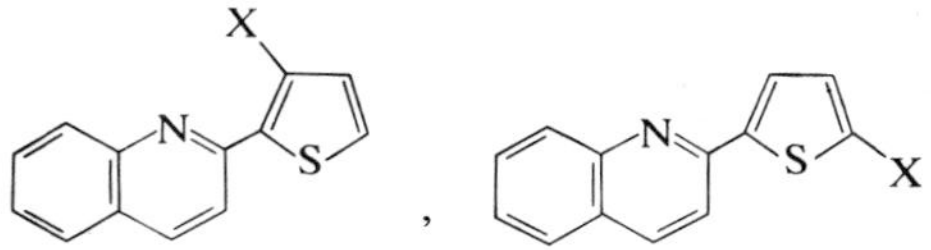 [structure] (45) , [structure] (8) $X = Si(CH_3)_3$	138
		BuLi/THF/ 0°/0.5 hr	,,	(−) (85) $X = Si(CH_3)_3$	138
$C_{13}H_{10}N_2$	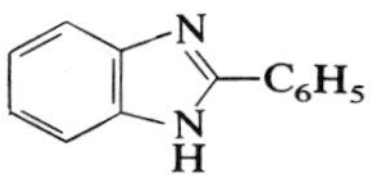 [structure] $-C_6H_5$	BuLi/ether	CO_2	2-(o-Carboxyphenyl)benzimidazole (41)	606

TABLE XXXIX. ARYLCARBIMINES (ORTHO BETA) (*Continued*)

Formula	Compound Lithiated	Conditions	Substrate	Product and Yield (%)	Refs.
$C_{13}H_{15}ClN_2$	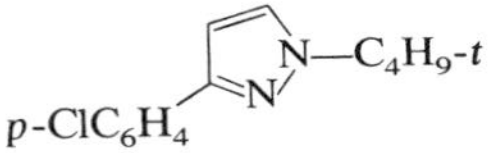 p-ClC_6H_4, N—C_4H_9-*t*	BuLi/ether/ 0°	o-ClC_6H_4CN (H_3O^+)	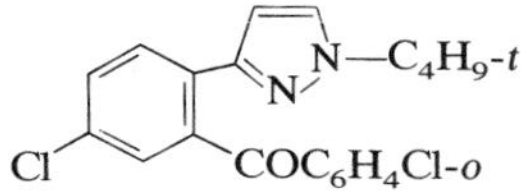 Cl, N—C_4H_9-*t*, COC_6H_4Cl-*o* (50)	23
$C_{13}H_{15}ClN_2O$	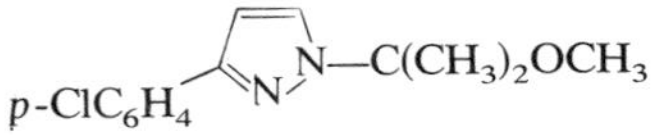 p-ClC_6H_4, N—$C(CH_3)_2OCH_3$	*s*-BuLi/ ether/ −70°/5 min	o-$ClC_6H_4CON(CH_3)_2$ (H_3O^+)	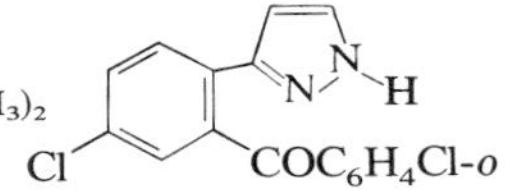 Cl, N—H, COC_6H_4Cl-*o* (40)	23
$C_{14}H_{17}NO_2$	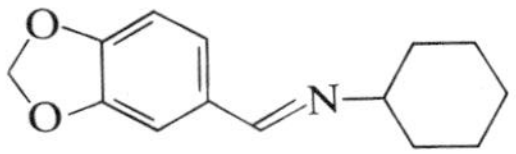 O, O, N	BuLi/THF/ −78°/ 15 min	D_2O	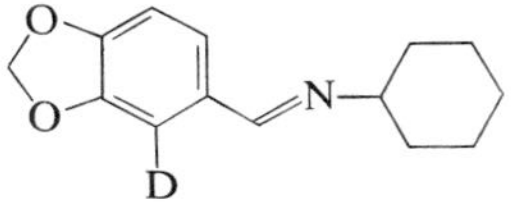 O, O, N, D (100)	212
		,,	I_2	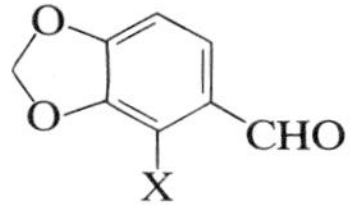 X = I (60)	212
		,,	CO_2 (H_3O^+)	O, O, CHO, X; X = CO_2H (54)	212
		,,	CH_3I (H_3O^+)	X = CH_3 (61)	212
		,,	$ClCO_2CH_3$ (H_3O^+)	X = CO_2CH_3 (68)	212
		,,	$CH_2{=}CHCH_2Br$ (CuI) (H_3O^+)	X = $CH_2CH{=}CH_2$ (72)	212
$C_{14}H_{19}N$	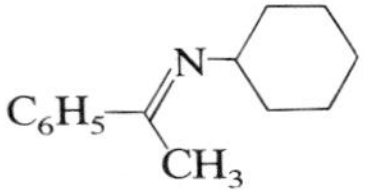 C_6H_5, N, CH_3	BuLi/ether	$(C_6H_5)_2CO$	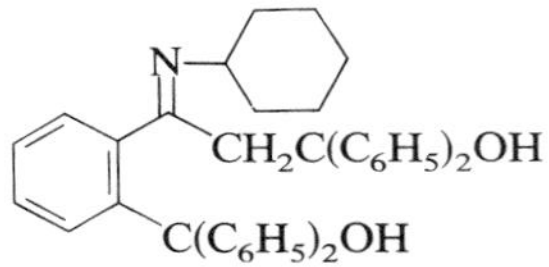 N, $CH_2C(C_6H_5)_2OH$, $C(C_6H_5)_2OH$ (16)	211

$C_{16}H_{10}N_2S$	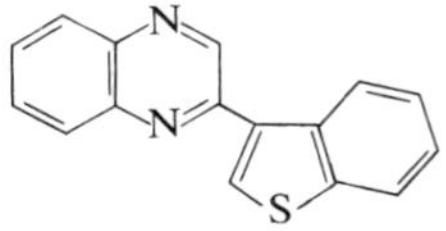	LDA/ether	CuI_2	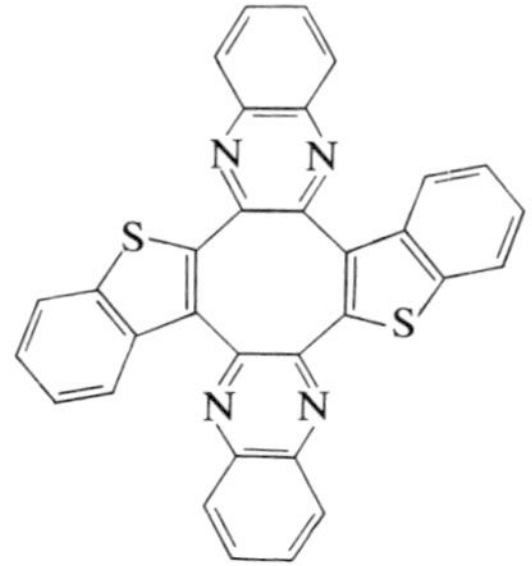 (61)	440
$C_{16}H_{17}N$	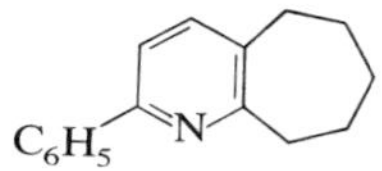	C_6H_5Li/ ether/ reflux/ 1 hr	D_2O	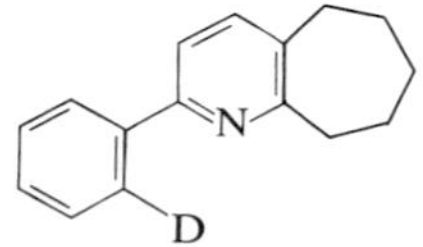 (I), 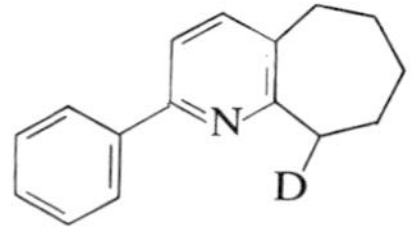 (II) (I:II, 1:0.4)	213
		C_6H_5Li/ ether/ reflux/ 5 hr	"	(I:II, 1:4.2)	213
$C_{17}H_{19}N$	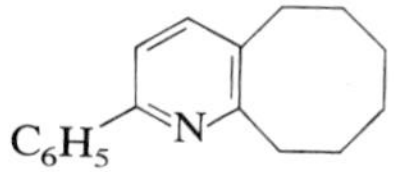	C_6H_5Li/ ether/ reflux/ 1 hr	D_2O	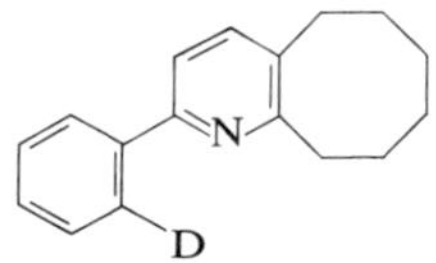 (III), 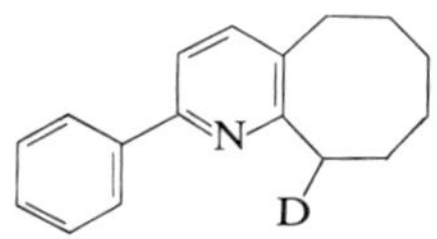 (IV) (III:IV, 1:0.5)	213

TABLE XXXIX. ARYLCARBIMINES (ORTHO, BETA) (*Continued*)

Formula	Compound Lithiated	Conditions	Substrate	Product and Yield (%)	Refs.
$C_{17}H_{19}N$ (*Contd.*)	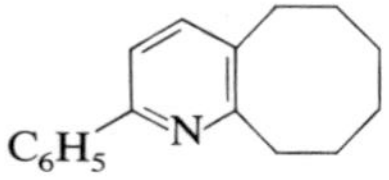 C_6H_5, N (structure)	C_6H_5Li/ether/ reflux 5 hr	"	(III:IV, 1:2.2)	213
$C_{18}H_{25}ClN_2O_3$	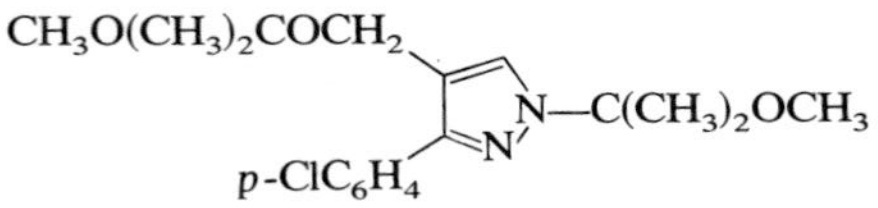 $CH_3O(CH_3)_2COCH_2$, N—$C(CH_3)_2OCH_3$, N, *p*-ClC_6H_4 (structure)	*s*-BuLi/ ether/ −50°–0°/ 1 hr	*o*-$ClC_6H_4CON(CH_3)_2$ (H_3O^+)	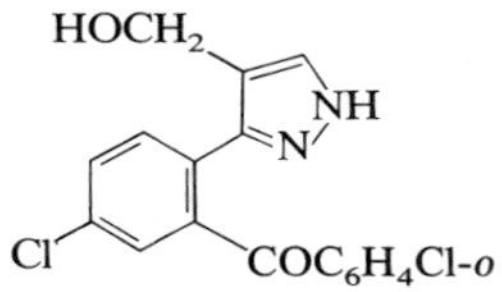 $HOCH_2$, NH, N, Cl, COC_6H_4Cl-*o* (structure) (64)	23

Note: References 360–607 are on pp. 355–360.

TABLE XL. N-ARYLPYRAZOLES (ORTHO)

Formula	Compound Lithiated	Conditions	Substrate	Product and Yield (%)	Refs.
$C_9H_8N_2$	N	BuLi/ether/0°–25°/2 hr	CO_2	1-Phenylpyrazole-5-carboxylic acid (39), 1-(*o*-carboxyphenyl)pyrazole (10)	88
		2 BuLi/ether/25°/7 hr	"	O N N (8), O N N CO_2H (26)	88
$C_{11}H_{12}N_2$	CH_3 N N CH_3 C_6H_5	BuLi/THF/−70°	CO_2	CH_3 N N CH_3 HO_2C (44)	155
$C_{11}H_{12}N_2O$	CH_3 N N OCH_3 C_6H_5	BuLi/THF/−70°	CO_2	CH_3 N N OCH_3 HO_2C (53)	155

TABLE XLI. ARYLNITRILES AND α, β-UNSATURATED NITRILES (ORTHO, BETA)

Formula	Compound Lithiated	Conditions	Substrate	Product and Yield (%)	Refs.
C_5H_3NS	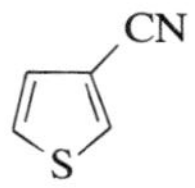 3-cyanothiophene (structure: thiophene ring with CN at 3-position)	BuLi/ether/ −70°/1 hr	CO_2	3-Cyano-2-thiophenecarboxylic acid (68)	271, 396
		BuLi/ether/ 0.5 hr	DMF	2-Formyl-3-thiophenecarboxamide (—)	396
C_5H_3NSe	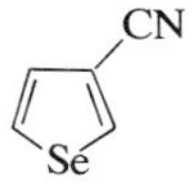 3-cyanoselenophene (structure: selenophene ring with CN at 3-position)	BuLi/ether/ 0.5 hr	CO_2	3-Cyano-2-selenophenecarboxylic acid (—)	396
		"	DMF	2-Formyl-3-selenophenecarboxamide (—)	396
C_7H_5ClN	m-ClC_6H_4CN	LTMP/THF/ −70°/1 hr	$(CH_3S)_2$	3-Chloro-2-(methylthio)benzonitrile (30)	24
$C_7H_8N_2$	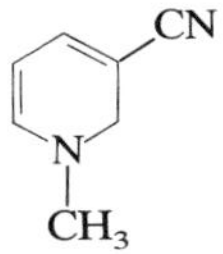 1-methyl-1,2-dihydropyridine-3-carbonitrile (structure: CN, N–CH_3)	LDA/THF/ −80°/0.5 hr	CH_3OD	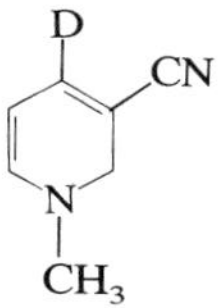 4-deuterio analogue (structure: D, CN, N–CH_3) (37)	218

Substrate	Conditions	Electrophile	Product	Yield (%)	Refs.
1-methyl-1,4-dihydropyridine-3-carbonitrile (structure)	LDA/THF/ −80°/20 hr	CH_3OD	2-X-1-methyl-1,4-dihydropyridine-3-carbonitrile (structure)	X = D (65)	218
	”	FSO_2OCH_3		X = CH_3 (77)	218
1-methyl-1,6-dihydropyridine-3-carbonitrile (structure)	LDA/THF/ −80°/1.5 hr	CH_3OD	2-X-1-methyl-1,6-dihydropyridine-3-carbonitrile (structure)	X = D (54)	218
	”	FSO_2OCH_3		X = CH_3 (72)	218

Note: References 360–607 are on pp. 355–360.

TABLE XLII. ALKYL VINYL ETHERS AND ALKYL ARYL ETHERS (ORTHO, BETA)

Formula	Compound Lithiated	Conditions	Substrate	Product and Yield (%)	Refs.
C_4H_7BrO	Br–CH=CH–OC_2H_5	BuLi/ether/ −78° to −50°/5 hr	Cyclopentanone (H_3O^+)	α-Bromo-$\Delta^{1,\alpha}$-cyclopentaneacetaldehyde (30)	174b
		,,	t-C_4H_9CHO	(E)-2-Bromo-1-ethoxy-4,4-dimethyl-1-penten-3-ol (58)	174b
C_4H_7ClO	Cl–CH=CH–OC_2H_5	BuLi/THF, hexane/ −100°	CO_2	(E)-2-Chloro-3-ethoxyacrylic acid (100)	174a
C_5H_6OS	3-methoxythiophene (OCH_3, S)	BuLi/ether/ 25°−reflux/ 2 hr	I_2	3-methoxy-2-X-thiophene (OCH_3, S, X): X = I (42)	567, 401
		BuLi/ether/ reflux/ 0.5 hr	CO_2	X = CO_2H (86)	402
		BuLi	DMF	X = CHO (83)	165
		,,	DMA	X = $COCH_3$ (32)	165
		BuLi/ether/ reflux/ 15 min	$(CH_3O)_2SO_2$	3-Methoxy-2-methylthiophene (I), 4-methoxy-2-methylthiophene (II) (I+II, 74) (I:II, 93:7)	145
		BuLi/reflux/ 20 min	$(n\text{-}C_4H_9O)_3B$ (H_2O_2)	3-methoxythiophen-2(5H)-one (OCH_3, S, =O) (24)	289

C_5H_6OSe	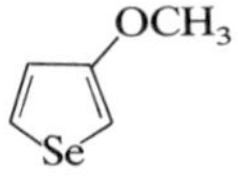 OCH_3, Se (structure)	BuLi	DMF	3-Methoxy-2-selenophenecarboxaldehyde (50)	165
		”	DMA	3-Methoxyselenophen-2-yl methyl ketone (40)	165
$C_5H_7NO_2$	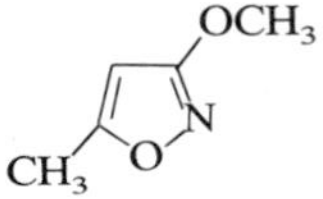 OCH_3, N, O, CH_3 (structure)	BuLi/THF/ −75°/1 hr	CO_2	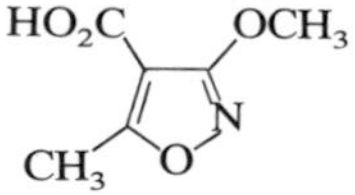 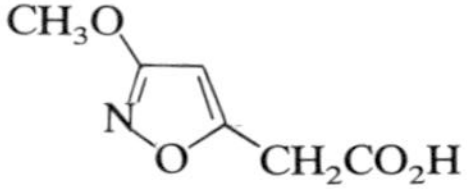 HO_2C, OCH_3, N, O, CH_3 (structure) (10), CH_3O, N, O, CH_2CO_2H (structure) (47)	498
C_6H_8OS	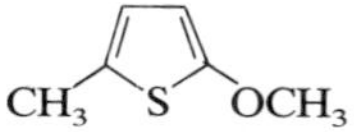 CH_3, S, OCH_3 (structure)	BuLi/ether/ 25°/24 hr	CO_2	2-Methoxy-5-methylthiophene-3-carboxylic acid (50)	231
C_7H_7FO	p-$FC_6H_4OCH_3$	BuLi/THF/ 27°/5 hr	CO_2	5-Fluoro-*o*-anisic acid (32)	19
C_7H_8O	$C_6H_5OCH_3$	BuLi/ether/ 2 hr	D_2O	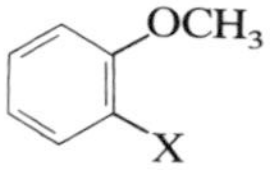 OCH_3, X (structure) X = D (30)	27
		BuLi/THF/ 24 hr	CO_2	X = CO_2H (51)	35
		BuLi/ether/ reflux	$CO_2(CH_2N_2)$	X = CO_2CH_3 (65)	50
		BuLi/ether/ reflux/ 12 hr	$CF_2{=}CCl_2$	X = $CF{=}CCl_2$ (41)	364
		BuLi/THF, ether/ 25°/6 hr	n-$C_{12}H_{25}Br$	X = $C_{12}H_{25}$-n (54)	499
$C_7H_8O_2$	m-$CH_3OC_6H_4OH$	BuLi/ether/ 25°/22 hr	CO_2	6-Hydroxy-*o*-anisic acid (7), 4-hydroxy-*o*-anisic acid (2)	233
$C_8H_7F_3O$	p-$CH_3OC_6H_4CF_3$	BuLi/ether/ 35°/21 hr	D_2O	3-*d*-α,α,α-Trifluoro-4-methoxytoluene (92)	19
		”	$(C_6H_5)_2CO$	(6-Methoxy-α,α,α-trifluoro-*m*-tolyl)diphenylmethanol (90)	19

TABLE XLII. ALKYL VINYL ETHERS AND ALKYL ARYL ETHERS (ORTHO, BETA) (*Continued*)

Formula	Compound Lithiated	Conditions	Substrate	Product and Yield (%)	Refs.
C_8H_8O	C_6H_5O–CH=CH_2	*s*-BuLi/THF/ 0°/2 hr	CH_3I	2-$CH_3C_6H_4O$–CH=CH_2 (20)	384
$C_8H_8O_2$	5-Methyl-1,3-benzodioxole (CH_3)	BuLi/ether/ 25°/1 hr	CO_2	4-X-6-CH_3-1,3-benzodioxole; X = CO_2H (20)	483
		"	$(C_6H_5)_2CO$	X = $C(C_6H_5)_2OH$ (50)	483
C_8H_9BrO	Br, CH_3, OCH_3 substituted benzene	C_6H_5Li/ ether/25°/ 18 hr, reflux/ 8 hr	Cyclopentanone	1-(3-Bromo-6-methoxy-*p*-tolyl)cyclopentanol (33)	221, 220
$C_8H_{10}O$	$C_6H_5OC_2H_5$	BuLi/THF/ 24 hr	CO_2	*o*-Ethoxybenzoic acid (42)	35
		BuLi/ether/ reflux/ 27 hr	$CO_2(CH_2N_2)$	Methyl *o*-ethoxybenzoate (63)	50
	m-$CH_3OC_6H_4CH_3$	BuLi/cyclo- hexane	CO_2	X, OCH_3, CH_3 (III), X, OCH_3, CH_3 (IV); (III:IV, 9:1) X = CO_2H	17

		BuLi/cyclo-hexane, TMEDA	"	(III:IV, 13:12)	$X = CO_2H$	17
		BuLi/ether/ reflux	$CO_2(CH_2N_2)$	(III+IV, 53) (III:IV, 3:2)	$X = CO_2CH_3$	50
	o-$CH_3OC_6H_4CH_3$	BuLi/cyclo-hexane/ reflux/10 hr	CO_2	2-Methoxy-*m*-toluic acid (V), 2-methoxyphenylacetic acid (VI) (V+VI, 57) (V:VI, 1:2)		62, 61
		BuLi/cyclo-hexane, TMEDA/ 25°/10 hr	"	(V+VI, 72) (V:VI, 3:1)		62
	p-$CH_3OC_6H_4CH_3$	BuLi/ether/ reflux/ 40 hr	CO_2	6-Methoxy-*m*-toluic acid (31)		61, 348
$C_8H_{10}OS$	m-$CH_3OC_6H_4SCH_3$	BuLi/ether/ reflux/ 4 hr	CO_2	6-(Methylthio)-*o*-anisic acid (46)		574
	p-$CH_3OC_6H_4SCH_3$	BuLi/ether/ reflux/ 4 hr	CO_2	5-(Methylthio)-*o*-anisic acid (50)		574
$C_8H_{10}O_2$	$C_6H_4(OCH_3)_2$-m	C_6H_5Li/ ether/ 2–3 days	Cl_2	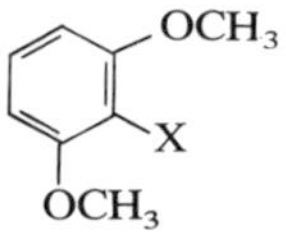 [structure: 2-X-1,3-dimethoxybenzene; labels OCH_3, X, OCH_3]	$X = OCl$ (39)	314
		C_6H_5Li/ ether/ 60 hr	Cu		X 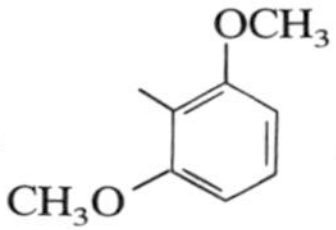 = [structure: 2,6-dimethoxyphenyl; labels OCH_3, CH_3O] (48)	219
		BuLi/ether/ 25°/70 hr	CuBr		$X = Cu$ (93)	501

TABLE XLII. ALKYL VINYL ETHERS AND ALKYL ARYL ETHERS (ORTHO, BETA) (*Continued*)

Formula	Compound Lithiated	Conditions	Substrate	Product and Yield (%)	Refs.
$C_8H_{10}O_2$ (*Contd.*)	$C_6H_4(OCH_3)_2$-*m*	C_6H_5Li/ ether/ 2–3 days	Br_2	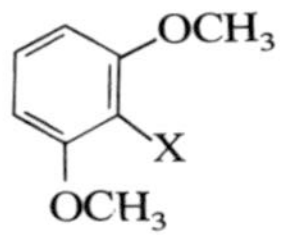 2,6-dimethoxyphenyl-X: X = Br (18)	314
		BuLi/ether	AgBr	X = Ag (21)	304
		C_6H_5Li/ether/ 2–3 days	I_2	X = I (80)	314
		BuLi/ether/ 25°/70 hr	$(CH_3)_2SnBr_2$	X = $Sn(CH_3)_2Br$ (87)	480
		C_6H_5Li	$B(OCH_3)_3$	X = $B(OH)_2$ (48)	502
		BuLi	$(CH_3)_3SnCl$	X = $Sn(CH_3)_3$ (80)	306
		—	$BrAuP(C_6H_5)_3$	X = $AuP(C_6H_5)_3$ (90)	307
		BuLi/ether/ reflux/2 hr	CO_2	X = CO_2H (70)	503
		C_6H_5Li/ether/ 2–3 days	BrCN	X = Br (46)	314
		"	ICN	X = I (46)	314
		C_6H_5Li/ether/ 25°/60 hr	CH_3I	X = CH_3 (95)	504
		C_6H_5Li/ether/ 2–3 days	$(SCN)_2$	X = CN (6), X = SCN (21)	314
		"	AcCl	X = Ac (14)	314
		"	Ethylene oxide	X = $(CH_2)_2OH$ (57)	314
		"	$Br(CH_2)_2Br$	X = Br (71)	314, 219
		"	$ClCH_2OCH_3$	X = CH_2OCH_3 (62)	314
		"	C_2H_5I	X = C_2H_5 (6)	314
		C_6H_5Li/25°/ 3 days	$(CH_3O)_2SO_2$	X = CH_3 (74–76)	505
		C_6H_5Li/ether 2–3 days	$BrCH_2CH{=}CH_2$	X = $CH_2CH{=}CH_2$ (88)	314

"	$\overset{O}{\triangle}CH_2Cl$	$X = CH_2\overset{O}{\triangle}$ (42),	314
		$X = CH_2CH(OH)CH_2Cl$ (18)	
BuLi/THF/ reflux/ 1 hr	$\overset{O}{\triangle}CH_3$	$X = CH_2CH(OH)CH_3$ (28)	506
C_6H_5Li/ether/ 2–3 days	CH_3COCH_3	$X = C(CH_3)_2OH$ (61)	314
C_6H_5Li/ether/ 60 hr	$\frac{1}{2}Br(CH_2)_3Br$	$X = (CH_2)_3C_6H_3(OCH_3)_2$-2,6 (42)	219
C_6H_5Li/ether/ 25°/2 days	$ClCH_2N(CH_3)_2$	$X = CH_2N(CH_3)_2$ (33)	507
C_6H_5Li/ether/ 2–3 days	$BrCH_2CH{=}CHCH_3$	$X = CH_2CH{=}CHCH_3$ (75)	314
"	$(CH_3)_2CHCHO$	$X = CH(OH)CH(CH_3)_2$ (76)	314
C_6H_5Li/ether/ 60 hr	$\frac{1}{2}Cl(CH_2)_4Cl$	$X = (CH_2)_4C_6H_3(OCH_3)_2$-2,6 (33)	219
"	$\frac{1}{2}Br(CH_2)_4Br$	$X = (CH_2)_4C_6H_3(OCH_3)_2$-2,6 (63)	219
"	$Cl(CH_2)_4Cl$	$X = (CH_2)_4Cl$ (45)	219
"	$Br(CH_2)_4Br$	$X = (CH_2)_4Br$ (50)	219

TABLE XLII. ALKYL VINYL ETHERS AND ALKYL ARYL ETHERS (ORTHO, BETA) (*Continued*)

Formula	Compound Lithiated	Conditions	Substrate	Product and Yield (%)	Refs.
$C_8H_{10}O_2$ (*Contd.*)	$C_6H_4(OCH_3)_2$-*m*	C_6H_5Li/ether/ 2–3 days	$(C_2H_5O)_2SO_2$	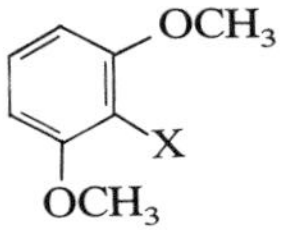 2,6-$(OCH_3)_2C_6H_3X$; X = C_2H_5 (18)	314
		"	$CH_2{=}C(CH_3)COCH_3$	X = $C(OH)(CH_3)C(CH_3){=}CH_2$ (35)	314
		"	$BrCH_2CH{=}C(CH_3)_2$	X = $CH_2CH{=}C(CH_3)_2$ (98)	314
		C_6H_5Li/ether/ 25°/2 days 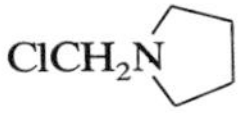 	$ClCH_2N$(pyrrolidine ring)	X = CH_2N(pyrrolidine ring) (55)	507
		"	$ClCH_2N$(morpholine ring)O	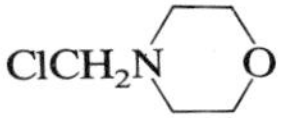 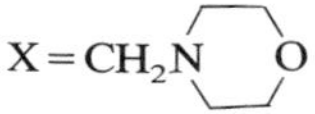 X = CH_2N(morpholine ring)O (55)	507
		C_6H_5Li/ether/ 60 hr	$\frac{1}{2}Br(CH_2)_5Br$	X 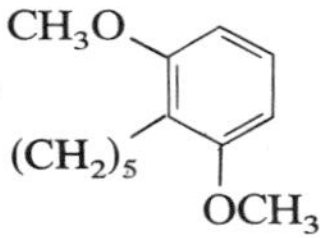 = $(CH_2)_5$-2,6-$(OCH_3)_2C_6H_3$ (58)	219
		C_6H_5Li/ether/ 25°/2 days	$ClCH_2N(C_2H_5)_2$	X = $CH_2N(C_2H_5)_2$ (68)	507
		C_6H_5Li/ether/ 2–3 days	$(CH_3)_2C{=}CHCOCH_3$	X = $C(OH)(CH_3)CH{=}C(CH_3)_2$ (59)	314
		"	$CH_3COC_4H_9$-*t*	X = $C(OH)(CH_3)C_4H_9$-*t* (28)	314
		C_6H_5Li/ether/ 25°/2 days 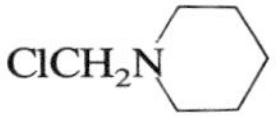 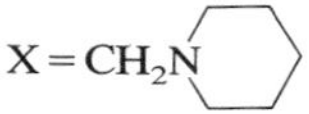	$ClCH_2N$(piperidine ring)	X = CH_2N(piperidine ring) (49)	507

C_6H_5Li/ether/ 60 hr	$\frac{1}{2}Br(CH_2)_6Br$	X 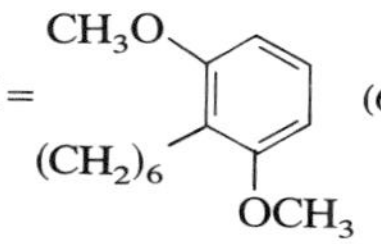 = CH_3O, $(CH_2)_6$, OCH_3 (64)	219
BuLi/ether/−5°	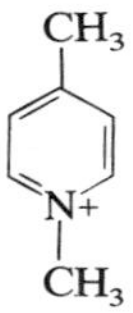 CH_3, N^+, CH_3 I^-	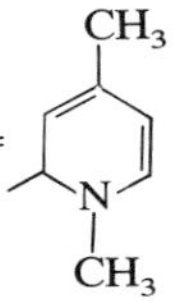 X = CH_3, N, CH_3 ("high")	508
C_6H_5Li/ether/ 2–3 days	$C_6H_5CH{=}CHCHO$	$X = CH(OH)CH{=}CHC_6H_5$ (79)	314
BuLi/THF/ −20°/2 hr	$n\text{-}C_{12}H_{25}Br$	$X = (C_{12}H_{25}\text{-}n$ (70)	499
t-BuLi	CO_2	OCH_3, X, OCH_3 (VII), X, OCH_3, OCH_3 (VIII)	
		(VII + VIII, 60) (VII : VIII, 19 : 1) $X = CO_2H$	36
BuLi/ether/ reflux	$CO_2(CH_2N_2)$	(VII + VIII, 60–75) (VII : VIII, 24 : 1) $X = CO_2CH_3$	50
C_6H_5Li/ether/ 2–3 days	$(CN)_2$	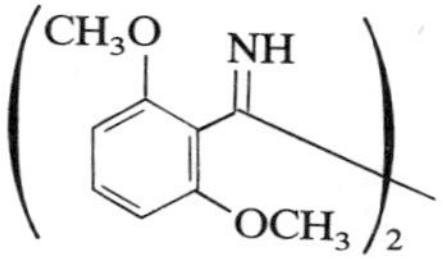 (CH_3O, NH, OCH_3)$_2$ (54)	314

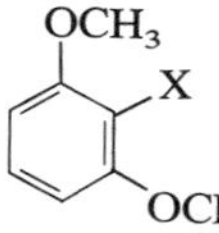

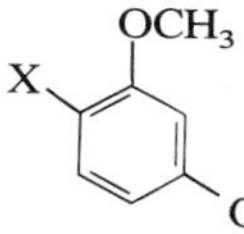

TABLE XLII. ALKYL VINYL ETHERS AND ALKYL ARYL ETHERS (ORTHO, BETA) (*Continued*)

Formula	Compound Lithiated	Conditions	Substrate	Product and Yield (%)	Refs.
$C_8H_{10}O_2$ (*Contd.*)	$C_6H_4(OCH_3)_2$-*o*	C_6H_5Li	$B(OCH_3)_3$	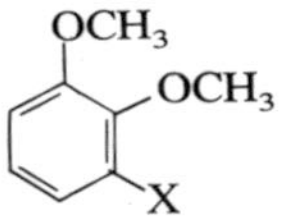 2,3-(OCH_3)$_2$C$_6$H$_3$X; X = $B(OH)_2$ (20)	502
		BuLi/ether/ 25°/24 hr	Cyclohexanone	X 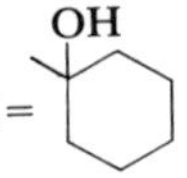 = 1-hydroxycyclohexyl (60)	509
		"	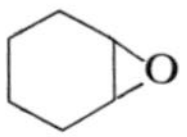 Cyclohexene oxide	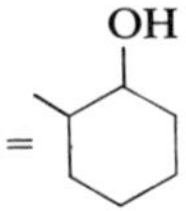 X = 2-hydroxycyclohexyl (70)	509
		BuLi/heptane, TMEDA/ 25°/20 hr	I_2	1,4-Diiodo-2,3-dimethoxybenzene (2)	510
	m-$CH_3OC_6H_4CH_2OH$	BuLi/hexane, TMEDA/ 60°/5 hr	$CO_2(CH_2N_2)$	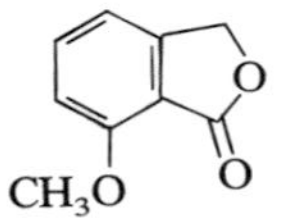 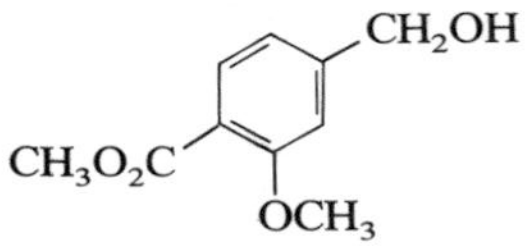 7-methoxyphthalide (CH_3O) (53), CH_3O_2C, OCH_3, CH_2OH benzene (6)	245
$C_8H_{12}OS$	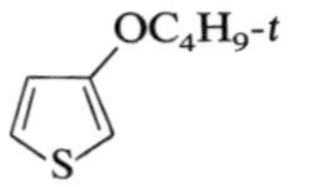 3-(OC_4H_9-*t*)thiophene	BuLi/ether/ reflux/ 0.5 hr	CO_2	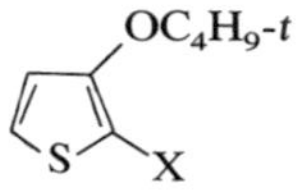 3-(OC_4H_9-*t*)-2-X-thiophene; X = CO_2H (62)	423
		BuLi/ether/ reflux/ 2 hr	$(CH_3O)_2SO_2$	X = CH_3 (87)	141

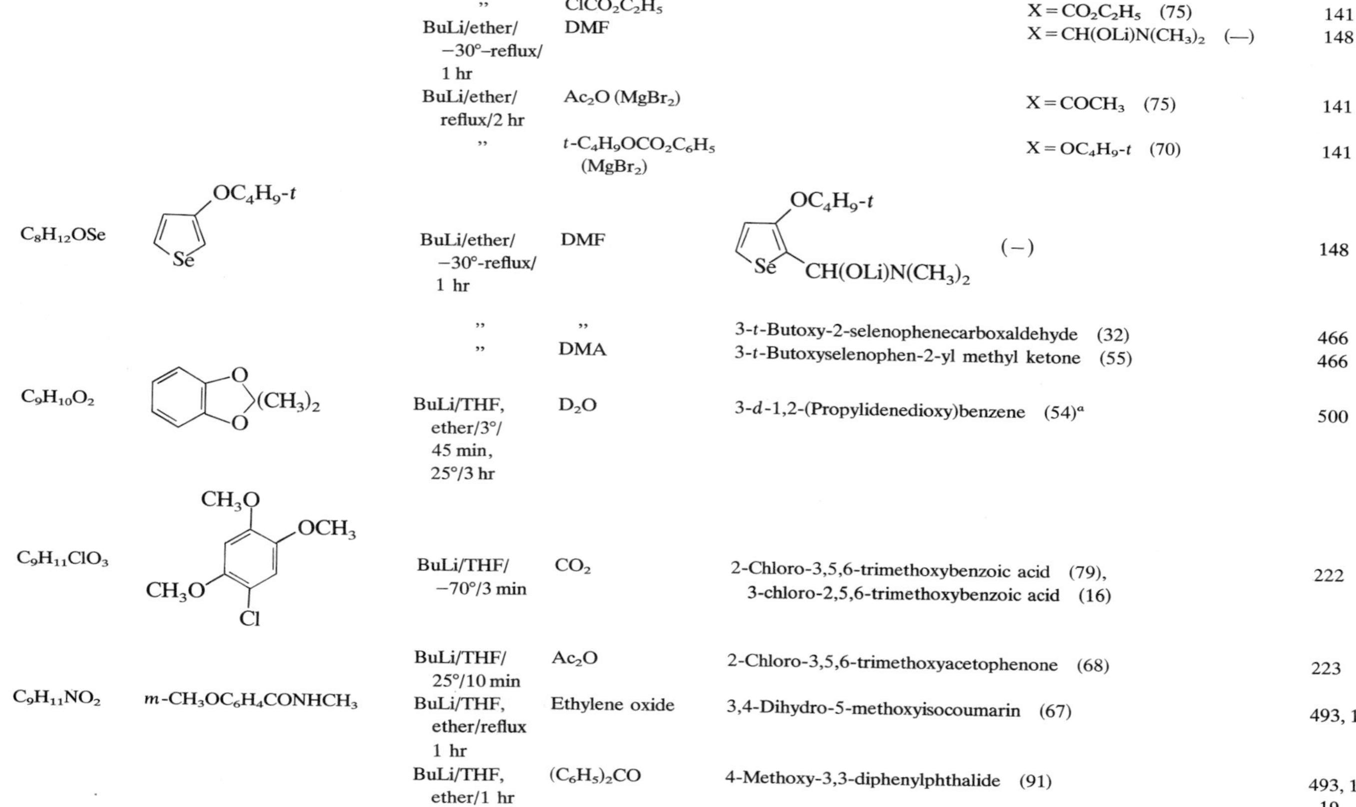

Formula	Substrate	Conditions	Electrophile	Product(s) and Yield(s) (%)	Refs.
		”	$ClCO_2C_2H_5$	$X = CO_2C_2H_5$ (75)	141
		BuLi/ether/ −30°–reflux/ 1 hr	DMF	$X = CH(OLi)N(CH_3)_2$ (—)	148
		BuLi/ether/ reflux/2 hr	Ac_2O ($MgBr_2$)	$X = COCH_3$ (75)	141
		”	t-$C_4H_9OCO_2C_6H_5$ ($MgBr_2$)	$X = OC_4H_9$-t (70)	141
$C_8H_{12}OSe$	3-(OC_4H_9-t)selenophene	BuLi/ether/ −30°-reflux/ 1 hr	DMF	3-(OC_4H_9-t)-2-[$CH(OLi)N(CH_3)_2$]selenophene (–)	148
		”	”	3-t-Butoxy-2-selenophenecarboxaldehyde (32)	466
		”	DMA	3-t-Butoxyselenophen-2-yl methyl ketone (55)	466
$C_9H_{10}O_2$	2,2-dimethyl-1,3-benzodioxole ($(CH_3)_2$)	BuLi/THF, ether/3°/ 45 min, 25°/3 hr	D_2O	3-d-1,2-(Propylidenedioxy)benzene (54)[a]	500
$C_9H_{11}ClO_3$	1-Cl-2,4,5-(CH_3O)$_3$-benzene	BuLi/THF/ −70°/3 min	CO_2	2-Chloro-3,5,6-trimethoxybenzoic acid (79), 3-chloro-2,5,6-trimethoxybenzoic acid (16)	222
		BuLi/THF/ 25°/10 min	Ac_2O	2-Chloro-3,5,6-trimethoxyacetophenone (68)	223
$C_9H_{11}NO_2$	m-$CH_3OC_6H_4CONHCH_3$	BuLi/THF, ether/reflux 1 hr	Ethylene oxide	3,4-Dihydro-5-methoxyisocoumarin (67)	493, 189
		BuLi/THF, ether/1 hr	$(C_6H_5)_2CO$	4-Methoxy-3,3-diphenylphthalide (91)	493, 189, 19

TABLE XLII. Alkyl Vinyl Ethers and Alkyl Aryl Ethers (Ortho, Beta) (*Continued*)

Formula	Compound Lithiated	Conditions	Substrate	Product and Yield (%)	Refs.
$C_9H_{12}O$	$C_6H_5OCH(CH_3)_2$	BuLi/THF/ 24 hr	CO_2	*o*-Isopropoxybenzoic acid (25)	35
	o-$C_2H_5C_6H_4OCH_3$	BuLi/ether/ reflux/ 60 hr	CO_2	3-Ethyl-*o*-anisic acid (32)	61
	3,5-$(CH_3)_2C_6H_3OCH_3$	BuLi/cyclo-hexane/ reflux/ 10 hr	CO_2	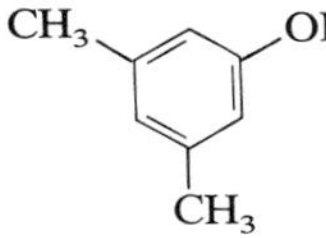 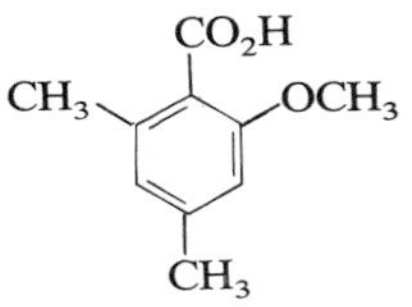 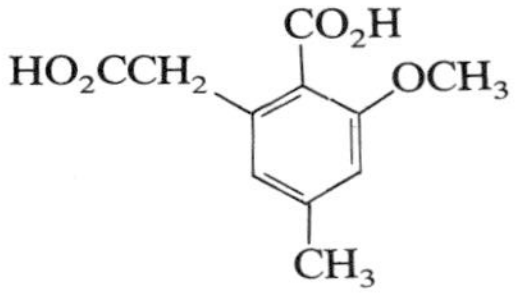 [structure: 3,5-dimethylphenol] (IX), [structure: 2-methoxy-4,6-dimethylbenzoic acid] (X), [structure: 2-carboxymethyl-6-methoxy-4-methylbenzoic acid] (XI) (IX+X+XI, 40) (IX:X:XI, 24:51:30)	17
		BuLi/ether	$(C_6H_5)_2CO$	2-Methoxy-4,6-dimethyl-α,α-diphenylbenzyl alcohol (80)	511
$C_9H_{12}O_2$	$C_6H_5OCH_2CH_2OCH_3$	BuLi/ether/ 2 hr	D_2O	*o*-*d*-2-Methoxyethoxybenzene (52), phenol (12), phenyl vinyl ether (8)	27
	m-$CH_3OC_6H_4CH(OH)CH_3$	BuLi/hexane, TMEDA/ 60°/5 hr*	CO_2 (CH_2N_2)	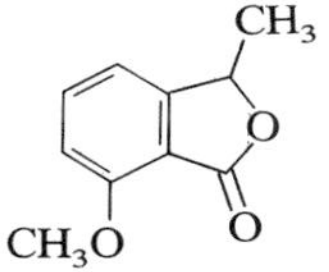 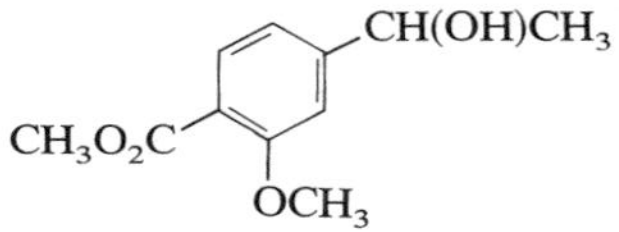 [structure: 3-methyl-7-methoxyphthalide] (62), [structure: methyl 2-methoxy-4-(1-hydroxyethyl)benzoate] (14)	245

Formula	Compound	Conditions	Electrophile	Product(s) (Yield, %)	Refs.
	3,4-$(CH_3O)_2C_6H_3CH_3$	BuLi/ether/ 0°/5 hr, 25° 5 hr	$Br(CH_2)_{10}Br$	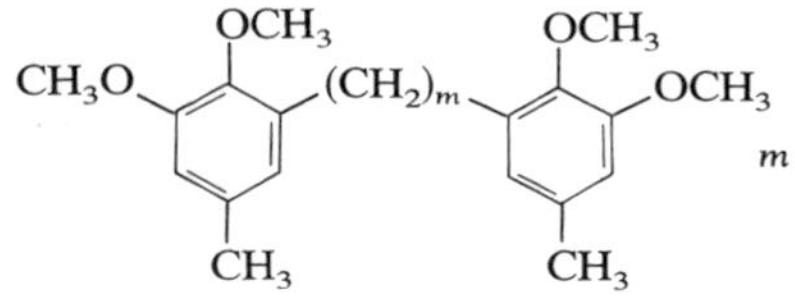 $m = 10$ (46)	512
		”	$I(CH_2)_{12}I$	$m = 12$ (46)	512
	3,5-$(CH_3O)_2C_6H_3CH_3$	BuLi/ether/ 0°/48 hr	I_2	4-Iodo-3,5-dimethoxytoluene (65)	513
		BuLi	CO_2	2,6-Dimethoxy-*p*-toluic acid (56)	514
$C_9H_{12}O_3$	1,2,3-$C_6H_3(OCH_3)_3$	BuLi/THF/ 25°/2 hr	D_2O	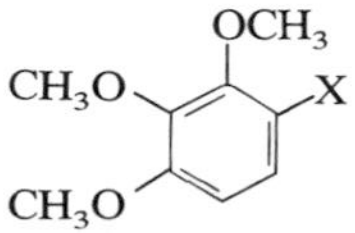 X = D (94)	224
		BuLi	CO_2	X = CO_2H (40)	514
		BuLi/THF/ 25°/2 hr	C_2H_5Br	X = C_2H_5 (56)	224
		”	n-$C_{12}H_{25}Br$	X = $C_{12}H_{25}$-n (55)	224
	1,3,5-$C_6H_3(OCH_3)_3$	BuLi/ether/ 25°/70 hr	CuBr	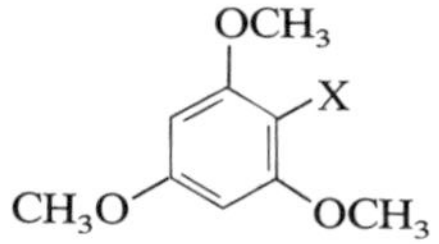 X = Cu (65)	501
		BuLi/ether	AgBr	X = Ag (21)	304
		BuLi/ether/ 25°/70 hr	$(CH_3)_2SnBr_2$	X = $Sn(CH_3)_2Br$ (95)	480
		BuLi/THF	$(CH_3)_3SnCl$	X = $Sn(CH_3)_3$ (—)	306
		—	$BrAuP(C_6H_5)_3$	X = $AuP(C_6H_5)_3$ (60)	307

TABLE XLII. ALKYL VINYL ETHERS AND ALKYL ARYL ETHERS (ORTHO, BETA) (*Continued*)

Formula	Compound Lithiated	Conditions	Substrate	Product and Yield (%)	Refs.
$C_9H_{12}O_3$ (*Contd.*)	3,5-$(CH_3O)_2C_6H_3CH_2OH$	BuLi/hexane, TMEDA/ 60°/5 hr*	$CO_2(CH_2N_2)$	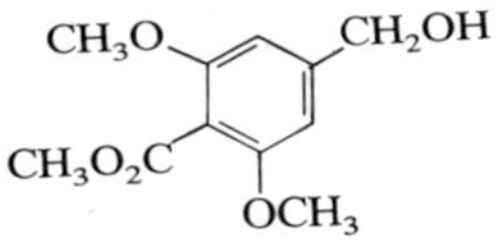 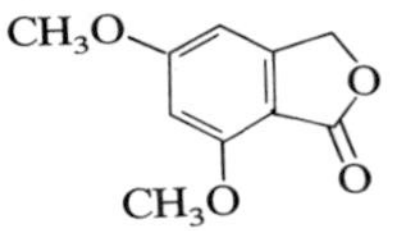 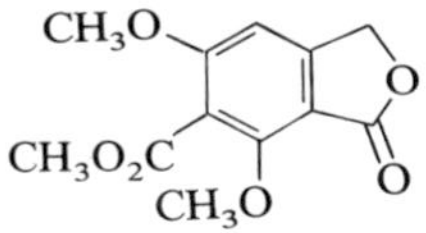 CH_3O, CH_2OH, CH_3O_2C, OCH_3 (22), CH_3O, O, CH_3O, O (8), CH_3O, O, CH_3O_2C, CH_3O, O (13)	245
$C_9H_{13}NO$	m-$CH_3OC_6H_4N(CH_3)_2$	BuLi/ether/ 35°/12 hr	$(C_6H_5)_2CO$	[2-(Dimethylamino)-6-methoxyphenyl]diphenylmethanol (71)	19
	o-$CH_3OC_6H_4N(CH_3)_2$	BuLi/ether/ 35°/12 hr	$(C_6H_5)_2CO$	[2-Methoxy-3-(dimethylamino)phenyl]diphenylmethanol (56)	19
	p-$CH_3OC_6H_4N(CH_3)_2$	BuLi/ether/ 35°/12 hr	D_2O	3-*d*-N,N-Dimethyl-*p*-anisidine (85)	19
		”	$(C_6H_5)_2CO$	[2-Methoxy-5-(dimethylamino)phenyl]diphenylmethanol (71)	19
$C_9H_{14}OS$	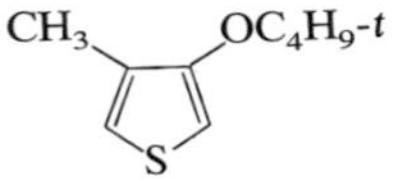 CH_3, OC_4H_9-*t*, S	BuLi/ether/ reflux/ 0.5 hr	*t*-$C_4H_9OCO_2C_6H_5$	2,3-Di-*t*-butoxy-4-methylthiophene (79)	147
$C_9H_{18}OS$	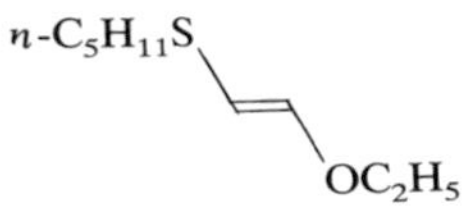 *n*-$C_5H_{11}S$, OC_2H_5	*t*-BuLi/THF −70°/1 hr	*n*-C_4H_9Br	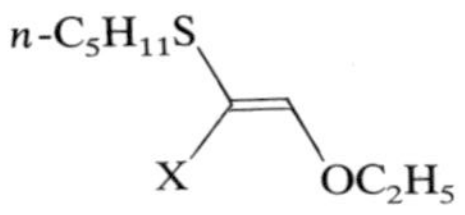 *n*-$C_5H_{11}S$, X, OC_2H_5 X = *n*-C_4H_9 (42)	127

Formula	Substrate	Conditions	Electrophile	Product(s) (Yield, %)	Refs.
		"	$n\text{-}C_4H_9I$	$X = n\text{-}C_4H_9$ (60)	127
		"	$n\text{-}C_6H_{13}CHO$	$X = CH(OH)C_6H_{13}\text{-}n$ (82)	127
		"	C_6H_5CHO	$X = CH(OH)C_6H_5$ (80)	127
$C_{10}H_7Cl_2NO_2$	3-CH_3O-5-($C_6H_3Cl_2$-2,6)isoxazole	BuLi/THF/ −70°/0.5 hr	CO_2	5-(2,6-Dichlorophenyl)-3-methoxy-4-isoxazolecarboxylic acid (75)	232
$C_{10}H_8ClNO_2$	3-CH_3O-5-(C_6H_4Cl-o)isoxazole	BuLi/THF/ −70°/0.5 hr	CO_2	5-(o-Chlorophenyl)-3-methoxy-4-isoxazolecarboxylic acid (62)	232
$C_{10}H_9NO_2$	3-CH_3O-5-(C_6H_5)isoxazole	BuLi/THF/ −70°/0.5 hr	I_2	4-Iodo-3-methoxy-5-phenylisoxazole (88)	232
		"	CO_2	3-Methoxy-5-phenyl-4-isoxazolecarboxylic acid (99)	232
$C_{10}H_{11}NO$	5-CH_3O-1-CH_3-indole	BuLi/ether/ reflux/ 13 hr	2-pyridinecarboxaldehyde (2-CHO-pyridine)	5-CH_3O-1-CH_3-2-X-indole (XII), 5-CH_3O-1-CH_3-6-X-indole (XIII), 5-CH_3O-1-CH_3-4-X-indole (XIV) $X = CH(OH)$-2-pyridyl (XII + XIII + XIV, 74) (XII : XIII : XIV, 4 : 5 : 1)	84
$C_{10}H_{12}O$	2,3,4,5-tetrahydro-1-benzoxepin	BuLi/hexane, TMEDA/ 25°/6 hr	CO_2	2,3,4,5-Tetrahydro-1-benzoxepin-9-carboxylic acid (73)	225

TABLE XLII. Alkyl Vinyl Ethers and Alkyl Aryl Ethers (Ortho, Beta) (*Continued*)

Formula	Compound Lithiated	Conditions	Substrate	Product and Yield (%)	Refs.
$C_{10}H_{12}OS$	$C_6H_5SCH=CHOC_2H_5$	t-BuLi/THF/ −70°/1 hr	D_2O	$C_6H_5SC(X)=CHOC_2H_5$ X = D (95)	127
		"	Ethylene oxide	X = $(CH_2)_2OH$ (60)	127
		"	Propylene oxide	X = $CH_2CH(OH)CH_3$ (55)	127
		"	n-C_4H_9I	X = n-C_4H_9 (55)	127
		"	$CH_3CH=CHCHO$	X = $CH(OH)CH=CHCH_3$ (78)	127
		"	Cyclopentanone	X = 1-hydroxycyclopentyl (78)	127
		"	C_6H_5CHO	X = $CH(OH)C_6H_5$ (75)	127
		"	n-$C_6H_{13}CHO$	X = $CH(OH)C_6H_{13}$-n (84)	127
$C_{10}H_{13}NO$	3,4-methylenedioxybenzyl-$N(CH_3)_2$ ($CH_2N(CH_3)_2$)	BuLi/ether/ 25°/1 hr	$(C_6H_5)_2CO$	[α-(Dimethylamino)-5,6-methylenedioxy)-*o*-tolyl]diphenylmethanol (60)	483
$C_{10}H_{14}O$	$C_6H_5OC_4H_9$-t	t-BuLi/cyclohexane/ reflux/ 17 hr	CO_2	*o*-*t*-Butoxybenzoic acid (82)	36
		BuLi/THF/ 24 hr	"	Salicylic acid (27)	35
$C_{10}H_{14}OS$	m-$CH_3OC_6H_4SC_3H_7$-i	BuLi/ether/ reflux/ 4 hr	CO_2	6-(Isopropylthio)-*o*-anisic acid (43)	574

	o-$CH_3OC_6H_4SC_3H_7$-i	BuLi/ether/ reflux/ 4 hr	CO_2	3-(Isopropylthio)-o-anisic acid (45)	574
	p-i-$C_3H_7OC_6H_4SCH_3$	BuLi/ether/ reflux/ 4 hr	CO_2	2-Isopropoxy-5-(methylthio)benzoic acid (38)	574
$C_{10}H_{14}O_2$	$C_6H_4(OC_2H_5)_2$-m	BuLi/THF	$(CH_3)_3SnCl$	(2,6-Diethoxyphenyl)trimethylstannane (73)	306
	m-$CH_3OC_6H_4C(CH_3)_2OH$	BuLi/hexane, TMEDA/ −60°/5 hr	$CO_2(CH_2N_2)$	7-Methoxy-3,3-dimethylphthalide (46)	245
$C_{10}H_{14}O_3$	3,5-$(CH_3O)_2C_6H_3CH$-$(OH)CH_3$	BuLi/hexane TMEDA/ 60°/5 hr	$CO_2(CH_2N_2)$	[structure: CH_3O, CH_3, O, O, CH_3O] (8), [structure: CH_3O, $CH(OH)CH_3$, CH_3O_2C, OCH_3] (24)	245
$C_{10}H_{14}O_4$	[structure: CH_3O, CH_3O, OCH_3, OCH_3]	BuLi/THF/ 25°	CO_2	1,2,3,6-Tetramethoxybenzoic acid (61)	514
	[structure: CH_3O, OCH_3, CH_3O, OCH_3]	BuLi/THF/25°/ 10 min	CO_2	[structure: CH_3O, OCH_3, CH_3O, X, OCH_3] (XV) $X=CO_2H$ (78)	223, 222, 514
		BuLi/THF, ether/ 25°/0.5 hr	CH_3I	$X=CH_3$ (75)	513
		BuLi/THF/ 25°/10 min	DMF	$X=CHO$ (62)	223

TABLE XLII. ALKYL VINYL ETHERS AND ALKYL ARYL ETHERS (ORTHO, BETA) (*Continued*)

Formula	Compound Lithiated	Conditions	Substrate	Product and Yield (%)	Refs.
$C_{10}H_{14}O_4$ (*Contd.*)	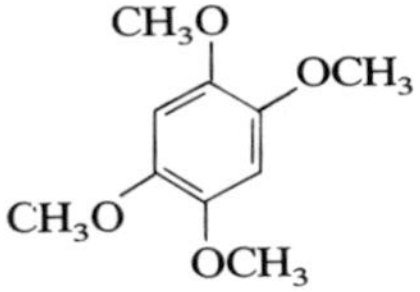 1,2,4,5-tetramethoxybenzene (CH_3O, OCH_3, CH_3O, OCH_3)	BuLi/THF/ 25°/40 min	$ClCH_2N(CH_3)_2$	$X = CH_2N(CH_3)_2$ (55)	507
		BuLi/THF/ 25°/10 min	Ac_2O	$X = Ac$ (64)	223
		BuLi/THF/ 25°/40 min	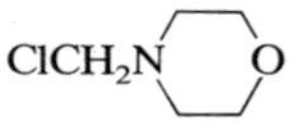 $ClCH_2N$(morpholine)O	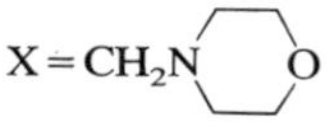 $X = CH_2N$(morpholine)O (64)	507
		BuLi/THF/ 25°/10 min	C_6H_5COCl	$X = COC_6H_5$ (88)	223
		BuLi/THF, ether/ 25°/0.5 hr	I_2	XV, 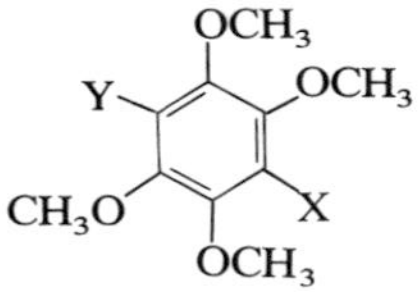 (Y, OCH_3, OCH_3, X, OCH_3, CH_3O-substituted benzene) (70) $X = I$ (9) $X = Y = I$	513
		4 BuLi/THF, ether/25°/ 15 min	I_2, CH_3I	(—) (39) $X = I$, $Y = CH_3$	513
		5 BuLi/ether/ reflux/ 38 hr	CO_2	(—) (35) $X = Y = CO_2H$	222, 514

Formula	Compound	Conditions	Electrophile	Product(s) (yield, %)	Refs.
		2 BuLi/THF/ 25°	$(CH_3O)_2SO_2$	(—) (86) $X = Y = CH_3$	514
$C_{10}H_{14}O_5$	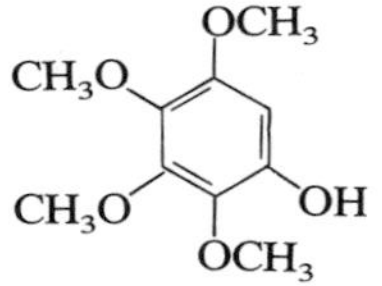 OCH_3, CH_3O, CH_3O, OH, OCH_3	BuLi/THF/ 25°/10 min	Ac_2O	2-Hydroxy-3,4,5,6-tetramethoxyacetophenone (85)	223
$C_{10}H_{15}NO$	*m*-$CH_3OC_6H_4CH_2N(CH_3)_2$	BuLi/ether	Ethylene oxide	[2-(Dimethylamino)methyl]-6-methoxyphenethyl alcohol (—)	189
		BuLi/ether/ 27°/2 hr	$(C_6H_5)_2CO$	[α-(Dimethylamino)-6-methoxy-*o*-tolyl]diphenylmethanol (79)	19, 45, 189
	o-$CH_3OC_6H_4CH_2N(CH_3)_2$	BuLi/ether/ 27°/2 hr	$(C_6H_5)_2CO$	[α-(Dimethylamino)-2-methoxy-*m*-tolyl]diphenylmethanol (58), [α-(Dimethylamino)-3-methoxy-*o*-tolyl]diphenylmethanol (<5)	19
	p-$CH_3OC_6H_4CH_2N(CH_3)_2$	BuLi/ether, TMEDA 15 hr	D_2O	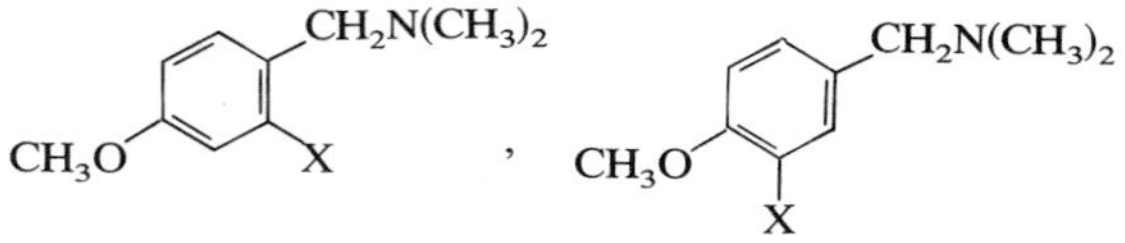 $CH_2N(CH_3)_2$, CH_3O, X , $CH_2N(CH_3)_2$, CH_3O, X (18) (48) $X = D$	19
		BuLi/ether, TMEDA/ 2 hr	$(C_6H_5)_2CO$	(7) (55) $X = C(C_6H_5)_2OH$	19, 180
$C_{11}H_{10}O$	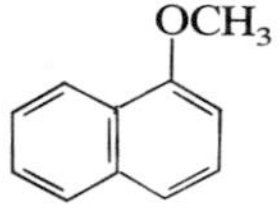 OCH_3	BuLi/ether/ reflux/ 20 hr	CO_2	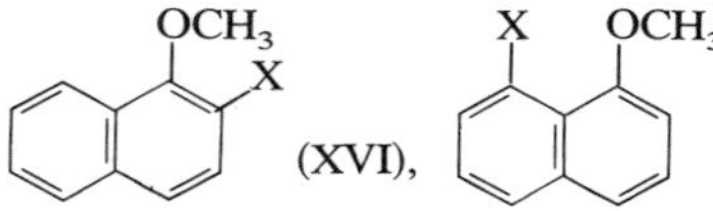 OCH_3, X (XVI), X, OCH_3 (XVII) (XVI + XVII, 29) (XVI : XVII, 65 : 35) $X = CO_2H$	226, 227

TABLE XLII. ALKYL VINYL ETHERS AND ALKYL ARYL ETHERS (ORTHO, BETA) (*Continued*)

Formula	Compound Lithiated	Conditions	Substrate	Product and Yield (%)	Refs.
$C_{11}H_{10}O$ (*Contd.*)	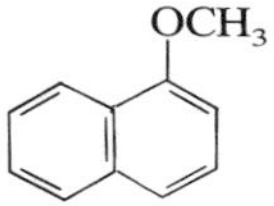 1-methoxynaphthalene (OCH_3)	BuLi/hexane, TMEDA/ 25°/2 hr	CO_2 (CH_2N_2)	(XVI + XVII, 60) (XVI : XVII, 99 : 0.3) X = CO_2CH_3	192
		t-BuLi/cyclo-hexane	"	(XVI + XVII, 35) (XVI : XVII, 1 : 99)	192
		BuLi/ether/ reflux/ 18 hr	$C_6H_5N(CH_3)CHO$	(37) (9) X = CHO	229, 515
		BuLi/THF/25°/ 2.5 hr	C_6H_5CN	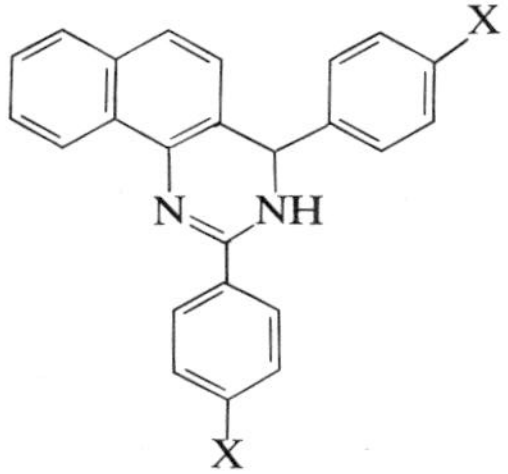 [structure: X, N, NH, X] (35) X = H	516
		"	*p*-ClC_6H_4CN	(31) X = Cl	516
	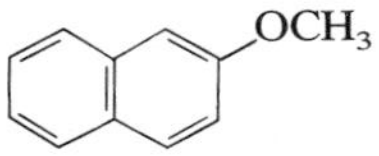 2-methoxynaphthalene (OCH_3)	BuLi/ether	CO_2	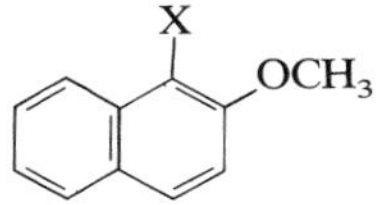 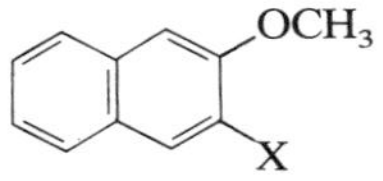 [structure: X, OCH_3] (XVIII), [structure: OCH_3, X] (XIX) (XVIII + XIX, 70) (XVIII : XIX, 7 : 43) X = CO_2H	17

Formula	Substrate	Conditions	Electrophile	Product(s)	(Yield)		Ref.
		BuLi/ether/ reflux/18 hr	$C_6H_5N(CH_3)CHO$	(7)	(65)	X = CHO	229, 515
		BuLi/ether	Ethylene oxide	(—)	(55)	$X = (CH_2)_2OH$	230
		”	$BrCH_2CH{=}CH_2$	(—)	(57)	$X = CH_2CH{=}CH_2$	230
		BuLi/THF/ 25°/2.5 hr	C_6H_5CN	(—)	(90)	$X = C(=NH)C_6H_5$	516
$C_{11}H_{11}NO$		BuLi/ether/ 0°/1 hr	Ethylene oxide, HBr		(4)		559
$C_{11}H_{11}NO_2$		BuLi	$C_6H_5CN\ (H_3O^+)$		(–)		478
		BuLi/ether/ 0°/45 min	$C_6H_5N(CH_3)CHO$	2,4-Dimethoxy-3-quinolinecarboxaldehyde	(68)		517, 515
		BuLi/THF/ −70°/0.5 hr	CO_2	5-Ethoxy-3-phenyl-4-isoxazolecarboxylic acid	(81)		232
$C_{11}H_{14}O$		BuLi/ether/ reflux/30 hr	O_2	2,2-Dimethyl-9-chromanol	(18)		518
		”	CO_2	2,2-Dimethyl-9-chromancarboxylic acid	(30)		518
		BuLi/hexane, TMEDA/ 25°/6 hr	CO_2	2,3,4,5-Tetrahydro-7-methyl-1-benzoxepin-9-carboxylic acid	(59)		225

TABLE XLII. ALKYL VINYL ETHERS AND ALKYL ARYL ETHERS (ORTHO, BETA) (*Continued*)

Formula	Compound Lithiated	Conditions	Substrate	Product and Yield (%)	Refs.
$C_{11}H_{14}O_2$	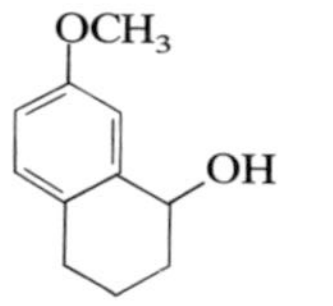 OCH_3, OH	BuLi/hexane, TMEDA/ 60°/5 hr	$CO_2(CH_2N_2)$	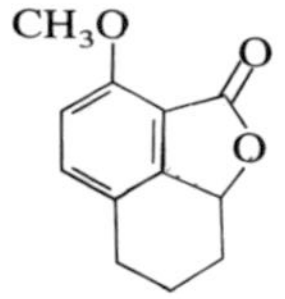 CH_3O, O, O (80)	245
$C_{11}H_{16}O$	$C_6H_5OCH(C_2H_5)_2$	BuLi/THF/ 24 hr	CO_2	*o*-(1-Ethylpropoxy)benzoic acid (17)	35
	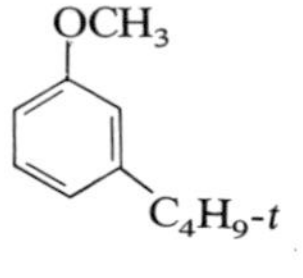 OCH_3, C_4H_9-*t*	*t*-BuLi/ cyclo- hexane	CO_2	4-*t*-Butyl-*o*-anisic acid (XX), 6-*t*-butyl-*o*-anisic acid (XXI) (XX+XXI, 73) (XX:XXI, 91:9)	17
$C_{11}H_{16}O$	*o*-*t*-$C_4H_9C_6H_4OCH_3$	BuLi/ether, TMEDA/ reflux/1 hr	$(CH_3)_3SiCl$	(3-*t*-Butyl-2-methoxyphenyl) trimethylsilane (29)	519
		"	$(C_6H_5)_2CO$	3-*t*-Butyl-2-methoxy-α,α-diphenylbenzyl alcohol (25)	519
$C_{11}H_{16}O_2$	3,5-$(CH_3O)_2C_6H_3C_3H_7$-*n*	C_6H_5Li/ether/ reflux/ 20 hr	CO_2	2,6-Dimethoxy-4-propylbenzoic acid (48)	520
$C_{11}H_{16}O_3$	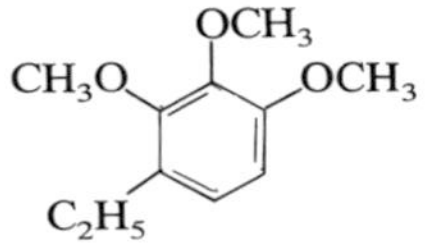 OCH_3, CH_3O, OCH_3, C_2H_5	BuLi/THF/ 25°/2 hr	C_2H_5Br	1,5-Diethyl-2,3,4-trimethoxybenzene (13)	224
$C_{11}H_{17}NO$	*p*-$CH_3OC_6H_4(CH_2)_2N(CH_3)_2$	BuLi/ether/ 27°/28 hr	D_2O	3-*d*-4-Methoxy-N,N-dimethylphenethylamine (72)	19

		BuLi/ether/ 27°/32 hr	$(C_6H_5)_2CO$	[2-Methoxy-5-(dimethylamino)ethylphenyl]-α,α-diphenylbenzyl alcohol (60)	19
$C_{11}H_{17}NO_2$	3,4-$(CH_3O)_2C_6H_3CH_2$-$N(CH_3)_2$	BuLi/ether/ 0°/3 hr	I_2	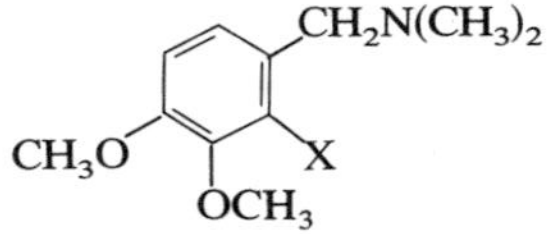 X = I (77)	23
		”	Ethylene oxide	X = $(CH_2)_2OH$ (60)	23
		”	CH_3CHO	X = $CH(OH)CH_3$ (65)	23
		”	CH_3OCH_2Cl	X = CH_2OCH_3 (66)	23
$C_{12}H_{12}O_2$	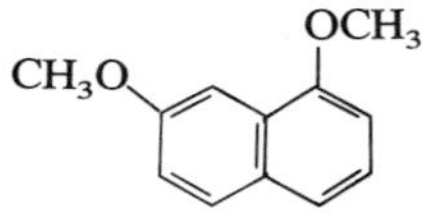	BuLi/ether/ reflux/ 9 hr	CO_2	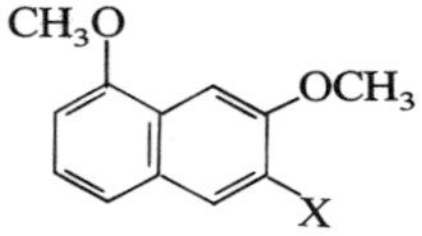 X = CO_2H (72)	521
		”	$(CH_3O)_2SO_2$	X = CH_3 (75)	521
		”	$C_6H_5N(CH_3)CHO$	X = CHO (84)	521
$C_{12}H_{13}NO_2$	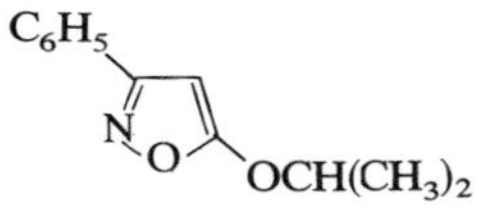	BuLi/THF/ −70°/0.5 hr	CO_2	5-Isopropoxy-3-phenyl-4-isoxazolecarboxylic acid (77)	232
$C_{12}H_{13}NO_3$	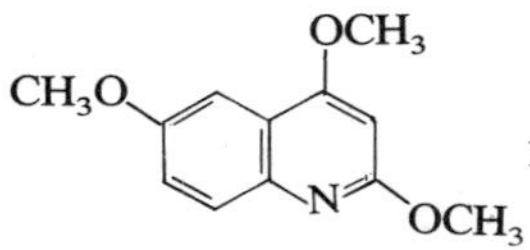	BuLi/ether/ 0°/3 hr, 25°/ “overnight”	Ethylene oxide	2,4,6-Trimethoxy-3-quinolinethanol (26)	559
		BuLi/ether/ 0°/1.5 hr	$C_6H_5N(CH_3)CHO$	2,4,6-Trimethoxy-3-quinoline-carboxaldehyde (58)	517

TABLE XLII. ALKYL VINYL ETHERS AND ALKYL ARYL ETHERS (ORTHO, BETA) (*Continued*)

Formula	Compound Lithiated	Conditions	Substrate	Product and Yield (%)	Refs.
$C_{12}H_{13}NO_3$ (*Contd.*)	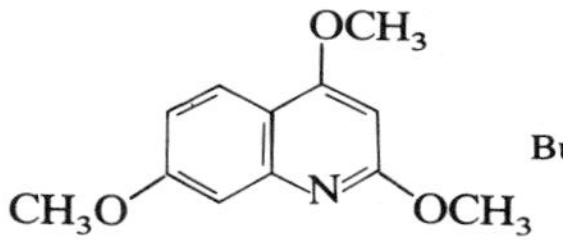 [structure: 2,4,7-trimethoxyquinoline; labels OCH$_3$, CH$_3$O, N, OCH$_3$]	BuLi/ether/ 0°/1.5 hr	$C_6H_5N(CH_3)CHO$	2,4,7-Trimethoxy-3-quinolinecarboxaldehyde (53)	517
	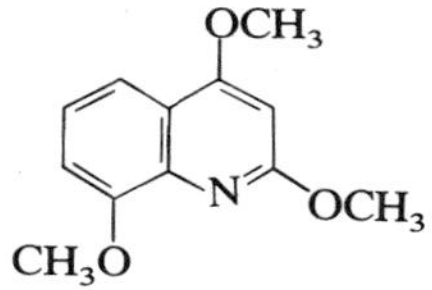 [structure: 2,4,8-trimethoxyquinoline; labels OCH$_3$, N, OCH$_3$, CH$_3$O]	BuLi/ether/ 0°/3 hr, 25°/ "overnight"	Ethylene oxide	2,4,8-Trimethoxy-3-quinolinethanol (14)	517
		BuLi/ether/ 0°/20 min	$C_6H_5N(CH_3)CHO$	2,4,8-Trimethoxy-3-quinolinecarboxaldehyde (40)	559
$C_{12}H_{14}O_2$	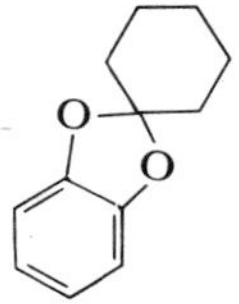 [structure: spiro[1,3-benzodioxole-2,1'-cyclohexane]]	BuLi/THF, ether/3°/ 45 min, 25°/3 hr	D_2O	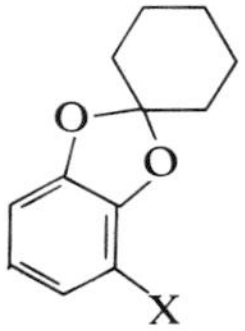 [structure: 4-X-spiro[1,3-benzodioxole-2,1'-cyclohexane]] X = D (92)[b]	500
		BuLi/THF, ether/0°/ 5 hr, 25°/ 5 hr	$\frac{1}{2}Br(CH_2)_{10}Br$	X = $(CH_2)_{10}$–[spiro[1,3-benzodioxole-2,1'-cyclohexane]-4-yl] (76) 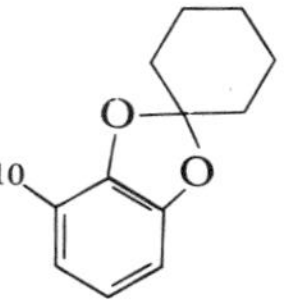	512
		BuLi/THF, ether/0°/ 1 hr, 25°/ 2.5 hr	n-$C_{12}H_{25}Br$	X = $C_{12}H_{25}$-n (71)	499

		BuLi/THF, ether/ 0°–25°/7 hr	$Br(CH_2)_{11}OTHP$	$X = (CH_2)_{11}OTHP$ (84)	500
$C_{12}H_{16}O_2$	OCH_3, OH, CH_3	BuLi/hexane, TMEDA/ 60°/5 hr	$CO_2(CH_2N_2)$	CH_3O, O, O, CH_3 (42), CH_3O_2C, OCH_3, OH, CH_3 (23)	245
	O, CH_3, OCH_3	BuLi/hexane, TMEDA/25°/ 6 hr	CO_2	2,3,4,5-Tetrahydro-5-methoxy-7-methyl-1-benzoxepin-9-carboxylic acid (65)	225
$C_{12}H_{16}O_3$	OCH_3, OTHP	BuLi/ether/ 25°/22 hr	CO_2 (H_3O^+)	6-Hydroxy-*o*-anisic acid (26)	233
$C_{12}H_{18}O$	CH_3, OCH_3, C_4H_9-t	BuLi/cyclo-hexane/ reflux/ 10 hr	CO_2	CH_3, OCH_3, CO_2H, C_4H_9-*t* (XXII) (12), CO_2H, CH_3, OCH_3, C_4H_9-*t* (XXIII) (24); CO_2H, HO_2CCH_2, OCH_3, C_4H_9-*t* (XXIV) (63)	17

TABLE XLII. ALKYL VINYL ETHERS AND ALKYL ARYL ETHERS (ORTHO, BETA) (*Continued*)

Formula	Compound Lithiated	Conditions	Substrate	Product and Yield (%)	Refs.
$C_{12}H_{18}O$ (*Contd.*)	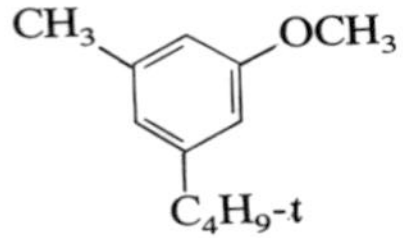 3-CH_3-5-OCH_3-C_6H_3-C_4H_9-*t*	BuLi/cyclohexane, TMEDA/25°	CO_2	XXII (3), XXIII (97), XXIV (0)	17
$C_{12}H_{19}NO$	*p*-$CH_3OC_6H_4(CH_2)_3N(CH_3)_2$	BuLi/ether/ 27°/24 hr	D_2O	3-(3-*d*-4-Methoxyphenyl)-N,N-dimethylpropylamine (70)	19
		"	$(C_6H_5)_2CO$	[5-[(3-Dimethylamino)propyl]-2-methoxyphenyl]-α,α-diphenylbenzyl alcohol (71)	19
$C_{12}H_{20}O_2S$	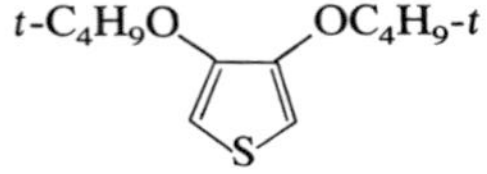 3,4-di(*t*-C_4H_9O)thiophene	BuLi/ether/ 20°/0.5 hr	$(CH_3O)_2SO_2$	3,4-Di-*t*-butoxy-2-methylthiophene (91)	147
		BuLi/ether/ −20°–reflux 1 hr	DMF	3,4-Di-*t*-butoxy-2-thiophenecarboxaldehyde (50)	148
$C_{12}H_{20}O_2Se$	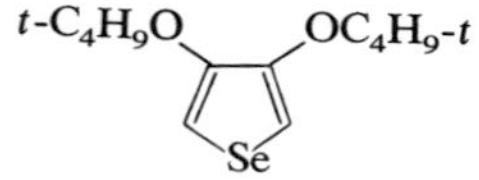 3,4-di(*t*-C_4H_9O)selenophene	BuLi/ether/ −20°–reflux/ 1 hr	DMF	3,4-Di-*t*-butoxy-2-selenophenecarboxaldehyde (5)	148
$C_{13}H_{10}O$	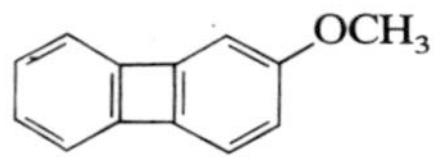 2-OCH_3-biphenylene	BuLi	$B(OC_4H_9\text{-}n)_3$, (H_2O_2)	2-Methoxy-1-biphenylenol (61)	522
$C_{13}H_{15}NO_2$	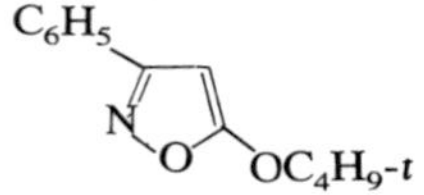 3-C_6H_5-5-OC_4H_9-*t*-isoxazole	BuLi/THF/ −70°/0.5 hr	CO_2	5-*t*-Butoxy-3-phenyl-4-isoxazolecarboxylic acid (75)	232

$C_{13}H_{17}NO_3$	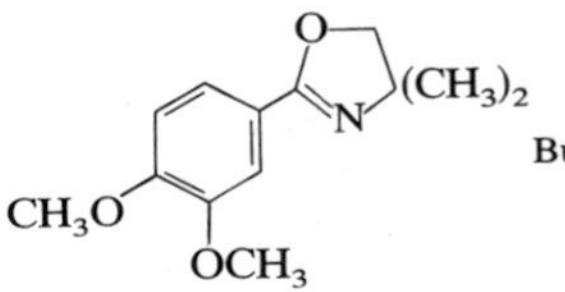	BuLi/THF/ −45°/1.5 hr	D_2O	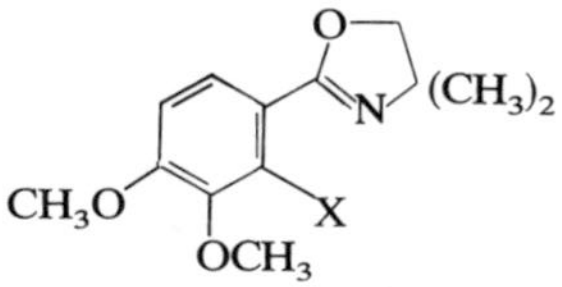 X = D (95)	204
		”	$(CH_3S)_2$	X = SCH_3 (92)	204
		”	N-Chloro-succinimide	X = Cl (90)	204
$C_{13}H_{20}O_2$	3,5-$(CH_3O)_2C_6H_3C_5H_{11}$-*n*	BuLi/ether, hydrocarbon/ 25°/16 hr	$C_6H_5N(CH_3)CHO$	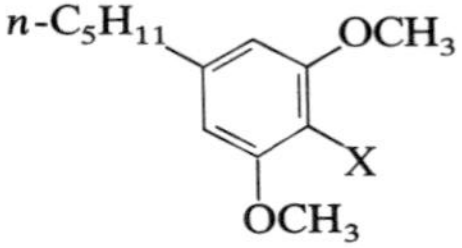 X = CHO (83)	523
		—	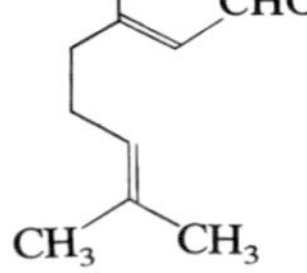 (Pyridine/TosOH)	X = 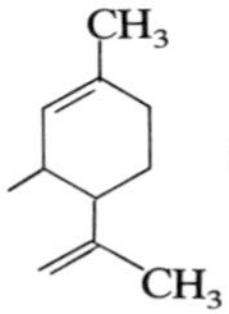 (−)	524
$C_{13}H_{22}O_2S$	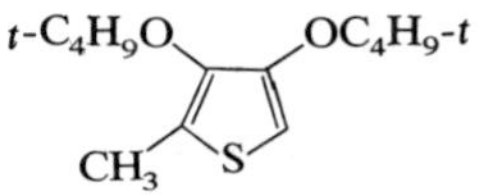	BuLi/ether/ 20°/0.5 hr	$(CH_3O)_2SO_2$	3,4-Di-*t*-butoxy-2,5-dimethylthiophene (81)	147
$C_{14}H_{17}NO$	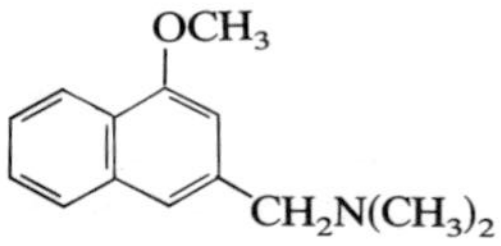	BuLi/ether/ 25°	CO_2 (CH_2N_2)	Methyl [3-(dimethylamino)methyl]-1-methoxy-2-naphthoate (56)	185

TABLE XLII. ALKYL VINYL ETHERS AND ALKYL ARYL ETHERS (ORTHO, BETA) (*Continued*)

Formula	Compound Lithiated	Conditions	Substrate	Product and Yield (%)	Refs.
$C_{14}H_{17}NO$ (*Contd.*)	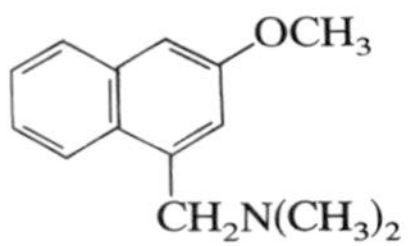 [structure: OCH_3, $CH_2N(CH_3)_2$]	BuLi/ether/ 25°	CO_2 (CH_2N_2)	Methyl [1-(dimethylamino)methyl]-3-methoxy-2-naphthoate (39)	185
$C_{14}H_{17}NO_2$	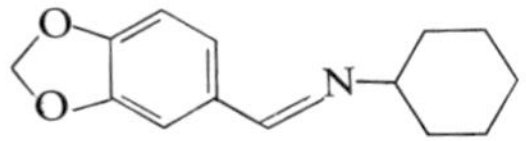 [structure: O, O, N]	BuLi/THF/ −78°/15 min	D_2O	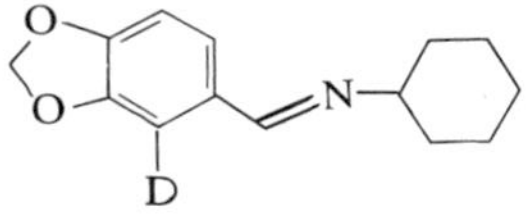 [structure: O, O, N, D] (100)	212
		”	I_2	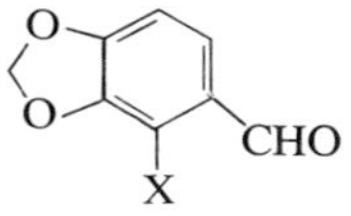 [structure: O, O, CHO, X] X = I (60)	212
		”	CO_2 (H_3O^+)	X = CO_2H (54)	212
		”	CH_3I (H_3O^+)	X = CH_3 (61)	212
		”	$ClCO_2CH_3$ (H_3O^+)	X = CO_2CH_3 (68)	212
		”	$CH_2{=}CHCH_2Br$ (CuI) (H_3O^+)	X = $CH_2CH{=}CH_2$ (72)	212
$C_{14}H_{18}O_2$	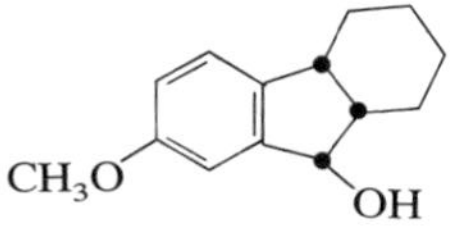 [structure: CH_3O, OH]	BuLi, $NaOC_4H_9$-*t*/ hexane/ 25°/1 hr	CO_2	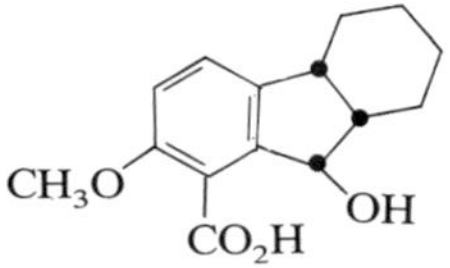 [structure: CH_3O, CO_2H, OH] (88)	525

Formula	Substrate	Conditions	Electrophile	Product (Yield %)	Ref.
$C_{15}H_{19}NO_2$	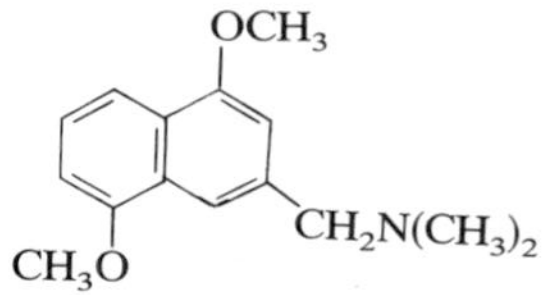 [structure: OCH_3, CH_3O, $CH_2N(CH_3)_2$]	BuLi/ether/25°	CO_2 (CH_2N_2)	Methyl [3-(dimethylamino)methyl]-1,5-dimethoxy-2-naphthoate (52)	185
$C_{15}H_{20}O_2$	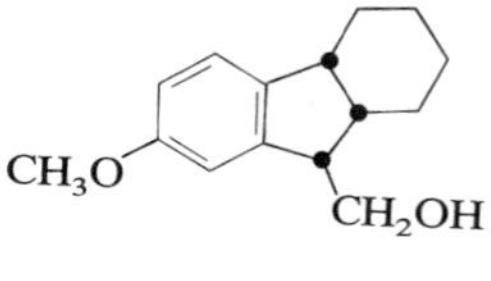 [structure: CH_3O, CH_2OH]	BuLi/hexane, TMEDA/ reflux/ 45 min	CO_2 (CH_2N_2)	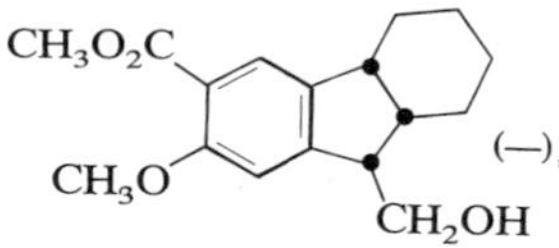 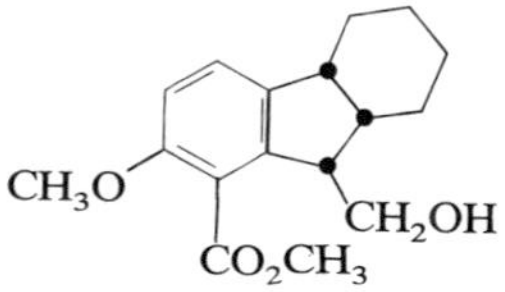 [structure: CH_3O_2C, CH_3O, CH_2OH] (—), [structure: CH_3O, CO_2CH_3, CH_2OH] (–)	525
$C_{16}H_{22}O_2$	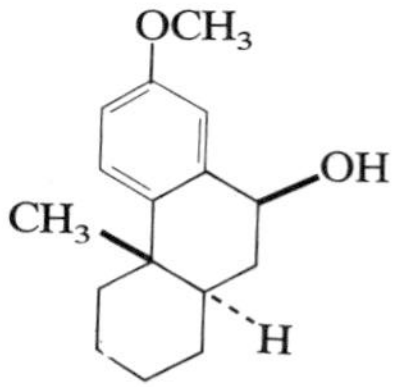 [structure: OCH_3, OH, CH_3, H]	BuLi/hexane, TMEDA/60°/ 5 hr	CO_2 (CH_2N_2)	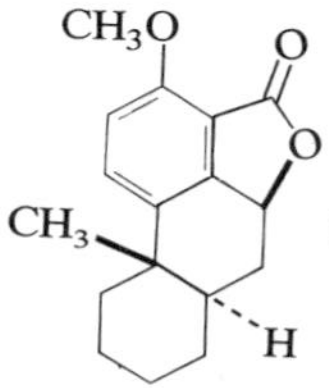 [structure: CH_3O, O, O, CH_3, H] (80)	245
$C_{17}H_{18}O$	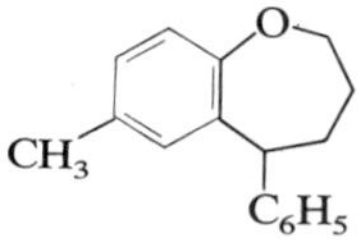 [structure: O, CH_3, C_6H_5]	BuLi/hexane, TMEDA/25°/ 6 hr	CO_2	2,3,4,5-Tetrahydro-7-methyl-5-phenyl-1-benzoxepin-9-carboxylic acid (48)	225
$C_{18}H_{26}O_6$	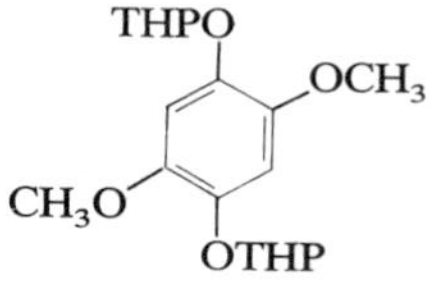 [structure: THPO, OCH_3, CH_3O, OTHP]	BuLi/THF/25°/ 45 min	CO_2 (H_3O^+)	2,5-Dihydroxy-3,6-dimethoxybenzoic acid (86)	223, 222

TABLE XLII. ALKYL VINYL ETHERS AND ALKYL ARYL ETHERS (ORTHO, BETA) (*Continued*)

Formula	Compound Lithiated	Conditions	Substrate	Product and Yield (%)	Refs.
$C_{18}H_{26}O_6$ (*Contd.*)	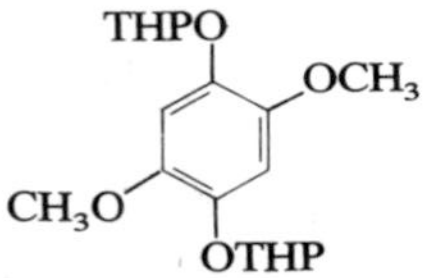 THPO OCH_3 CH_3O OTHP	BuLi/THF/25°	DMF	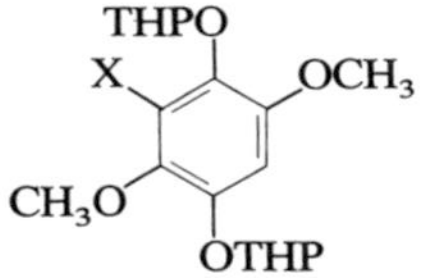 THPO X OCH_3 CH_3O OTHP X = CHO (74)	223
		BuLi/THF/25°/ 45 min	Ac_2O	X = Ac (82)	223
$C_{20}H_{34}O_2$	2,4-$(CH_3O)_2C_6H_3$-$C_{12}H_{25}$-*n*	BuLi/THF/ −20°/2 hr, 0°/0.5 hr	*n*-$C_{12}H_{25}Br$	1,3-Didodecyl-2,4-dimethoxybenzene (45)	499
$C_{20}H_{34}O_3$	3,4-$(CH_3O)_2C_6H_3$-$(CH_2)_{11}OCH_3$	BuLi/THF, ether/0°/ 5 hr, 25°/5 hr	½$Br(CH_2)_{10}Br$	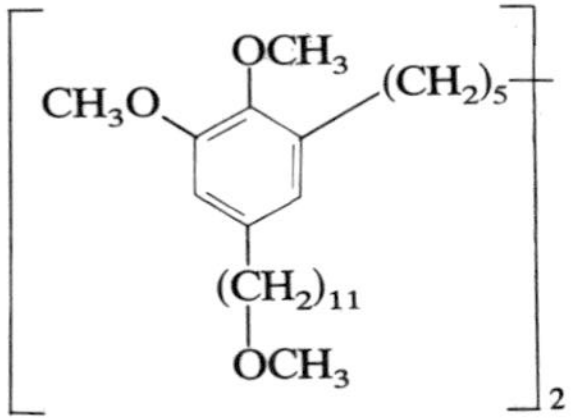 [OCH_3 CH_3O $(CH_2)_5$— $(CH_2)_{11}$ OCH_3]$_2$ (38)	512
$C_{21}H_{36}O_3$	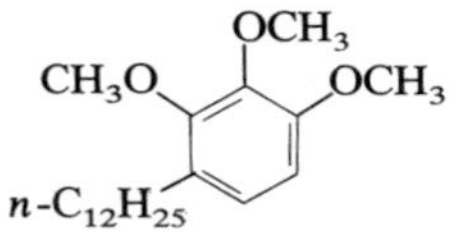 OCH_3 CH_3O OCH_3 *n*-$C_{12}H_{25}$	BuLi/THF, TMEDA/ 25°/3 hr	D_2O	1-*d*-5-Decyl-2,3,4-trimethoxybenzene (100)	224
		BuLi/THF/25°/ 2 hr	*n*-$C_{12}H_{25}Br$	1,5-Didodecyl-2,3,4-trimethoxybenzene (25)	224

$C_{26}H_{30}O_4$	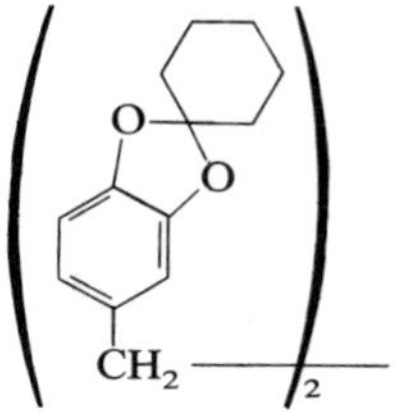 CH_2 (bis-structure, subscript 2)	2 BuLi/THF, ether/0°/ 5 hr, 25°/5 hr	$Br(CH_2)_{11}OTHP$ (H_3O^+)	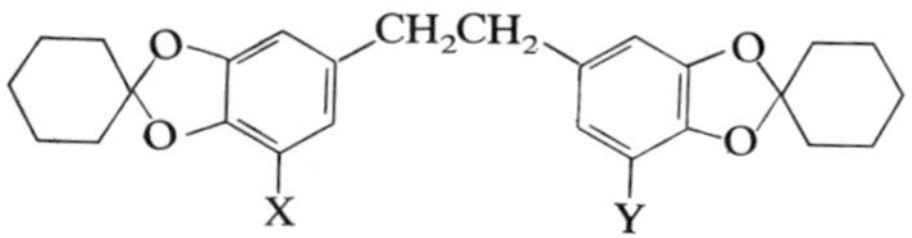 CH_2CH_2 (X, Y); X = Y = $(CH_2)_{11}OH$ (19), X = H, Y = $(CH_2)_{11}OH$ (47)	512
$C_{27}H_{46}O_2$	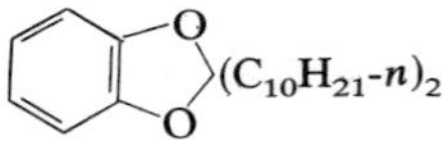 $(C_{10}H_{21}\text{-}n)_2$	BuLi/THF, ether/−15°–25°/2 hr	D_2O	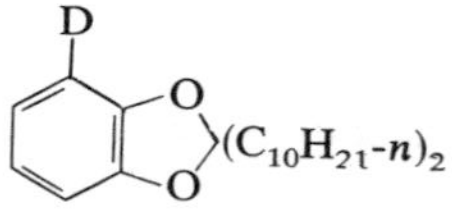 D; $(C_{10}H_{21}\text{-}n)_2$ (70)	500
$C_{28}H_{44}O_4$	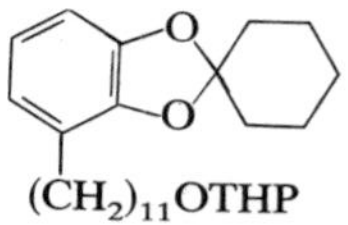 $(CH_2)_{11}OTHP$	BuLi/THF, ether/0°–25°/8 hr	$Br(CH_2)_{11}OTHP$	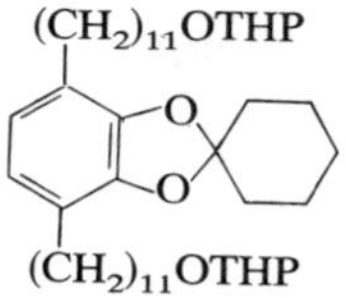 $(CH_2)_{11}OTHP$; $(CH_2)_{11}OTHP$ (61)	500

[a] The isolated yield was 54%; the deuterium incorporation was 30%.

[b] The isolated yield was 92%; the deuterium incorporation was 97%.

Note: References 360–607 are on pp. 355–360.

TABLE XLIII. ALKOXYALKYL ARYL ETHERS (ORTHO)

Formula	Compound Lithiated	Conditions	Substrate	Product and Yield (%)	Refs.
$C_8H_9ClO_2$	o-$ClC_6H_4OCH_2OCH_3$	BuLi/hexane, TMEDA/0°/ 0.5 hr	DMF	3-Chloro-2-(methoxymethoxy)-benzaldehyde (85)	234
$C_9H_{11}BrO_2$	OCH_2OCH_3, Br, CH_3	C_6H_5Li/ether, pet. ether/ 25°/24 hr	CO_2 (CH_2N_2)	Methyl 5-bromo-2-(methoxymethoxy)-p-toluate (I), methyl 3-bromo-6-(methoxymethoxy)-o-toluate (II) (I+II, 93) (I:II, 92:8)	526
$C_9H_{12}O_2$	m-$CH_3C_6H_4OCH_2OCH_3$	t-BuLi/pet. ether/0°/ 2 hr	CO_2 (CH_2N_2)	Methyl 2-(methoxymethoxy)-p-toluate (90)	526
	o-$CH_3C_6H_4OCH_2OCH_3$	BuLi/hexane, TMEDA/0°/ 0.5 hr	DMF	2-(Methoxymethoxy)-m-tolualdehyde (80)	234
	p-$CH_3C_6H_4OCH_2OCH_3$	BuLi/hexane, TMEDA/0°/ 0.5 hr	DMF	6-(Methoxymethoxy)-m-tolualdehyde (81)	234
	$C_6H_5O(CH_2)_2OCH_3$	BuLi/ether/ 2 hr	D_2O	2-d-(2-Methoxyethoxy)benzene (52) phenol (12), phenoxyethylene (8)	27
$C_{10}H_{14}O_4$	OCH_2OCH_3, OCH_2OCH_3	BuLi/ether/ 25°/24 hr	Ethylene oxide	2,3-Bis(methoxymethoxy)phenethyl alcohol (72)	527

Formula	Compound	Conditions	Electrophile	Product (Yield %)	Ref.
	1,4-Bis(methoxymethoxy)benzene (OCH_2OCH_3 / OCH_2OCH_3)	BuLi/ether/ 25°/24 hr	Ethylene oxide	2,5-Bis(methoxymethoxy)phenethyl alcohol (92)	527
$C_{11}H_{14}O_2$	Phenyl-OTHP	BuLi	CO_2 (H_3O^+)	Salicylic acid (52)	528
			Ethylene oxide (H_3O^+)	*o*-Hydroxyphenethyl alcohol (70)	235
$C_{12}H_{16}O_3$	3-Methoxyphenyl-OTHP ($OTHP$, OCH_3)	BuLi/ether/ 25°/22 hr	CO_2 (H_3O^+)	6-Hydroxy-*o*-anisic acid (26)	233
$C_{16}H_{18}O_4$	4,4'-Bis(methoxymethoxy)biphenyl (OCH_2OCH_3 / OCH_2OCH_3)	BuLi/ether/ 25°/5 min	Ethylene oxide	4,4'-Bis(methoxymethoxy)biphenyl, 3-X (OCH_2OCH_3, X / OCH_2OCH_3): $X = (CH_2)_2OH$ (28)	527
		”	CH_3CHO	$X = CH(OH)CH_3$ (11)	527

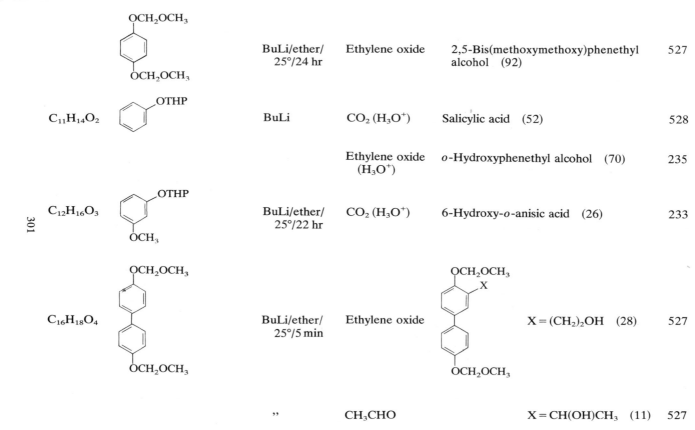

TABLE XLIII. ALKOXYALKYL ARYL ETHERS (ORTHO) (*Continued*)

Formula	Compound Lithiated	Conditions	Substrate	Product and Yield (%)	Refs.
$C_{16}H_{22}O_4$	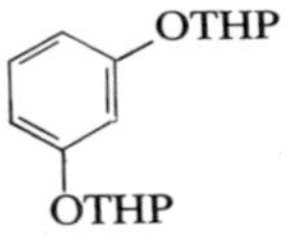 OTHP OTHP	BuLi	CO_2 (H_3O^+)	2,6-Dihydroxybenzoic acid (60)	528
	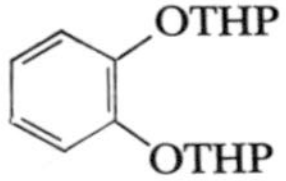 OTHP OTHP	BuLi/ether/ reflux/20 hr	CO_2 (H_3O^+)	2,3-Dihydroxybenzoic acid (48)	528
	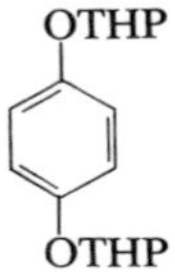 OTHP OTHP	BuLi	CO_2 (H_3O^+)	2,5-Dihydroxybenzoic acid (65)	528, 235
$C_{18}H_{26}O_6$	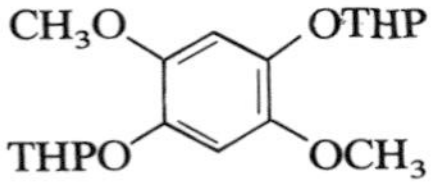 CH_3O OTHP THPO OCH_3	BuLi/THF/ 25°/45 min	CO_2 (H_3O^+)	2,5-Dihydroxy-3,6-dimethoxybenzoic acid (86)	223, 222
		BuLi/THF/ 25°	DMF	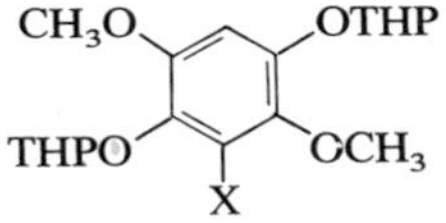 CH_3O OTHP THPO OCH_3 X X = CHO (74)	223
		BuLi/THF/ 25°/45 min	Ac_2O	X = $COCH_3$ (82)	223

Note: References 360–607 are on pp. 355–360.

TABLE XLIV. DIARYL ETHERS AND CONDENSED DIARYL ETHERS (ORTHO)

Formula	Compound Lithiated	Conditions	Substrate	Product and Yield (%)	Refs.
$C_{12}H_8O$	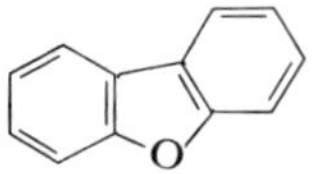	BuLi/ether	Br_2	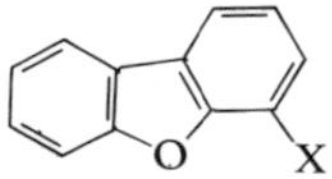 X = Br (17)	249
		”	O_2	X = OH (45)	288
		BuLi/THF, ether/ −55°–0°/1 hr	CO_2	X = CO_2H (86)	529, 560
		BuLi/ether/25°/ 38 hr	CH_3ONH_2	X = NH_2 (79)	286
		—	$(CH_3)_2N$–CH=CH–NO_2	X = CH=CHNO_2 (56)	285
		BuLi/ether/ reflux/17 hr	$(C_6H_5)_3SiCl$	X = $Si(C_6H_5)_3$ (63)	452
$C_{12}H_8OS$	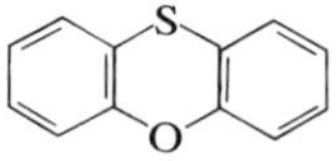	BuLi/ether/ reflux/24 hr	O_2	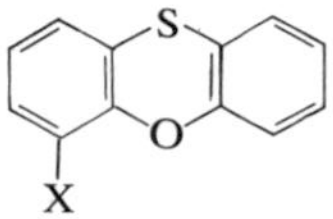 (16) X = OH	288
		BuLi	CO_2	(37) X = CO_2H	237
		BuLi/ether/reflux	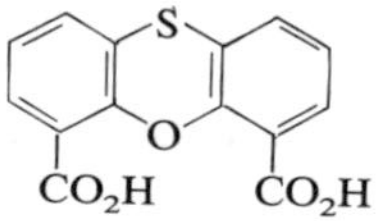 ”	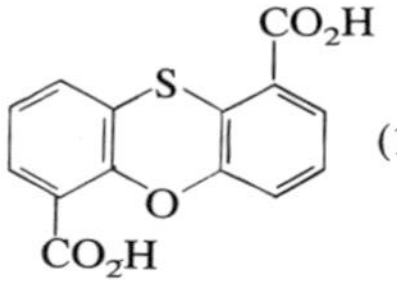 CO_2H, CO_2H (35), CO_2H, CO_2H (1–9)	236, 237

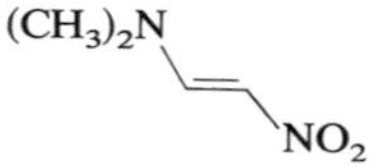

TABLE XLIV. DIARYL ETHERS AND CONDENSED DIARYL ETHERS (ORTHO) (*Continued*)

Formula	Compound Lithiated	Conditions	Substrate	Product and Yield (%)		Refs.
$C_{12}H_9ClO$	p-$ClC_6H_4OC_6H_5$	C_6H_5Li/ether/ 20°/26 hr*	$(C_6H_5)_2SiCl_2$	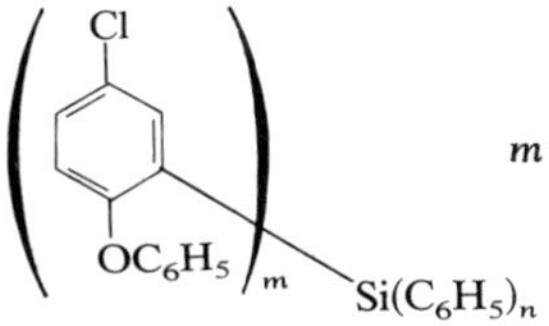 $(4\text{-}Cl\text{-}2\text{-}(OC_6H_5)C_6H_3)_m Si(C_6H_5)_n$	$m = n = 2$ (7)	243
		,,	$(C_6H_5)_3SiCl$		$m = 1$, $n = 3$ (38)	243
$C_{12}H_9NO$	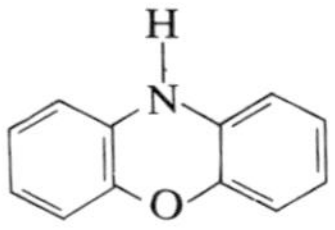 phenoxazine	BuLi/ether/25°/ 110 hr	CO_2	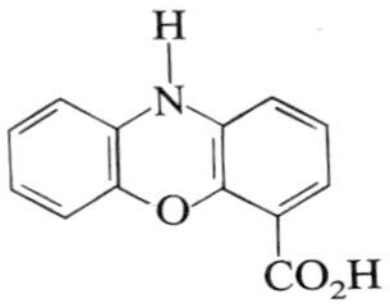 phenoxazine-4-carboxylic acid (CO_2H) (5)		238
$C_{12}H_{10}O$	$(C_6H_5)_2O$	BuLi/THF/25°/ 2 hr	Ethylene oxide	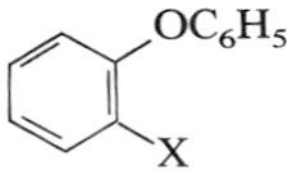 $2\text{-}X\text{-}C_6H_4OC_6H_5$	$X = (CH_2)_2OH$ (93)	23
		BuLi/ether/reflux/ 24 hr	$(C_6H_5)_3SiCl$		$X = Si(C_6H_5)_3$ (67)	243
		BuLi/ether/reflux/ 72 hr	CO_2	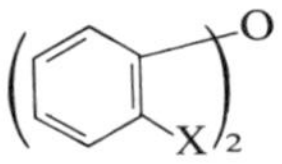 $(2\text{-}X\text{-}C_6H_4)_2O$	$X = CO_2H$ (23)	239
		,,	$(CH_3)_3SiCl$		$X = Si(CH_3)_3$ (60)	242

BuLi/THF, ether/ reflux/5 hr	$(CH_3)_2SiCl_2$	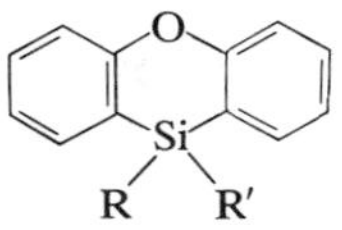 [10-silaphenoxazine-type structure: dibenzoxasiline with Si(R)(R′)]	$R = R' = CH_3$ (32)	240, 239
,,	$C_6H_5SiH_3$		$R = H$, $R' = C_6H_5$ (26)	241
,,	$(C_6H_5)_2SiH_2$		$R = R' = C_6H_5$ (20)	241
BuLi/ether/reflux/ 72 hr	$(C_6H_5)_2SiCl_2$		$R = R' = C_6H_5$ (34)	239
,,	$(C_6H_5CH_2)_2SiCl_2$		$R = R' = CH_2C_6H_5$ (52)	242
,,	$(n\text{-}C_{12}H_{25})_2SiCl_2$		$R = R' = n\text{-}C_{12}H_{25}$ (17)	242
,,	$SiCl_4$	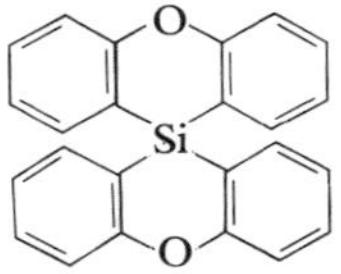 [spiro bis(dibenzoxasiline) structure] (25)		239
BuLi/ether/reflux/ 24 hr	$C_6H_5SiCl_3$	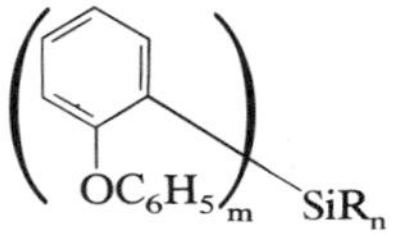 $(o\text{-}C_6H_5OC_6H_4)_m SiR_n$	$m = 3$, $n = 1$, $R = C_6H_5$ (14)	243
,,	$(C_6H_5)_2SiCl_2$		$m = n = 2$, $R = C_6H_5$ (40)	243
,,	$(n\text{-}C_{12}H_{25})_2SiCl_2$		$m = n = 2$, $R = n\text{-}C_{12}H_{25}$ (53)	242

TABLE XLIV. Diaryl Ethers and Condensed Diaryl Ethers (Ortho) (*Continued*)

Formula	Compound Lithiated	Conditions	Substrate	Product and Yield (%)		Refs.
$C_{14}H_{12}O$	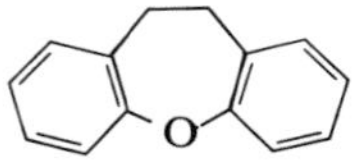 O	BuLi/THF/25°/ 7 hr	CH_2O	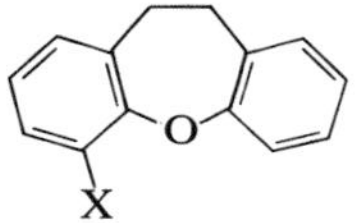 O, X	$X = CH_2OH$ (56)	23
		”	Ethylene oxide		$X = (CH_2)_2OH$ (52)	23
		BuLi/ether/reflux/ 68 hr	CO_2 (CH_2N_2)	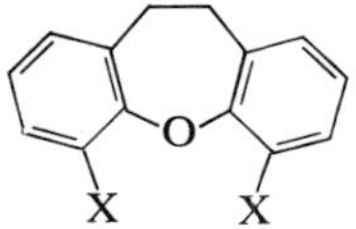 O, X, X	$X = CO_2CH_3$ (30)	530
		”	CH_2O		$X = CH_2OH$ (55)	530
$C_{14}H_{13}NO$	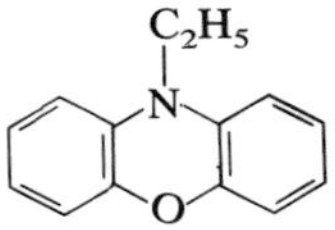 C_2H_5, N, O	BuLi/reflux/42 hr	CO_2	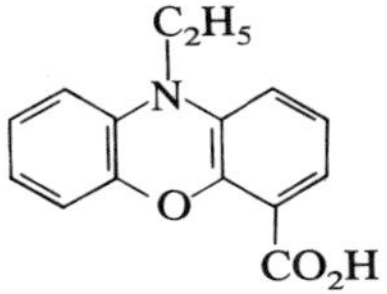 C_2H_5, N, O, CO_2H (35)		238
$C_{14}H_{14}O_2$	$C_6H_5O(CH_2)_2OC_6H_5$	BuLi/ether/25°/ 1.5 hr	CO_2	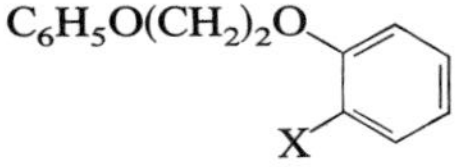 $C_6H_5O(CH_2)_2O$, X	$X = CO_2H$ (14)	243
		”	$(C_6H_5)_3SiCl$		$X = Si(C_6H_5)_3$ (7)	243
$C_{15}H_{18}OSi$	p-$(CH_3)_3SiC_6H_4OC_6H_5$	BuLi/THF, ether/ 15 hr	$(C_6H_5)_2SiCl_2$	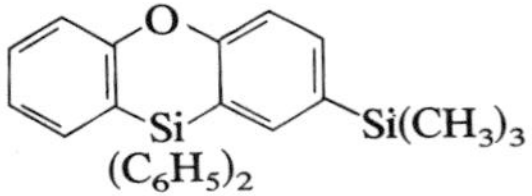 O, Si, $(C_6H_5)_2$, $Si(CH_3)_3$ (17)		241

$C_{18}H_{13}NO$	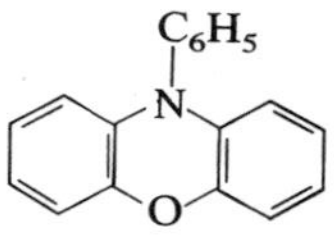 10-phenylphenoxazine (C_6H_5 on N)	BuLi/ether/reflux/ 23 hr	CO_2	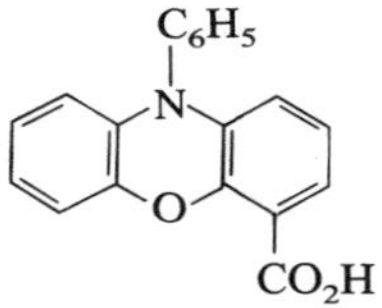 10-phenylphenoxazine-4-carboxylic acid (C_6H_5 on N, CO_2H)	(4)	238
$C_{20}H_{14}O$	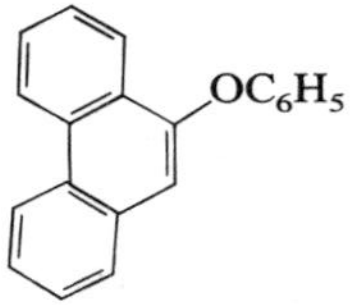 9-phenoxyphenanthrene (OC_6H_5)	C_6H_5Li/ether/ 25°/6 days	CO_2	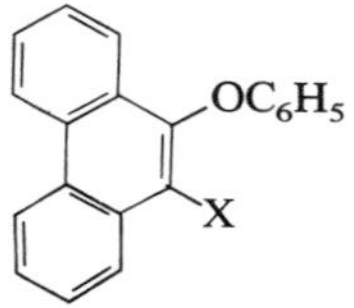 9-phenoxy-10-X-phenanthrene (OC_6H_5, X)	$X = CO_2H$ (46)	531
		"	CH_3I		$X = CH_3$ (15)	531
		"	$(C_6H_5)_2CO$		$X = C(C_6H_5)_2OH$ (59)	531

Note: References 360–607 are on pp. 355–360.

TABLE XLV. Alkyl Aralkyl Ethers and Aralkyl Alcohols (Ortho, Beta)

Formula	Compound Lithiated	Conditions	Substrate	Product and Yield (%)	Refs.
C_6H_8OS	3-(CH_2OCH_3)thiophene	BuLi/ether/25°/ 10 hr	CO_2	3-(CH_2OCH_3)-2-X-thiophene, X = CO_2H (89)	142
		BuLi/ether/25°–35°/1–1.5 hr	$(CH_3S)_2$	X = SCH_3 (61)	134, 43
		BuLi	DMF	X = CHO (72)	43, 134
		BuLi/ether/25°–35°/1–1.5 hr	$(C_6H_5)_2CO$	X = $C(C_6H_5)_2OH$ (46)	134
$C_6H_9NO_2$	3-(CH_2OCH_3)-5-CH_3-isoxazole	BuLi/THF/−60°/ 1 hr	CH_3I	3-(CH_2OCH_3)-4-CH_3-5-CH_3-isoxazole (I), 3-(CH_2OCH_3)-5-C_2H_5-isoxazole (II) (I + II, 90) (I : II, 1 : 4)	186
C_7H_8O	$C_6H_5CH_2OH$	BuLi/pet. ether, TMEDA/ reflux/11 hr	I_2	2-X-$C_6H_4CH_2OH$, X = I (58)	244b
		"	CH_2O	X = CH_2OH (70)	244b
		"	CH_3I	X = CH_3 (30)	244b
		"	n-C_4H_9Br	X = C_4H_9-n (21)	244b
		"	n-C_4H_9Cl	X = C_4H_9-n (55)	244b
		"	Cyclohexanone	X = 1-hydroxycyclohexyl (71)	244b
		"	C_6H_5CHO	X = $CH(OH)C_6H_5$ (95)	244b

Formula	Substrate	Conditions	Electrophile	Product (yield, %)	Ref.
		”	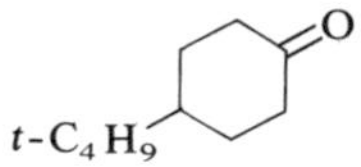 t-C_4H_9 (cyclohexanone, =O)	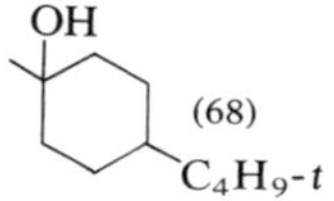 X = (OH, C_4H_9-t) (68)	244b
		”	CO_2	Phthalide (50)	244b
$C_7H_8O_2S$	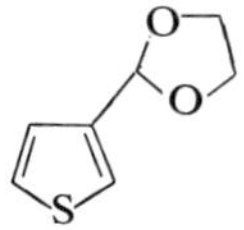 (3-thienyl, 1,3-dioxolane; S, O, O)	BuLi/ether/ reflux/20 min	“S” + $ClCH_2CO_2CH_3$ (NaOEt)	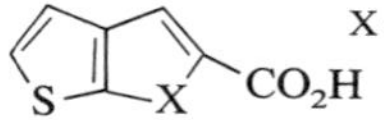 (S, X, CO_2H) X = S (94)	143
		BuLi/ether/25°– reflux/15 min	“Se” + $ClCH_2CO_2CH_3$ (NaOEt)	X = Se (48)	245
		BuLi/ether/ reflux/15 min	CO_2 (H_3O^+)	3-Formyl-2-thiophenecarboxylic acid (78)	412
		BuLi/ether/−70°	C_6H_5CN (H_3O^+)	2-Benzoyl-3-thiophenecarboxaldehyde (16)	413
$C_7H_8O_2Se$	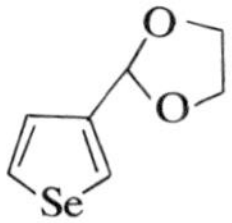 (3-selenienyl, 1,3-dioxolane; Se, O, O)	BuLi/ether/ reflux/0.5 hr	“Se” + $ClCH_2CO_2CH_3$ (H_3O^+, $NaOC_2H_5$)	Selenolo[2,3-*b*]selenophene-2-carboxylic acid (45)	575
$C_7H_8O_3$	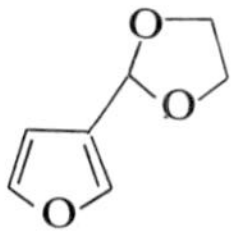 (3-furyl, 1,3-dioxolane; O, O, O)	BuLi/ether/−10°– reflux/15 min	CO_2 (H_3O^+)	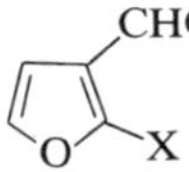 (CHO, O, X) X = CO_2H (50)	376
		”	$C_2H_5CON(CH_3)_2$ (H_3O^+)	X = COC_2H_5 (40)	376
		BuLi/ether/0°/ 0.5 hr	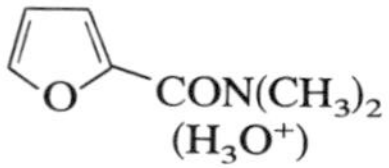 (2-furyl, O) $CON(CH_3)_2$ (H_3O^+)	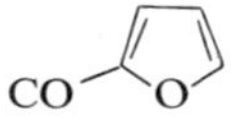 X = CO (2-furyl, O) (41)	113

TABLE XLV. Alkyl Aralkyl Ethers and Aralkyl Alcohols (Ortho, Beta) (*Continued*)

Formula	Compound Lithiated	Conditions	Substrate	Product and Yield (%)	Refs.
$C_7H_8O_3$ (*Contd.*)	[structure: O, O, O]	,,	[structure: S] $CON(CH_3)_2$ (H_3O^+)	[structure: CHO, O, X] $X = CO$–[structure: S] (30)	113
		BuLi/ether/0°/ 0.5 hr	$C_6H_5CON(CH_3)_2$ (H_3O^+)	$X = COC_6H_5$ (61)	113
		BuLi/ether/−30°–25°/0.5 hr	$(n\text{-}C_4H_9O)_3B$ (H_3O^+)	$X = B(OH)_2$ (36)	115, 114
$C_8H_{10}O$	$C_6H_5CH(CH_3)OH$	BuLi/pet. ether, TMEDA/reflux/ 11 hr	C_6H_5CHO	α^1-Methyl-α^2-phenyl-*o*-xylene-α^1,α^2-diol (88)	244b
$C_8H_{10}O_2$	*m*-$CH_3OC_6H_4CH_2OH$	BuLi/hexane, TMEDA/60°/ 5 hr	CO_2 (CH_2N_2)	[structure: CH_3O, O, O] (53), [structure: CH_3O_2C, CH_2OH, OCH_3] (6)	245
$C_8H_{10}O_2S$	[structure: CH_3, O, O, S]	BuLi/ether/25°/ 45 min, reflux/15 min	C_6H_5CN (H_3O^+)	3-Acetyl-2-benzoylthiophene (27)	413
$C_9H_{12}O$	$C_6H_5CH(C_2H_5)OH$	BuLi/pet. ether, TMEDA/reflux/ 11 hr	C_6H_5CHO	α^1-Ethyl-α^2-phenyl-*o*-xylene-α^1,α^2-diol (82)	244b

Formula	Substrate	Conditions	Electrophile	Product(s) and Yield(s) (%)	Refs.
	$C_6H_5C(CH_3)_2OH$	BuLi/pet. ether, TMEDA/reflux/ 11 hr	C_6H_5CHO	α^1,α^1-Dimethyl-α^2-phenyl-*o*-xylene-α^1,α^2-diol (86)	244b
$C_9H_{12}O_2$	*m*-$CH_3OC_6H_4CH(OH)CH_3$	BuLi/hexane, TMEDA/60°/* 5 hr	CO_2 (CH_2N_2)	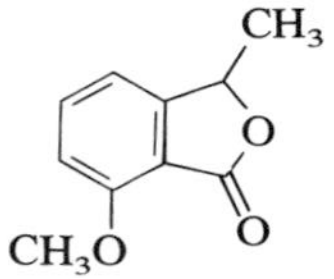 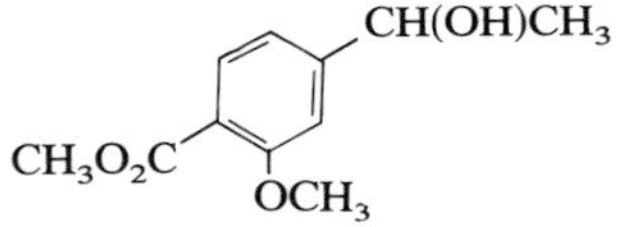 CH_3; O; O; CH_3O (62),	245
				$CH(OH)CH_3$; CH_3O_2C; OCH_3 (14)	
$C_9H_{12}O_3$	3,5-$(CH_3O)_2C_6H_3CH_2OH$	BuLi/hexane, TMEDA/60°/* 5 hr	CO_2 (CH_2N_2)	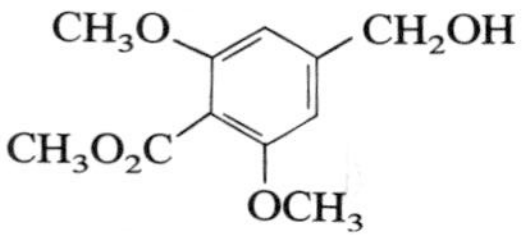 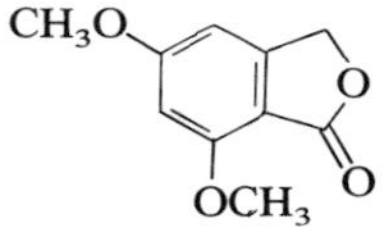 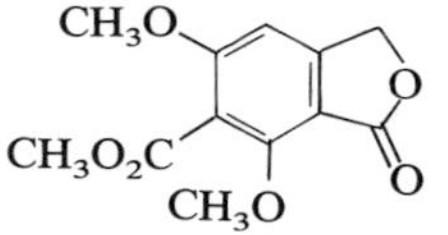 CH_3O; CH_2OH; CH_3O_2C; OCH_3 (22),	245
				CH_3O; O; OCH_3; O (8),	
				CH_3O; O; CH_3O_2C; CH_3O; O (13)	
$C_9H_{13}NO$	$C_6H_5CH(OH)CH_2NHCH_3$	BuLi/ether/25°/ 24 hr	$(CH_3)_3SiCl$	α-[(Methylamino)methyl]-*o*-(trimethylsilyl)benzyl alcohol (21)	175

TABLE XLV. Alkyl Aralkyl Ethers and Aralkyl Alcohols (Ortho, Beta) (*Continued*)

Formula	Compound Lithiated	Conditions	Substrate	Product and Yield (%)	Refs.
$C_9H_{14}O_2S$	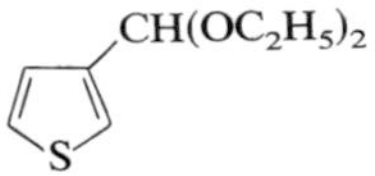 3-$CH(OC_2H_5)_2$-thiophene	BuLi/ether/−30°–reflux/20 min	"S" + $ClCH_2CO_2CH_3$ (H_3O^+)	Methyl[(3-formyl-2-thienyl)thio]acetate (77)	427
		BuLi	DMF (H_3O^+)	2,3-Thiophenedicarboxaldehyde (85)	428
$C_9H_{14}O_2Se$	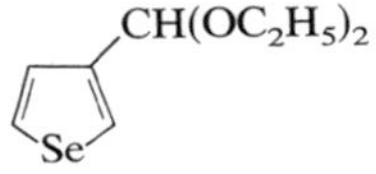 3-$CH(OC_2H_5)_2$-selenophene	BuLi/ether/reflux	I_2	3-(Diethoxymethyl)-2-iodoselenophene (—)	396
		BuLi/ether/−40°/1 hr	DMF	2,3-Selenophenedicarboxaldehyde, 3-(diethyl acetal) (80)	283
$C_{10}H_{11}NO$	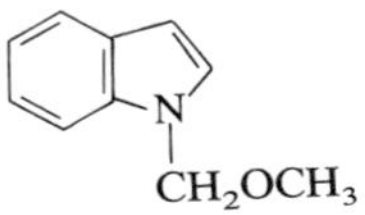 1-(CH_2OCH_3)-indole	*t*-BuLi/ether/25°/1 hr	CO_2	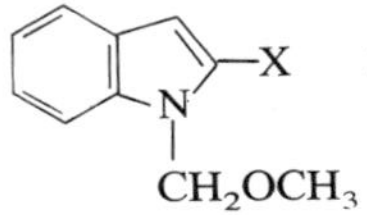 1-(CH_2OCH_3)-2-X-indole, X = CO_2H (80)	85
		"	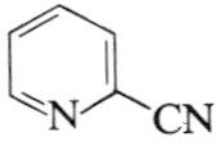 2-pyridyl-CN (H_3O^+)	X 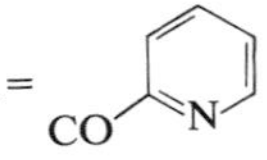 = CO-(2-pyridyl) (56)	85
		"	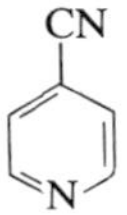 4-pyridyl-CN (H_3O^+)	X 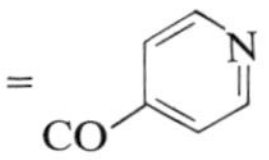 = CO-(4-pyridyl) (56)	85
		"	C_6H_5CN (H_3O^+)	X = COC_6H_5 (84)	85
		"	C_6H_5CHO	X = $CH(OH)C_6H_5$ (40)	85
		"	$C_6H_5N(CH_3)CHO$	X = CHO (46)	85
		"	*p*-$CH_3OC_6H_4CN$ (H_3O^+)	X = $COC_6H_4OCH_3$-*p* (70)	85

$C_{10}H_{14}O_2$	m-$CH_3OC_6H_4C(CH_3)_2OH$	BuLi/hexane, TMEDA/60°/ 5 hr	CO_2 (CH_2N_2)	7-Methoxy-3,3-dimethylphthalide (46)	245
$C_{10}H_{14}O_3$	3,5-$(CH_3O)_2C_6H_3CH(OH)CH_3$	BuLi/hexane, TMEDA/60°/ 5 hr	CO_2 (CH_2N_2)	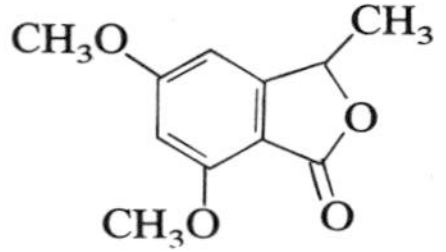 (8), 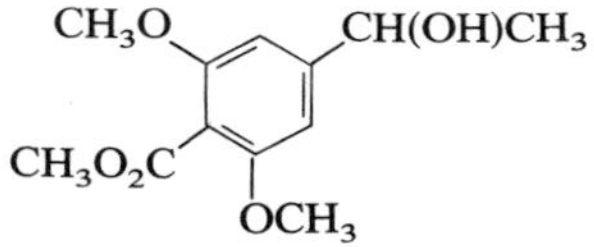 (24)	245
$C_{10}H_{15}NO$	$C_6H_5CH(OH)CH_2N(CH_3)_2$	BuLi/ether/25°/ 24 hr	CH_3I	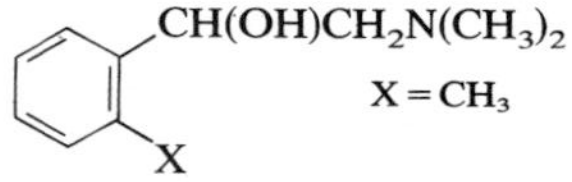 $X = CH_3$ (47)	175
		”	CH_2O	$X = CH_2OH$ (33)	175
		”	$(CH_3)_3SiCl$	$X = Si(CH_3)_3$ (61)	175
		”	$(C_6H_5)_2CO$	$X = C(C_6H_5)_2OH$ (48)	175
$C_{11}H_{10}O_2S_2$	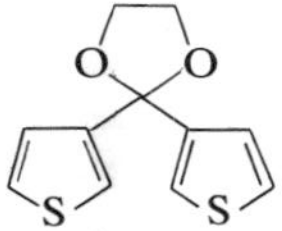	BuLi/ether/−35° to −20°/20 min	“S” (H_3O^+)	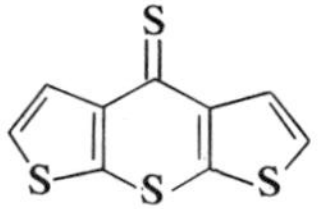 (35), 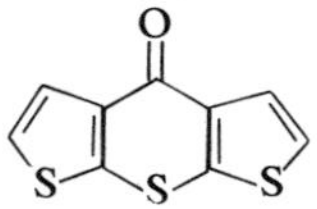 (9), 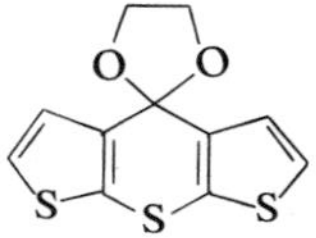 (6)	436

TABLE XLV. ALKYL ARALKYL ETHERS AND ARALKYL ALCOHOLS (ORTHO, BETA) (*Continued*)

Formula	Compound Lithiated	Conditions	Substrate	Product and Yield (%)	Refs.
$C_{11}H_{10}O_2S_2$ (*Contd.*)	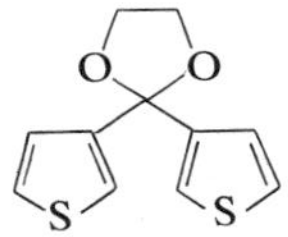	BuLi/ether/−35°	I_2	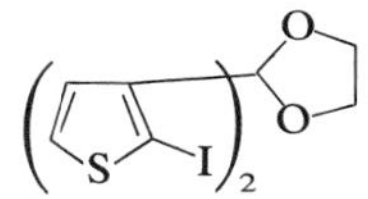 (56)	294
		BuLi/ether/ reflux/15 min	CO_2 (H_3O^+)	3,3′-Carbonyldi-2-thiophenecarboxylic acid (93)	437
$C_{11}H_{14}ClNO$	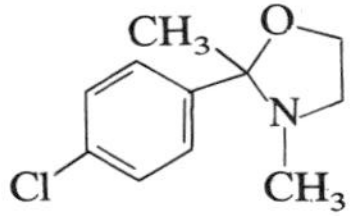	BuLi/ether/0°/ 4 hr	$(CH_3S)_2$	2-[4-Chloro-2-(methylthio)phenyl]-2,3-dimethyloxazolidine (80)	23
$C_{11}H_{14}O_2$	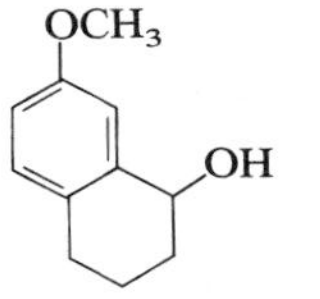	BuLi/hexane, TMEDA/60°/ 5 hr	CO_2 (CH_2N_2)	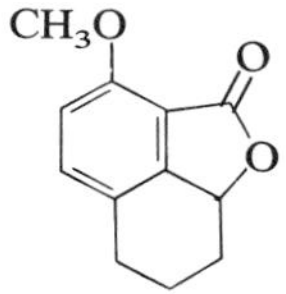 (80)	245
$C_{11}H_{16}O$	$C_6H_5CH(C_4H_9$-*n*$)OH$	BuLi/pet. ether, TMEDA/reflux/ 11 hr	C_6H_5CHO	α^1-Butyl-α^2-phenyl-*o*-xylene-α^1,α^2-diol (92)	244b
$C_{11}H_{17}NO_2$	*p*-$CH_3OC_6H_4CH(OH)CH_2N(CH_3)_2$	BuLi/ether/25°/ 24 hr	$(C_6H_5)_2CO$	α^1-[(Dimethylamino)methyl]-4-methoxy-α^2,α^2-diphenyl-*o*-xylene-α^1,α^2-diol (48)	175
$C_{12}H_{16}O_2$	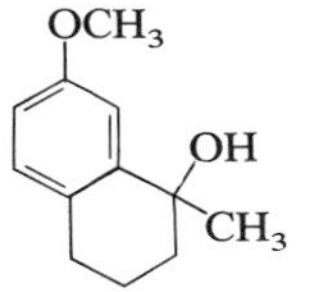	BuLi/hexane, TMEDA/60°/ 5 hr	CO_2 (CH_2N_2)	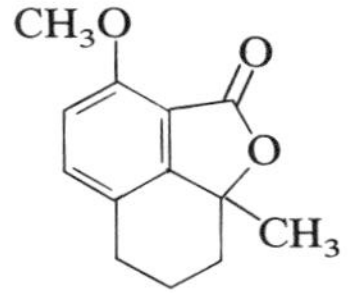 (42),	245

Formula	Substrate	Conditions	Electrophile	Product(s) (% yield)	Refs.
				CH_3O_2C, OCH_3, OH, CH_3 (23)	
$C_{13}H_{11}FO_2S$	o-FC_6H_4, O, O, S	BuLi/ether/0°/ 2 hr	CH_3CHO	S, $CH(OH)CH_3$, O, O, C_6H_4F-o (60)	23
$C_{13}H_{12}O_2S$	C_6H_5, O, O, S	BuLi/ether/ reflux/2 hr	CO_2 (H_3O^+)	3-Benzoyl-2-thiophenecarboxylic acid (43)	413
$C_{14}H_{15}ClN_2O$	p-ClC_6H_4, N, N—THP	BuLi/ether/0°/ 0.5 hr	CH_2O	X, p-ClC_6H_4, N, N—THP; X = CH_2OH (73)	23
		"	t-C_4H_9NCO	X = $CONHC_4H_9$-t (63)	23
		"	$(C_6H_5S)_2$	X = SC_6H_5 (80)	23
$C_{14}H_{18}O_2$	CH_3O, OH	BuLi, $NaOC_4H_9$-t/ hexane/25°/ 1 hr	CO_2	CH_3O, CO_2H, OH (88)	525
$C_{14}H_{24}O_4S$	$(C_2H_5O)_2CH$, $CH(OC_2H_5)_2$, S	BuLi/ether/−40°/ 1 hr	DMF	2,3,4-Thiophenetricarboxaldehyde, 3,4-bis(diethyl acetal) (30)	283

TABLE XLV. ALKYL ARALKYL ETHERS AND ARALKYL ALCOHOLS (ORTHO, BETA) (*Continued*)

Formula	Compound Lithiated	Conditions	Substrate	Product and Yield (%)	Refs.
$C_{14}H_{24}O_4Se$	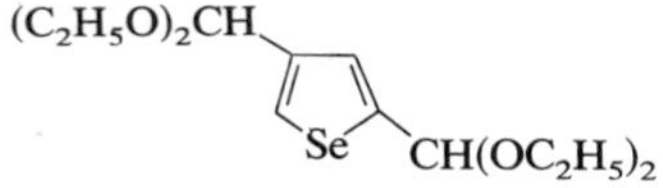 $(C_2H_5O)_2CH$, Se, $CH(OC_2H_5)_2$	BuLi/ether/−50°/ 5 hr	DMF	2,3,5-Selenophenetricarboxaldehyde, 3,5-bis(diethyl acetal) (40)	283
	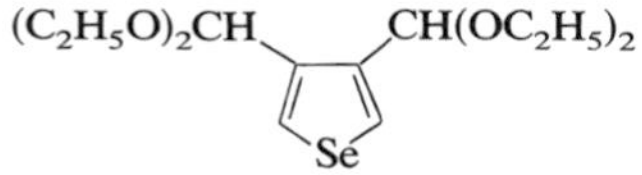 $(C_2H_5O)_2CH$, $CH(OC_2H_5)_2$, Se	BuLi/ether/−40°/ 1 hr	DMF	2,3,4-Selenophenetricarboxaldehyde, 3,4-bis(diethyl acetal) (40)	283
$C_{15}H_{20}O_2$	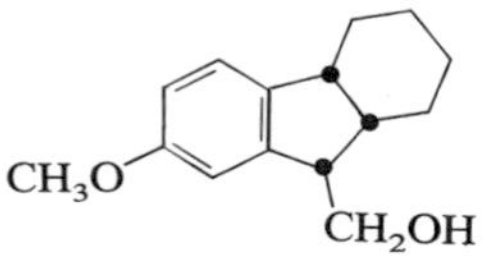 CH_3O, CH_2OH	BuLi/hexane, TMEDA/ reflux/45 min	$CO_2(CH_2N_2)$	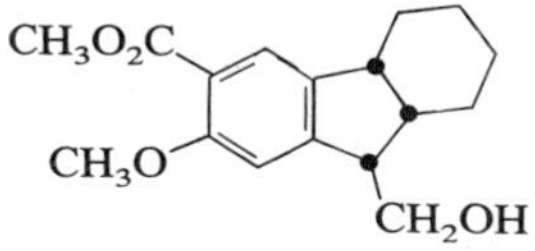 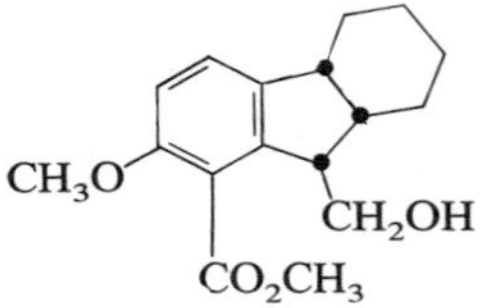 CH_3O_2C, CH_3O, CH_2OH (−),	525
				CH_3O, CH_2OH, CO_2CH_3 (−)	
$C_{16}H_{22}O_2$	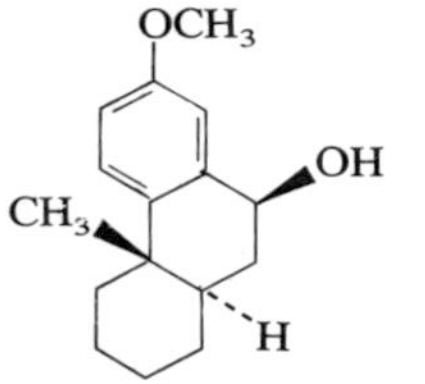 OCH_3, OH, CH_3, H	BuLi/hexane, TMEDA/60°/ 5 hr	$CO_2(CH_2N_2)$	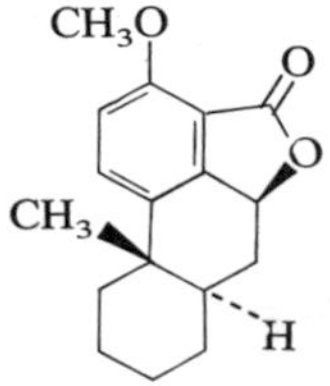 CH_3O, O, O, CH_3, H (80)	245

$C_{19}H_{34}O_6S$	$(C_2H_5O)_2CH$, $CH(OC_2H_5)_2$, $CH(OC_2H_5)_2$ (thiophene, S)	BuLi/"long time"	DMF	2,3,4,5-Thiophenetetracarboxaldehyde, 2,3,4-tris(diethyl acetal) (—)	441
$C_{19}H_{34}O_6Se$	$(C_2H_5O)_2CH$, $CH(OC_2H_5)_2$, $CH(OC_2H_5)_2$ (selenophene, Se)	BuLi/"long time"	DMF	2,3,4,5-Selenophenetetracarboxaldehyde, 2,3,4-tris(diethyl acetal) ("good")	441
$C_{20}H_{18}O$	$(C_6H_5)_3C(OCH_3)$	C_6H_5Li/ether, toluene/ reflux/24 hr	H_2O	(fluorene, X, C_6H_5) X = H (17)	532
		BuLi/ether, toluene/ reflux/24 hr	CO_2	X = CO_2H (13)	532
		BuLi/ether/ reflux/24 hr	CO_2	3,3-Diphenylphthalide (—)	532
$C_{26}H_{22}O$	$[(C_6H_5)_2CH]_2O$	BuLi/ether, benzene/25°/ 50 hr	CO_2	(benzene, $CH(C_6H_5)OCH(C_6H_5)_2$, CO_2H) (3)	533

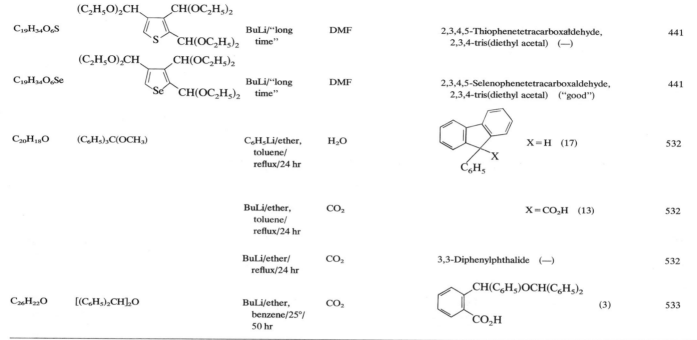

Note: References 360–607 are on pp. 355–360.

TABLE XLVI. α-Silyloxystyrenes (Ortho)

Formula	Compound Lithiated	Conditions	Substrate	Product and Yield (%)	Refs.
$C_{11}H_{16}OSi$	$C_6H_5C(=CH_2)OSi(CH_3)_3$	BuLi/hexane, TMEDA/−30° −25°/24 hr	CH_3Br (H_3O^+)	Acetophenone (8), 2′-methylacetophenone (34), 2′-methylpropiophenone (26), 2′-methylisobutyrophenone (26)	228
		”	$(CH_3)_3SiCl$	2-$(CH_3)_3SiC_6H_4C(=CH_2)OSi(CH_3)_3$ (90)	228
$C_{12}H_{18}OSi$	4-$CH_3C_6H_4C(=CH_2)OSi(CH_3)_3$	BuLi/hexane, TMEDA/−30° −25°/24 hr	$(CH_3)_3SiCl$	4-CH_3-2-$XC_6H_3C(=CH_2)OX$ (71), 4-CH_3-2-$XC_6H_3C(=CHX)OX$ (10), 4-$XCH_2C_6H_4C(=CH_2)OX$ (10); $X = Si(CH_3)_3$	228

TABLE XLVII. ARYLCARBOXYLIC ACIDS, ESTERS, AND DIARYLKETONES (ORTHO, BETA)

Formula	Compound Lithiated	Conditions	Substrate	Product and Yield (%)	Refs.
$C_4H_3NO_2S$	4-Isothiazolecarboxylic acid (CO_2H)	BuLi/THF/−65°/ 15 min*	CO_2	4,5-Isothiazoledicarboxylic acid (15)	154
$C_5H_4O_2S$	3-Thiophenecarboxylic acid (CO_2H)	LDA/THF or ether, HMPA/ −70°	$(CH_3)_3SiCl$	2-(Trimethylsilyl)-3-thiophenecarboxylic acid (—)	246
$C_5H_5NO_2S$	3-Methyl-4-isothiazolecarboxylic acid (CH_3, CO_2H)	BuLi/THF/−70°/ 15 min	Br_2	3-Methyl-5-X-4-isothiazolecarboxylic acid (CH_3, CO_2H, X) X = Br (52)	154
		BuLi/THF/−65°/ 15 min*	CO_2	X = CO_2H (29)	154
		,,	DMF	X = CHO (25)	154
$C_9H_{10}O_2$	$C_6H_5CO_2C_2H_5$	LTMP/THF/ −78°*	(—)	C_6H_4(CO_2R)(COC_6H_5) (I) (44) R = C_2H_5	53
$C_{10}H_{12}O_2$	$C_6H_5CO_2C_3H_7$-*n*	LTMP/THF/−78°	(—)	I (46) R = C_3H_7-*n*	53
	$C_6H_5CO_2C_3H_7$-*i*	LTMP/THF/−78°	(—)	I (10) R = C_3H_7-*i*	53
$C_{11}H_{14}O_2$	$C_6H_5CO_2C_4H_9$-*n*	LTMP/THF/−78°	(—)	I (38) R = C_4H_9-*n*	53
$C_{13}H_{10}O$	$(C_6H_5)_2CO$	LTMP/THF/−78°	(—)	3-Hydroxy-1,1,3-triphenylphthalan (80)	53

TABLE XLVIII. ALKYL ARYL SULFIDES (ORTHO, BETA)

Formula	Compound Lithiated	Conditions	Substrate	Product and Yield (%)	Refs.
$C_5H_6S_2$	3-(SCH_3)thiophene	BuLi/ether/ reflux/0.5 hr	CO_2	3-(Methylthio)-2-thiophenecarboxylic acid (70)	248
$C_6H_8S_2$	2-CH_3-5-(SCH_3)thiophene	BuLi	CO_2	5-Methyl-2-(methylthio)-3-thiophene-carboxylic acid (27)	404
C_7H_8S	$C_6H_5SCH_3$	BuLi/5 min	CO_2 (CH_2N_2)	SCH_3-C_6H_4-CO_2CH_3 *o*+*m* (I), $C_6H_5SCH_2CO_2CH_3$ (II) (I+II, 30) (I:II, 37:63)	58
$C_8H_{10}OS$	*m*-$CH_3OC_6H_4SCH_3$	BuLi/ether/ reflux/4 hr	CO_2	6-(Methylthio)-*o*-anisic acid (46)	574
$C_8H_{10}S$	$C_6H_5SC_2H_5$	BuLi/15 hr	CO_2(CH_2N_2)	SC_2H_5-C_6H_4-CO_2CH_3 *o* (III), *m* (IV), *p* (V) (III+IV+V, 27) (III:IV:V, 55:19:5)	58

$C_8H_{12}S_2$	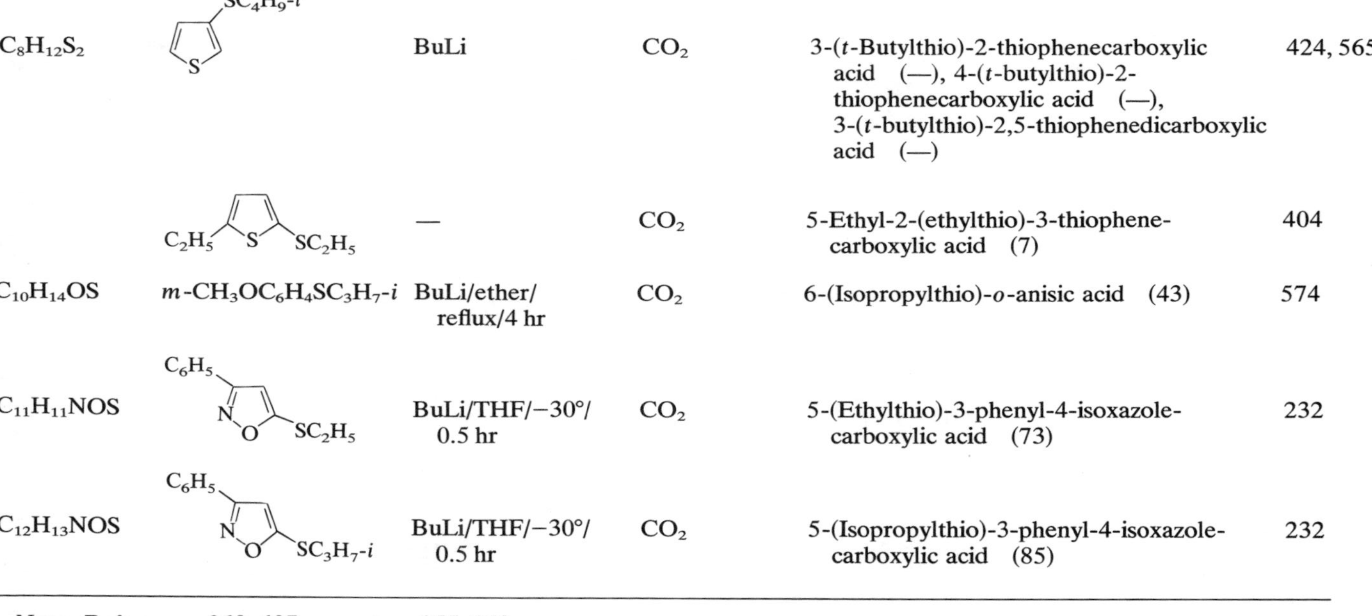 3-(SC_4H_9-*t*)thiophene	BuLi	CO_2	3-(*t*-Butylthio)-2-thiophenecarboxylic acid (—), 4-(*t*-butylthio)-2-thiophenecarboxylic acid (—), 3-(*t*-butylthio)-2,5-thiophenedicarboxylic acid (—)	424, 565
	2-C_2H_5-5-SC_2H_5-thiophene	—	CO_2	5-Ethyl-2-(ethylthio)-3-thiophenecarboxylic acid (7)	404
$C_{10}H_{14}OS$	*m*-$CH_3OC_6H_4SC_3H_7$-*i*	BuLi/ether/reflux/4 hr	CO_2	6-(Isopropylthio)-*o*-anisic acid (43)	574
$C_{11}H_{11}NOS$	3-C_6H_5-5-SC_2H_5-isoxazole	BuLi/THF/−30°/0.5 hr	CO_2	5-(Ethylthio)-3-phenyl-4-isoxazolecarboxylic acid (73)	232
$C_{12}H_{13}NOS$	3-C_6H_5-5-SC_3H_7-*i*-isoxazole	BuLi/THF/−30°/0.5 hr	CO_2	5-(Isopropylthio)-3-phenyl-4-isoxazolecarboxylic acid (85)	232

Note: References 360–607 are on pp. 355–360.

TABLE XLIX. Diaryl Sulfides and Condensed Diaryl Sulfides (Ortho, Beta)

Formula	Compound Lithiated	Conditions	Substrate	Product and Yield (%)		Refs.
$C_8H_6S_3$	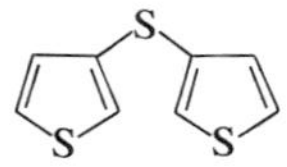	BuLi/ether/ reflux/1 hr	$CuCl_2$	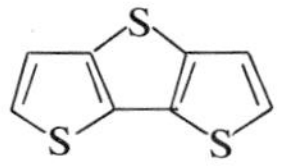 (52)		144
$C_{12}H_8OS$	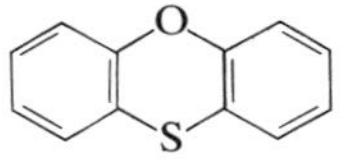	BuLi/ether/ reflux/24 hr	O_2	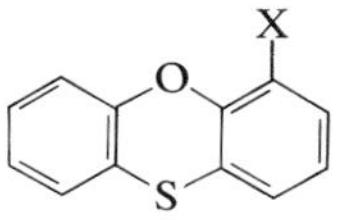	X = OH (16)	288
		BuLi	CO_2		X = CO_2H (37)	237
		BuLi/ether/ reflux	"	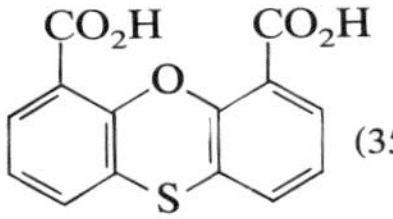 CO_2H CO_2H (35), CO_2H CO_2H (1–9)	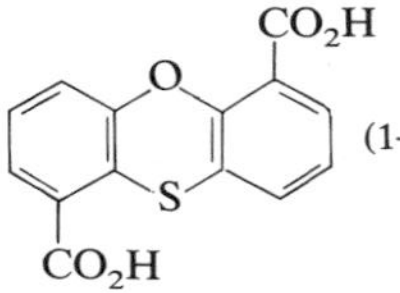	236, 237
$C_{12}H_8S$	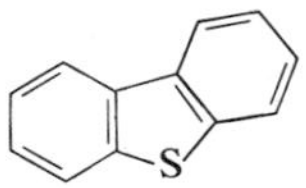	BuLi/ether	Br_2	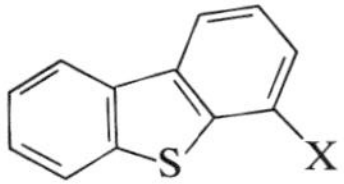	X = Br (77)	152, 249
		BuLi/THF, ether/ 25°/5 hr	CO_2		X = CO_2H (60)	560
		BuLi/ether/ reflux/3 hr	$(C_6H_5)_3SiCl$		X = $Si(C_6H_5)_3$ (42)	452
$C_{12}H_8S_2$	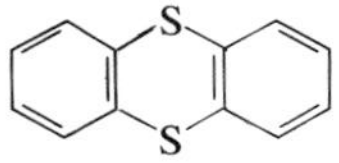	BuLi/ether/ reflux/22 hr	O_2	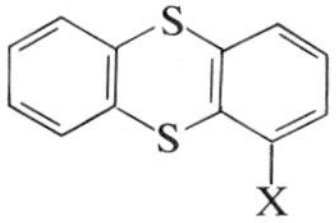	X = OH (3)	288

		BuLi/ether/ 25°/40 hr	CO_2	$X = CO_2H$ (28)	250
		”	CH_3ONH_2	$X = NH_2$ (26)	250
		”	$(n\text{-}C_4H_9O)_3B$ (H_3O^+)	$X = B(OH)_2$ (22)	250
		”	$(C_6H_5)_3SiCl$	$X = Si(C_6H_5)_3$ (10)	250
$C_{12}H_{10}S$	$(C_6H_5)_2S$	BuLi/ether	$(CH_3)_3SiCl$	Trimethyl[*o*-(phenylthio)phenyl]silane (I), trimethyl[*m*-(phenylthio)phenyl]silane (II) (I : II, 9 : 1)	252
		”	”	I (23)	251
$C_{13}H_{11}NS$	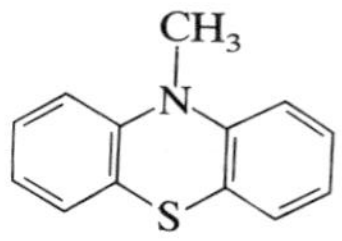	BuLi/ether/ 30 hr	CO_2	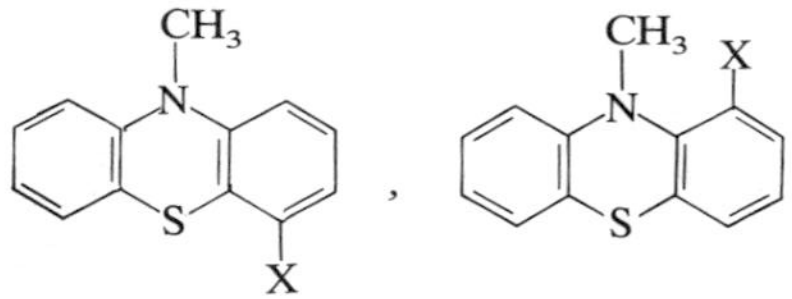 (III) (16), (IV) (14) $X = CO_2H$	177, 183a
		”	CH_2O	III $X = CH_2OH$ (10)	177
		”	Ethylene oxide	” $X = (CH_2)_2OH$ (10)	177
		”	$(CH_3O)_2SO_2$	” $X = CH_3$ (25)	177
		”	CH_3CO_2Li	” $X = COCH_3$ (23)	177
		”	$C_2H_5CO_2Li$	” $X = COC_2H_5$ (17)	177
		”	$C_6H_5CO_2Li$	” $X = COC_6H_5$ (20)	177
		”	$C_6H_5N(CH_3)CHO$	” $X = CHO$ (10)	177
$C_{14}H_{13}NS$	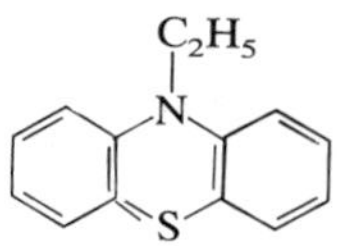	BuLi/ether/30 hr	CO_2	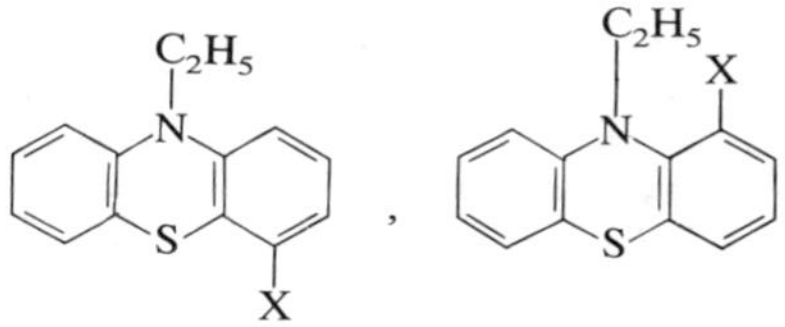 (V) (14), (VI) (13) $X = CO_2H$	177, 183a
		”	CH_3CO_2Li	V $X = COCH_3$ (13)	177

TABLE XLIX. Diaryl Sulfides and Condensed Diaryl Sulfides (Ortho, Beta) (*Continued*)

Formula	Compound Lithiated	Conditions	Substrate	Product and Yield (%)	Refs.
$C_{14}H_{13}NS$ (*Contd.*)	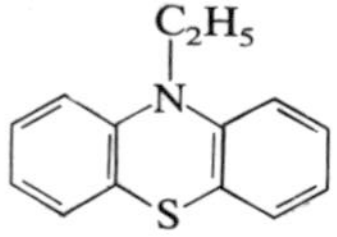 10-ethylphenothiazine (structure; N–C_2H_5)	"	$C_6H_5CO_2Li$	" $X = COC_6H_5$ (18)	183a, 177
		"	$(C_6H_5)_2CO$	" $X = C(C_6H_5)_2OH$ (55)	178
$C_{16}H_{10}S$	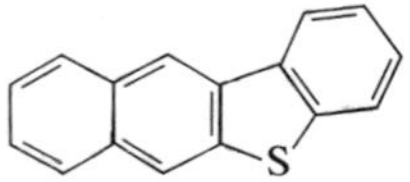 benzo[b]naphtho[2,3-d]thiophene (structure)	BuLi	CO_2	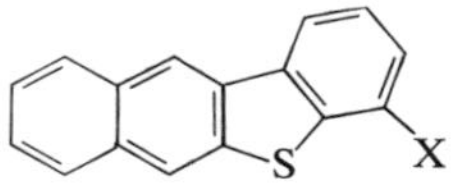 (structure with X) $X = CO_2H$ (51)	534
		"	$(CH_3O)_2SO_2$	$X = CH_3$ (70)	534
$C_{19}H_{24}N_2S$	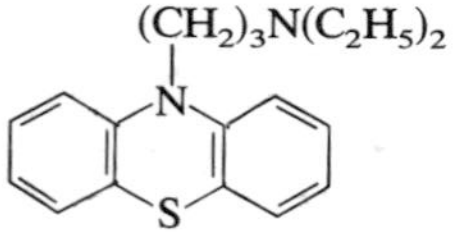 phenothiazine (structure; N–$(CH_2)_3N(C_2H_5)_2$)	BuLi	$(C_6H_5)_2CO$	10-[(3-Diethylamino)propyl]-α,α-diphenyl-4-phenothiazinemethanol (8)	178

Note: References 360–607 are on pp. 355–360.

TABLE L. SULFONES (ORTHO, BETA)

Formula	Compound Lithiated	Conditions	Substrate	Product and Yield (%)	Refs.
$C_8H_{12}O_2S_2$	2-$SO_2C_4H_9$-*t*-thiophene	BuLi/THF/ −20°/9.5 hr	CO_2	5-(*t*-Butylsulfonyl)-2,4-thiophenedicarboxylic acid (68)	136
		"	DMF	5-(*t*-Butylsulfonyl)-2,4-thiophenedicarboxaldehyde (59)	136
	3-$SO_2C_4H_9$-*t*-thiophene	BuLi	CO_2	3-(*t*-Butylsulfonyl)-2,4-thiophenedicarboxylic acid (42), 3-(*t*-butylsulfonyl)-2,5-thiophenecarboxylic acid (8), 3-(*t*-butylsulfonyl)-2-thiophenecarboxylic acid (4)	565
$C_9H_8N_2O_2S$	1-$SO_2C_6H_5$-imidazole	*t*-BuLi/THF/ −20°/10 min	D_2O	1-$SO_2C_6H_5$-2-X-imidazole: X = D (100)	573
		BuLi/THF/ −10°/15 min	I_2	X = I (71)	573
		t-BuLi/THF/ −20°/0.5 hr	CH_2O	X = CH_2OH (10)	573
		BuLi/THF/ 0°/5 min	Cyclohexanone	X = 1-hydroxycyclohexyl (HO-cyclohexyl) (15)	573
		t-BuLi/THF/ 0°/10 min	C_6H_5CHO	X = $CH(OH)C_6H_5$ (18)	573
$C_{10}H_{14}O_2S$	$C_6H_5SO_2C_4H_9$-*t*	BuLi	—	2-X-$C_6H_4SO_2C_4H_9$-*t*: X = Li (—)	572
		BuLi/THF/ −10°/3 hr	CO_2	X = CO_2H (45)	535

TABLE L. SULFONES (ORTHO, BETA) (*Continued*)

Formula	Compound Lithiated	Conditions	Substrate	Product and Yield (%)	Refs.
$C_{10}H_{14}O_2S$ (*Contd.*)	$C_6H_5SO_2C_4H_9$-*t*	RLi/20°/20 hr	CO_2	HO$_2$C, R, CO$_2$H (benzene ring) $R = C_2H_5$ (42); $R = C_4H_9$-*n* (44)	535
		"Excess" BuLi	"	*t*-C_4H_9, SO$_2$H, CO$_2$H (biphenyl) (40)	572
$C_{11}H_{16}O_2S$	*p*-$CH_3C_6H_4SO_2C_4H_9$-*t*	BuLi	—	CH$_3$, Li, SO$_2$C$_4$H$_9$-*t* (benzene ring) (−)	572
		"Excess" BuLi	CO_2	CH$_3$, SO$_2$H, *t*-C$_4$H$_9$, CO$_2$H, CH$_3$ (biphenyl) (5)	572

$C_{12}H_8O_2S$	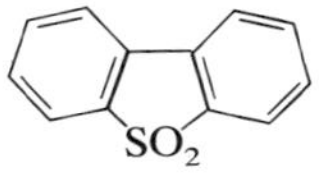	BuLi/ether/ −30°	CO_2	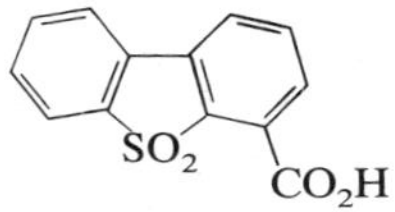	(18)	536
		3 BuLi	”	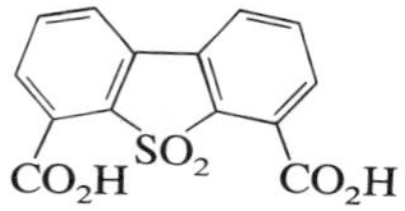	(20)	536
$C_{12}H_8O_2S_2$	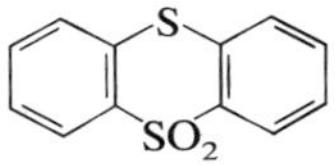	BuLi/ether/ −70°/16 hr	CO_2	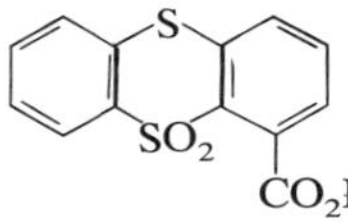	(41)	250
		C_6H_5Li/ether/ 0°/1 hr, 25°/ 3 hr	”	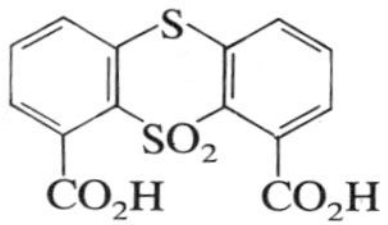	(54)	250
$C_{12}H_8O_3S$	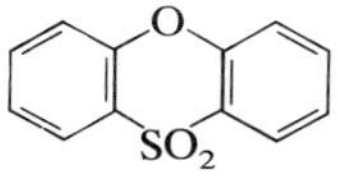	BuLi/ether/ 0°–25°	CO_2	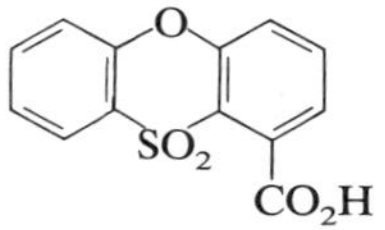	(46)	537
		BuLi/ether/ −45° to −10°	”	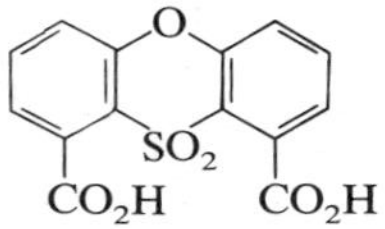	(52)	236
$C_{12}H_9BrO_2S$	$C_6H_5SO_2C_6H_4Br$-*m*	BuLi/ether/0°	CO_2	2-Bromo-6-(phenylsulfonyl)benzoic acid	(54)	69
	$C_6H_5SO_2C_6H_4Br$-*p*	BuLi/ether/0°	CO_2	5-Bromo-2-(phenylsulfonyl)benzoic acid	(52)	69

TABLE L. SULFONES (ORTHO, BETA) (*Continued*)

Formula	Compond Lithiated	Conditions	Substrate	Product and Yield (%)	Ref.
$C_{12}H_{10}O_2S$	$(C_6H_5)_2SO_2$	BuLi/ether/−40° to −30°/1 hr	CO_2	2-X-$C_6H_4SO_2C_6H_5$ $X = CO_2H$ (90)	538, 69, 536
		BuLi/ether/−40° to −30°/1 hr	$C_6H_5SO_2F$	$X = SO_2C_6H_5$ (73)	538
		,,	$C_6H_5SO_2Cl$	$X = Cl$ (90)	538
		,,	$(C_6H_5)_2CO$	$X = C(C_6H_5)_2OH$ (63)	538
		4 BuLi/ether/ −30°	CO_2	$(2-HO_2CC_6H_4)_2SO_2$ (22)	536
		BuLi/ether	$SiCl_4$	spiro-bis(dibenzo-Si,SO₂ ring) (−)	539
		,,	$(CH_3)_2SiCl_2$	$R_2Si(C_6H_4)_2SO_2$ $R = CH_3$ (24)	539
		,,	$(C_6H_5)_2SiCl_2$	$R = C_6H_5$ (8)	539

$C_{12}H_{20}O_4S_3$	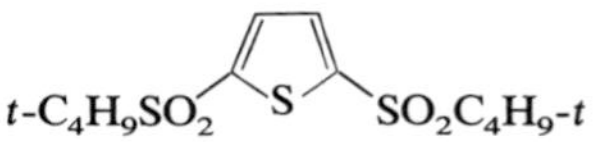 t-$C_4H_9SO_2$ S $SO_2C_4H_9$-t	BuLi/−40° to −20°	CO_2	2,5-Bis(t-butylsulfonyl)thiophene-3-carboxylic acid (I) (41)	540
		"Excess" BuLi	"	I, 2,5-Bis(t-butylsulfonyl)thiophene-3,4-dicarboxylic acid (II)	540
				(I + II, —) (I : II, —)	
$C_{13}H_{12}O_2S$	$C_6H_5SO_2C_6H_4CH_3$-p	BuLi/ether/0°/1.5 hr, 25°/1 hr	CO_2	6-(Phenylsulfonyl)-m-toluic acid (III), o-(p-toluenesulfonyl)benzoic acid (IV) (III + IV, 60) (III : IV, 3 : 2)	541
$C_{14}H_{11}NO_2S$	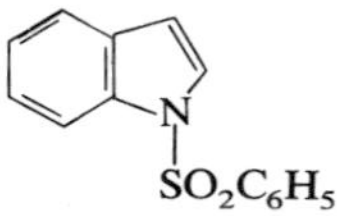 N–$SO_2C_6H_5$	t-BuLi/THF/−12°–25°/20 min		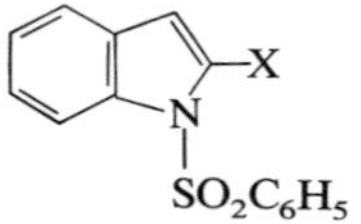 X, N–$SO_2C_6H_5$; X = CO_2H (63)	85
		"	$ClCO_2C_2H_5$	X = $CO_2C_2H_5$ (75)	85
		"	N CHO	X = CH(OH) N (32)	85
		"	COCl N	X = CO N (60)	85
		"	C_6H_5CHO	X = $CH(OH)C_6H_5$ (55)	85
		"	C_6H_5COCl	X = COC_6H_5 (65)	85
		"	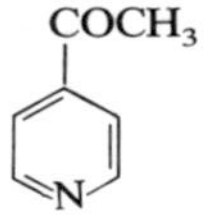 $COCH_3$ N	X = $C(CH_3)$ OH 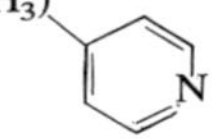 N (35)	85

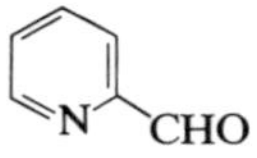

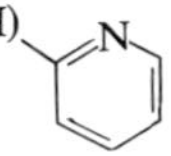

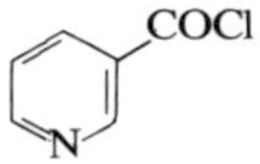

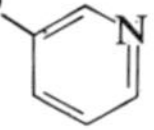

TABLE L. SULFONES (ORTHO, BETA) (*Continued*)

Formula	Compound Lithiated	Conditions	Substrate	Product and Yield (%)		Refs.
$C_{14}H_{11}NO_2S$ (*Contd.*)	1-(phenylsulfonyl)indole, $SO_2C_6H_5$	*t*-BuLi/THF/−12°–25°/20 min	$C_6H_5COCH_3$	2-X-1-($SO_2C_6H_5$)indole	$X = C(CH_3)(OH)C_6H_5$ (64)	85
		"	*p*-$CH_3OC_6H_4CHO$		$X = CH(OH)C_6H_4OCH_3$-*p* (65)	85
		"	*p*-$CH_3OC_6H_4COCH_3$		$X = C(CH_3)(OH)C_6H_4OCH_3$-*p* (35)	85
		"	2-cyanopyridine (H_3O^+)	2-X-1H-indole (N–H)	X = CO-(2-pyridyl) (36)	85
		"	4-cyanopyridine (H_3O^+)		X = CO-(4-pyridyl) (26)	85
		"	C_6H_5CN (H_3O^+)		$X = COC_6H_5$ (30)	85
		"	3-pyridyl-$CO_2C_2H_5$		X = CO-(3-pyridyl) (22)	85

Formula	Substrate	Conditions	Electrophile	Product (Yield, %)	Ref.
		”	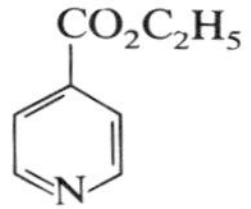 $CO_2C_2H_5$ (4-pyridyl)	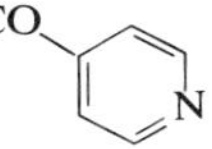 $X = CO$–(4-pyridyl) (31)	85
		”	$C_6H_5CO_2C_2H_5$	$X = COC_6H_5$ (26)	85
$C_{14}H_{14}ClNO_2S$	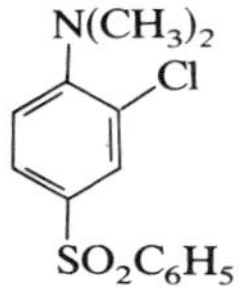 $N(CH_3)_2$, Cl, $SO_2C_6H_5$	BuLi/ether/−80°/ 24 hr	CO_2	2-Chloro-3-(dimethylamino)-6-(phenylsulfonyl)-benzoic acid (82)	542
$C_{14}H_{16}O_2S$	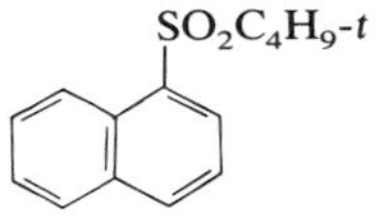 $SO_2C_4H_9$-*t* (1-naphthyl)	BuLi/ether/ −70°/7 hr	CO_2	1-(*t*-Butylsulfonyl)-2-naphthoic acid (47)	253
	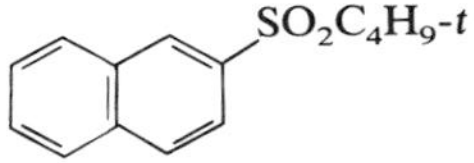 $SO_2C_4H_9$-*t* (2-naphthyl)	BuLi	CO_2	1-Butyl-2-naphthoic acid (14), 2-(*t*-butylsulfonyl)-1-naphthoic acid (13)	543
$C_{15}H_{13}NO_3S$	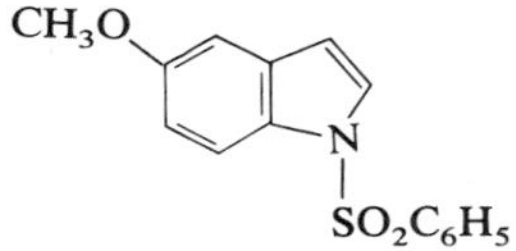 CH_3O, N–$SO_2C_6H_5$ (indole)	*t*-BuLi/THF/0°– 25°/45 min	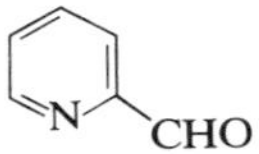 CHO (2-pyridyl)	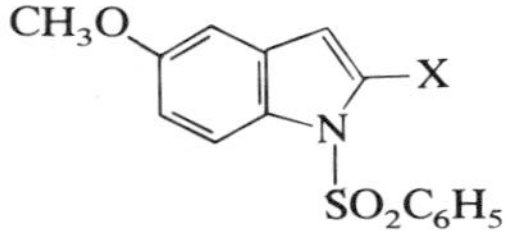 CH_3O, X, N–$SO_2C_6H_5$; $X = CH(OH)$–(2-pyridyl) (59) 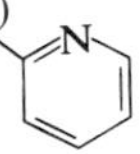	84
		”	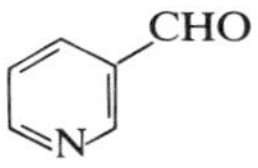 CHO (3-pyridyl)	$X = CH(OH)$–(3-pyridyl) (84) 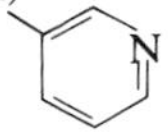	84

TABLE L. SULFONES (ORTHO, BETA) (*Continued*)

Formula	Compound Lithiated	Conditions	Substrate	Product and Yield (%)	Refs.
$C_{15}H_{13}NO_3S$ (*Contd.*)	6-CH_3O-1-($SO_2C_6H_5$)indole	*t*-BuLi/THF/0°–25°/45 min	2-pyridine-CHO	6-CH_3O-1-($SO_2C_6H_5$)-2-X-indole, X = CH(OH)-2-pyridyl (62)	84
		"	3-pyridine-CHO	X = CH(OH)-3-pyridyl (70)	84
		"	4-pyridine-$COCH_3$	X = C(CH_3)(OH)-4-pyridyl (58)	84
$C_{16}H_{12}O_2S$	2-$SO_2C_6H_5$-naphthalene	BuLi/ether/−70°	$Cl_2P(O)OC_2H_5$	cyclic P(O)OC_2H_5 / SO_2 product (25)	577
$C_{16}H_{18}O_2S$	$C_6H_5SO_2C_6H_4C_4H_9$-*t*-*p*	BuLi/ether/0°	CO_2	*o*-(*p*-*t*-Butylphenylsulfonyl)benzoic acid (69), 3-*t*-butyl-6-(phenylsulfonyl)benzoic acid (31)	544
$C_{20}H_{14}O_2S$	[2-naphthyl]$_2SO_2$	BuLi/ether/−70°	$Cl_2P(O)OC_2H_5$	cyclic P(O)OC_2H_5 / SO_2 product (25)	577

Note: References 360–607 are on pp. 355–360.

TABLE LI. ARYLSULFONAMIDES (ORTHO, BETA)

Formula	Compound Lithiated	Conditions	Substrate	Product and Yield (%)	Refs.
$C_7H_8ClNO_2S$	*p*-$ClC_6H_4SO_2NHCH_3$	BuLi/THF/ −10°/1 hr	$CH_2{=}CHCHO$	4-Cl-2-(SO_2NHCH_3)$C_6H_3CH(OH)CH{=}CH_2$ (>90)	24
$C_7H_9NO_2S$	$C_6H_5SO_2NHCH_3$	BuLi/THF/0°/ 15–20 min	CO_2	2-X-$C_6H_4SO_2NHCH_3$, $X = CO_2H$ (49)	545
		BuLi/THF/ −10°/3 hr	$CH_2{=}CHCHO$	$X = CH(OH)CH{=}CH_2$ (>80)	24
		BuLi/THF/0°/ 15–20 min	Cyclohexanone	X = 1-hydroxycyclohexyl (72)	545
		BuLi/THF/0°	C_6H_5NCO	$X = CONHC_6H_5$ (67)	496
		BuLi/THF/0°/ 15–20 min	$C_6H_5COCH_3$	$X = C(OH)(CH_3)C_6H_5$ (58)	545
		"	$(C_6H_5)_2CO$	$X = C(C_6H_5)_2OH$ (82)	545
		BuLi/THF/0°/ 0.5 hr	C_6H_5CN	cyclic benzo-fused SO_2–N–CH_3, C(C_6H_5)(NH_2) (64)	201
$C_8H_{11}NO_2S$	$C_6H_5SO_2N(CH_3)_2$	BuLi/THF/0°/ 0.5 hr	CO_2	2-X-$C_6H_4SO_2N(CH_3)_2$, $X = CO_2H$ (75)	254

TABLE LI. ARYLSULFONAMIDES (ORTHO, BETA) (*Continued*)

Formula	Compound Lithiated	Conditions	Substrate	Product and Yield (%)	Refs.
$C_8H_{11}NO_2S$ (*Contd.*)	$C_6H_5SO_2N(CH_3)_2$	"	C_6H_5CN	X = C(=NH)C$_6$H$_5$ (79)	254
		"	C_6H_5NCO	X = CONHC$_6$H$_5$ (57)	254
		"	$(C_6H_5)_2CO$	X = C(C$_6$H$_5$)$_2$OH (82)	254
$C_9H_{17}NO_2S_2Si$	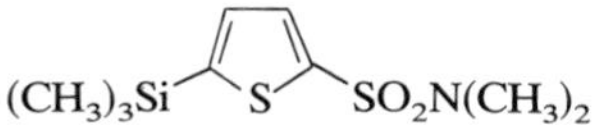 $(CH_3)_3Si$–[thiophene, S]–$SO_2N(CH_3)_2$	BuLi/ether/0°/ 6 hr	CO_2	2-(Dimethylsulfamoyl)-5-(trimethylsilyl)-3-thiophenecarboxylic acid (10)	135
$C_{10}H_{14}ClNO_2S$	p-$ClC_6H_4SO_2NHC_4H_9$-t	BuLi/THF/25°/ 1–2 hr	CO_2	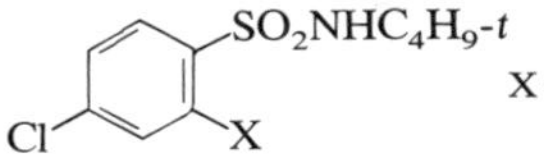 [structure: Cl, $SO_2NHC_4H_9$-t, X] X = CO$_2$H (25)	255
		BuLi/THF/ −30°/1 hr	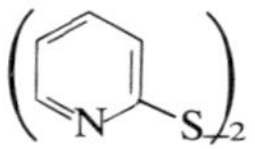 [structure: (N, S)$_2$]	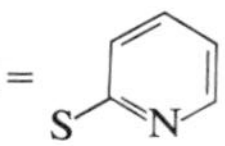 X = [structure: S, N] (32)	24
		"	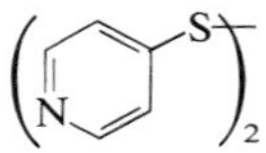 [structure: (N, S)$_2$]	X = S–[structure: N] (30)	24
		"	$(C_6H_5S)_2$	X = SC$_6$H$_5$ (50)	24
$C_{10}H_{14}FNO_2S$	p-$FC_6H_4SO_2NHC_4H_9$-t	BuLi/THF/25°/ 1–2 hr	CO_2	2-(t-Butylsulfamoyl)-5-fluorobenzoic acid (35)	255
$C_{10}H_{15}NO_2S$	$C_6H_5SO_2NHC_4H_9$-t	BuLi/THF/25°/ 1–2 hr	CO_2	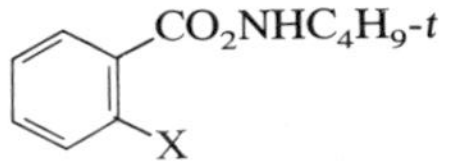 [structure: $CO_2NHC_4H_9$-t, X] X = CO$_2$H (48)	255
		BuLi/THF/0°/	$CH_2{=}CHCHO$	X = CH(OH)CH=CH$_2$ (>80)	24
		"	t-C_4H_9NCO	X = CONHC$_4$H$_9$-t (>80)	24
$C_{11}H_{17}NO_2S$	$C_6H_5SO_2N(CH_3)C_4H_9$-t	BuLi/ether/0°/ 1 hr	DMF	N-t-Butyl-o-formyl-N-methylbenzenesulfonamide (72)	24

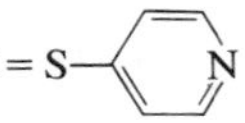

	p-$CH_3C_6H_4SO_2NHC_4H_9$-t	BuLi/THF/25°/ 1–2 hr	CO_2	6-(t-Butylsulfamoyl)-m-toluic acid (28)	255
$C_{11}H_{17}NO_3S$	p-$CH_3OC_6H_4SO_2NHC_4H_9$-t	BuLi/THF/25°/ 1–2 hr	CO_2	6-(t-Butylsulfamoyl)-m-anisic acid (27)	255
$C_{11}H_{21}NO_2S_2Si$	$(CH_3)_3Si$ 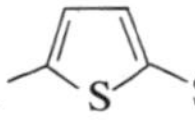 $SO_2N(C_2H_5)_2$	BuLi/ether, hexane, TMEDA/ −30°/6 hr	CO_2	2-(Diethylsulfamoyl)-5-(trimethylsilyl)-3-thiophenecarboxylic acid (44)	135
$C_{12}H_{11}NO_2S$	$C_6H_5SO_2NHC_6H_5$	BuLi/THF/0°/ 15–20 min	CO_2	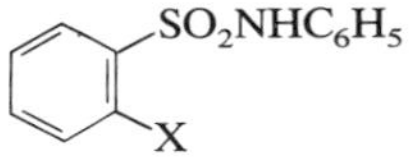 X = CO_2H (22)	545
		BuLi/THF/0°/ 0.5 hr	C_6H_5CN	X = $C(=NH)C_6H_5$ (64)[a]	201
		BuLi/THF/0°	C_6H_5NCO	X = $CONHC_6H_5$ (53)	496
		BuLi/THF/0°/ 15–20 min	$C_6H_5COCH_3$	X = $C(CH)_3(OH)C_6H_5$ (48)	545
		"	$(C_6H_5)_2CO$	X = $C(C_6H_5)_2OH$ (50)	545
$C_{13}H_{11}NO_3S$	$C_6H_5SO_2NHCOC_6H_5$	BuLi/THF/ −78°/4 hr	—	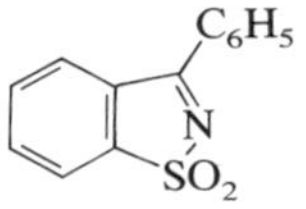 (9)	546
$C_{13}H_{13}NO_2S$	$C_6H_5SO_2N(CH_3)C_6H_5$	CH_3Li/THF/ 25°/1–2 hr	H_2O	N-Methyl-o-(phenylsulfonyl)aniline (89)	256
		"	C_6H_5CHO	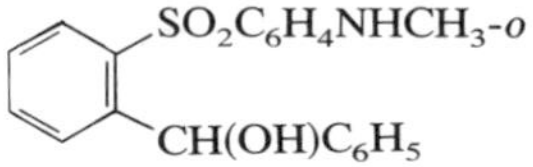 (85)	256
	p-$CH_3C_6H_4SO_2NHC_6H_5$	BuLi/THF	$(C_6H_5)_2CO$	α^2-Hydroxy-α^2,α^2-diphenyl-2,4-xylenesulfonanilide (80)	545

TABLE LI. ARYLSULFONAMIDES (ORTHO, BETA) (*Continued*)

Formula	Compound Lithiated	Conditions	Substrate	Product and Yield (%)	Refs.
$C_{14}H_{15}NO_2S$	$C_6H_5SO_2N(C_2H_5)C_6H_5$	CH_3Li/THF/ 25°/1–2 hr	H_2O	N-Ethyl-*o*-(phenylsulfonyl)aniline (40)	256
	o-$CH_3C_6H_4SO_2N(CH_3)C_6H_5$	"	H_2O	N-Methyl-*o*-(*o*-tolylsulfonyl)aniline (50)	256
	p-$CH_3C_6H_4SO_2N(CH_3)C_6H_5$	"	H_2O	N-Methyl-*o*-(*p*-tolysulfonyl)aniline (52)	256,257
$C_{14}H_{15}NO_3S$	*p*-$CH_3OC_6H_4SO_2N(CH_3)C_6H_5$	BuLi/THF/25°/ 1-2 hr	H_2O	*o*-(*p*-Methoxyphenylsufonyl)-N-methylaniline (45)	256
$C_{14}H_{17}NO_2S$	$SO_2NHC_4H_9$-*t*	BuLi/THF/25°/ 1–2 hr	CO_2	8-(*t*-Butylsulfamoyl)-1-naphthoic acid (14)	255
	$SO_2NHC_4H_9$-*t*	BuLi/THF/25°/ 1–2 hr	CO_2	2-(*t*-Butylsulfamoyl)-1-naphthoic acid (30)	255
$C_{15}H_{15}NO_2S$	SO_2 N—C_6H_5 $(CH_3)_2$	BuLi/THF/25°	H_2O	SO_2 N H $(CH_3)_2$ ("good")	257
$C_{15}H_{17}NO_2S$	*p*-$CH_3C_6H_4SO_2N(C_2H_5)C_6H_5$	CH_3Li/THF/ 25°/1–2 hr	H_2O	N-Ethyl-*o*-(*p*-tolylsulfonyl)aniline (81)	256
$C_{15}H_{17}NO_3S$	*p*-$CH_3C_6H_4SO_2N(CH_3)C_6H_4OCH_3$-*p*	"		N-Methyl-2-(*p*-tolylsulfonyl)-*p*-anisidine (61)	256
	p-$CH_3OC_6H_4SO_2N(C_2H_5)C_6H_5$	CH_3Li/THF/ 25°/1–2 hr	H_2O	N-Ethyl-*o*-(*p*-methoxyphenylsulfonyl)aniline (57)	256
$C_{15}H_{17}NO_4S$	*p*-$CH_3OC_6H_4SO_2N(CH_3)C_6H_4$-$OCH_3$-*p*	BuLi/THF/ 25°/1–2 hr	H_2O	2-(*p*-Methoxyphenylsulfonyl)-N-methyl-*p*-anisidine (53)	256
$C_{15}H_{18}N_2O_2S$	*p*-$(CH_3)_2NC_6H_4SO_2N(CH_3)C_6H_5$	BuLi/THF/25°/ 1–2 hr	H_2O	*o*-(*p*-Dimethylaminophenylsulfonyl)-N-methylaniline (54)	256

$C_{16}H_{17}NO_2S$	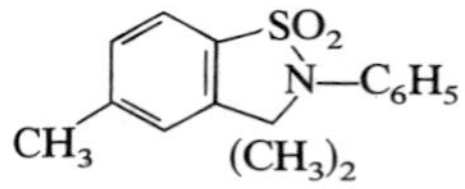 [structure]: SO_2, $N—C_6H_5$, CH_3, $(CH_3)_2$	BuLi/THF/25° H_2O	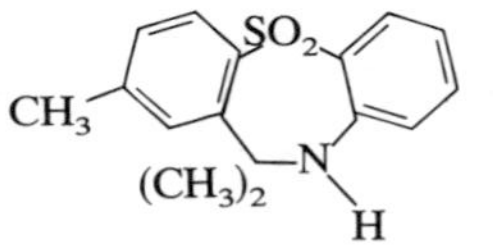 [structure]: SO_2, CH_3, $(CH_3)_2$, N, H ("good")	257
$C_{18}H_{15}NO_2S$	$C_6H_5SO_2N(C_6H_5)_2$	BuLi/THF/25°/ 1–2 hr H_2O	N-Phenyl-*o*-(phenylsulfonyl)aniline (86)	256
$C_{19}H_{17}NO_2S$	*p*-$CH_3C_6H_4SO_2N(C_6H_5)_2$	C_6H_5Li/ether/ 34° H_2O	N-Phenyl-*p*-(tolylsulfonyl)aniline (48)	258
$C_{26}H_{21}NO_2S$	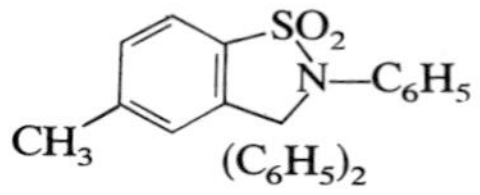 [structure]: SO_2, $N—C_6H_5$, CH_3, $(C_6H_5)_2$	BuLi/THF/25° H_2O	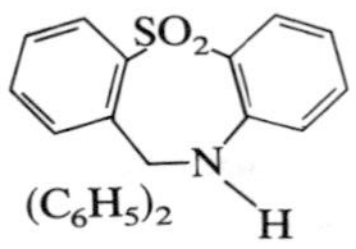 [structure]: SO_2, $(C_6H_5)_2$, N, H ("good")	257

[a] The product is a mixture of open-chain and ring isomers.

Note: References 360–607 are on pp. 355–360.

TABLE LII. Aryl Fluorides and Vinyl Fluorides (Ortho, Beta)

Formula	Compound Lithiated	Conditions	Substrate	Product and Yield (%)	Refs.
C_2HClF_2	$CF_2{=}CClH$	BuLi/ether/−100° to −78°/0.5–1 hr	CH_3COCH_3	2-Chloro-3-methylcrotonic acid (15)	172
		”	CF_3COCF_3	$CF_2{=}CClX$ $X = C(CF_3)_2OH$ (56)	172
		”	CH_3COCF_3	$X = C(CF_3)(OH)CH_3$ (61)	172
		”	$(C_2H_5)_3SiCl$	$X = Si(C_2H_5)_3$ (10)	173
C_2HF_3	$CF_2{=}CFH$	BuLi/ether/−100° to −78°/0.5–1 hr	CH_3COCH_3	2-Fluoro-3-methylcrotonic acid (30)	172
		”	CO_2	$CF_2{=}CFX$ $X = CO_2H$ (57)	172
		”	CF_3COCF_3	$X = C(CF_3)_2OH$ (63)	172
		”	$(C_2H_5)_3SiCl$	$X = Si(C_2H_5)_3$ (79)	173
C_4HF_5	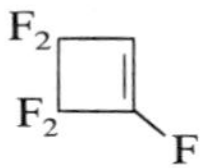	CH_3Li/ether/−70°	CH_3CHO	2,3,3,4,4-Pentafluoro-α-methyl-1-cyclobutene-1-methanol (23)	263
C_4H_3FS	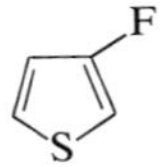	BuLi/ether/reflux/ 15 min	CO_2	3-Fluoro-2-thiophenecarboxylic acid (75)	385
C_5HF_7	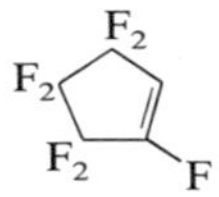	CH_3Li/ether/−70°	CH_3CHO	2,3,3,4,4,5,5-Heptafluoro-α-methyl-1-cyclopentene-1-methanol (42)	263
C_6HF_5	Pentafluorobenzene	BuLi/THF/−70°/ 5 min	CO_2	 $X = CO_2H$ (82)	547

		BuLi/ether/−55°/ 2 hr	CF_3COCF_3		$X = C(CF_3)_2OH$ (79)	548
C_6HF_9	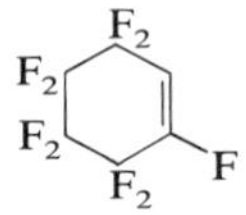	CH_3Li/ether/−70°	Br_2	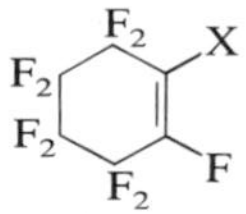	X = Br (46)	263
		”	I_2		X = I (59)	263
		”	CO_2		$X = CO_2H$ (77)	263
		”	CH_3CHO		$X = CH(OH)CH_3$ (63)	263
$C_6H_2F_4$	1,2,3,4-Tetrafluoro-benzene	BuLi/THF/−70°/ 0.5 hr	$HgCl_2$	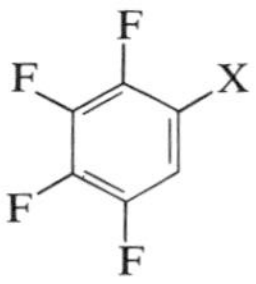	X = Hg 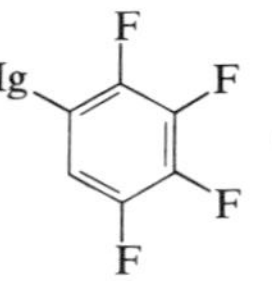 (90)	262
		”	CuI		X = 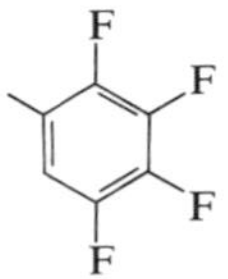 (69)	262
		”	CO_2		$X = CO_2H$ (—)	262
		”	$(CH_3)_2SiHCl$		$X = SiH(CH_3)_2$ (64)	262
		”	$(CH_3)_3SiCl$		$X = Si(CH_3)_3$ (82)	262
	1,2,4,5-Tetrafluoro-benzene	BuLi/ether/−65°/ 2 hr	CO_2	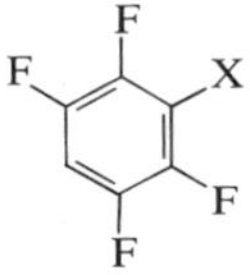 (I)	$X = CO_2H$ (85)	277

TABLE LII. Aryl Fluorides and Vinyl Fluorides (Ortho, Beta) (*Continued*)

Formula	Compound Lithiated	Conditions	Substrate	Product and Yield (%)	Refs.
$C_6H_2F_4$ (*Contd.*)	1,2,4,5-Tetrafluoro-benzene	BuLi/THF/−65°/ 2 hr	CF_3COCF_3	(I) (11),  [structure: ring with F, F, X, F, F, X] (61) X = $C(CF_3)_2OH$	548
$C_6H_2F_4S$	2,3,5,6-Tetrafluoro-benzenethiol	BuLi/THF/−70°/ 1 hr	CO_2	2,3,5,6-Tetrafluoro-4-mercaptobenzoic acid (77)	549
$C_6H_3F_3$	1,3,5-Trifluoro-benzene	BuLi	$(CH_3)_3SiCl$	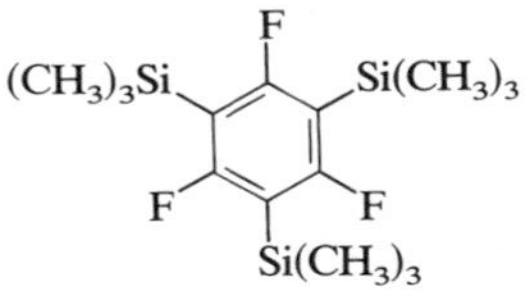 [structure: ring with F, $Si(CH_3)_3$, F, $Si(CH_3)_3$, F, $(CH_3)_3Si$] (—)	550
$C_6H_3F_4N$	2,3,5,6-Tetrafluoro-aniline	BuLi/THF/−70°/ 3 hr	CO_2	4-Amino-2,3,5,6-tetrafluorobenzoic acid (37)	549
C_6H_4ClF	p-FC_6H_4Cl	BuLi/THF/−70°/ 4 hr	$(CH_3S)_2$	5-Chloro-2-fluorophenyl methyl sulfide (>90)	23
$C_6H_4F_2$	*m*-Difluorobenzene	BuLi/THF/−65°	CO_2	2,6-Difluorobenzoic acid (88)	549, 551
		BuLi/THF	$(CH_3)_3SnCl$	(2,6-Difluorophenyl)trimethylstannane (60)	306
	o-Difluorobenzene	BuLi/THF, hexane/−50°	$FClO_3$	1,2,3-Trifluorobenzene (—)	551
		"	CO_2	2,3-Difluorobenzoic acid (74)	551
C_6H_5F	C_6H_5F	BuLi/THF/−50°/ 7 hr	CO_2	2-Fluorobenzoic acid (60)	260

C_7HF_7	CF_3, F, F, F, F	BuLi/THF/−70°/ 1 hr	CO_2	CF_3, F, F, F, F, X	$X = CO_2H$ (77)	549
		BuLi/THF/−60°/ 0.5 hr	CF_3COCF_3		$X = C(CF_3)_2OH$ (61)	548
$C_7H_2F_4O_2$	CO_2H, F, F, F, F	BuLi/THF/−65°/ 45 min	Cl_2	CO_2H, F, F, F, F, X	$X = Cl$ (—)	277
		,,	"S"		$X = SH$ (69)	277
		,,	CO_2		$X = CO_2H$ (94)	277
$C_7H_4F_4$	2,3,5,6-Tetrafluoro- toluene	BuLi/THF/−70°/ 1 hr	CO_2	CH_3, F, F, F, F, X	$X = CO_2H$ (88)	549
		BuLi/THF/−60°/ 0.5 hr	CF_3COCF_3		$X = C(CF_3)_2OH$ (91)	548
C_7H_7F	o-$FC_6H_4CH_3$	BuLi/THF/−50°/ 7 hr	CO_2	2-Fluoro-m-toluic acid (3)		260
	p-$FC_6H_4CH_3$	BuLi/THF/−50°/ 7 hr	CO_2	6-Fluoro-m-toluic acid (58)		260
$C_{10}H_7F$	1-Fluoronaphthalene	BuLi/THF/−50°/ 7 hr	CO_2	1-Fluoro-2-naphthoic acid (30)		260
	2-Fluoronaphthalene	BuLi/THF/−60°/ 6.5 hr	CO_2	2-Fluoro-1-naphthoic acid (II), 3-fluoro-2-naphthoic acid (III) (II+III, 70) (II:III, 54:46)		261

TABLE LII. ARYL FLUORIDES AND VINYL FLUORIDES (ORTHO, BETA) (*Continued*)

Formula	Compound Lithiated	Conditions	Substrate	Product and Yield (%)	Refs.
$C_{12}H_2F_8$	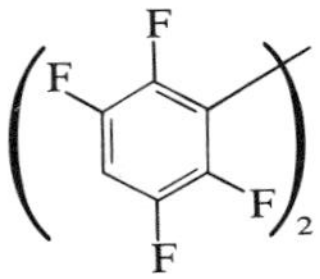 [structure]	BuLi/THF/−70°/ 1 hr	CO_2	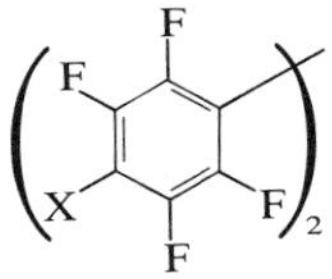 [structure] X = CO_2H (97)	549
		BuLi/THF/−60°/ 1 hr	CF_3COCF_3	X = $C(CF_3)_2OH$ (74)	548
$C_{12}H_7F_4N$	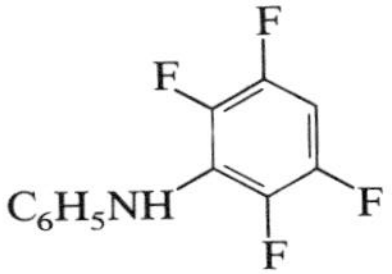 [structure]	BuLi	CO_2	4-Anilino-2,3,5,6-tetrafluorobenzoic acid (56)	552
$C_{14}H_9F$	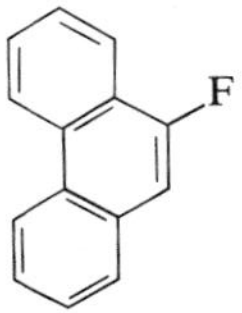 [structure]	BuLi/THF/−50°/ 7 hr	Br_2	9-Bromo-10-fluorophenanthrene (42)	531
		"	CO_2	10-Fluoro-9-phenanthrenecarboxylic acid (65)	531

Note: References 360–607 are on pp. 355–360.

TABLE LIII. ARYL CHLORIDES AND VINYL CHLORIDES (ORTHO, BETA)

Formula	Compound Lithiated	Conditions	Substrate	Product and Yield (%)	Refs.
C_2HCl_3	$CCl_2{=}CClH$	BuLi/THF, ether, pet. ether/−100°/50 min	CO_2	Trichloroacrylic acid (81)	167, 168
$C_2H_2Cl_2$	$CCl_2{=}CH_2$	BuLi/THF/−110°/ 68 min	CO_2	Chloropropiolic acid (86)	167
		BuLi/THF/−110°/ 40 min	Br_2	(Z)-1-Bromo-1,2-dichloroethylene (26)	167, 168
		"	CO_2	(E)-2,3-Dichloroacrylic acid (99)	167
C_3H_2ClNS		BuLi/THF/−65°/ 15 min*	CO_2	4-Chloro-5-isothiazolecarboxylic acid (68)	154
		"	DMF	4-Chloro-5-isothiazolecarboxaldehyde (65)	154
$C_4H_2Cl_2Se$		LDA/ether/−70°/2 hr	CO_2	2,5-Dichloro-3-selenophenecarboxylic acid (78)	291
		LDA/ether/−70°/4 hr	$(CH_3Se)_2$	2,5-Dichloro-3-(methylselenyl)selenophene (51)	291
C_4H_3ClO		LDA/THF/−80°/2.5 hr	$(CH_3)_2C{=}CHCH_2Br$	3-Chloro-2-(3-methyl-2-butenyl)furan (41)	566
C_4H_4ClNS		BuLi/THF/−65°/ 15 min*	CO_2	4-Chloro-3-methyl-5-isothiazolecarboxylic acid (75)	154
		"	DMF	4-Chloro-3-methyl-5-isothiazolecarboxaldehyde (47)	154

TABLE LIII. ARYL CHLORIDES AND VINYL CHLORIDES (ORTHO, BETA) (*Continued*)

Formula	Compound Lithiated	Conditions	Substrate	Product and Yield (%)	Refs.
C_4H_5ClO	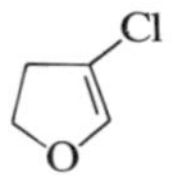	BuLi/THF/−78°	CH_3I	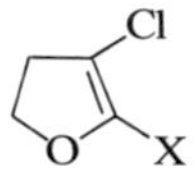 $X = CH_3$ (74)	266
		"	CH_3COCl	X 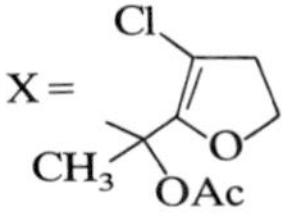 = (78)	266
		"	n-C_4H_9I	$X = C_4H_9$-n (62)	266
		"	C_6H_5CHO	$X = CH(OH)C_6H_5$ (77)	266
$C_4H_7ClO_2$	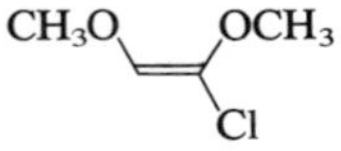	s-BuLi/THF/−100°/ 0.5 hr	$HgCl_2$	Bis[(Z)-2-chloro-(1,2-dimethoxyvinyl)]mercury (−)	600
		"	CO_2	(E)-3-Chloro-2,3-dimethoxyacrylic acid (45)	600
$C_5H_2Cl_3N$	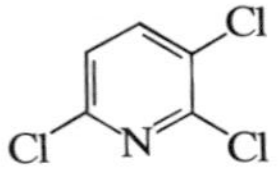	BuLi/ether/−70°/ 45 min, −20°/2 hr	$(CH_3O)_2SO_2$	2,3,6-Trichloro-4-methylpyridine (45)	265
C_5H_7ClO	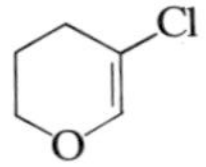	BuLi/THF/25°/2 hr	CH_3I	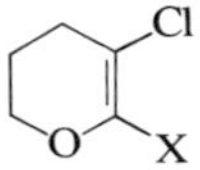 $X = CH_3$ (>65)	562
		"	C_2H_5I	$X = C_2H_5$ (>65)	562
		"	n-C_4H_9I	$X = C_4H_9$-n (>65)	562
C_6HCl_5	Pentachlorobenzene	BuLi/THF/−65°	CO_2	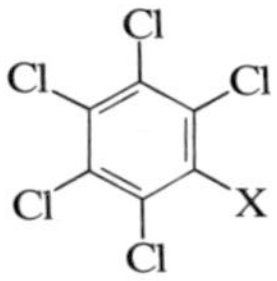 $X = CO_2H$ (91)	553

		BuLi/THF/−70°	$(CH_3)_3SiCl$	X = $Si(CH_3)_3$ (95)	264, 554
		BuLi/THF/−78°/3 hr	$(CH_3)_3SnCl$	X = $Sn(CH_3)_3$ (73)	554
$C_6H_2Cl_4$	1,2,3,4-Tetrachlorobenzene	CH_3Li/THF/−70°	$(CH_3)_3SiCl$	Trimethyl(2,3,6-trichlorophenyl)silane (7), trimethyl(2,3,4,5-tetrachlorophenyl)silane (93)	264
	1,2,4,5-Tetrachlorobenzene	BuLi/ether/−65°	CO_2	1-X-2,3,5,6-tetrachlorobenzene (48), 1,4-di-X-2,3,5,6-tetrachlorobenzene (27); X = CO_2H	553
		2 BuLi/THF	”	(30) (43) X = CO_2H	553
		BuLi/THF/−78°	$(CH_3)_2SiHCl$	(58) X = $SiH(CH_3)_2$	554
		BuLi/THF, ether/ −78°/5 hr	$(CH_3)_3SiCl$	(24) (56) X = $Si(CH_3)_3$	554
		BuLi/THF/−78°	$C_6H_5(CH_3)_2SiCl$	(—) (67) X = $Si(CH_3)_2C_6H_5$	554
		”	$CH_3(C_6H_5)_2SiCl$	(—) (62) X = $Si(C_6H_5)_2CH_3$	554
		”	$(CH_3)_3SnCl$	(—) (32) X = $Sn(CH_3)_3$	554
$C_6H_3Cl_3$	1,3,5-Trichlorobenzene	3 BuLi/THF/−65°	$(CH_3)_3SiCl$	(2,4,6-Trichloro-*s*-phenenyl)tris[trimethylsilane] (38)	555
		2 BuLi/THF/−65°/ 0.5 hr	”	(2,4,6-Trichloro-*m*-phenylene)bis[trimethylsilane] (44)	555
$C_7H_3Cl_2F_3$	2,4-dichloro-1-(CF_3)benzene	BuLi/THF/−50°/ 1 hr	CO_2	2,6-Dichloro-α,α,α-trifluoro-*m*-toluic acid (—)	556
C_7H_5ClN	*m*-ClC_6H_4CN	LTMP/THF/−70°/ 1 hr	$(CH_3S)_2$	3-Chloro-2-(methylthio)benzonitrile (30)	24

TABLE LIII. Aryl Chlorides and Vinyl Chlorides (Ortho, Beta) (*Continued*)

Formula	Compound Lithiated	Conditions	Substrate	Product and Yield (%)	Refs.
$C_7H_6Cl_2$		BuLi/THF/−50°	$C_6H_5SO_2F$	(2,6-Dichloro-*m*-tolyl)phenylsulfone (—)	557
		”	C_6H_5COCl	2,6-Dichloro-3-methylbenzophenone (—)	557
C_9H_6ClNO		BuLi/THF/−70°/ 0.5 hr	I_2	X = I (73)	232
		”	CO_2	X = CO_2H (80)	232
		”	CH_3I	X = CH_3 (74)	232
$C_9H_8Cl_2O_2$		BuLi/THF/−50°	C_6H_5COCl (H_3O^+)	2,4-Dichloro-3-benzoylbenzaldehyde (—)	557
$C_9H_{11}ClO_3$		BuLi/THF/−70°/ 3 min	CO_2	2-Chloro-3,5,6-trimethoxybenzoic acid (79), 3-chloro-2,5,6-trimethoxybenzoic acid (16)	222
		BuLi/THF/25°/ 10 min	Ac_2O	2-Chloro-3,5,6-trimethoxyacetophenone (68)	223

$C_{11}H_{14}ClNO$	m-$ClC_6H_4CONHC_4H_9$-t	BuLi/THF/−70°	$(CH_3S)_2$	N-t-Butyl-3-chloro-2-(methylthio)benzamide (31)	24
$C_{14}H_4ClNO_2S$	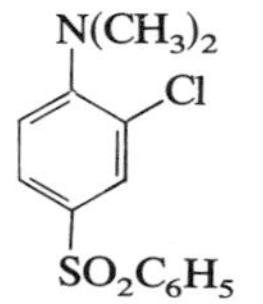	BuLi/ether/−80°/ 24 hr	CO_2	2-Chloro-3-(dimethylamino)-6-(phenylsulfonyl)-benzoic acid (82)	542
$C_{18}H_{18}Cl_2O_2$	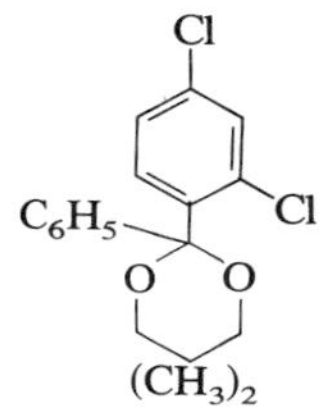	BuLi/THF/−50°	CO_2 (H_3O^+)	2,6-Dichloro-3-benzoylbenzoic acid (—)	557

Note: References 360–607 are on pp. 355–360.

TABLE LIV. ARYL BROMIDES (ORTHO, BETA)

Formula	Compound Lithiated	Conditions	Substrate	Product and Yield (%)	Refs.
C_3H_2BrNS	4-bromoisothiazole (structure)	BuLi/THF/−65°/ 15 min*	CO_2	4-Bromo-5-isothiazolecarboxylic acid (70)	154
		"	DMF	4-Bromo-5-isothiazolecarboxaldehyde (73)	154
$C_3H_3BrN_2$	4-bromopyrazole (structure)	C_6H_5Li/ether/25°/ 2 hr	CO_2	4-Bromopyrazole-5-carboxylic acid (35)	87
$C_4H_2Br_2S$	2,5-dibromothiophene (structure)	LDA/THF or ether/−70°	$(CH_3)_3SiCl$	(2,5-Dibromo-3-thienyl)trimethylsilane (—)	246
C_4H_3BrO	3-bromofuran (structure)	LDA/THF/−80°/ 2.5 hr	CH_2O	3-bromo-2-X-furan (structure): $X = CH_2OH$ (52)	566
		"	CH_3OCH_2Cl	$X = CH_2OCH_3$ (49)	566
		LDA/THF or ether/−70°	$(CH_3)_3SiCl$	$X = Si(CH_3)_3$ (—)	246
		LDA/THF/ −80°/2.5 hr	$(CH_3)_2C{=}CHCH_2Br$	$X = CH_2CH{=}C(CH_3)_2$ (66)	566
C_4H_3BrS	3-bromothiophene (structure)	C_6H_5Li/ether/ "overnight"	CO_2	3-Bromo-2-thiophenecarboxylic acid (72)	48

Formula	Substrate	Conditions	Electrophile	Product(s) (Yield, %)	Refs.
C_4H_4BrNS	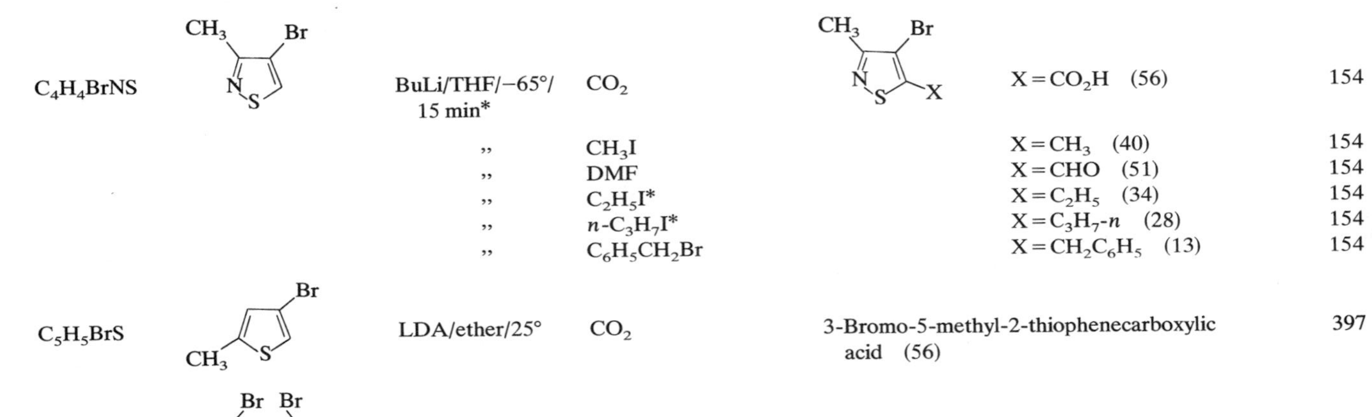	BuLi/THF/−65°/ 15 min*	CO_2	X = CO_2H (56)	154
		”	CH_3I	X = CH_3 (40)	154
		”	DMF	X = CHO (51)	154
		”	C_2H_5I*	X = C_2H_5 (34)	154
		”	n-C_3H_7I*	X = C_3H_7-n (28)	154
		”	$C_6H_5CH_2Br$	X = $CH_2C_6H_5$ (13)	154
C_5H_5BrS		LDA/ether/25°	CO_2	3-Bromo-5-methyl-2-thiophenecarboxylic acid (56)	397
$C_8H_4Br_2S_2$		EtLi/THF/−70°/ 2 min	CO_2	4,4′-Dibromo-[3,3′-bithiophene]-5-carboxylic acid (51)	417
$C_{12}H_9BrO_2S$	$C_6H_5SO_2C_6H_4Br$-m	BuLi/ether/0°	CO_2	2-Bromo-6-(phenylsulfonyl)benzoic acid (54)	69

Note: References 360–607 are on pp. 355–360.

TABLE LV. Aryl Iodides (Beta)

Formula	Compound Lithiated	Conditions	Substrate	Product and Yield (%)	Refs.
C_3H_2INS	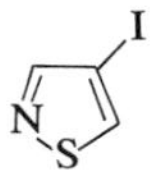	BuLi/THF/ −65°/15 min*	DMF	4-Iodo-5-isothiazolecarboxaldehyde (33)	154
C_4H_3IS	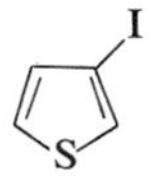	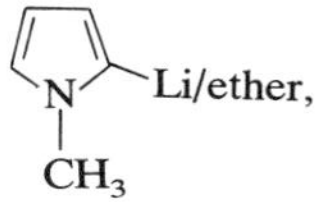 N-methyl-2-pyrrolyl-Li/ether, CH_3 TMEDA/ −10°/0.5 hr	CO_2	3-Iodo-2-thiophenecarboxylic acid (80), 4-iodo-2-thiophenecarboxylic acid (20)	68
C_4H_4INS	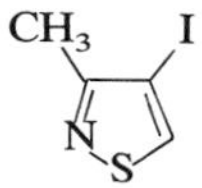 CH_3 I	BuLi/THF/ −65°/15 min*	CO_2	4-Iodo-3-methyl-5-isothiazolecarboxylic acid (58)	154
		”	DMF	4-Iodo-3-methyl-5-isothiazolecarboxaldehyde (68)	154

TABLE LVI. (Trifluoromethyl)benzenes (Ortho)

Formula	Compound Lithiated	Conditions	Substrate	Product and Yield (%)	Refs.
$C_7H_5F_3$	$C_6H_5CF_3$	BuLi	$CO_2(CH_2N_2)$	Methyl α,α,α-trifluoro-*o*-toluate (73), methyl α,α,α-trifluoro-*m*-toluate (26)	50
$C_8H_4F_6$	*m*-$CF_3C_6H_4CF_3$	BuLi/ether/ 20°/1 hr	CO_2	2,4-Bis(trifluoromethyl)benzoic acid (I), 2,6-bis(trifluoromethyl)benzoic acid (II) (I+II, 85) (I:II, 3:2)	49, 267, 268, 269
		BuLi/ether, TMEDA/25°	”	(II) (30)	558
		BuLi/THF	$(CH_3)_3SnCl$	[2,6-Bis(trifluoromethyl)phenyl] trimethylstannane (30), [2,4-bis(trifluoromethyl)phenyl] trimethyl stannane (—)	306
	p-$CF_3C_6H_4CF_3$	BuLi/ether/25° 1.5 hr	CO_2	2,5-Bis(trifluoromethyl)benzoic acid (87)	49, 268, 269
$C_{10}H_9F_6N$	CF_3 $(CH_3)_2N$ CF_3	BuLi/ether, hexane/reflux	CO_2	4-(Dimethylamino)-2,6-bis-(trifluoromethyl)benzoic acid (76)	558
$C_{10}H_{12}F_3N$	*m*-$CF_3C_6H_4CH_2$-$N(CH_3)_2$	BuLi/ether/ 25°/1 hr	$(C_6H_5)_2CO$	[α^6-(Dimethylamino)-$\alpha^2,\alpha^2,\alpha^2$-trifluoro-2,6-xylyl]diphenylmethanol (72)	45

Note: References 360–607 are on pp. 355–360.

TABLE LVII. Arylphosphine Oxides and Imides (Ortho, Beta)

Formula	Compound Lithiated	Conditions	Substrate	Product and Yield (%)		Refs.
$C_9H_9OPS_2$	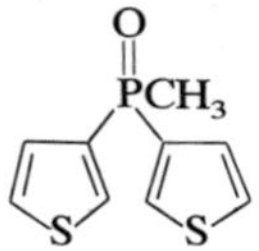 O, PCH_3, S, S	BuLi/THF/25°/ 1 hr	CO_2	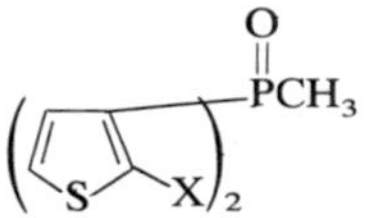 O, PCH_3, S, $X)_2$	$X = CO_2H$ (33)	270
		”	CH_3COCH_3		$X = C(CH_3)_2OH$ (40)	270
$C_{14}H_{11}OPS_2$	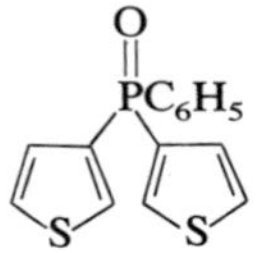 O, PC_6H_5, S, S	BuLi/THF/25°/ 1 hr	Br_2	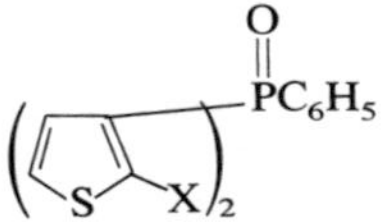 O, PC_6H_5, S, $X)_2$	$X = Br$ (65)	270
		”	CO_2		$X = CO_2H$ (75)	270
		”	$(CH_3)_3SiCl$		$X = Si(CH_3)_3$ (30)	270
		”	CH_3COCH_3		$X = C(CH_3)_2OH$ (45)	270
		”	C_6H_5CHO		$X = CH(OH)C_6H_5$ (70)	270
		”	$C_6H_5COCH_3$		$X = C(OH)(CH_3)C_6H_5$ (50)	270
		”	$(C_6H_5)_2CO$		$X = C(C_6H_5)_2OH$ (80)	270
		”	CH_3CO_2Et	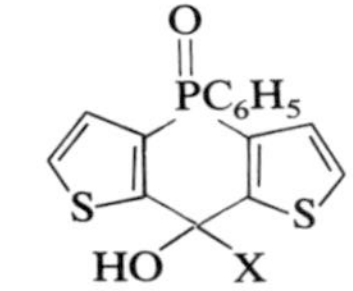 O, PC_6H_5, S, S, HO, X	$X = CH_3$ (35)	270

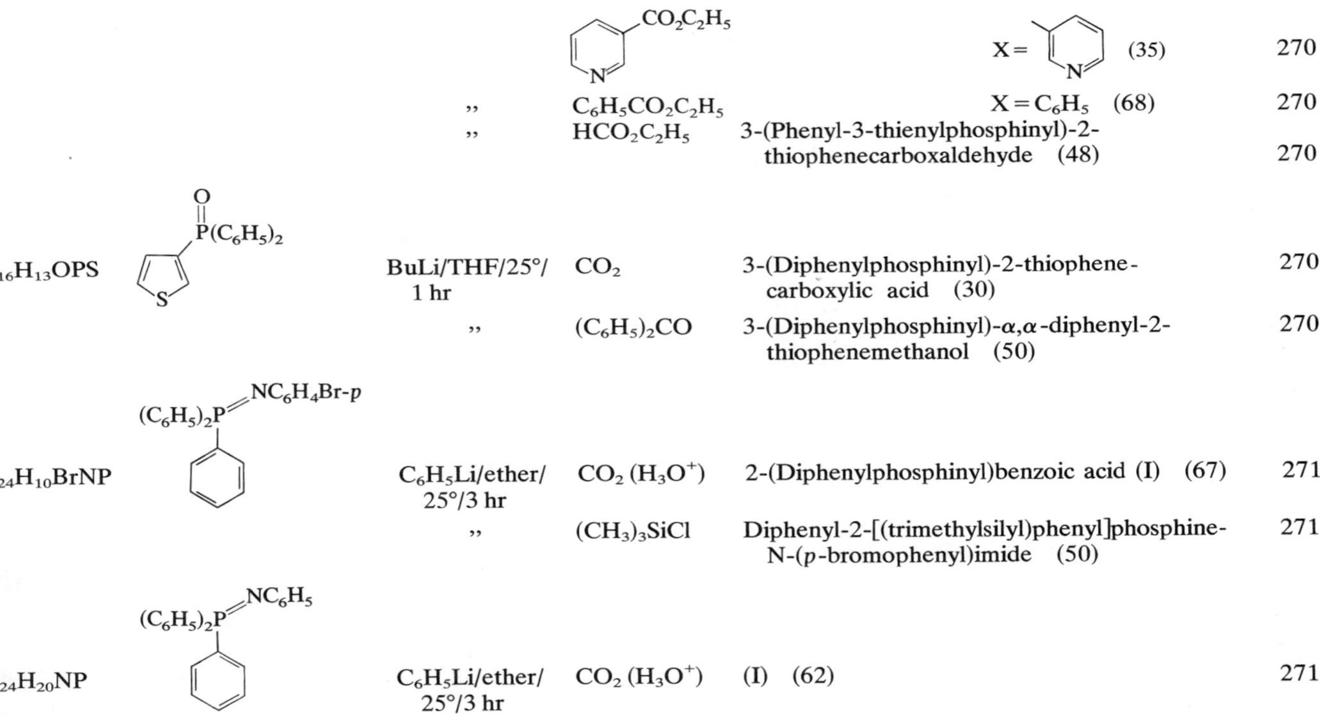

Formula	Structure	Conditions	Electrophile	Product (Yield %)	Ref.
			3-pyridyl-$CO_2C_2H_5$	X = 3-pyridyl (35)	270
		,,	$C_6H_5CO_2C_2H_5$	$X = C_6H_5$ (68)	270
		,,	$HCO_2C_2H_5$	3-(Phenyl-3-thienylphosphinyl)-2-thiophenecarboxaldehyde (48)	270
$C_{16}H_{13}OPS$	3-thienyl-$P(O)(C_6H_5)_2$	BuLi/THF/25°/ 1 hr	CO_2	3-(Diphenylphosphinyl)-2-thiophene-carboxylic acid (30)	270
		,,	$(C_6H_5)_2CO$	3-(Diphenylphosphinyl)-α,α-diphenyl-2-thiophenemethanol (50)	270
$C_{24}H_{10}BrNP$	$(C_6H_5)_2P(=NC_6H_4Br\text{-}p)C_6H_5$	C_6H_5Li/ether/ 25°/3 hr	CO_2 (H_3O^+)	2-(Diphenylphosphinyl)benzoic acid (I) (67)	271
		,,	$(CH_3)_3SiCl$	Diphenyl-2-[(trimethylsilyl)phenyl]phosphine-N-(*p*-bromophenyl)imide (50)	271
$C_{24}H_{20}NP$	$(C_6H_5)_2P(=NC_6H_5)C_6H_5$	C_6H_5Li/ether/ 25°/3 hr	CO_2 (H_3O^+)	(I) (62)	271

TABLE LVIII. SELENIDES (ORTHO, BETA)

Formula	Compound Lithiated	Conditions	Substrate	Product and Yield (%)	Refs.
C_5H_6SSe	3-($SeCH_3$)thiophene	BuLi/2 hr	CO_2	3-(Methylselenyl)-2-thiophenecarboxylic acid (I), 4-(methylselenyl)-2-thiophenecarboxylic acid (II) (I+II, 75) (I:II, 56:44)	140
$C_8H_{12}SSe$	3-(SeC_4H_9-*n*)thiophene	BuLi/2 hr	CO_2	3-(SeC_4H_9-*n*)-2-X-thiophene (III), 4-(SeC_4H_9-*n*)-2-X-thiophene (IV) (III+IV, 79) (III:IV, 6:4) X = CO_2H	140
		BuLi/ether/−30°	"S" + CH_3I	(III:IV, 7:3) X = SCH_3	140
		"	"Se" + CH_3I	(III+IV, 67) X = $SeCH_3$	140
$C_{12}H_8Se$	dibenzoselenophene	BuLi/ether/ reflux	CO_2	4-(CO_2H)dibenzoselenophene (96)	273

REFERENCES TO TABLES I–LVIII

[360] Y. Leroux and C. Roman, *Tetrahedron Lett.*, **1973,** 2585.
[361] Y. L. Gol'dfarb, Y. L. Danyushevskii, and M. A. Vinogradova, *Dokl. Akad. Nauk SSSR*, **151,** 332 (1963) [*C.A.*, **59,** 8681c (1963)].
[362] Y. L. Gol'dfarb and V. P. Litvinov, *Izv. Akad. Nauk SSSR, Ser. Khim.*, **1964,** 2088 [*C.A.*, **62,** 9090f (1965)].
[363] W. E. Truce and E. Wellisch, *J. Am. Chem. Soc.*, **74,** 5177 (1952).
[364] K. Okuhara, *J. Org. Chem.*, **41,** 1487 (1976).
[365] S. Gronowitz, A. B. Hörnfeldt, and K. Pettersson, *Synth. Commun.*, **1973,** 213.
[366] D. J. Chadwick, J. Chambers, G. Meakins, and R. L. Snowden, *J. Chem. Soc., Perkin Trans. I*, **1975,** 523.
[367] M. J. Arco, M. H. Trammell, and J. D. White, *J. Org. Chem.*, **41,** 2075 (1976).
[368] J. Gombos, E. Haslinger, H. Zak, and U. Schmidt, *Monatsh. Chem.*, **106,** 219 (1975).
[369] R. A. Benkeser and R. B. Currie, *J. Am. Chem. Soc.*, **70,** 1780 (1948).
[370] H. Bredereck and R. Gompper, *Angew. Chem.*, **70,** 571 (1958).
[371] Y. L. Gol'dfarb and Y. L. Danyushevskii, *Izv. Akad. Nauk SSSR, Otd. Khim. Nauk*, **1963,** 540 [*C.A.*, **59,** 3861a (1963)].
[372] Y. A. Zhdanov, V. G. Alekseeva, and E. L. Korol, *Dokl. Akad. Nauk SSSR*, **225,** 1336 (1975) [*C.A.*, **84,** 150848f (1976)].
[373] G. R. Ziegler and G. S. Hammond, *J. Am. Chem. Soc.*, **90,** 513 (1968).
[374] S. Gronowitz and B. Holm, *Synth. Commun.*, **4,** 63 (1974).
[375] A. V. Koblik, T. I. Polyakova, B. A. Tertov, B. V. Mezhov, and G. N. Dorofeenko, *Zh. Org. Khim.*, **11,** 2153 (1975) [*C.A.*, **84,** 43782h (1976)].
[376] M. C. Zalushi, M. Robba, and M. Bonhomme, *Bull. Soc. Chim. Fr.*, **1970,** 1838.
[377] I. Vlattas, Ciba-Geigy Pharmaceutical Co., Summit, New Jersey, personal communication.
[378] V. G. Kul'nevich, Z. I. Zelikman, A. Shkrebets, B. A. Tertov, and M. M. Ketslakh, *Khim. Geterotsikl. Soedin.*, **1973,** 595 [*C.A.*, **79,** 42424e (1973)].
[379] P. Pastour and C. Plantard, *C.R. Acad. Sci., Paris, Ser. C*, **262,** 1539 (1966).
[380] E. N. Givens, L. G. Alexakos, and P. B. Venuto, *Tetrahedron*, **25,** 2407 (1969).
[381] S. Gronowitz and B. Uppström, *Acta Chem. Scand.*, **B29,** 441 (1975).
[382] T. A. Dobson and L. G. Humber, *J. Heterocycl. Chem.*, **12,** 591 (1975).
[383] D. Hoppe, *Angew. Chem., Int. Ed. Engl.*, **13,** 789 (1974).
[384] R. Muthukrishnan and M. Schlosser, *Helv. Chim. Acta*, **59,** 13 (1976).
[385] S. Gronowitz and U. Rosen, *Chem. Scr.*, **1,** 33 (1971) [*C.A.*, **75,** 20080e (1971)].
[386] S. Gronowitz and K. Halvarson, *Ark. Kemi*, **8,** 343 (1955) [*C.A.*, **49,** 10921f (1955)].
[387] S. Gronowitz and P. Moses, *Acta Chem. Scand.*, **16,** 155 (1960).
[388] Y. L. Gol'dfarb, V. P. Litvinov, and S. A. Ozolin, *Izv. Akad. Nauk SSSR, Ser. Khim.*, **1966,** 1432 [*C.A.*, **66,** 65409x (1967)].
[389] Y. L. Gol'dfarb and M. L. Kirmalova, *J. Gen. Chem. (USSR) (Engl. Transl.)*, **25,** 1321 (1955) [*C.A.*, **50,** 6422h (1956).]
[390] Y. L. Gol'dfarb, B. P. Fabrichnyi, and V. I. Rogovik, *Izv. Akad. Nauk SSSR, Ser. Khim.*, **1965,** 515 [*C.A.*, **63,** 562f (1965)].
[391] S. Gronowitz and J. Röe, *Acta Chem. Scand.*, **19,** 1741 (1965).
[392] P. L. Kelly, S. F. Thames, and J. E. McCleskey, *J. Heterocycl. Chem.*, **9,** 141 (1972).
[393] S. Gronowitz and R. Hakansson, *Ark. Kemi*, **17,** 73 (1960) [*C.A.*, **56,** 3497g (1962)].
[394] S. Gronowitz and B. Holm, *Chem. Scr.*, **2,** 245 (1972) [*C.A.*, **78,** 84156b (1973)].
[395] D. W. Adamson, *J. Chem. Soc.*, **1950,** 885.
[396] P. Dubus, B. Decroix, J. Morel, and P. Pastour, *Bull. Soc. Chim. Fr.*, **1976,** 628.
[397] S. Gronowitz and T. Frejd, *Acta Chem. Scand.*, **B30,** 485 (1976).
[398] W. R. Biggerstaff, H. Arzoumanian, and K. L. Stevens, *J. Med. Chem.*, **7,** 110 (1964).
[399] S. Gronowitz, P. Moses, A. B. Hörnfeldt, and R. Hakansson, *Ark. Kemi*, **17,** 165 (1961) [*C.A.*, **57,** 8528h (1962)].
[400] W. R. Biggerstaff and K. L. Stevens, *J. Org. Chem.*, **28,** 733 (1963).
[401] J. Pankiewicz, B. Decroix, and J. Morel, *C. R. Acad. Sci., Paris, Ser. C*, **281,** 39 (1975).
[402] S. Gronowitz, *Ark. Kemi*, **12,** 239 (1958) [*C.A.*, **52,** 20115d (1958)].

[403] Y. L. Gol'dfarb, M. L. Kirmalova, and M. A. Kalik, *Zh. Obshch. Khim.*, **29,** 2034 (1959) [*C.A.*, **54,** 8775h (1960)].

[404] Y. L. Gol'dfarb, M. A. Kalik, and M. L. Kirmalova, *Zh. Obsch. Khim.*, **29,** 3631 (1959) [*C.A.*, **54,** 19638h (1960)].

[405] S. Gronowitz and B. Gestblom, *Ark. Kemi*, **18,** 513 (1962) [*C.A.*, **57,** 3000c (1962)].

[406] E. Wiklund and R. Hakansson, *Chem. Scr.*, **3,** 220 (1973) [*C.A.*, **79,** 79495r (1973)].

[407] J. Sicé, *J. Org. Chem.*, **19,** 70 (1954).

[408] Y. L. Gol'dfarb, M. A. Kalik, and M. L. Kirmalova, *Metody Poluch. Khim. Reakt. Prep.*, **14,** 156 (1966) [*C.A.*, **67,** 11380h (1967)].

[409] A. Wiersema and S. Gronowitz, *Acta Chem. Scand.*, **24,** 2593 (1970).

[410] J. Skramstad, *Chem. Scr.*, **4,** 81 (1973) [*C.A.*, **79,** 77884e (1973)].

[411] A. Bugge, *Acta Chem. Scand.*, **23,** 1823 (1969).

[412] S. Gronowitz, B. Gestblom, and B. Mathiasson, *Ark. Kemi*, **20,** 407 (1963) [*C.A.*, **59,** 7459b (1963)].

[413] P. Pirson, A. Schonne, and L. Christiaens, *Bull. Soc. Chim. Belg.*, **79,** 575 (1970).

[414] U. Michael and S. Gronowitz, *Acta Chem. Scand.*, **22,** 1353 (1968).

[415] Y. L. Gol'dfarb, G. I. Gorushkina, and B. P. Federov, *Izv. Akad. Nauk SSSR, Otd. Khim. Nauk*, **1959,** 2021 [*C.A.*, **54,** 9879b (1960)].

[416] R. T. Hawkins and D. B. Stroup, *J. Org. Chem.*, **34,** 1173 (1969).

[417] R. Hakansson, *Acta Chem. Scand.*, **25,** 1313 (1971).

[418] H. Wynberg and A. Bantjes, *J. Am. Chem. Soc.*, **82,** 1447 (1960).

[419] S. Gronowitz and V. Vilks, *Ark. Kemi*, **21,** 191 (1963) [*C.A.*, **59,** 13919g (1963)].

[420] Y. L. Gol'dfarb, S. Ozolins, and V. P. Litvinov, *Zh. Obshch. Khim.*, **37,** 2220 (1967) [*C.A.*, **68,** 87218c (1968)].

[421] S. F. Thames and J. E. McCleskey, *J. Heterocycl. Chem.*, **3,** 104 (1966).

[422] V. P. Litvinov, T. Shchedrinskaya, P. A. Konstantinov, and Y. L. Gol'dfarb, *Khim. Geterotsikl. Soedin.*, **1975,** 492 [*C.A.*, **83,** 27273s (1975)].

[423] S. Gronowitz, *Ark. Kemi*, **16,** 363 (1960) [*C.A.*, **55,** 21092f (1961)].

[424] Y. A. Gol'dfarb, F. M. Stoyanovich, and G. B. Chermanova, *Izv. Akad. Nauk SSSR, Ser. Khim.*, **1973,** 2290 [*C.A.*, **80,** 47743n (1974)].

[425] Y. L. Gol'dfarb, B. P. Fabrichnyi, and V. I. Rogovik, *Izv. Akad. Nauk SSSR, Ser. Khim.*, **1963,** 2172 [*C.A.*, **60,** 9227b (1964)].

[426] Y. L. Gol'dfarb, B. P. Fabrichniyi, and V. I. Rogovik, *Izv. Akad. Nauk SSSR, Ser. Khim.*, **1963,** 2178 [*C.A.*, **60,** 9226g (1964)].

[427] Y. L. Gol'dfarb, S. Ozolins, and V. P. Litvinov, *Khim. Geterotsikl. Soedin.*, **1967,** 935 [*C.A.*, **69,** 2883p (1968)].

[428] P. Pastour, P. Savalle, and P. Eymery, *C. R. Acad. Sci., Paris, Ser. C*, **260,** 6130 (1965).

[429] R. Leardini, G. Martelli, P. Spagnolo, and M. Tiecco, *J. Chem. Soc., C*, **1970,** 1464.

[430] S. Gronowitz and H. Frostling, *Acta Chem. Scand.*, **16,** 1127 (1962).

[431] S. Gronowitz and H. Frostling, *Tetrahedron Lett.*, **1961,** 604.

[432] S. Gronowitz and S. Hagen, *Ark. Kemi*, **27,** 153 (1967) [*C.A.*, **67,** 99931x (1967)].

[433] Y. L. Gol'dfarb and M. L. Kirmalova, *Zh. Obshch. Khim.*, **29,** 897 (1959) [*C.A.*, **54,** 1485d (1960)].

[434] Y. L. Gol'dfarb and Y. B. Vol'kenshtein, *Izv. Akad. Nauk SSSR, Otd. Khim. Nauk*, **1960,** 2738 [*C.A.*, **55,** 14423g (1961)].

[435] D. W. H. MacDowell and A. T. Jeffries, *J. Org. Chem.*, **35,** 871 (1970).

[436] C. J. Grol, *Tetrahedron*, **30,** 3621 (1974).

[437] S. Gronowitz, J. E. Skramstad, and B. Eriksson, *Ark. Kemi*, **28,** 99 (1967) [*C.A.*, **69,** 2886s (1968)].

[438] D. W. H. MacDowell and T. B. Patrick, *J. Org. Chem.*, **32,** 2441 (1967).

[439] D. W. H. MacDowell and A. T. Jeffries, *J. Org. Chem.*, **36,** 1053 (1971).

[440] T. Kauffmann and R. Otter, *Angew. Chem.*, **88,** 513 (1976).

[441] J. Morel, C. Paulmier, and P. Pastour, *C. R. Acad. Sci., Paris, Ser. C*, **269,** 37 (1969).

[442] H. Wynberg, J. De Wit, and H. J. M. Sinnige, *J. Org. Chem.*, **35,** 711 (1970).

[443] L. H. Klemm, E. E. Klopfenstein, R. Zell, D. R. McCoy, and R. A. Klemm, *J. Org. Chem.*, **34,** 347 (1969).

[444] L. H. Klemm and R. E. Merrill, *J. Heterocycl. Chem.*, **11,** 355 (1974).

[445] R. P. Dickinson and B. Iddon, *J. Chem. Soc., C*, **1971,** 182.
[446] T. V. Shchedrinskaya, V. P. Litvinov, P. A. Konstantinov, Y. L. Gol'dfarb, and E. G. Ostapenko, *Khim. Geterotsikl. Soedin.*, **1973,** 1026 [*C.A.*, **79,** 126198v (1973)].
[447] S. Gronowitz, J. Rehnö, and J. Sandström, *Acta Chem. Scand.*, **24,** 304 (1970).
[448] E. N. Karaulova, D. S. Meilanova, and G. D. Gal'pern, *Zh. Obshch. Khim.*, **30,** 3292 (1960) [*C.A.*, **55,** 19892a (1961)].
[449] J. E. Banfield, W. Davies, N. W. Gamble, and S. Middleton, *J. Chem. Soc.*, **1956,** 4791.
[450] B. Iddon, H. Suschitzky, and D. S. Taylor, *J. Chem. Soc., Perkin Trans. I*, **1974,** 2505.
[451] W. E. Parham and B. Gadsby, *J. Org. Chem.*, **25,** 234 (1960).
[452] R. H. Meen and H. Gilman, *J. Org. Chem.*, **20,** 73 (1954).
[453] N. B. Chapman, C. G. Hughes, and R. M. Scrowston, *J. Chem. Soc., C*, **1971,** 463.
[454] V. P. Litvinov, Y. L. Gol'dfarb, V. V. Zelentsov, L. G. Bogdanova, and N. N. Petukhova, *Khim. Geterotsikl. Soedin.*, **1975,** 486 [*C.A.*, **83,** 96909r (1975)].
[455] K. Clarke, G. Rawson, and R. M. Scrowston, *J. Chem. Soc., C*, **1969,** 537.
[456] R. P. Dickinson and B. Iddon, *J. Chem. Soc., C*, **1970,** 2592.
[457] D. E. Horning and J. M. Muchowski, *Can. J. Chem.*, **52,** 2950 (1974).
[458] R. Breslow and E. McNelis, *J. Am. Chem. Soc.*, **81,** 3080 (1959).
[459] M. Erne and H. Erlenmeyer, *Helv. Chim. Acta*, **31,** 652 (1948).
[460] A. I. Meyers and G. N. Knaus, *J. Am. Chem. Soc.*, **95,** 3408 (1973).
[461] P. Jutzi and H. J. Hoffmann, *Chem. Ber.*, **106,** 594 (1973).
[462] P. Jutzi, H. J. Hoffmann, K. Beier, and K. H. Wyes, *J. Organomet. Chem.*, **82,** 209 (1974).
[463] S. Gronowitz and T. Frejd, *Acta Chem. Scand.*, **B30,** 313 (1976).
[464] J. Morel, C. Paulmier, D. Semard, and P. Pastour, *C. R. Acad. Sci., Paris, Ser. C*, **270,** 825 (1970).
[465] G. Hinrio, J. Morel, and P. Pastour, *Ann. Chim.* (*Paris*), **10,** 37 (1975).
[466] J. Morel, C. Paulmier, and P. Pastour, *J. Heterocycl. Chem.*, **9,** 355 (1972).
[467] C. Paulmier and P. Pastour, *Bull. Soc. Chim. Fr.*, **1966,** 4021.
[468] H. Böhme and W. Stammberger, *Arch. Pharm.* (*Weinheim*), **305,** 392 (1972).
[469] G. Köbrich, W. E. Breckoff, and W. Drischel, *Justus Liebigs Ann. Chem.*, **704,** 51 (1967).
[470] G. Köbrich and W. Drischel, *Angew. Chem., Int. Ed. Engl.*, **4,** 74 (1965).
[471] G. Köbrich and W. Drischel, *Tetrahedron*, **22,** 2621 (1966).
[472] J. H. Boyer and R. F. Reinisch, *J. Am. Chem. Soc.*, **82,** 2218 (1960).
[473] M. Sorm and S. Nespurek, *Collect. Czech. Chem. Commun.*, **40,** 3459 (1975).
[474] C. V. Greco and B. P. O'Reilly, *J. Heterocycl. Chem.*, **7,** 1433 (1970).
[475] G. Van Koten and J. G. Noltes, *Chem. Commun.*, **1972,** 940.
[476] G. Pifferi and R. Monguzzi, *J. Heterocycl. Chem.*, **9,** 1445 (1972).
[477] Z. Horii, H. Hakusui, T. Shigeuchi, M. Hanaoka, and T. Momose, *Yakugaku Zasshi*, **92,** 503 (1972) [*C.A.*, **77,** 84156b (1973)].
[478] A. C. Ranade, R. S. Mali, R. M. Gidwani, and H. R. Deshpande, *Chem. Ind.* (*London*), **1977,** 310.
[479] M. Cugnon de Sevricourt and M. Robba, *Bull. Soc. Chim. Fr.*, **1977,** 142.
[480] G. Van Koten, C. A. Schaap, and J. G. Noltes, *J. Organomet. Chem.*, **99,** 157 (1975).
[481] G. Van Koten and J. G. Noltes, *J. Am. Chem. Soc.*, **98,** 5393 (1976).
[482] B. H. Bhike and V. P. Gupta, *Indian J. Chem.*, **15B**, 30 (1977).
[483] A. C. Ranade, R. S. Mali, S. R. Bhide, and S. R. Mehta, *Synthesis*, **1976,** 123.
[484] M. Julia, M. Duteil, and J. Y. Lallemand, *J. Organomet. Chem.*, **102,** 239 (1975).
[485] G. Van Koten and J. G. Noltes, *J. Organomet. Chem.*, **104,** 127 (1976).
[486] C. C. Eaborn, P. Golborn, and R. Taylor, *J. Organomet. Chem.*, **10,** 171 (1967).
[487] C. L. Mao, I. T. Barnish, and C. R. Hauser, *J. Heterocycl. Chem.*, **6,** 475 (1969).
[488] W. J. Houlihan, U.S. Pat. 3,745,165 (1973) [*C.A.*, **79,** 66201d (1973)].
[489] W. J. Houlihan, U.S. Pat. 3,879,418 (1975) [*C.A.*, **83,** 58651t (1975)].
[490] W. J. Houlihan, U.S. Pat. 3,838,174 (1974) [*C.A.*, **81,** 151979z (1974)].
[491] N. S. Narasimhan and R. S. Mali, *Synthesis*, **1975,** 797.
[492] W. J. Houlihan, U.S. Pat. 3,767,673 (1973) [*C.A.*, **80,** 59882u (1974)].
[493] N. S. Narasimhan and B. H. Bhide, *Tetrahedron*, **27,** 6171 (1971).
[494] N. S. Narasimhan and B. H. Bhide, *Chem. Commun.*, **1970,** 1552.
[495] B. H. Bhide and H. J. Parekh, *Chem. Ind.* (*London*), **1974,** 733.

[496] H. Watanabe, C. L. Mao, and C. R. Hauser, *J. Org. Chem.*, **34,** 1786 (1969).
[497] B. Cederlund, R. Lantz, A. B. Hörnfeldt, O. Thorstad, and K. Undheim, *Acta Chem. Scand.*, **B31,** 198 (1977).
[498] K. Bowden, G. Crank, and W. J. Ross, *J. Chem. Soc., C,* **1968,** 172.
[499] W. Vetter, G. Schill, and C. Zürcher, *Chem. Ber.*, **107,** 424 (1974).
[500] J. Boeckmann and G. Schill, *Chem. Ber.*, **110,** 703 (1977).
[501] G. Van Koten, A. J. Leusink, and J. G. Noltes, *J. Organomet. Chem.*, **85,** 105 (1975).
[502] Q. Q. Dang, *Bull. Soc. Chim. Fr.*, **1973,** 767.
[503] R. Levine and J. R. Sommers, *J. Org. Chem.*, **39,** 3559 (1974).
[504] A. K. Tanaka, A. Kobayashi, and K. Yamashita, *Agric. Biol. Chem.*, **37,** 669 (1973) [*C.A.*, **79,** 53136m (1973)].
[505] J. P. Lambooy, *J. Am. Chem. Soc.*, **78,** 771 (1956).
[506] M. E. Lewellyn and D. S. Tarbell, *J. Org. Chem.*, **39,** 1755 (1974).
[507] H. Böhme and U. Bomke, *Arch. Pharm.*, **303,** 779 (1970).
[508] R. K. Razdan, H. G. Pars, B. A. Zitko, V. V. Kane, and W. R. Thompson, *Tetrahedron Lett.*, **1973,** 1623.
[509] E. D. Bergmann, R. Pappo, and D. Ginsburg, *J. Chem. Soc.*, **1950,** 1369.
[510] B. G. Pring, *Acta Chem. Scand.*, **27,** 3873 (1973).
[511] N. S. Narasimhan and S. P. Chandrachood, *Indian J. Chem.*, **11,** 1192 (1973).
[512] G. Schill and K. Murjahn, *Chem. Ber.*, **104,** 3587 (1971).
[513] J. Kalamar, E. Steiner, E. Charollais, and T. Posternak, *Helv. Chim. Acta*, **57,** 2368 (1974).
[514] I. W. Mathison, R. C. Gueldner, and D. M. Carroll, *J. Pharm. Sci.*, **57,** 1820 (1968).
[515] N. S. Narasimhan and R. S. Mali, *Tetrahedron Lett.*, **1973,** 843.
[516] W. J. Houlihan and A. J. Pieroni, *J. Heterocycl. Chem.*, **10,** 405 (1973).
[517] N. S. Narasimhan and R. S. Mali, *Tetrahedron*, **30,** 4153 (1974).
[518] M. Hallet and R. Huls, *Bull. Soc. Chim. Belg.*, **61,** 33 (1952).
[519] D. W. Slocum and B. P. Koonsvitsky, *J. Org. Chem.*, **38,** 1675 (1973).
[520] J. K. K. Lam and M. V. Sargent, *J. Chem. Soc., Perkin Trans. I,* **1974,** 1417.
[521] R. A. Barnes and W. M. Bush, *J. Am. Chem. Soc.*, **81,** 4705 (1959).
[522] M. Sato, S. Ebine, and J. Tsunetsugu, *Tetrahedron Lett.*, **1977,** 855.
[523] M. Cushman and N. Castagnoli, Jr., *J. Org. Chem.*, **39,** 1546 (1974).
[524] R. Mechoulam and Y. Gaoni, *J. Am. Chem. Soc.*, **87,** 3273 (1965).
[525] H. O. House, T. M. Bare, and W. E. Hanners, *J. Org. Chem.* **34,** 2209 (1969).
[526] R. C. Ronald, *Tetrahedron Lett.*, **1975,** 3973.
[527] R. Stern, J. English, Jr., and H. G. Cassidy, *J. Am. Chem. Soc.*, **79,** 5792 (1957).
[528] W. E. Parham and E. L. Anderson, *J. Am. Chem. Soc.*, **70,** 4187 (1948).
[529] H. Gilman and R. D. Gorsich, *J. Org. Chem.*, **22,** 687 (1957).
[530] B. A. Hess, Jr., A. S. Bailey, B. Bartusek, and V. Boekelheide, *J. Am. Chem. Soc.*, **91,** 1665 (1969).
[531] G. Wittig, W. Uhlenbrock, and P. Weinhold, *Chem. Ber.*, **95,** 1692 (1962).
[532] H. Gilman, W. J. Meikle, and J. W. Morton, Jr., *J. Am. Chem. Soc.*, **74,** 6282 (1952).
[533] D. Y. Curtin and M. J. Fletcher, *J. Org. Chem.*, **19,** 352 (1954).
[534] V. P. Litvinov, D. D. Gverdtsiteli, and E. D. Lubuzh, *Izv. Akad. Nauk SSSR, Ser. Khim.*, **1972,** 79 [*C.A.*, **77,** 34226h (1972)].
[535] F. M. Stoyanovich and B. P. Fedorov, *Angew. Chem., Int. Ed. Engl.*, **5,** 127 (1966).
[536] H. Gilman and D. L. Esmay, *J. Am. Chem. Soc.*, **75,** 278 (1953).
[537] D. A. Shirley and E. A. Lehto, *J. Am. Chem. Soc.*, **77,** 1841 (1955).
[538] G. Köbrich, *Chem. Ber.*, **92,** 2981 (1959).
[539] K. Oita and H. Gilman, *J. Org. Chem.*, **22,** 336 (1957).
[540] F. M. Stoyanovich, G. B. Chermanova, Y. L. Gol'dfarb, and R. G. Karpenko, *Izv. Akad. Nauk SSSR, Ser. Khim.*, **1976,** 1583 [*C.A.*, **85,** 177163m (1976)].
[541] W. E. Truce and O. L. Norman, *J. Am. Chem. Soc.*, **75,** 6023 (1953).
[542] H. E. Zieger and G. Wittig, *J. Org. Chem.*, **27,** 3270 (1962).
[543] Y. L. Gol'dfarb and F. M. Stoyanovich, *Izv. Akad. Nauk SSSR, Ser. Khim.*, **1975,** 1588 [*C.A.*, **83,** 178626h (1975)].

[544] D. A. Shirley and E. A. Lehto, *J. Am. Chem. Soc.*, **79,** 3481 (1957).
[545] H. Watanabe, R. L. Gay, and C. R. Hauser, *J. Org. Chem.*, **33,** 900 (1968).
[546] R. A. Abramovitch, E. M. Smith, M. Humber, B. Purtschert, P. C. Srinivasan, and G. M. Singer, *J. Chem. Soc., Perkin Trans. I,* **1974,** 2589.
[547] R. J. Harper, Jr., E. J. Soloski, and C. Tamborski, *J. Org. Chem.*, **29,** 2385 (1964).
[548] C. Tamborski, W. H. Burton, and L. W. Breed, *J. Org. Chem.*, **31,** 4229 (1966).
[549] C. Tamborski and E. J. Soloski, *J. Org. Chem.*, **31,** 746 (1966).
[550] S. S. Dua and H. Gilman. *J. Organomet. Chem.*, **64,** *C*-1 (1974).
[551] A. M. Roe, R. A. Burton, and D. R. Reavill, *Chem. Commun.*, **1965,** 582.
[552] R. Koppong, *J. Fluorine Chem.*, **5,** 323 (1975) [*C.A.*, **83,** 42934b (1975)].
[553] C. Tamborski, E. J. Soloski, and C. E. Dills, *Chem. Ind.* (*London*), **1965,** 2067.
[554] K. Shiina, T. Brennan, and H. Gilman, *J. Organomet. Chem.*, **11,** 471 (1968).
[555] I. Haiduc and H. Gilman, *J. Organomet. Chem.*, **12,** 394 (1968).
[556] W. J. Houlihan, U.S. Pat. 3,823,134 (1974) [*C.A.*, **81,** 77692g (1974)].
[557] W. J. Houlihan, U.S. Pat. 3,898,275 (1975) [*C.A.*, **83,** 147305a (1975)].
[558] D. E. Grocock, T. K. Jones, G. Hallas, and J. D. Hepworth, *J. Chem. Soc., C,* **1971,** 3305.
[559] N. S. Narasimhan, M. V. Paradkar, and R. H. Alurkar, *Tetrahedron,* **27,** 1351 (1971).
[560] H. Gilman and S. Gray, *J. Org. Chem.*, **23,** 1476 (1958).
[561] U. Schmidt, J. Gombos, E. Haslinger, and H. Zak, *Chem. Ber.*, **109,** 2628 (1976).
[562] O. Riobe, A. Lebouc, J. Delaunay, and M. H. Normant, *C. R. Acad. Sci., Paris, Ser. C,* **284,** 281 (1977).
[563] J. E. Baldwin, O. W. Lever, and N. R. Tzodikov, *J. Org. Chem.*, **41,** 2874 (1976).
[564] J. Beraud and J. Metzger, *Bull. Soc. Chim. Fr.*, **1962,** 2072.
[565] Y. L. Gol'dfarb, G. B. Chermanova, and F. M. Stoyanovich, *Tezisy Dokl. Nauchn. Sess. Khim. Tekhnol. Org. Soedin. Sery Sernistykh Neftei,* **1974,** 178 [*C.A.*, **86,** 43476c (1977)].
[566] N. D. Ly and M. Schlosser, *Helv. Chim. Acta,* **60,** 2085 (1977).
[567] P. Netchitailo, J. Morel, and J. C. Halle, *J. Chem. Res.*, (*S*), **1977,** 196.
[568] F. Fringuelli, S. Gronowitz, A. B. Hörnfeldt, I. Johnson, and A. Taticchi, *Acta Chem. Scand.*, **B30,** 605 (1976).
[569] A. S. Fletcher and K. Swaminathan, *J. Chem. Soc., Perkin Trans. I,* **1977,** 1881.
[570] D. W. H. MacDowell and F. H. Ballas, *J. Org. Chem.*, **42,** 3717 (1977).
[571] R. M. Sandifer, C. F. Beam, M. Perkins, and C. R. Hauser, *Chem. Ind.* (*London*), **1977,** 231.
[572] F. M. Stoyanovich, R. G. Karpenko, G. I. Gorushkina, and Y. L. Gol'dfarb, *Tetrahedron,* **28,** 5017 (1972).
[573] R. J. Sundberg, *J. Heterocycl. Chem.*, **14,** 517 (1977).
[574] S. Cabiddu, S. Melis, P. P. Piras, and M. Secci, *J. Organomet. Chem.*, **132,** 321 (1977).
[575] S. Gronowitz, A. Konar, and A. B. Hörnfeldt, *Chem. Scr.*, **10,** 159 (1976) [*C.A.*, **87,** 201377 (1977)].
[576] R. K. Boeckman, Jr., and K. J. Bruza, *Tetrahedron Lett.*, **1977,** 4187.
[577] A. R. Acherakar, V. N. Gogte, and B. D. Tilak, *Indian J. Chem., Sec. B,* **15,** 408 (1977).
[578] K. Smith and K. Swaminathan, *J. Chem. Soc., Chem. Commun.*, **1976,** 387.
[579] R. R. Fraser and P. R. Hubert, *Can. J. Chem.*, **52,** 185 (1974).
[580] J. C. Chen, *Org. Prep. Proced. Int.*, **8,** 91 (1976).
[581] D. J. Chadwick, *J. Chem. Soc., Chem. Commun.*, **1974,** 790.
[582] E. T. Holmes and H. R. Snyder, *J. Org. Chem.*, **29,** 2155 (1964).
[583] F. E. Ziegler and E. B. Spitzner, *J. Am. Chem. Soc.*, **92,** 3492 (1970).
[584] F. E. Ziegler and E. B. Spitzner, *J. Am. Chem. Soc.*, **95,** 7146 (1973).
[585] H. R. Snyder, F. Verbanac, and D. B. Bright, *J. Am. Chem. Soc.*, **74,** 3243 (1952).
[586] D. A. Shirley and P. W. Alley, *J. Am. Chem. Soc.*, **79,** 4922 (1957).
[587] A. M. Roe, *J. Chem. Soc.*, **1963,** 2195.
[588] D. S. Noyce, G. T. Stowe, and W. Wong, *J. Org. Chem.*, **39,** 2301 (1974).
[589] E. F. Godefroi and J. J. H. Geenan, *J. Med. Chem.*, **18,** 530 (1975).
[590] H. Ogura and H. Takahashi, *J. Org. Chem.*, **39,** 1374 (1974).
[591] P. A. Alley and D. A. Shirley, *J. Org. Chem.*, **23,** 1791 (1958).
[592] B. A. Tertov, N. A. Ivankova, and A. M. Simonov, *Zh. Obshch. Khim.*, **32,** 2989 (1962) [*C.A.*, **58,** 9048b (1963)].

[593] R. A. Abramovitch, M. Saha, E. M. Smith, and R. T. Coutts, *J. Am. Chem., Soc.*, **89,** 1537 (1967).
[594] R. A. Abramovitch, E. M. Smith, E. E. Knaus, and M. Saha, *J. Org. Chem.*, **37,** 1690 (1972).
[595] R. A. Abramovitch, J. Campbell, E. E. Knaus, and A. Silhankova, *J. Heterocycl. Chem.*, **9,** 1367 (1972).
[596] R. A. Abramovitch and E. E. Knaus, *J. Heterocycl. Chem.*, **6,** 989 (1969).
[597] T. Kauffmann, B. Greving, J. König, A. Mitschker, and A. Woltermann, *Angew. Chem., Int. Ed. Engl.*, **14,** 713 (1975).
[598] A. B. Levy and S. J. Schwartz, *Tetrahedron Lett.*, **1976,** 2201.
[599] P. H. M. Schreurs, J. Meijer, P. Vermeer, and L. Brandsma, *Tetrahedron Lett.*, **1976,** 2387.
[600] B. R. O'Connor, *J. Org. Chem.*, **33,** 1991 (1968).
[601] E. M. Dexheimer and L. Spialter, *J. Organomet. Chem.*, **107,** 229 (1976).
[602] Y. Leroux and R. Mantione, *J. Organomet. Chem.*, **30,** 295 (1971).
[603] Y. Leroux and R. Mantione, *Tetrahedron Lett.*, **1971,** 591.
[604] R. R. Schmidt and J. Talbiersky, *Synthesis*, **1977,** 869.
[605] R. H. Everhardus, H. G. Eeuwhorst, and L. Brandsma, *J. Chem. Soc., Chem. Commun.*, **1977,** 801.
[606] A. C. Ranade and J. Gopal, *Chem. Ind.* (*London*), **1978,** 582.
[607] A. I. Belyashova, N. N. Zatsepina, E. N. Malysheva, A. F. Pozharskii, L. P. Smirnova, and I. F. Tupitsyn, *Khim. Geterotsikl. Soedin*, **1977,** 1544 [*C.A.*, **88,** 73917f (1978)].

CHAPTER 2

INTRAMOLECULAR REACTIONS OF DIAZOCARBONYL COMPOUNDS

Steven D. Burke and Paul A. Grieco

University of Pittsburgh, Pittsburgh, Pennsylvania

CONTENTS

ACKNOWLEDGMENT

We thank Mrs. Barbara Hunt for typing the manuscript.

INTRODUCTION

Since the demonstration by Stork and Ficini in 1961 that unsaturated diazoketones undergo intramolecular cyclization to form cyclopropanes (Eq. 1),[1] the intramolecular reactions of α-diazocarbonyl compounds have been extensively studied under thermal, catalytic, and photochemical conditions. Intramolecular cyclization of α-carbonyl carbenes and carbenoids has found widespread application to the synthesis of theoretically interesting compounds such as bullvalene,[2] twistane,[3] bridged annulenes,[4,5] and barbaralone,[2] as well as syntheses of natural products such as sabinene,[6] sirenin,[7–12] α-chamigrene,[13] and phyllocladene.[14] The reaction has also allowed the construction of several intriguing polycyclic systems that were unattainable by alternative methods.[15]

O, CHN_2 — Cu bronze, cyclohexane, reflux → O (Eq. 1)

A recent review has covered the reactions of diazoacetic esters with alkenes, alkynes, heterocyclics, and aromatic compounds.[16] Unsaturated alkoxycarbonyl carbenes (:$CHCO_2R$) and unsaturated carbonyl carbenes (:CHCOR) (where R contains either aromatic or olefinic groups), which are considered to be intermediates in the absence of catalyst, have been the subject of several other reviews.[17–20] A review concerning photo-

[1] G. Stork and J. Ficini, *J. Am. Chem. Soc.*, **83,** 4678 (1961).

[2] W. von E. Doering, B. M. Ferrier, E. T. Fossel, J. H. Hartenstein, M. Jones, Jr., G. Klumpp, R. M. Rubin, and M. Saunders, *Tetrahedron*, **23,** 3943 (1967).

[3] M. Tichý, *Tetrahedron Lett.*, **1972,** 2001.

[4] E. Vogel, A. Vogel, H.-K. Kübbeler, and W. Sturm, *Angew. Chem., Int. Ed. Engl.*, **9,** 514 (1970).

[5] E. Vogel and H. Reel, *J. Am. Chem. Soc.*, **94,** 4388 (1972).

[6] O. P. Vig, M. S. Bhatia, K. C. Gupta, and K. L. Matta, *J. Indian Chem. Soc.*, **46,** 991 (1969).

[7] J. J. Plattner, U. T. Bhalerao, and H. Rapoport, *J. Am. Chem. Soc.*, **91,** 4933 (1969).

[8] P. A. Grieco, *J. Am. Chem. Soc.*, **91,** 5660 (1969).

[9] E. J. Corey, K. Achiwa, and J. A. Katzenellenbogen, *J. Am. Chem. Soc.*, **91,** 4318 (1969).

[10] K. Mori and M. Matsui, *Tetrahedron Lett.*, **1969,** 4435.

[11] U. T. Bhalerao, J. J. Plattner, and H. Rapoport, *J. Am. Chem. Soc.*, **92,** 3429 (1970).

[12] K. Mori and M. Matsui, *Tetrahedron*, **26,** 2801 (1970).

[13] J. D. White, S. Torii, and J. Nogami, *Tetrahedron Lett.*, **1974,** 2879.

[14] A. Tahara, M. Shimagaki, S. Ohara, and T. Nakata, *Tetrahedron Lett.*, **1973,** 1701.

[15] For example, (a) S. Masamune, *J. Am. Chem. Soc.*, **86,** 735 (1964): (b) W. von E. Doering and M. Pomerantz, *Tetrahedron Lett.*, **1964,** 961.

[16] V. Dave and E. W. Warnhoff, *Org. Reactions*, **18,** 217 (1970).

[17] W. Kirmse, *Prog. Org. Chem.*, **6,** 164 (1964).

[18] W. Kirmse, *Carbene Chemistry*, Academic Press, New York, 1971.

[19] M. Jones, Jr. and R. A. Moss, *Carbenes*, Wiley, New York, 1973.

[20] (a) B. Capon, M. J. Perkins, and C. W. Rees, *Organic Reaction Mechanisms*—1966, Interscience, New York, 1967, p. 279. (b) B. Capon and C. W. Rees, *Organic Reaction Mechanisms*—1970, Interscience, New York, 1971, p. 391, (c) R. A. Butler and M. J. Perkins, *Organic Reaction Mechanisms*—1973, Interscience, New York, 1975, p. 203.

chemically, thermally, and catalytically induced Wolff rearrangements of α-diazocarbonyl compounds has recently appeared.[21] However, to date no comprehensive review has appeared which demonstrates the synthetic potential of α-diazocarbonyl insertion and addition reactions.

In reviewing the reactions of α-diazocarbonyl compounds we have chosen not to include certain peripheral topics such as intermolecular reactions, intramolecular dimerization of bis(α-carbonyl carbenoids) to give diacyl cycloolefins (Eq. 2),[22] and intramolecular trimerizations of tris(α-keto carbenoids), affording triacylcyclopropanes (Eq. 3).[23] Also excluded are base-catalyzed reactions of α-diazocarbonyl compounds[24] and certain ring-contraction reactions involving C–C bond insertion of intermediate α-ketocarbenes (Eq. 4).[25]

$(CH_2)_n(COCHN_2)_2$ —[Cu(acetylacetonate)$_2$, benzene, reflux]→ (Eq. 2)

COCHN$_2$ / COCHN$_2$ / COCHN$_2$ —[$C_6H_5SCH_3$, $CuSO_4$, xylene, reflux]→ (Eq. 3)

=N$_2$ → [:] → (Eq. 4)

The topics included in this report are the intramolecular reactions of α-carbonyl carbenes and carbenoids with olefinic and aromatic unsaturation, with C–H bonds, and with C–C and N–H single bonds. Also included are acid-catalyzed cyclization reactions of α-diazoketones.

MECHANISM

The intramolecular reactions of α-carbonyl carbenes and carbenoids elaborated in this review are referred to as additions if the divalent center

[21] H. Meier and K.-P. Zeller, *Angew. Chem., Int. Ed. Engl.*, **7**, 32 (1975).

[22] (a) J. Font, J. Valls, and F. Serratosa, *Tetrahedron*, **30**, 455 (1974); (b) J. Font, F. Serratosa, and J. Valls, *Chem. Commun.*, **1970**, 721.

[23] J. Font, F. López, and F. Serratosa, *Tetrahedron Lett.*, **1972**, 2589.

[24] T. L. Burkoth, *Tetrahedron Lett.*, **1969**, 5049.

[25] G. Stork and R. P. Szajewski, *J. Am. Chem. Soc.*, **96**, 5787 (1974).

is reacting with olefinic or aromatic unsaturation, and as insertions if the carbenic center is reacting with a C–C, C–H, or N–H single bond. Although mechanistic studies based on intramolecular reactions of α-diazocarbonyl compounds are relatively few, analogous intermolecular reactions have been rather extensively studied, and it is from these reports that much of the contemporary thought on the mechanism of the intramolecular reactions has been inferred.[16–20]

Photochemical, catalytic, or thermal conditions have been employed in the decomposition of α-diazocarbonyl compounds. Catalytic conditions are most commonly used. They generate a carbene complex (as opposed to a free carbene generated by photochemical decomposition). The carbene complex is referred to in this report as a carbenoid species. The catalytic decompositions are carried out under homogeneous or heterogeneous conditions with a wide variety of catalysts, usually copper or a copper salt. However, soluble complexes of copper and palladium have also been used.

Several investigations have strongly implicated a carbene–metal–olefin complex in the homogeneous copper-catalyzed decomposition of α-diazocarbonyl compounds.[26–29] The effects of varying the steric and electronic requirements of (trialkyl and triarylphosphite)copper(I) chloride catalysts have been studied in terms of the intermolecular addition of ethyl diazoacetate to olefins.[26,27] Early work considered both intermolecular and intramolecular additions of α-carbonyl carbenoids to olefins using soluble copper(II) chelates.[28,29] Strong evidence has been provided for a carbene–copper–olefin complex as an intermediate by utilizing an optically active copper(II) complex to induce asymmetry in reaction products (Eqs. 5 and 6).[29] Although interpretations differ slightly

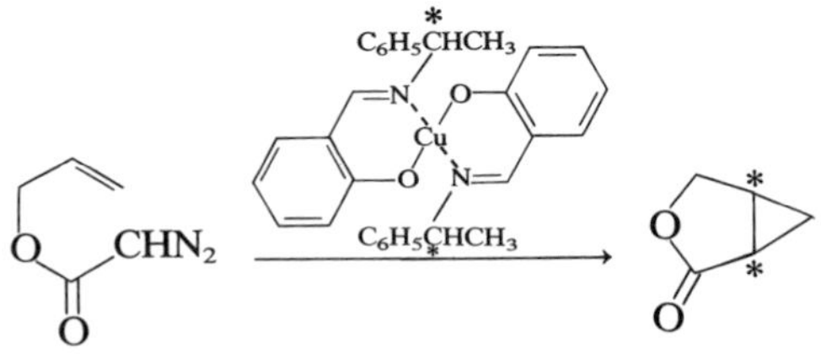

$C_6H_5\overset{*}{C}HCH_3$ … $C_6H_5\overset{*}{C}HCH_3$ (Cu chelate catalyst) CHN$_2$ → (Eq. 5)

C_6H_5, CHN$_2$ → ---C_6H_5 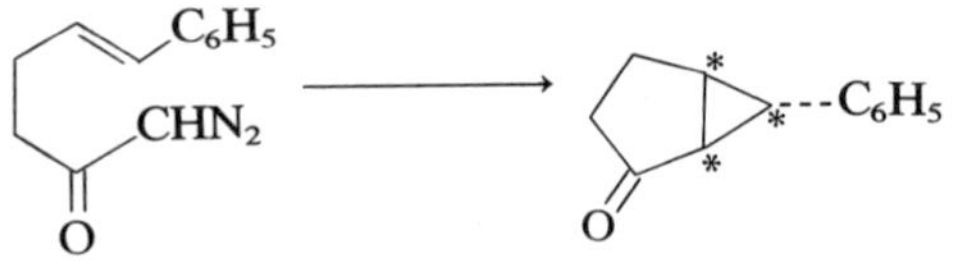 (Eq. 6)

[26] W. R. Moser, *J. Am. Chem. Soc.*, **91,** 1135 (1969).
[27] W. R. Moser, *J. Am. Chem. Soc.*, **91,** 1141 (1969).
[28] H. Nozaki, S. Moriuti, H. Takaya, and R. Noyori, *Tetrahedron Lett.*, **1966,** 5239.
[29] H. Nozaki, H. Takaya, S. Moriuti, and R. Noyori, *Tetrahedron*, **24,** 3655 (1968).

on the exact nature of the intermediate complexes, it is proposed that intermediates are formed where the carbene acts as a ligand coordinated to the copper by donation of its electron pair. Concomitant d_π–p_π back donation from the metal to the empty p orbital of the carbene carbon results in a stabilization of the complex. Analogous carbenoid complexes have been proposed as intermediates in the catalytic decomposition of α-diazocarbonyls with di-μ-chlorodi-π-allylpalladium (**1**)[30] and bis-(acetylacetonato)copper(II) (**2**).[28,29,31–34]

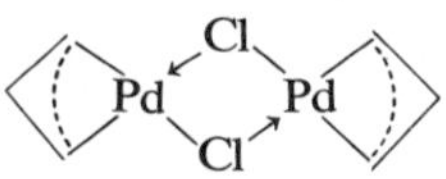

Cl Pd Pd Cl

1

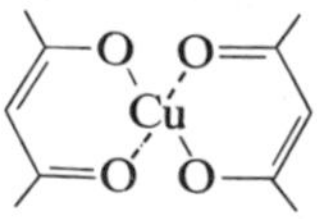

O O Cu O O

2

The carbenoid intermediates generated by the decomposition of α-diazocarbonyl compounds in the presence of copper or its salts generally show reduced reactivity (hence greater selectivity), as evidenced by a lack of intermolecular C–H insertion reactions which are prevalent in thermally or photochemically generated free carbenes. One example of this type of behavior is the photolysis of allyl diazoacetate, which gave only the intermolecular insertion product.[35] In order to obtain the olefin addition product copper catalysis was required. The enhanced selectivity of the modified carbenes obtained through catalysis explicitly allows the intramolecular reactions to be of synthetic importance.

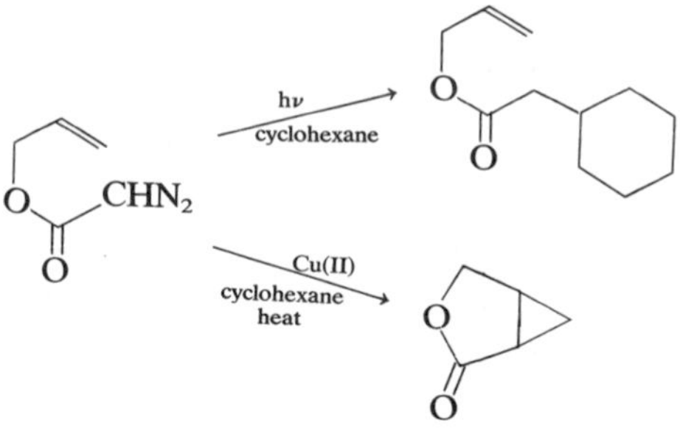

O CHN$_2$ O

$h\nu$ cyclohexane

O O

Cu(II) cyclohexane heat

O O

It has been suggested that the intramolecular insertion reactions of photolytically generated alkoxycarbonyl carbenes are realized only to a

[30] R. K. Armstrong, *J. Org. Chem.*, **31,** 618 (1966).
[31] M. Takebayashi, T. Ibata, H. Kohara, and B. H. Kim, *Bull. Chem. Soc. Jpn.*, **40,** 2392 (1967).
[32] M. Takebayashi, T. Ibata, H. Kohara, and K. Ueda, *Bull. Chem. Soc. Jpn.*, **42,** 2338 (1969).
[33] M. Takebayashi, T. Ibata, and K. Ueda, *Bull. Chem. Soc. Jpn.*, **43,** 1500 (1970).
[34] K. Ueda, T. Ibata, and M. Takebayashi, *Bull. Chem. Soc. Jpn.*, **45,** 2779 (1972).
[35] W. Kirmse and H. Dietrich, *Chem. Ber.*, **98,** 4027 (1965).

slight extent because of dominant conformational effects.[36,37] For example, the low yields observed in the intramolecular C–H insertion reactions of the carbenes generated from *t*-butyl- and *t*-pentyldiazoacetates[38] can be attributed to the inability of the O-alkyl moieties to attain sufficient proximity to the carbenic center. In support of this premise, it has been demonstrated that N,N-dialkyldiazoamides (Eq. 7) undergo principally lactam formation via intramolecular C–H insertion of the photolytically generated carbene.[36,37] In this reaction one of the alkyl groups must be held in close proximity to the divalent center by the planar amide bond.

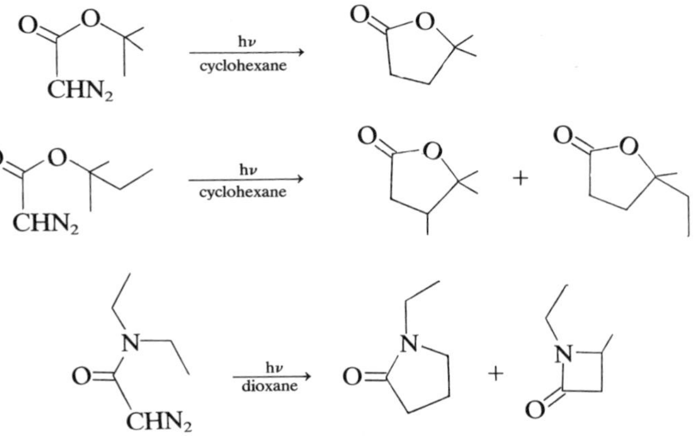

CHN_2 —hν, cyclohexane→

CHN_2 —hν, cyclohexane→ +

CHN_2 —hν, dioxane→ + (Eq. 7)

It has also been observed that the intramolecular C–H insertion reaction proceeds with retention of configuration.[39]

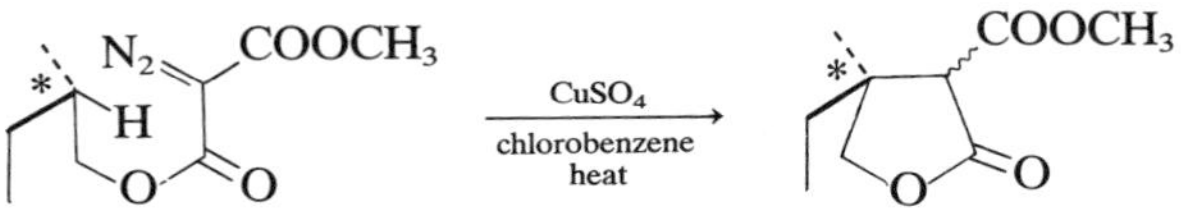

N_2 $COOCH_3$ —$CuSO_4$, chlorobenzene, heat→ $COOCH_3$

As elaborated in the following section, the intramolecular addition reactions of unsaturated alkoxycarbonyl carbenoids and unsaturated ketocarbenoids are stereospecific.[40] Thus the intramolecular addition of an α-diazocarbonyl compound to a suitably substituted olefinic site allows for the simultaneous construction of three asymmetric centers. The observed stereospecificity of the process has been applied to several syntheses of the sesquiterpene hormone sirenin (**3**)[7–12] as well as to other bicyclo[*x*.1.0] systems.

[36] R. R. Rando, *J. Am. Chem. Soc.*, **92,** 6706 (1970).
[37] R. R. Rando, *J. Am. Chem. Soc.*, **94,** 1629 (1972).
[38] W. Kirmse, H. Dietrich, and H. W. Bücking, *Tetrahedron Lett.*, **1967,** 1833.
[39] H. Ledon, G. Linstrumelle, and S. Julia, *Tetrahedron Lett.*, **1973,** 25.
[40] G. Stork and M. Gregson, *J. Am. Chem. Soc.*, **91,** 2372 (1969), footnote 6.

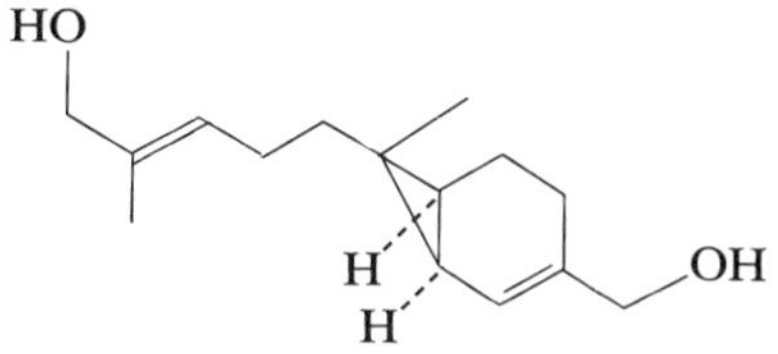

3

SCOPE AND LIMITATIONS

Reactions involving intramolecular additions[1] and insertions[41a] of α-carbonyl carbenes and carbenoids have found widespread use in the relatively short time since their initial application. The early applications are exemplified in syntheses of compounds of theoretical interest. Highly strained compounds containing the tricyclo[2.1.0.0^{2,5}]pentane carbon skeleton were obtained via olefin addition of photolytically and catalytically decomposed diazoketones.[42–44] A large number of more recent applications were directed toward total syntheses of the following natural products containing cyclopropane rings: 3-carone,[45] aristolone,[46] thujopsene,[47] sabinene,[6] sirenin,[7–12,48] and several others. Studies have also been directed at intramolecular C–H insertion of carbenes derived from α-diazoamides to afford a new synthetic approach to β-lactam antibiotics.[49–55]

The following discussion elaborates some of these areas of interest, including consideration of the factors involved in the relative success or failure of the intramolecular insertion and addition reactions. Examples are chosen which illustrate a general principle or have some especially interesting characteristic features. The applications are organized into sections corresponding to the table headings.

[41] (a) F. Greuter, J. Kalvoda, and O. Jeger, *Proc. Chem. Soc. (London)*, **1958,** 349; (b) T. Nakata and A. Tahara, *Tetrahedron Lett.*, **1976,** 1515.

[42] S. Masamune, *J. Am. Chem. Soc.*, **86,** 735 (1964).

[43] J. Trotter, C. S. Gibbons, N. Nakatsuka, and S. Masamune, *J. Am. Chem. Soc.*, **89,** 2792 (1967).

[44] W. von E. Doering and M. Pomerantz, *Tetrahedron Lett.*, **1964,** 961.

[45] F. Medina and A. Manjarrez, *Tetrahedron*, **20,** 1807 (1964).

[46] E. Piers, R. W. Britton, and W. de Waal, *Can. J. Chem.*, **47,** 831 (1969).

[47] K. Mori, O. Ohki, A. Kobayashi, and M. Matsui, *Tetrahedron*, **26,** 2815 (1970).

[48] E. J. Corey and K. Achiwa, *Tetrahedron Lett.*, **1970,** 2245.

[49] E. J. Corey and A. M. Felix, *J. Am. Chem. Soc.*, **87,** 2518 (1965).

[50] R. H. Earle, Jr., D. T. Hurst, and M. Viney, *J. Chem. Soc., (C)*, **1969,** 2093.

[51] G. Lowe and J. Parker, *J. Chem. Soc., Chem. Commun.*, **1971,** 577.

[52] D. M. Brunwin, G. Lowe, and J. Parker, *J. Chem. Soc., Chem. Commun.*, **1971,** 865.

[53] D. M. Brunwin, G. Lowe, and J. Parker, *J. Chem. Soc., (C)*, **1971,** 3756.

[54] D. M. Brunwin and G. Lowe, *J. Chem. Soc., Chem. Commun.*, **1972,** 589.

[55] B. T. Golding and D. R. Hall, *J. Chem. Soc., Chem. Commun.*, **1973,** 293.

Diazoketone Additions to Olefins

The ketocarbene or ketocarbenoid addition reaction with a suitably situated olefinic moiety within the same molecule is by far the most widely encountered application. Since the prototype (Eq. 1),[1] reactions of this type have come into utilization for a wide range of synthetic goals.

For the synthesis of bicyclo [x.1.0] compounds it was found that the controlling factor in determining the usefulness of this method was the proximity of the olefin to the divalent center, whereas the nature of the substitution on the double bond had little effect.[56]

This method proved useful for the construction (albeit in low yield) of strained polycyclic compounds, notably the tricyclo[$2.1.0.0^{2,5}$]pentane system. Interestingly, one synthesis is based on a photochemical decomposition of the diazoketone (Eq. 8),[42,43] whereas the other is based on catalytic carbenoid generation (Eq. 9).[44]

R, O, CHN_2 —hν, tetrahydrofuran→ R, R, O (Eq. 8)

$R = C_6H_5$, n-C_3H_7

O, CHN_2 —Cu powder, hexane, reflux→ O (Eq. 9)

The low yields associated with the conversions indicated in Eqs. 8 and 9 are undoubtedly attributable to fragmentation of the initially formed tricyclic systems. It has recently been observed[57–60a] that catalytic decomposition of β, γ-unsaturated diazoketones of this type can lead to rearranged γ, δ-unsaturated carboxylic acid derivatives in a process termed the vinylogous Wolff rearrangement.[59] Mechanistic pathways for this process have been proposed.[60a] This method provides a synthetic alternative to the Claisen rearrangement.

O, CHN_2 —catalyst, CH_3OH→ $COOCH_3$

[56] M. M. Fawzi and C. D. Gutsche, *J. Org. Chem.*, **31,** 1390 (1966).

[57] J. P. Lokensgard, J. O'Dea, and E. A. Hill, *J. Org. Chem.*, **39,** 3355 (1974).

[58] H. E. Zimmerman and R. D. Little, *J. Am. Chem. Soc.*, **96,** 4623 (1974).

[59] A. B. Smith, III, *J. Chem. Soc., Chem. Commun.*, **1974,** 695.

[60] (a) A. B. Smith, B. H. Toder, and S. J. Branca, *J. Am. Chem. Soc.*, **98,** 7456 (1976); (b) C. Iwata, M. Yamada, Y. Shinoo, K. Kobayashi, and H. Okada, *J. Chem. Soc., Chem. Commun.*, **1977,** 888.

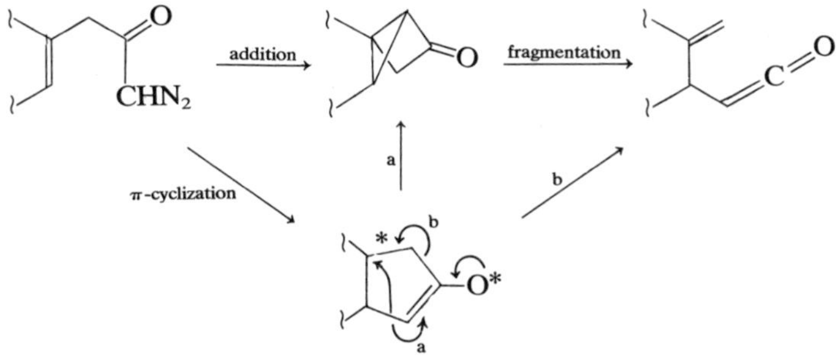

addition
fragmentation
π-cyclization
a
b

Other theoretically intriguing compounds prepared by intramolecular carbenoid addition are barbaralone (Eq. 10),[2] twistane (**4**),[3] a [4.4.4]propellane (**5**),[61] and a bridged [14]annulene (Eq. 11).[4] Hydrocarbons **4** and **5** are of course not directly attainable by this method, but their syntheses do proceed through cyclopropyl ketones obtained via intramolecular diazoketone–olefin addition.

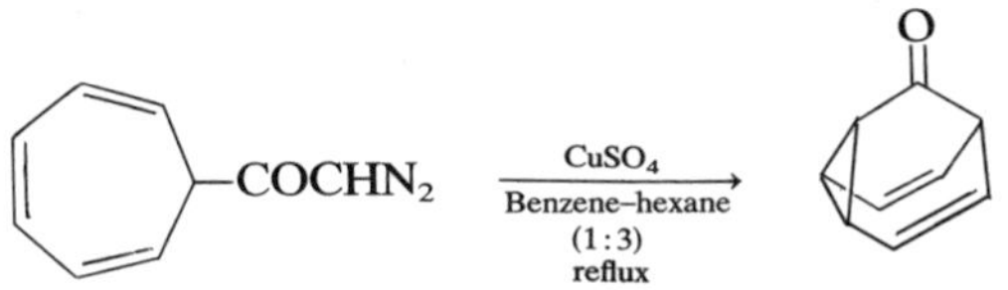

$COCHN_2$ $\xrightarrow[\text{Benzene–hexane (1:3) reflux}]{CuSO_4}$ (Eq. 10)

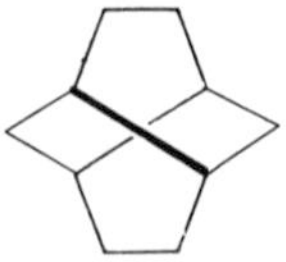

4 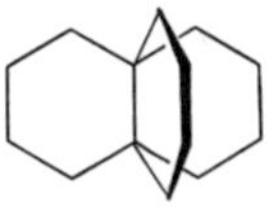 **5**

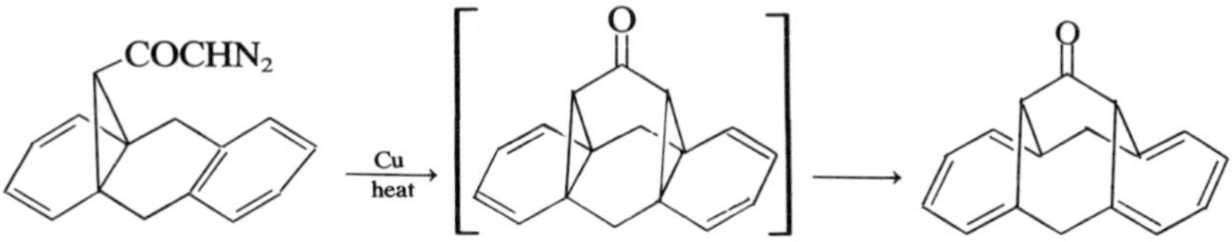

$COCHN_2$ $\xrightarrow[\text{heat}]{Cu}$ (Eq. 11)

Many natural products have been obtained through synthetic application of this type of reaction. The stereospecific nature of the addition is put to good use in achieving the proper stereochemistry about the three-membered ring in natural products that contain a cyclopropyl ring. Examples of compounds synthesized in this manner are sirenin (**3**) (p. 370),[7–12]

[61] J. Altman, D. Becker, D. Ginsburg, and H. J. E. Loewenthal, *Tetrahedron Lett.*, **1967,** 757.

$COCHN_2$

Cu bronze, cyclohexane, reflux

6

3

(Ref. 8)

OH

OH

O

10

7

O

$COCHN_2$

Cu powder, $CuSO_4$, cyclohexane, reflux

O

8

N_2CHCO

Cu

tetrahydrofuran, heat

O

9

aristolone (**7**),[45] thujopsene (**8**),[47,62a] longicyclene (**9**),[63–64] and sesquicarene (**10**).[65–69]

In several applications the cyclopropyl ketone resulting from intramolecular cyclization of an olefinic ketocarbenoid is subsequently cleaved by hydrogenolysis,[70–75] protonolysis,[70,72,74,76–78] lithium/ammonia reduction,[70,74,79] or Lewis-acid treatment.[80–83] The spiro sesquiterpenoids, α-chamigrene (**11**)[13] and epihinesol (**12**),[77,84] as well as the tetracyclic diterpenes kaurene (**13**)[14] and phyllocladene (**14**),[14] have been synthesized in this manner.

This strategy involving intramolecular olefinic ketocarbenoid addition followed by regio- and stereospecific cleavage of one of the conjugated cyclopropane bonds has been applied to the synthesis of many compounds containing the bicyclo[3.2.1]octane[71,72,75,76,78,79] and bicyclo-[2.2.2]octane[76,78] ring systems in addition to phyllocladene and kaurene. The two-step, intramolecular, angular-alkylation sequence takes on particular importance in that it has allowed the construction of skeletal analogs of the plant-growth-regulatory gibberellin diterpenes [*e.g.*, gibberellic acid (**15**)] and of the sesquiterpenoid helminthosporins [*e.g.*,

[62] (a) P. L. Andersen, *Diss. Abstr., B*, **28** (1), 91 (1967); (b) S. J. Branca, R. L. Lock, and A. B. Smith, III, *J. Org. Chem.*, **42,** 3165 (1977).

[63] S. C. Welch and R. L. Walters, *Synth. Commun.*, **3,** 15 (1973).

[64] S. C. Welch and R. L. Walters, *J. Org. Chem.*, **39,** 2665 (1974).

[65] E. J. Corey and K. Achiwa, *Tetrahedron Lett.*, **1969,** 1837.

[66] K. Mori and M. Matsui, *Tetrahedron Lett.*, **1969,** 2729.

[67] R. M. Coates and R. M. Freidinger, *J. Chem. Soc., Chem. Commun.*, **1969,** 871.

[68] R. M. Coates and R. M. Freidinger, *Tetrahedron*, **26,** 3487 (1970).

[69] O. P. Vig, O. P. Chugh, R. C. Anand, and M. S. Bhatia, *J. Indian Chem. Soc.*, **47,** 506 (1970).

[70] S. K. Dasgupta and A. S. Sarma, *Tetrahedron Lett.*, **1968,** 2983.

[71] S. K. Dasgupta, R. Dasgupta, S. R. Ghosh, and U. R. Ghatak, *J. Chem. Soc., Chem. Commun.*, **1969,** 1253.

[72] P. N. Chakrabortty, R. Dasgupta, S. K. Dasgupta, S. R. Ghosh, and U. R. Ghatak, *Tetrahedron*, **28,** 4653 (1972).

[73] D. Becker and H. J. E. Loewenthal, *Isr. J. Chem.*, **10,** 375 (1972).

[74] S. K. Dasgupta and A. S. Sarma, *Tetrahedron*, **29,** 309 (1973).

[75] U. R. Ghatak, P. C. Chakraborti, B. C. Ranu, and B. Sanyal, *J. Chem. Soc., Chem. Commun.*, **1973,** 548.

[76] D. J. Beames and L. N. Mander, *J. Chem. Soc., Chem. Commun.*, **1969,** 498.

[77] M. Mongrain, J. Lafontaine, A. Bélanger, and P. Deslongchamps, *Can. J. Chem.*, **48,** 3273 (1970).

[78] D. J. Beames, J. A. Halleday, and L. N. Mander, *Aust. J. Chem.*, **25,** 137 (1972).

[79] L. N. Mander, R. H. Prager, and J. V. Turner, *Aust. J. Chem.*, **27,** 2645 (1974).

[80] G. Stork and P. A. Grieco, *J. Am. Chem. Soc.*, **91,** 2407 (1969).

[81] G. Stork and M. Marx, *J. Am. Chem. Soc.*, **91,** 2371 (1969).

[82] G. Stork and P. A. Grieco, *Tetrahedron Lett.*, **1971,** 1807.

[83] (a) G. Stork and M. Gregson, *J. Am. Chem. Soc.*, **91,** 2373 (1969); (b) P. Deslongchamps, J. Lafontaine, L. Ruest, and P. Soucy, *Can. J. Chem.*, **55,** 4117 (1977).

[84] P. M. McCurry, Jr., *Tetrahedron Lett.*, **1971,** 1845.

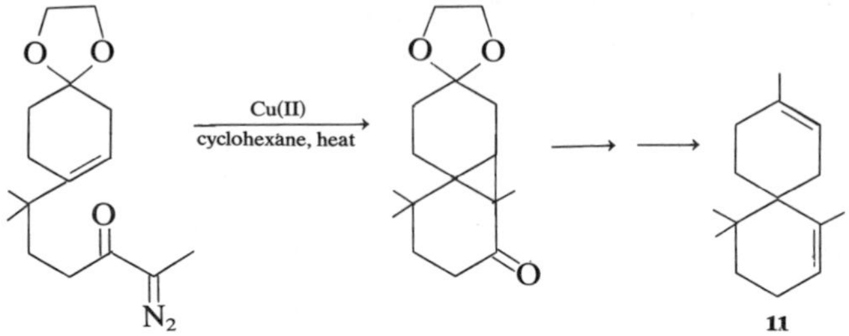

Cu(II)
cyclohexane, heat

11

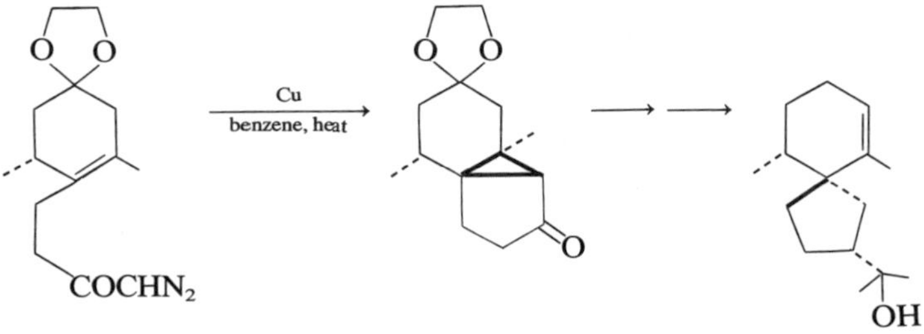

Cu
benzene, heat

$COCHN_2$

12

$COCHN_2$

$CuSO_4$
cyclohexane,
heat

H
CO_2Me

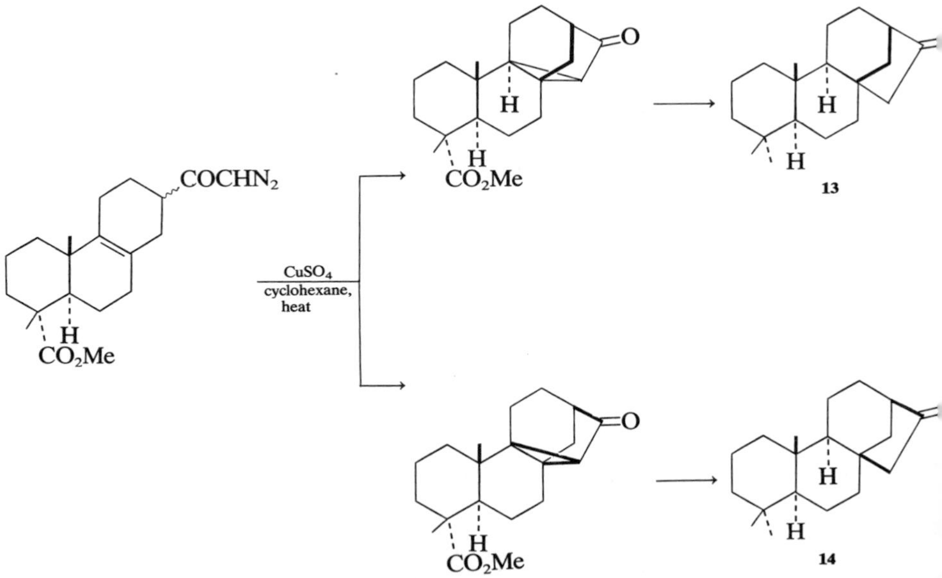

13

14

helminthosporic acid (**16**)] which possess gibberellin-like activity. Exemplary of this application are the tricyclic compounds **17** and **18**.[79]

An interesting application to the synthesis of bicyclo[3.2.1]octane and bicyclo[2.2.2]octane systems uses a retrograde Michael reaction to effect regiocontrolled cleavage of the cyclopropane.[78] The cyclopropane bond

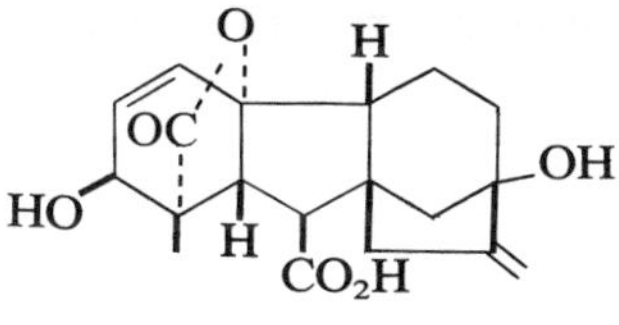

O H OC HO H CO_2H OH

15

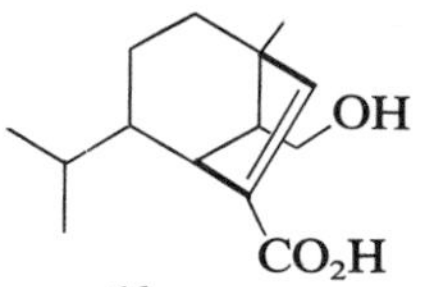

OH CO_2H

16

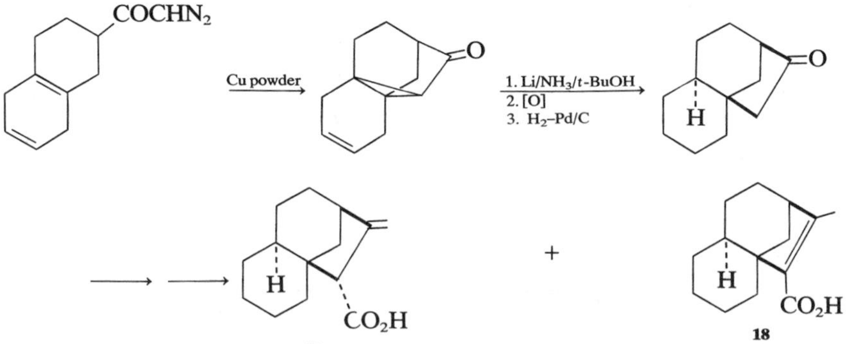

$COCHN_2$ Cu powder O 1. $Li/NH_3/t$-BuOH 2. [O] 3. H_2–Pd/C O H

H CO_2H

17

+

H CO_2H

18

cleaved in Eqs. 12 and 13 is the one that is activated by both carbonyl groups. Conceptually the sequence allows for intramolecular γ alkylation of an incipient α,β-unsaturated ketone.

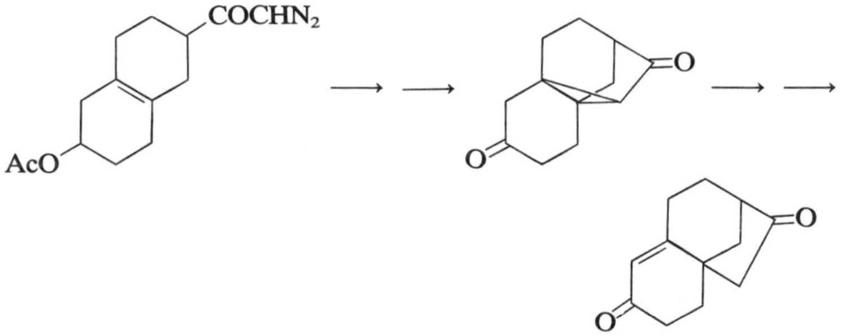

$COCHN_2$ AcO O O O O

(Eq. 12)

$COCHN_2$ AcO O O O O

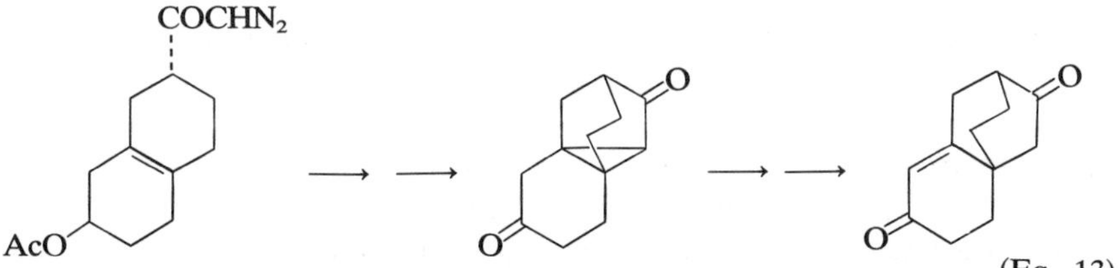

(Eq. 13)

The intramolecular olefinic ketocarbenoid addition reaction provided ready access to a series of compounds which were used to demonstrate a Lewis-acid-catalyzed cleavage of an acylcyclopropane with simultaneous participation of a suitably situated olefinic or aromatic center, leading to ring formation.[80–83a] Illustrative of this study, which also nicely demonstrates the stereospecificity of the internal diazoketone addition,[40] is the sequence shown in Eqs. 14 and 15.[83a]

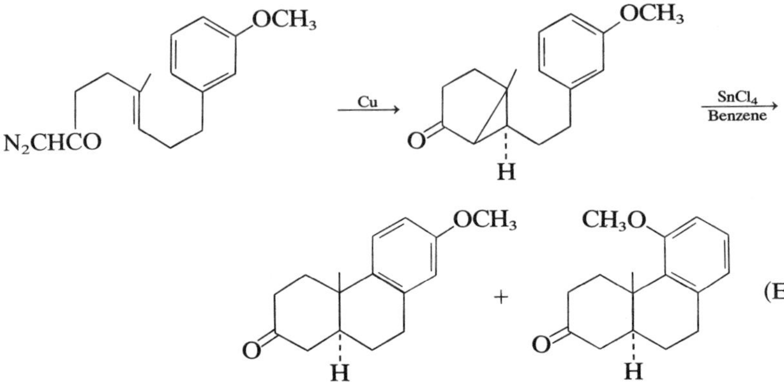

(Eq. 14)

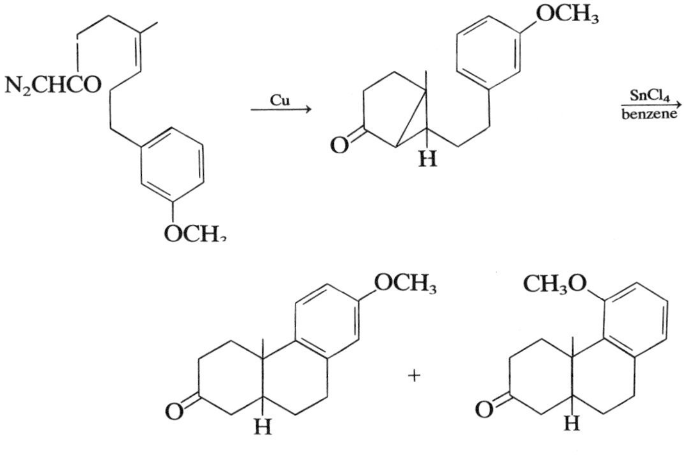

(Eq. 15)

In another intramolecular addition of ketocarbenoids to unsaturated centers, bicyclo[5.3.0]decatrienones have been synthesized[85–87] and subsequently converted to azulenes.[87] The sequence outlined in Eq. 16 is representative.[87]

The unstable norcaradiene resulting from addition of the ketocarbenoid to the 1,2 position of the aromatic ring isomerizes by ring opening and subsequent 1,5-hydrogen shift to the trienone (**19**).

[85] A. Costantino, G. Linstrumelle, and S. Julia, *Bull. Soc. Chim. Fr.*, **1970,** 907.

[86] A. Costantino, G. Linstrumelle, and S. Julia, *Bull. Soc. Chim. Fr.*, **1970,** 912.

[87] L. T. Scott, *J. Chem. Soc., Chem. Commun.*, **1973,** 882.

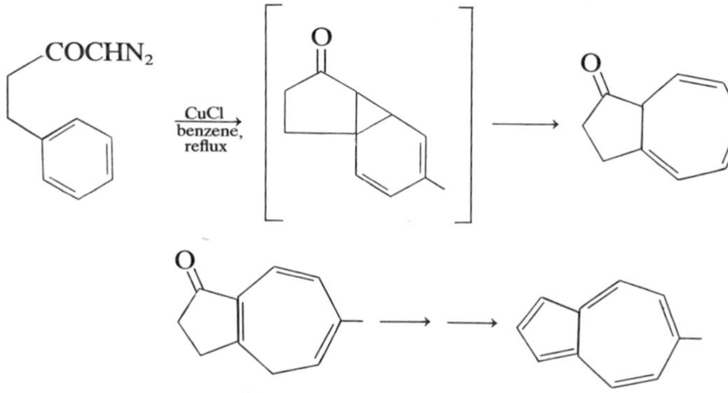

$COCHN_2$

CuCl
benzene,
reflux

19

(Eq. 16)

Finally, an interesting example has been reported involving the intramolecular addition of a ketocarbenoid to a furan ring.[88] The intermediate heterocyclic ring system thus formed undergoes ring opening to give the ketoaldehyde **20**.

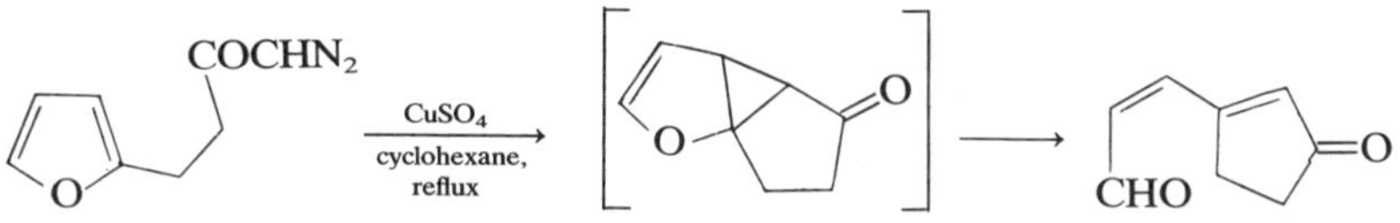

$COCHN_2$

$CuSO_4$
cyclohexane,
reflux

CHO

20

Diazoester Additions to Olefins

The addition of alkoxycarbonyl carbenoids to internally situated olefinic sites has not been used as extensively as olefinic ketocarbenoid additions. However, the synthetic potential of this type of reaction is clear.

The decomposition of allyl diazoacetate is catalyzed by various copper reagents, and the bicyclic lactone (**21**) is formed in an optimium yield of 52%.[35] Similar applications provided syntheses of three bicyclic lactone esters in yields of 50–55%.[89,90]

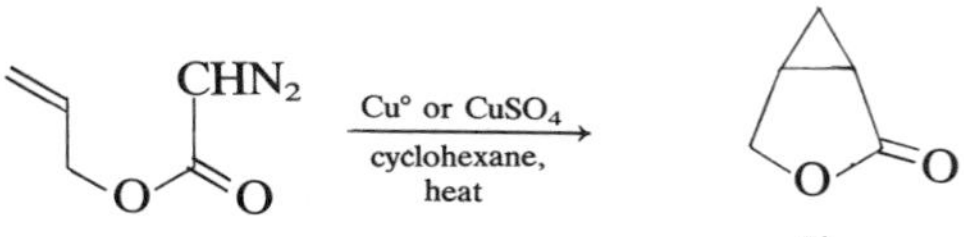

CHN_2

Cu° or $CuSO_4$
cyclohexane,
heat

21

[88] M. N. Nwaji and O. S. Onyiriuka, *Tetrahedron Lett.*, **1974,** 2255.

[89] G. Cannic, G. Linstrumelle, and S. Julia, *Bull. Soc. Chim. Fr.*, **1968,** 4913.

[90] S. Julia, G. Cannic, and G. Linstrumelle, *C. R. Acad. Sci. Paris, Ser. C,* **264,** 1890 (1967).

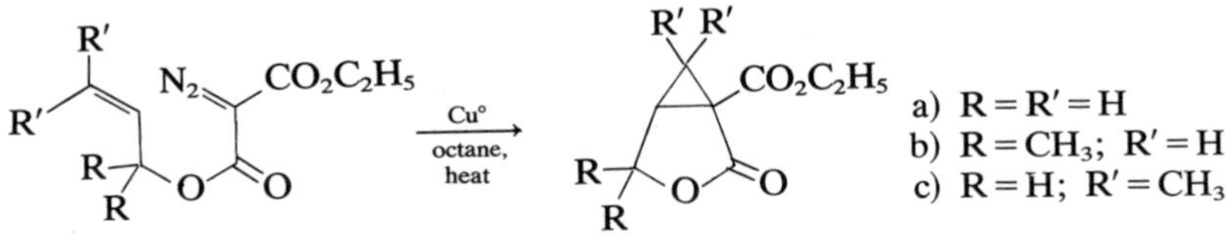

a) R = R′ = H
b) R = CH_3; R′ = H
c) R = H; R′ = CH_3

In an extension of this method to the construction of tetracyclic lactones such as **22** and **23** it was found that the intramolecular addition reaction is complicated by the formation of maleate and fumarate esters, formally carbene dimer products.[91] This type of behavior was also observed elsewhere,[35] implying that alkoxycarbonyl carbenoids are not as selective in their reactions as the analogous ketocarbenoids.

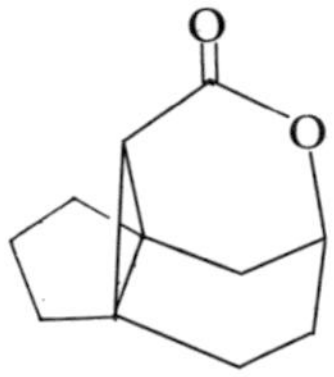

22 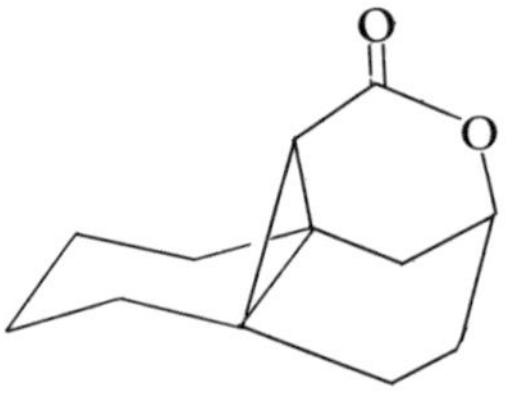 **23**

More recently it was found that the intramolecular addition reactions of carbenoids derived from mixed diazomalonates yielded lactones of 1-hydroxymethyl-7-carboxycycloheptatrienes.[92,93a]

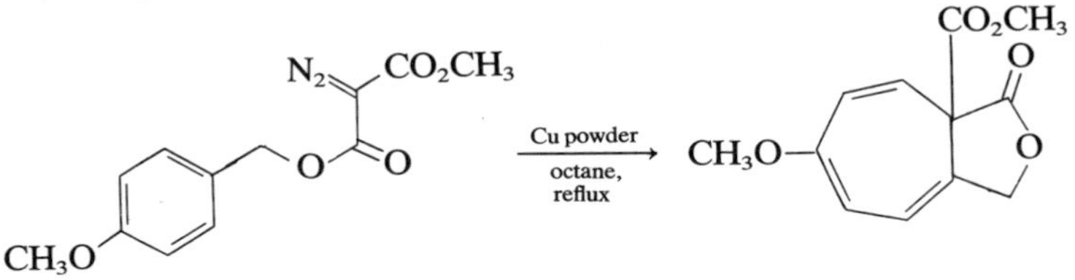

The copper-catalyzed decomposition of *trans, trans*-farnesyl diazoacetate (**24**) provided a stereoselective route to racemic presqualene alcohol through bicyclic lactone **25**.[94]

An interesting application has been made to the synthesis of a promising intermediate **26** for prostanoid synthesis.[95] The demonstration that cleavage of the cyclopropane can be achieved via homoconjugate addition of an organocopper (Gilman) reagent to compound **26** suggests a stereocontrolled route to prostaglandins.

[91] H. O. House and C. J. Blankley, *J. Org. Chem.*, **33,** 53 (1968).

[92] H. Ledon, G. Cannic, G. Linstrumelle, and S. Julia, *Tetrahedron Lett.*, **1970,** 3971.

[93] (a) H. Ledon, G. Linstrumelle, and S. Julia, *Tetrahedron*, **29,** 3609 (1973); (b) F. Zutterman, P. De Clercq, and M. Vandewalle, *Tetrahedron Lett.*, **1977,** 3191.

[94] R. M. Coates and W. H. Robinson, *J. Am. Chem. Soc.*, **93,** 1785 (1971).

[95] E. J. Corey and P. L. Fuchs, *J. Am. Chem. Soc.*, **94,** 4014 (1972).

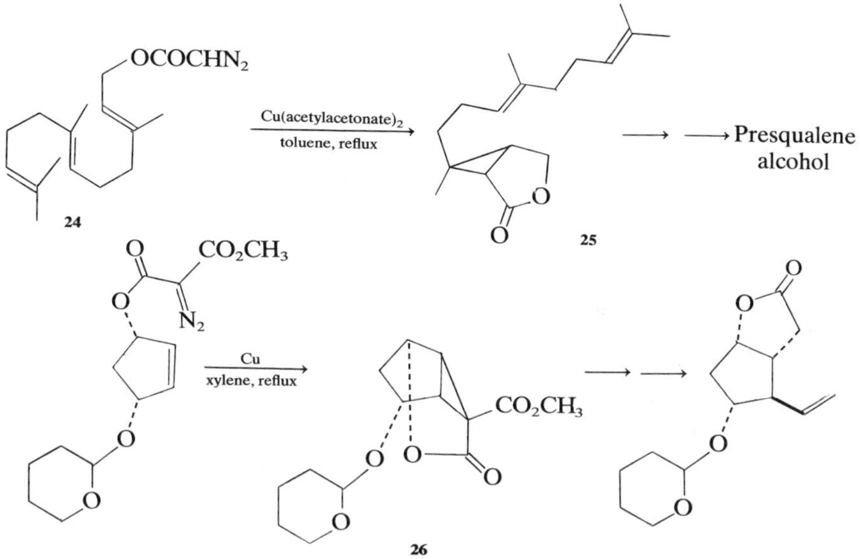

$OCOCHN_2$

24

Cu(acetylacetonate)$_2$, toluene, reflux

25

⟶ ⟶ Presqualene alcohol

CO_2CH_3 N_2

Cu, xylene, reflux

CO_2CH_3

26

⟶ ⟶

Interest in certain cytotoxic natural products containing α-methylene-γ-butyrolactones prompted the sequence shown in Eq. 17.[96] A similar study led to the synthesis of spiro lactone **27** and spiro ketone **28**.[97]

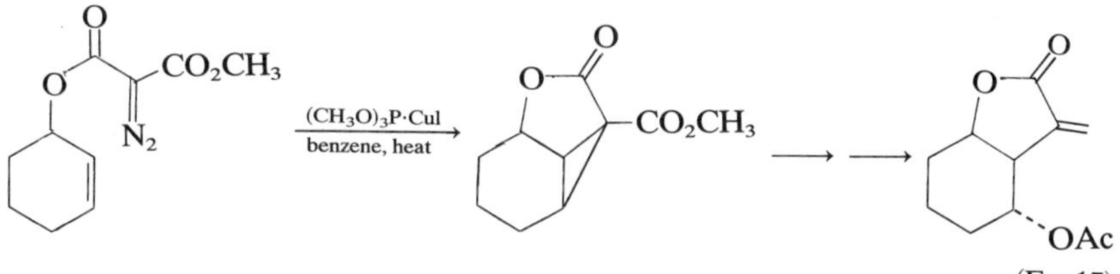

CO_2CH_3 N_2

$(CH_3O)_3P{\cdot}CuI$, benzene, heat

CO_2CH_3

⟶ ⟶ OAc

(Eq. 17)

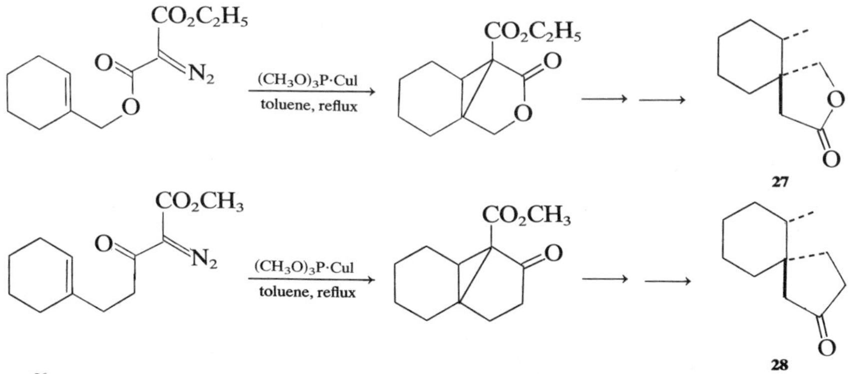

$CO_2C_2H_5$ N_2

$(CH_3O)_3P{\cdot}CuI$, toluene, reflux

$CO_2C_2H_5$

⟶ ⟶

27

CO_2CH_3 N_2

$(CH_3O)_3P{\cdot}CuI$, toluene, reflux

CO_2CH_3

⟶ ⟶

28

[96] F. E. Ziegler, A. F. Marino, O. A. C. Petroff, and W. L. Studt, *Tetrahedron Lett.*, **1974**, 2035.

[97] R. D. Clark and C. H. Heathcock, *Tetrahedron Lett.*, **1975**, 529.

The stereocontrolled generation of acyclic side chains of natural products (*e.g.*, steroids and prostaglandins) has recently been reported,[98–100] utilizing the nucleophilic ring opening of appropriately substituted acylcyclopropanes.

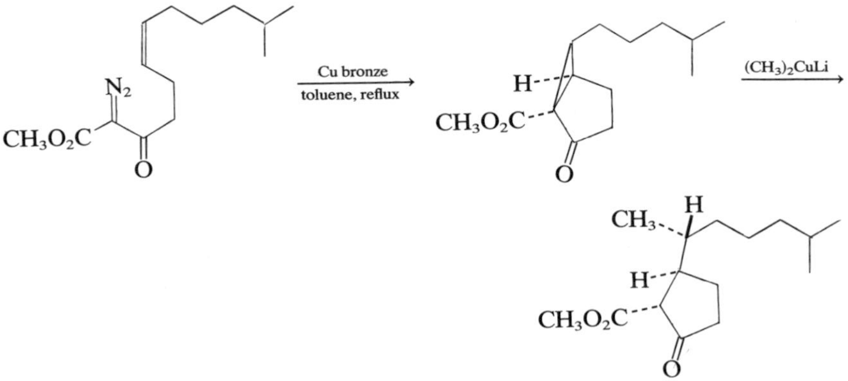

N_2 CH_3O_2C O —(Cu bronze / toluene, reflux)→ H CH_3O_2C O —($(CH_3)_2CuLi$)→ CH_3 H H CH_3O_2C O

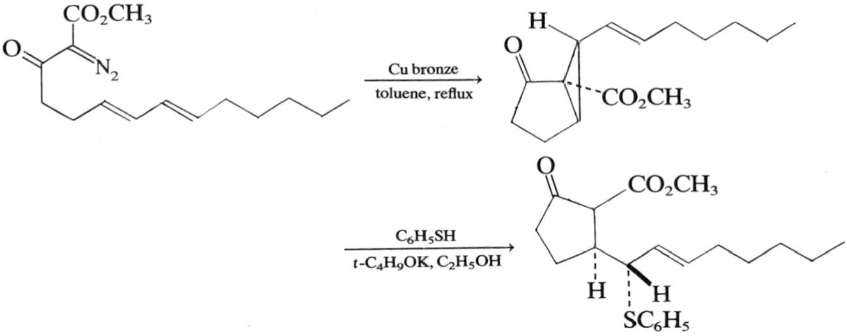

CO_2CH_3 O N_2 —(Cu bronze / toluene, reflux)→ H O CO_2CH_3 —(C_6H_5SH / $t\text{-}C_4H_9OK$, C_2H_5OH)→ O CO_2CH_3 H H SC_6H_5

Diazocarbonyl Insertions into C–H Bonds

Although the majority of the synthetic intramolecular reactions of α-diazocarbonyl compounds have relied on conditions that supress insertion into C–H bonds in favor of intramolecular cycloaddition of the carbene or carbenoid to an olefinic moiety, many important and otherwise difficult synthetic transformations have been based on intramolecular C–H insertion of α-diazocarbonyl compounds.

Initial observation of intramolecular cyclization via such a C–H insertion reaction was made when 21-diazo-5α-pregnan-20-one (**29**) was decomposed in refluxing toluene in the presence of copper(I) oxide.[41]

[98] B. M. Trost, D. F. Taber, and J. B. Alper, *Tetrahedron Lett.*, **1976,** 3857.

[99] D. F. Taber, *J. Am. Chem. Soc.*, **99,** 3513 (1977).

[100] K. Kondo, T. Umenoto, Y. Takahatake, and D. Tunemoto, *Tetrahedron Lett.*, **1977,** 113.

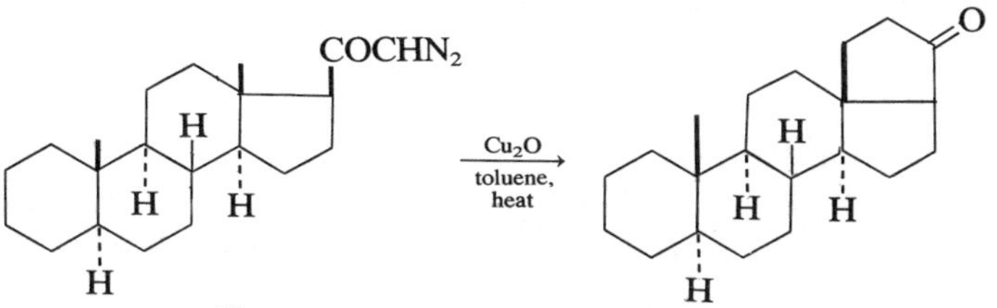

COCHN$_2$ H H H H 29 $\xrightarrow[\text{toluene, heat}]{Cu_2O}$ O H H H H

Several examples demonstrate that in geometrically rigid systems intramolecular C–H insertion is a highly favored process. Indicative of this is the study in which diazocamphor was catalytically decomposed to give cyclocamphanone (**30**) in high yield with no evidence of intermolecular addition or insertion products.[101] Contrast this with the behavior of 3-diazobicyclo[2.2.2]octan-2-one (**31**) which, upon copper-catalyzed decomposition in benzene, gives mainly the Wolff rearrangement product with only 5% formation of compound **32.**[102] Another illustration of geometrically favored intramolecular attack at an unactivated C–H bond is furnished by the 1-substituted adamantane derivative, which gives the tetracyclic ketone **33** in 70–80% yield.[103]

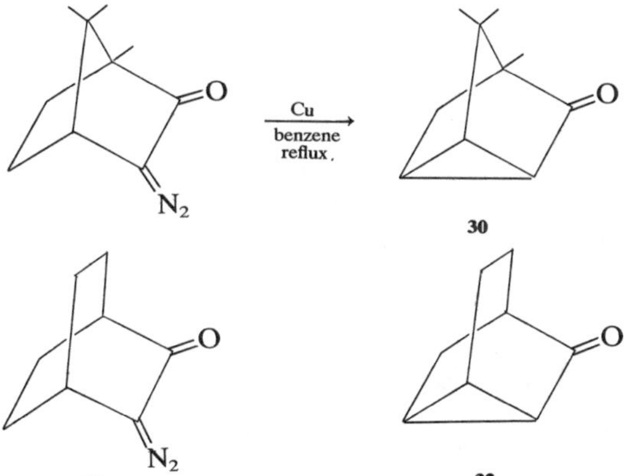

O N$_2$ $\xrightarrow[\text{benzene reflux}]{Cu}$ O 30

O N$_2$ 31 O 32

COCHN$_2$ $\xrightarrow[\text{toluene, heat}]{CuSO_4}$ O 33

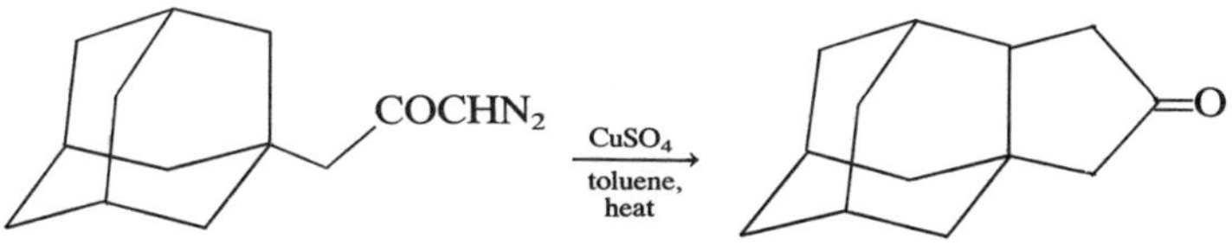

[101] P. Yates and S. Danishefsky, *J. Am. Chem. Soc.*, **84,** 879 (1962); J. Bredt and W. Holz, *J. Prakt. Chem.*, **95,** 133 (1917).

[102] P. Yates and R. J. Crawford, *J. Am. Chem. Soc.*, **88,** 1562 (1966).

[103] J. K. Chakrabarti, S. S. Szinai, and A. Todd, *J. Chem. Soc.*, (C), **1970,** 1303.

In some instances intramolecular C–H insertion has been observed when diazoketones are decomposed under Wolff rearrangement conditions.[104–107] For example, anomalous Wolff rearrangement products **34** and **35** are formed, *inter alia*, from the corresponding diazoketone.[104] Similarly, one of the products of the attempted Wolff rearrangement of diazoketone **36** is the C–H insertion product **37**, which was obtained in 22% yield.[105]

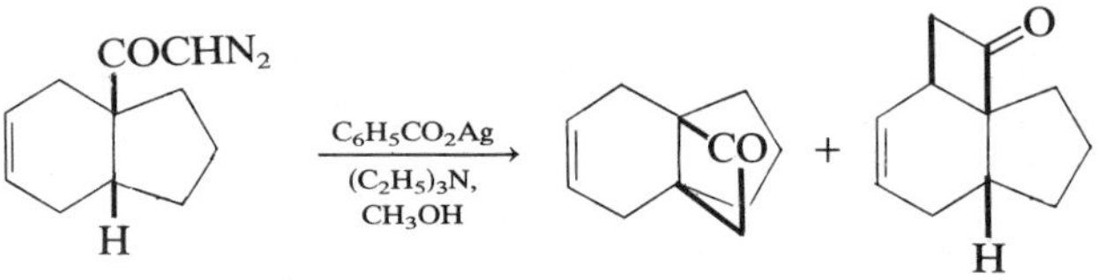

$COCHN_2$ H —$C_6H_5CO_2Ag$, $(C_2H_5)_3N$, CH_3OH→ CO + O H

34 **35**

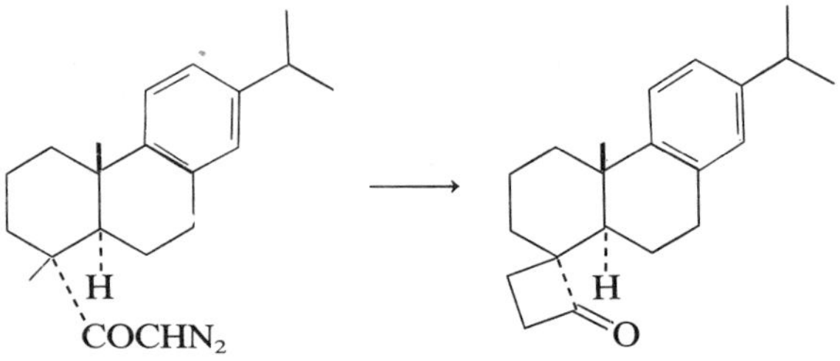

H $COCHN_2$ → H O

36 **37**

A general study of intramolecular C–H insertion reactions of diazoketones catalyzed by copper(II) sulfate in cyclohexane is summarized in Eq. 18.[105]

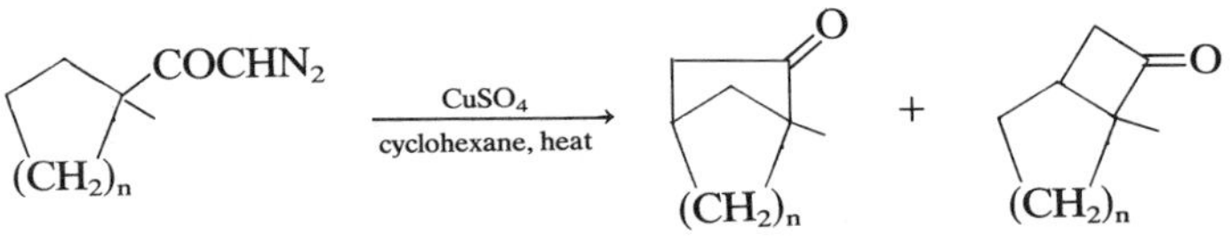

$COCHN_2$ $(CH_2)_n$ —$CuSO_4$, cyclohexane, heat→ O $(CH_2)_n$ + O $(CH_2)_n$ (Eq. 18)

A B

n	% yield A	% yield B
1	(−)	(−)
2	19	3
3	62	(−)
4	72	9

[104] H. O. House, S. G. Boots, and V. K. Jones, *J. Org. Chem.*, **30,** 2519 (1965).
[105] E. Wenkert, B. L. Mylari, and L. L. Davis, *J. Am. Chem. Soc.*, **90,** 3870 (1968).
[106] J. P. Tresca, J. L. Fourrey, J. Polonsky, and E. Wenkert, *Tetrahedron Lett.*, **1973,** 895.
[107] S. Wolff and W. C. Agosta, *J. Org. Chem.*, **38,** 1694 (1973).

Several transannular C–H insertions have been reported in which monocyclic α-diazoketones are converted to *cis*-fused bicyclic ketones.[108]

CuO, petroleum ether (60–90°), heat → (31%)

CuO, petroleum ether (60–90°), heat → (52%)

CuO, petroleum ether (60–90°), heat → (13%)

Applications have also been made to functionalization of diterpenoids. The isopropyl group of dehydroabietic acid was functionalized by intramolecular C–H insertion of the C–12 diazomethyl ketone **38.**[109] A partial synthesis of the tetracyclic diterpene isohibaene **39** (p. 382) uses as the key step a regiospecific C–H ketocarbenoid insertion.[110] The observed insertion is toward the tertiary C_8–H bond, whereas insertion into the secondary C_{11}–H (axial) bond is not observed. The key step in the total

N_2CH, O, H, CO_2CH_3 — **38** — Cu_2O, cyclohexane, heat → O, H, CO_2CH_3

[108] M. Regitz and J. Rüter, *Chem. Ber.*, **102,** 3877 (1969).

[109] R. C. Cambie and R. A. Franich, *J. Chem. Soc., Chem. Commun.*, **1969,** 725.

[110] M. Kitadani, K. Ito, and A. Yoshikoshi, *Bull. Chem. Soc. Jpn.*, **44,** 3431 (1971).

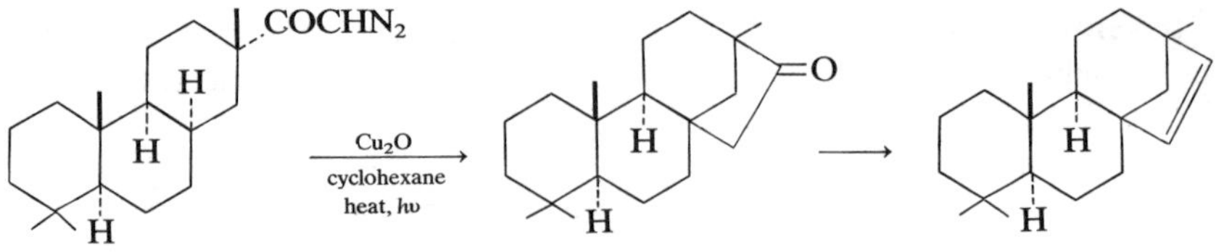

COCHN$_2$ H H H Cu$_2$O cyclohexane heat, $h\nu$ H H O H H

39

synthesis of atisine, veatchine, and gibberellin-A_{15} involves an intramolecular angular alkylation based on the regioselective and stereospecific intramolecular α-ketocarbenoid insertion reaction illustrated in Eq. 19.[111–113]

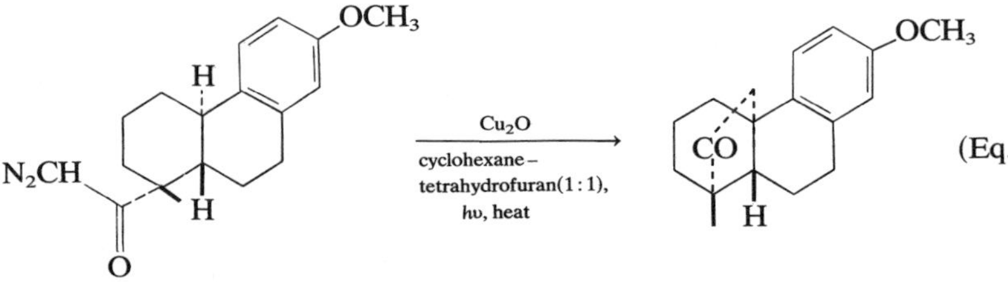

OCH$_3$ H N$_2$CH H O Cu$_2$O cyclohexane–tetrahydrofuran(1:1), $h\nu$, heat OCH$_3$ CO H (Eq. 19)

The intramolecular C–H insertion reactions of α-diazoamides have been shown to give fused and monocyclic β-lactams.[36,37,49–55] The initial demonstration of this type of reaction was made in the synthesis of methyl-6-phenylpenicillinate (**41**).[49] The ease with which carbenes derived from α-diazo amides undergo intramolecular C–H insertion, in contrast to the behavior of nonrigid alkoxycarbonyl carbenes,[38] has been attributed to a conformational effect of the planar amide bond forcing the C–H bond close to the divalent center.[36,37] The photochemical decomposition of N,N-diethyldiazoacetamide gives the two possible intramolecular C–H insertion products unaccompanied by any intermolecular insertion products (Eq. 7).[36,37] Further applications have been made to the synthesis of nuclear analogs of the penicillin–cephalosporin antibiotics leading to compounds **42,**[52,53] **43,**[54] **44,**[54] and **45.**[55]

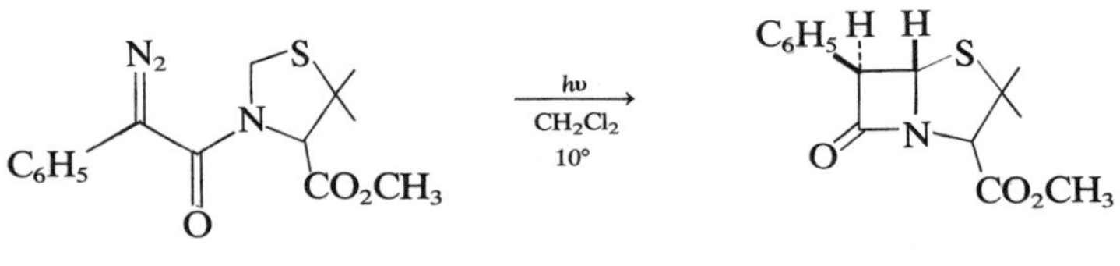

N$_2$ S N C$_6$H$_5$ O CO$_2$CH$_3$ $h\nu$ CH$_2$Cl$_2$ 10° C$_6$H$_5$ H H S N O CO$_2$CH$_3$

40 **41**

[111] U. R. Ghatak and S. Chakrabarty, *J. Am. Chem. Soc.*, **94,** 4756 (1972).

[112] U. R. Ghatak and S. Chakrabarty, *J. Org. Chem.*, **41,** 1089 (1976).

[113] S. Chakrabarty, J. K. Ray, D. Mukherjee, and U. R. Ghatak, *Synth. Commun.*, **5,** 275 (1975).

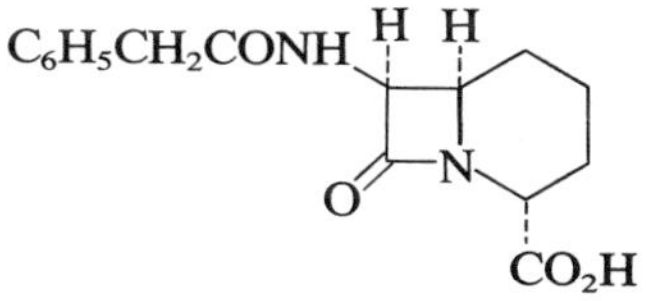

42 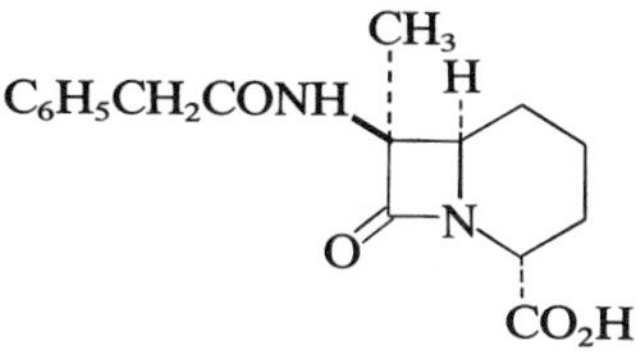 **43**

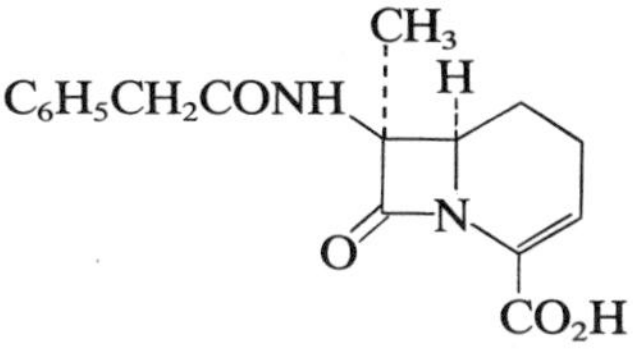

44 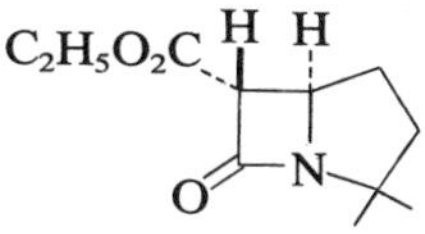 **45**

The copper-catalyzed decomposition of mixed diazomalonates has been used to prepare γ-butyrolactones.[39,114] It was found that chlorobenzene is a superior solvent for intramolecular C–H insertions of this type.

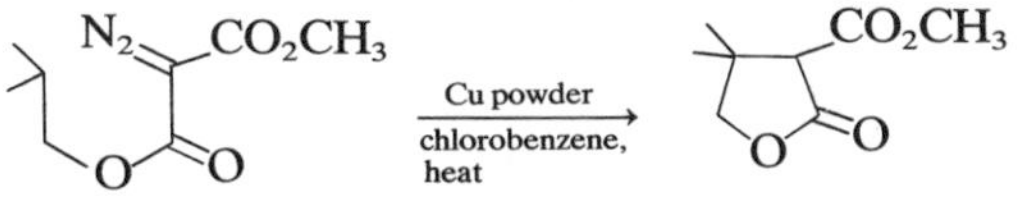

Cu powder, chlorobenzene, heat

Acid-Catalyzed Diazoketone Cyclizations

A facet of diazocarbonyl chemistry that was observed some 30 years ago[115,116] and is continuing to generate a great deal of interest is the acid-catalyzed cyclization of diazomethyl ketones through π participation of suitably situated olefinic bonds or aromatic groups.[117–121] The reaction probably proceeds via initial protonation (Brönsted-acid catalysis) or complexing (Lewis-acid catalysis) of the diazocarbonyl functionality, followed by displacement of nitrogen from the resultant diazonium species by π-bond participation.[122] This method promises to be of great synthetic utility for carbocyclic ring annulation, as a single product is frequently obtained in essentially quantitative yield.

[114] H. Ledon, G. Linstrumelle, and S. Julia, *Bull. Soc. Chim. Fr.*, **1973,** 2071.

[115] J. W. Cook and R. Schoental, *J. Chem. Soc.*, **1945,** 288.

[116] M. S. Newman, G. Eglinton, and H. M. Grotta, *J. Am. Chem. Soc.*, **75,** 349 (1953).

[117] R. Malherbe, N. T. T. Tam, and H. Dahn, *Helv. Chim. Acta*, **55,** 245 (1972).

[118] (a) W. F. Erman and L. C. Stone, *J. Am. Chem. Soc.*, **93,** 2821 (1971); (b) T. Miyashi, H. Kawamoto, and T. Mukai, *Tetrahedron Lett.*, **1977,** 4623.

[119] D. J. Beames, T. R. Klose, and L. N. Mander, *J. Chem. Soc., Chem. Commun.*, **1971,** 773.

[120] D. J. Beames and L. N. Mander, *Aust. J. Chem.*, **24,** 343 (1971).

[121] A. B. Smith and R. Karl Dieter, *J. Org. Chem.*, **42,** 396 (1977).

[122] D. J. Beames and L. N. Mander, *Aust. J. Chem.*, **27,** 1257 (1974).

This method was applied to construct the skeleton of the sesquiterpene α-patchoulane.[118a] The one-step, acid-catalyzed cyclization of diazomethyl ketone **46** proceeded under mild conditions to give two cyclization products, **47** and **48.** Other applications generate compounds containing the bicyclo[3.2.1]octane system by means of olefinic π-bond participation.[72,119,123–125] In this way intermediates are readily accessible for prospective total synthesis of a large class of tetracyclic diterpenoids. This use is illustrated by the preparation of compounds **49,**[119] **50,**[124] and **51.**[125]

$CO_2C_2H_5$ $COCHN_2$ **46** $\xrightarrow[ClCH_2CH_2Cl,\ 0°–27°]{BF_3 \cdot Et_2O}$ **47** ($CO_2C_2H_5$, O) + **48** ($CO_2C_2H_5$, O)

CH_3O, $COCHN_2$ $\xrightarrow[CH_3NO_2]{HBF_4}$ **49** (CH_3O, =O)

CH_3O, $COCHN_2$ $\xrightarrow[0°]{CF_3CO_2H}$ **50** (CH_3O, =O)

CH_3O, $OCOCF_3$, $COCHN_2$ $\xrightarrow{1.\ CF_3CO_2H,\ CH_2Cl_2,\ 0°;\ 2.\ Hydrolysis}$ **51** (CH_3O, OH, =O)

Aromatic π-bond participation has also been applied to the generation of compounds containing the bicyclo [3.2.1] system.[119,126,127] Examples of

[123] L. N. Mander, J. V. Turner, and B. G. Coombe, *Aust. J. Chem.*, **27,** 1985 (1974).
[124] T. R. Klose and L. N. Mander, *Aust. J. Chem.*, **27,** 1287 (1974).
[125] D. J. Beames, L. N. Mander, and J. V. Turner, *Aust. J. Chem.*, **27,** 1977 (1974).
[126] D. J. Beames, T. R. Klose, and L. N. Mander, *Aust. J. Chem.*, **27,** 1269 (1974).
[127] D. W. Johnson and L. N. Mander, *Aust. J. Chem.*, **27,** 1277 (1974).

this are the two diketones shown in the following reactions[126,127]:

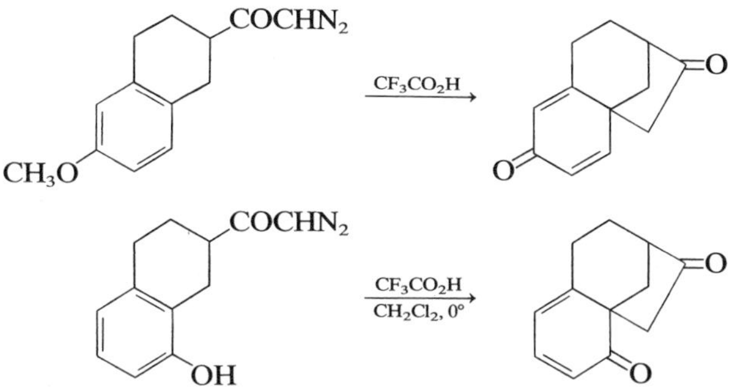

CH_3O … $COCHN_2$ —CF_3CO_2H→

$COCHN_2$ … OH —CF_3CO_2H, CH_2Cl_2, 0°→

A further application was made to the synthesis of norhelminthosporin analogs via **52** and **53.**[123]

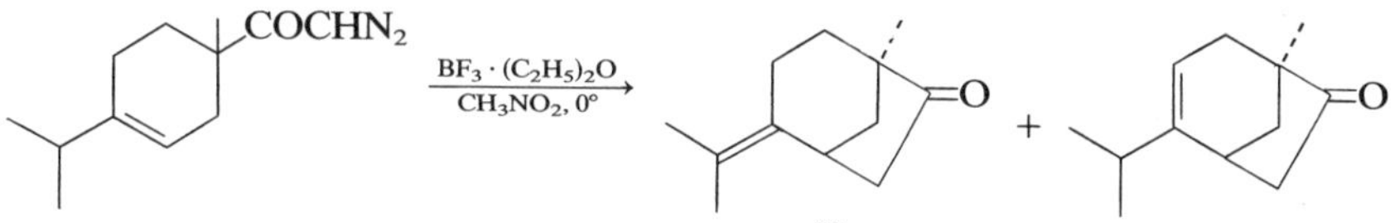

$COCHN_2$ —$BF_3 \cdot (C_2H_5)_2O$, CH_3NO_2, 0°→ **52** + **53**

Compound **54,** containing the bicyclo[2.2.2]octane ring system, has also been prepared in fair yield by this acid-catalyzed cyclization procedure.[120,126]

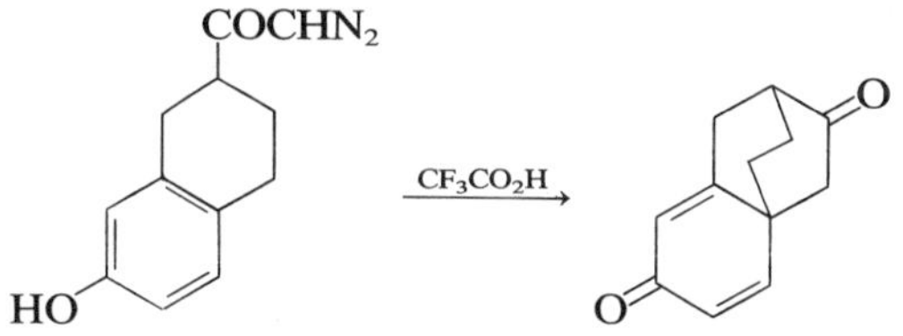

$COCHN_2$ … HO —CF_3CO_2H→ **54**

It has been shown that angularly fused cyclobutanones are available through acid-catalyzed intramolecular C alkylation of β,γ-unsaturated α-diazomethyl ketones.[128]

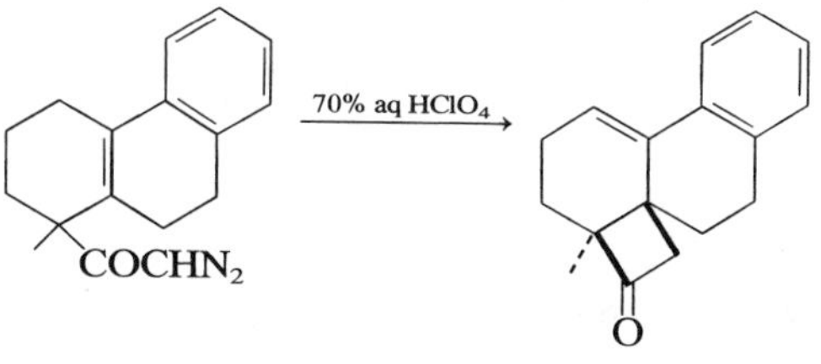

$COCHN_2$ —70% aq $HClO_4$→

[128] U. R. Ghatak and B. Sanyal, *J. Chem. Soc., Chem. Commun.*, **1974,** 876.

Spirodienones and products derived from a dienonephenol rearrangement have also been prepared by acid-catalyzed cyclization of aromatic diazoketones.[122]

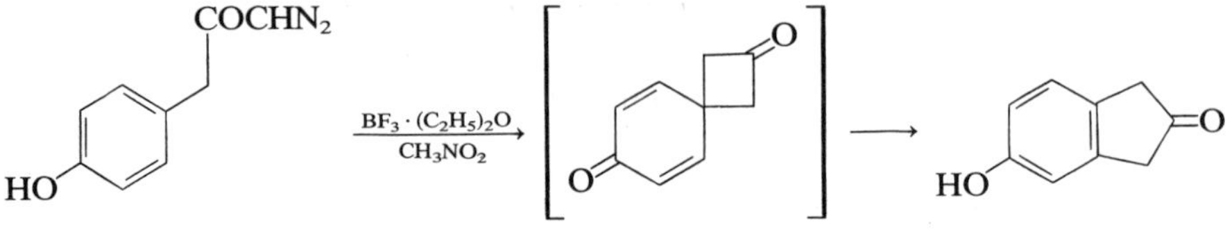

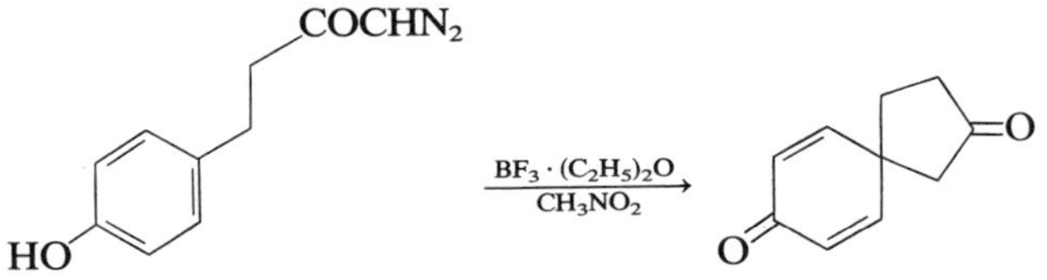

A synthesis of semibullvalene has been reported in which the key step is the acetolysis of 4-*endo*-diazoacetylbicyclo[3.1.0]hexene (**55**) to give the tricyclic ketoacetate.[129]

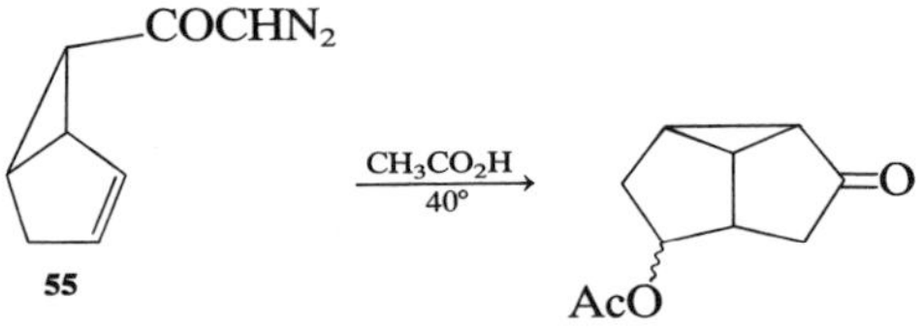

This acid-catalyzed cyclization method has been developed into a cyclopentenone annulation sequence. An example is the conversion of diazoketone **56** to the α,β-unsaturated cyclopentenone.[130]

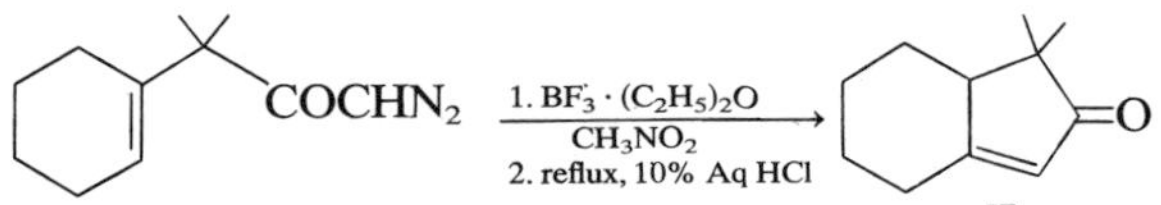

Similarly, the cyclization of acyclic β,γ-unsaturated diazoketones provided a synthesis of cyclopentenones, *e.g.*, *cis*-jasmone.[131]

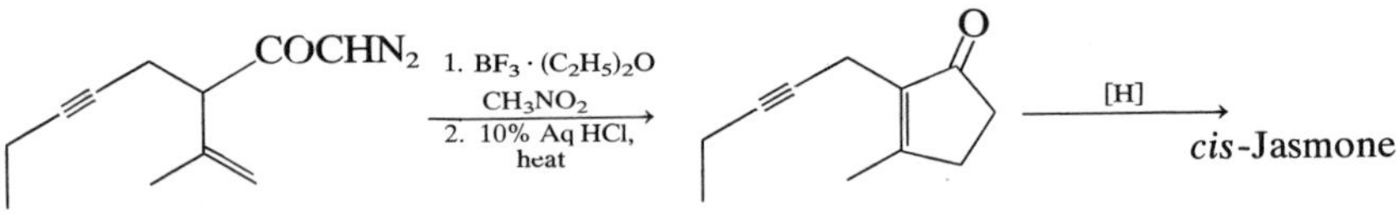

[129] R. Malherbe, *Helv. Chim. Acta*, **56**, 2845 (1973).

[130] A. B. Smith, III, *J. Chem. Soc., Chem. Commun.*, **1975**, 274.

[131] A. B. Smith, III, S. J. Branca, and B. H. Toder, *Tetrahedron Lett.*, **1975**, 4225.

It has been proposed that the transformation of the type **56** → **57** involves initial complexation of the Lewis acid to the carbonyl oxygen, producing an intermediate which can, in either a stepwise or concerted process, lose nitrogen and undergo a π cyclization.[121,122,130,131] Appropriately functionalized diazoketones of this type have been demonstrated to undergo Lewis-acid-catalyzed polyene cyclization via internal trapping of the initially formed carbonium ion (Eq. 20).[121]

$O\bar{B}F_3$ N_2^+ → $O\bar{B}F_3$

$COCHN_2$ $\xrightarrow[CH_3NO_2]{BF_3 \cdot (C_2H_5)_2O}$ O (Eq. 20)

Miscellaneous

Among the reactions that do not fall into the preceding categories is the insertion of an α-carbonyl carbene or carbenoid into a C–C single bond to give an olefin (Eq. 21)[132a] or a new ring (Eq. 22).[101]

N_2 $CO_2C_2H_5$ $\xrightarrow[\text{cyclopentane}]{h\nu}$ $CO_2C_2H_5$ (Eq. 21)

O N_2 $\xrightarrow[\text{benzene, heat}]{\text{Cu bronze}}$ O (Eq. 22)

The catalytic decomposition of bisdiazoketones has been shown to yield the 3,3′-spirobis(bicyclo[3.1.0]hexane) system.[133,134]

$COCHN_2$ N_2CHCO $\xrightarrow[\text{dioxane, } 5\text{–}15°]{\text{Di-}\mu\text{-dichloro-}\pi\text{-diallypalladium}}$ O O + O O (6:1)

[132] (a) U. Schöllkopf, D. Hoppe, N. Rieber, and V. Jacobi, *Justus Liebigs Ann. Chem.*, **730,** 1 (1969); (b) R. Malherbe and H. Dahn, *Helv. Chim. Acta*, **60,** 2539 (1977).

[133] S. Bien and D. Ovadia, *J. Org. Chem.*, **35,** 1028 (1970).

[134] S. Bien and D. Ovadia, *J. Chem. Soc., Perkin Trans. I*, **1974,** 333.

There has been one report in which a vinylogous α-diazo ester was used for an intramolecular cyclopropanation reaction.[48] This reaction was applied to the following synthesis of sirenin:

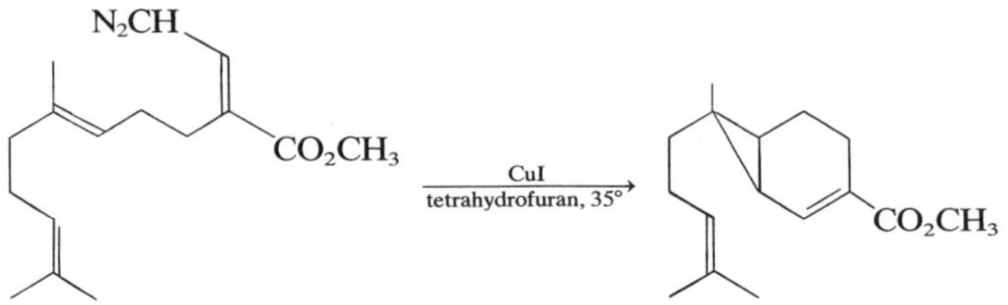

Recently an example of a diazoester insertion reaction into an N–H bond employing rhodium catalysis has been reported.[135]

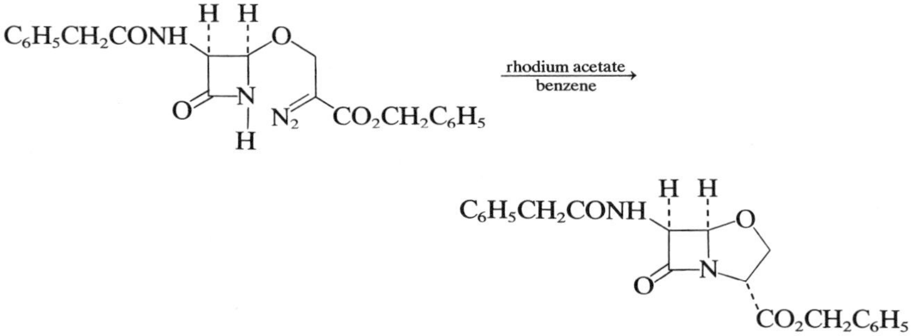

Copper(II)-catalyzed decomposition of penicillin-derived diazoketones in aprotic solvents has been shown to result in the formation of tricyclic ketones.[136]

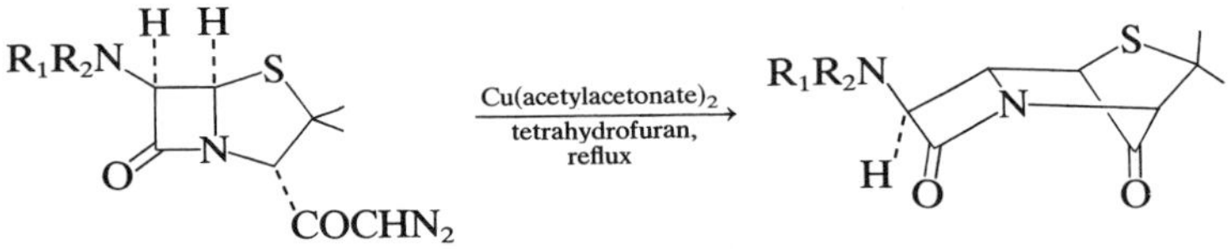

Finally, the acid-catalyzed decomposition of *o*-substituted diazoacetophenones possessing an oxygen functionality at the *ortho* position leads to coumaranones in high yield.[137–139]

[135] L. D. Cama, R. A. Firestone, and B. G. Christensen, Abstracts of the 10th Middle Atlantic Regional Meeting of the American Chemical Society, Philadelphia, Pa., Feb. 23–26, 1976, Abstract J-16.

[136] I. Ernest, *Tetrahedron*, **33,** 547 (1977).

[137] E. R. Marshall, J. A. Kuck, and R. C. Elderfield, *J. Org. Chem.*, **7,** 444 (1942).

[138] P. Pfeiffer, and E. Endres, *Chem. Ber.* **84,** 247 (1951).

[139] A. J. Bose and P. Yates, *J. Am. Chem. Soc.*, **74,** 4703 (1952).

COCHN$_2$ OCH$_3$ $\xrightarrow{H^+}$ O O

EXPERIMENTAL PROCEDURES

Tricyclo[2.2.1.0^{2,6}]heptan-3-one (Nortricyclanone).[140] A solution of crude 3-cyclopenten-1-yl diazomethylketone (from 23.9 g of the corresponding acid chloride) in 100 mL of tetrahydrofuran was added through a high-dilution head over a 4.5-hour period to a stirred slurry of 25 g of Cu powder in 700 mL of boiling tetrahydrofuran. The mixture was stirred at reflux for another 2 hours, filtered, dried over magnesium sulfate, and concentrated on a steam bath. Distillation of the residue under reduced pressure gave 12.9 g (65%) of nortricyclanone (bp 72–79°/29 mm), which was 95% pure by glpc analysis on a Carbowax 20M column. Redistillation in a vacuum-jacketed Vigreux column (3:1 reflux ratio) gave material with bp 75–77°/15 mm, n_D^{25} 1.4864, and a 2,4-dinitrophenylhydrazine derivative, mp 185–187° with decomposition.

Tricyclo[2.2.2.0^{2,6}]octan-3-one.[104] To a cold (5°) suspension of the anhydrous sodium salt from 3.77 g (30 mmol) of Δ^3-cyclohexene-1-carboxylic acid in 30 mL of benzene containing 0.8 mL of pyridine was added, with stirring, 15 g (0.12 mol) of oxalyl chloride. The resulting mixture was stirred for 2 hours at 5°, filtered, and concentrated. The residue was dissolved in 30 mL of benzene. This solution of the crude acid chloride was added to a cold (0–5°) ethereal solution of 60 mmol of diazomethane. The solution was stirred for 2 hours at 0–5° and 12 hours at room temperature, then filtered, dried, and concentrated, leaving 4.88 g of the crude diazoketone as a yellow liquid.

To a suspension of 400 mg of copper bronze in 250 mL of refluxing cyclohexane was added dropwise over a 1-hour period 3.79 g (25 mmol) of the diazoketone in 50 mL of cyclohexane. After stirring at reflux for 24 hours the mixture was filtered and concentrated, leaving 2.95 g of yellow liquid, which upon short-path distillation gave 2.34 g of a yellow liquid (bp 120–130°/20–25 mm). Chromatography of this material on 60 g of Florisil (elution with ether–petroleum ether mixtures) followed by distillation gave 1.85 g of tricyclo[2.2.2.0^{2,6}]octan-3-one as a colorless liquid, bp 100–108°/17 mm, n_D^{25} 1.5086, which solidified on standing, mp 41–43°. Short-path distillation (150°/25 mm) of the residue from the above distillation gave another 300 mg of product for a total yield of 2.15 g (57%).

[140] W. von E. Doering, E. T. Fossel, and R. L. Kaye, *Tetrahedron*, **21**, 25 (1965).

Tetracyclo[3.3.1.0^{2,8}.0^{4,6}]nonan-3-one (Triasteranone).[141] To a solution of *endo*-7-diazoacetyl-Δ^3-norcarene (from 2.76 g of the corresponding acid) in 1.2 L of *n*-hexane was added 19.2 g of activated copper powder. The suspension was refluxed for 12 hours, filtered, concentrated to a volume of 100 mL, and shaken with 150 mL of water. The water phase was washed with 100 mL of *n*-hexane and then with four portions (100 mL each) of chloroform. The combined organic layers were dried over sodium sulfate, and the solvent was removed under reduced pressure, yielding 1.73 g of crude triasteranone. Recrystallization from *n*-hexane and sublimation at 60°/1 mm gave 1.21 g (45%) of pure product, mp 74.5°; infrared (carbon tetrachloride) cm^{-1}: 1672 (C=O).

8-Tetracyclo[4.3.0.0^{2,4}.0^{3,7}]nonen-5-one.[142] A solution of 7-diazoacetylnorbornadiene (870 mg) in 10 mL of dry tetrahydrofuran was added slowly to a stirred suspension of copper powder in 20 mL of refluxing tetrahydrofuran. The mixture was heated at reflux for 1.8 hours, then set aside at room temperature overnight. The catalyst was filtered, and approximately 1 mL of water was added to the filtrate. After a few minutes the solution was diluted with ether, extracted with water and saturated aqueous sodium bicarbonate solution, and then dried and evaporated. This procedure removed two impurities observed in the glpc analysis of the crude product. Further purification was effected by passage through a column of 15 g of silica gel (elution with 25% ether–pentane), giving 577 mg (54%) of pure cyclopropyl ketone as a colorless, clear liquid; infrared (film) cm^{-1}: 1764, 1750 (C=O).

1,5,6-Triphenyltricyclo[3.1.0.0^{2,6}]hexan-3-one.[143] In a dry, three-necked flask equipped with a reflux condenser, dropping funnel, and a nitrogen inlet tube were placed 30 mL of benzene and 0.9 g of copper powder. The mixture was heated to reflux. A solution of 1-(1,2,3-triphenylcycloprop-2-enyl)-3-diazopropan-2-one (0.992 g) in 30 mL of benzene was added dropwise over a 10-minute period, and the mixture was refluxed for 1 hour, cooled, filtered, and concentrated. Chromatography with benzene on Fisher alumina (80–120 mesh) gave two fractions: 0.51 g (57%) with mp 155.5–156° and 0.068 g (7%) with mp 148–156°. Recrystallization from ethyl acetate provided an analytical sample, mp 155–157°; infrared (KBr) cm^{-1}: 1750 (C=O).

Tricyclo[3.3.1.0^{2,8}]nona-3,6-dien-9-one (Barbaralone).[2] Cycloheptatrien-7-yl diazomethyl ketone (from 14 g of the acid chloride) was dissolved in 80 mL of anhydrous benzene and 80 mL of dry hexane. This

[141] H. Musso and U. Biethan, *Chem. Ber.*, **100**, 119 (1967).

[142] R. M. Coates and J. L. Kirkpatrick, *J. Am. Chem. Soc.*, **92**, 4883 (1970).

[143] A. S. Monahan, *J. Org. Chem.*, **33**, 1441 (1968).

solution was divided into two equal parts, and each was added dropwise over a 45-minute period to a vigorously stirred, boiling suspension of 16 g of anhydrous cupric sulfate in 160 mL of *n*-hexane under nitrogen. The mixture was refluxed for an additional hour. The supernatant liquid was decanted, and the residue was washed with acetone. The solutions were combined, concentrated to about 80 mL by distillation through a Vigreux column, and steam distilled. The distillate was extracted with three 100-mL portions of ether, and the ethereal extracts were dried over magnesium sulfate, concentrated to 50 mL, and cooled to −70°. The crystals were collected by filtration at −70°. Recrystallization from 30 mL of pentane at −70° and again at −10 to −20° followed by vacuum sublimation gave 1.6–3.5 g of barbaralone, mp 40–48°. Recrystallization from water gave colorless needles, mp 53.5°; oxime, mp 130–131°; thiosemicarbazone, mp 165°; semicarbazone, mp 205–207° (dec).

7-*endo*-Methyl-7-(4-methyl-3-pentenyl)bicyclo[4.1.0]heptan-2-one.[11] *trans*-6,10-Dimethyl-5,9-decadienyl diazomethyl ketone obtained from 2.1 g of the corresponding acid was dissolved in 1.1 mL of dry cyclohexane containing 3.0 g of anhydrous copper sulfate and refluxed for 2 hours with stirring. The mixture was then filtered and the cyclohexane removed under reduced pressure. The residue was dissolved in ether, washed with aqueous sodium bicarbonate and aqueous sodium chloride solutions, dried, and the ether was evaporated. Chromatography on 250 g of silica gel (elution with benzene, benzene–ethyl acetate mixtures) gave 1.35 g (66%) of 7-*endo*-methyl-7-(4-methyl-3-pentenyl)bicyclo[4.1.0]heptan-2-one; infrared (film) cm^{-1}: 1685 (C=O). A gas chromatogram (5% Carbowax 20M, 10 ft×0.25 in., 165°) of the oil gave one major peak with a retention time of 14 min (60 mL/min, He).

1-Carbethoxy-3-oxa-6,6-dimethylbicyclo[3.1.0]hexan-2-one.[89] To a cold (0–5°) solution of 10 g (0.05 mol) of ethyl 3-methyl-2-butenyl malonate in ether (approximately 60 mL) containing 9.8 g of tosyl azide was added dropwise 5 mL of diethylamine. After the addition was complete, the mixture was stirred for 15 minutes at 0° and 2 hours at ambient temperature. Addition of approximately 200 mL of pentane, followed by filtration, removed the tosyl azide. The filtrate was dried over sodium sulfate and concentrated to give 11 g of the diazomalonate.

The crude diazomalonate (11 g) was dissolved in 700 mL of octane and added dropwise to a suspension of 20 g of copper powder in 300 mL of octane at reflux. After 3 hours at reflux the suspension was filtered, and the solvent was removed by distillation, leaving 5 g (50%) of 1-carbethoxy-3-oxa-6,6-dimethyltricyclo[3.1.0]hexan-2-one, bp 100–105° (0.5 mm); $n_D^{22} = 1.4670$; infrared (film) cm^{-1}: 1785, 1735 (C=O).

3-Hydroxytricyclo[4.4.1.0]undecylcarboxylic Acid Lactone.[91] A cold (0°) solution of 1.01 g (3.89 mmol) of the acid chloride of the *p*-toluene sulfonylhydrazone of glyoxylic acid in 10 mL of methylene chloride was treated successively with 577 mg (3.79 mmol) of $\Delta^{9(10)}$-2-octalol and with a solution of 397 mg (3.95 mmol) of triethylamine in 2 mL of methylene chloride, and was then stirred for 30 minutes at room temperature. After a second 404-mg (3.98-mmol) portion of triethylamine was added, the mixture was stirred for 1 hour at room temperature and then concentrated under reduced pressure. A solution of the residue in benzene was filtered through a 12-g column of Florisil to give 783 mg (89%) of the crude diazo ester. A solution of this material in hexane, when cooled to dry-ice temperatures, deposited the diazoester as yellow prisms, mp 29–32°; infrared (carbon tetrachloride) cm^{-1}: 2100 (C=N=N), 1700 (C=O); ultraviolet (*n*-hexane), nm max(ϵ): 217 (8100), 245 (12, 700).

A solution of 2.2 g (10 mmol) of the diazo ester in 125 mL of cyclohexane was added to a suspension of 3.1 g of cuprous oxide in 125 mL of cyclohexane over a 12.5-hour period. Filtration and concentration of the filtrate gave 2.58 g of crude product, which was chromatographed on Florisil. The early fractions (eluted with 5% ether in hexane) were followed by fractions eluted with 50% ether in hexane, affording 434 mg of material containing the lactone. Repetition of the chromatography and crystallization from an ether–hexane mixture gave the pure lactone as white needles, mp 54–55.5°; infrared (carbon tetrachloride): 1735 (C=O), 1715 (shoulder).

Adamantocyclopentan-2-one.[103] A solution of 1-(adamantan-1-yl)-3-diazopropan-2-one (3.8 g) in 100 mL of dry toluene was added dropwise during 5–6 hours to 400 mL of refluxing dry toluene containing 0.8 g of cupric sulfate. The mixture was refluxed for an additional 2 hours, filtered, and washed with water, 5 *N* sodium hydroxide solution, and again with water. The organic layer was dried over sodium sulfate and evaporated under reduced pressure, yielding approximately 2.8 g of a thick, light-brown oil. Glpc analysis of this oil showed a main peak consisting of 70–80% of the desired material. This peak was collected by preparative glc (6 ft x 0.25 in. glass column, 3% JRX, temperature programmed from 60 to 200°, 6°/min) and concentrated to give crystalline adamantocyclopentan-2-one; mp 74–76°; semicarbazone, mp 234–236°.

1,4,5,6,7,7*a*-Hexahydro-(2H)-inden-2-one.[144] A rapidly stirred solution containing 390 mg (2.38 mmol) of the α-diazoketone derived from 1-cyclohexeneacetic acid and 50 mL of dry nitromethane (distilled and stored over 4-Å molecular sieves) was treated dropwise with 355 μL (1.2

[144] A. B. Smith, III, and S. J. Branca, University of Pennsylvania, unpublished observations.

eq) of boron trifluoride etherate. After the evolution of nitrogen had ceased (≈10 min), 20 mL of 10% (v/v) aqueous hydrochloric acid was then added, and the resulting mixture was heated at reflux under nitrogen for 1 hour, cooled, and added to a 1:1 mixture of brine and ethyl acetate. The organic phase was then separated, washed with 5% (w/v) aqueous sodium bicarbonate solution, water, brine, and dried over anhydrous magnesium sulfate. Removal of the solvent under vacuum yielded 340 mg of an oil which contained 162 mg (50%) of the title enone. An analytical sample obtained by glpc (10 ft X 0.375 in. 25% Carbowax 20M, column temperature 205°C) gave the following spectral data: infrared (carbon tetrachloride) cm^{-1}: 1715 (C=O), 1630 (C=C).

TABULAR SURVEY

The five tables that follow contain reported examples of reactions discussed in this chapter. Tables I and II list examples of the intramolecular addition of diazoketones and diazoesters to olefins, respectively, and Table III lists examples of the intramolecular insertion of diazocarbonyl compounds into C–H bonds. Table IV deals with the acid-catalyzed cyclization of diazoketones. Table V lists miscellaneous reactions of diazocarbonyl compounds.

The examples are arranged in the tables according to increasing carbon number. Products appearing in brackets represent initial cyclopropane adducts that were not isolated. Throughout all the tables the reaction conditions, when available, have been summarized. Product ratios and/or percentage yields are recorded; a dash indicates that no yield was given in the reference.

The literature has been reviewed through December 1977.

TABLE I. Intramolecular Addition of Diazoketones to Olefins

Reactant	Conditions	Product(s)	Yield (%)	Refs.
C_6 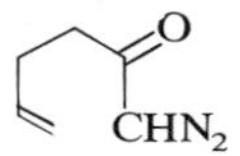 O, CHN_2	$CuSO_4$, cyclohexane, reflux	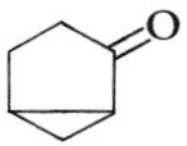 O	59	56
C_7 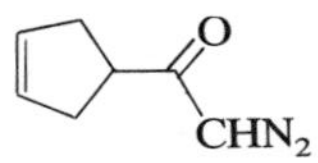 O, CHN_2	Cu powder, tetrahydrofuran, reflux	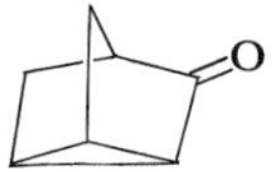 O	65	140
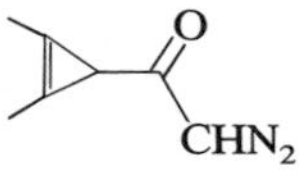 O, CHN_2	Cu powder, hexane, reflux	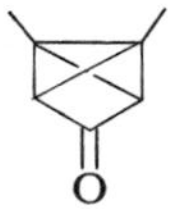 O	(*ca.* 1)	44
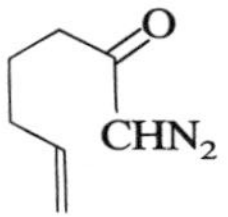 O, CHN_2	Cu bronze, cyclohexane, reflux	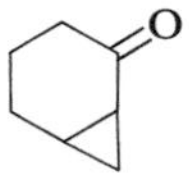 O	(*ca.* 50)	1
	$CuSO_4$, cyclohexane, reflux	"	37	56
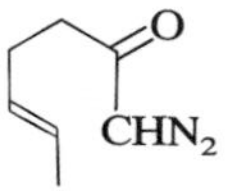 O, CHN_2	Cu powder, cyclohexane, reflux	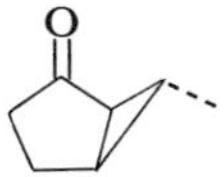 O	—	158
	$CuSO_4$, cyclohexane, reflux	"	30	159
O, CHN_2	$CuSO_4$, cyclohexane, reflux	O	30	159

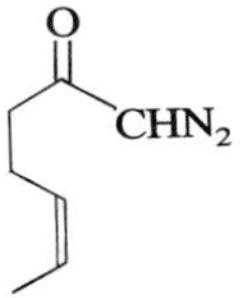

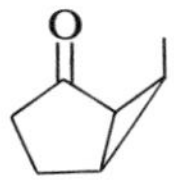

	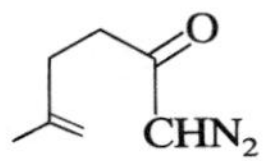	Cu catalyst	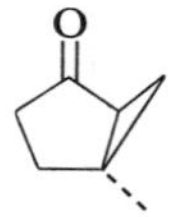	—	160
C_8	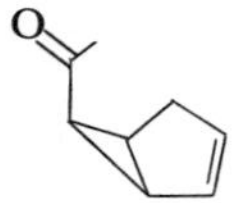	Tetrahydrofuran, *hν*	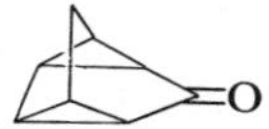	Major	161
		Cu powder, benzene	”	30	162
	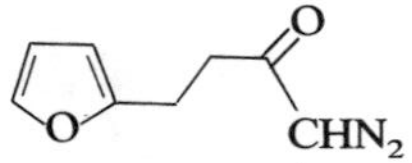	$CuSO_4$, cyclohexane, reflux	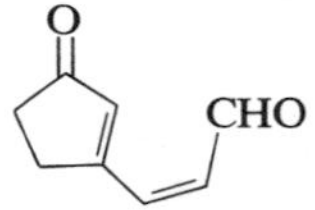	60	88
	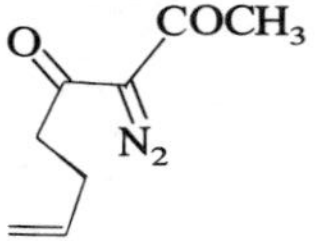	$CuSO_4$, benzene, reflux		37	145
	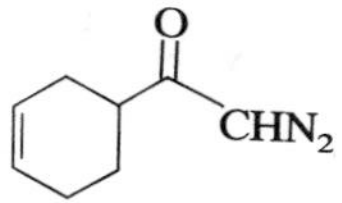	Cu powder, hexane, reflux	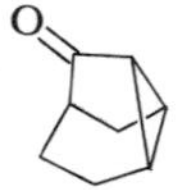	31	140
		Cu bronze, cyclohexane, reflux	”	57	104
		CuCl, cyclohexane, reflux	”	53	163
	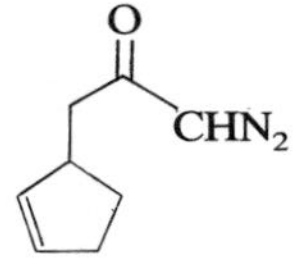	$CuSO_4$, cyclohexane, reflux		53	164

TABLE I. INTRAMOLECULAR ADDITION OF DIAZOKETONES TO OLEFINS (*Continued*)

Reactant	Conditions	Product(s)	Yield (%)	Refs.
C_8 (*Contd.*) 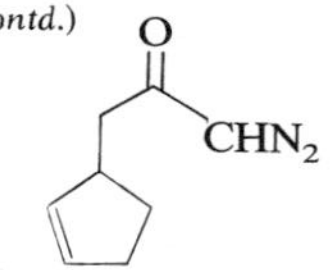				
O, CHN_2	"	"	67	159
O, CDN_2	"	O, D	31	164
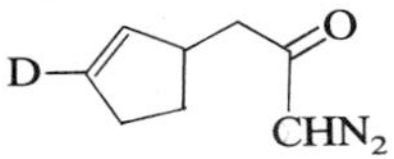 D, O, CHN_2	$CuSO_4$, cyclohexane, reflux	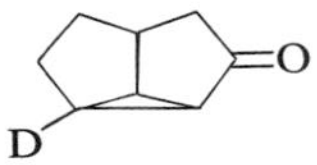 O, D	41	164
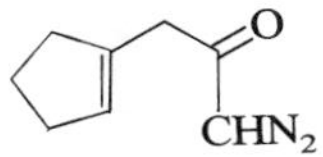 O, CHN_2	Cu(acetylacetonate)$_2$	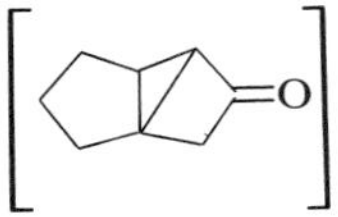 [O]	—	57
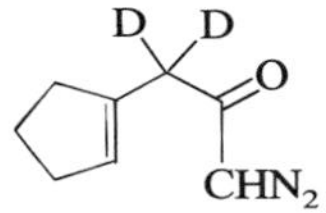 D D, O, CHN_2	Cu(acetylacetonate)$_2$	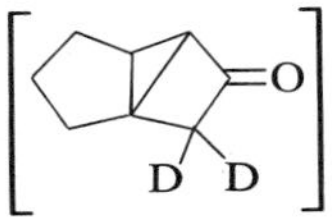 [O, D D]	—	57
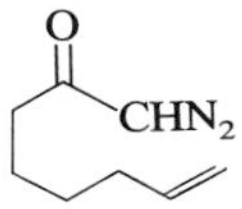 O, CHN_2	$CuSO_4$, cyclohexane, reflux	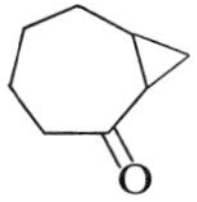 O	*ca.* 3	56

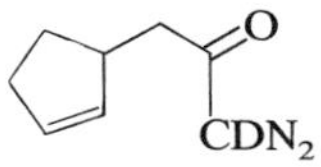

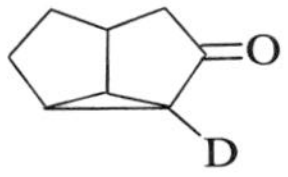

Diazoketone	Conditions	Product	Yield (%)	Ref.
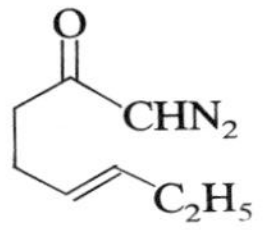 O, CHN_2, C_2H_5	Cu powder, cyclohexane, reflux	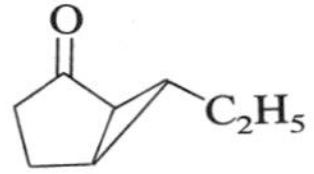 O, C_2H_5	—	158, 165
	$CuSO_4$, cyclohexane, reflux	”	—	159
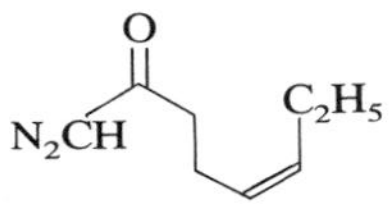 O, N_2CH, C_2H_5	Cu powder, cyclohexane, reflux	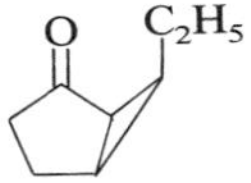 O, C_2H_5	—	158
	$CuSO_4$, cyclohexane, reflux	”	—	158
O, CHN_2	Cu powder, cyclohexane, reflux	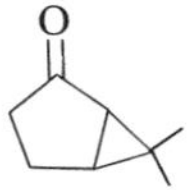 O	35–42	167, 168
C_9				
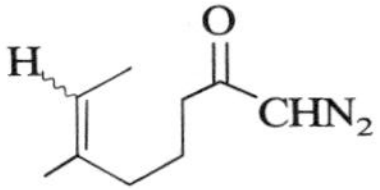 H, O, CHN_2	$CuSO_4$	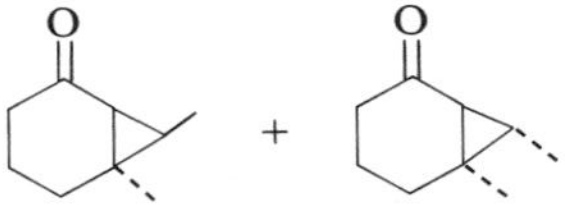 O + O	—	166
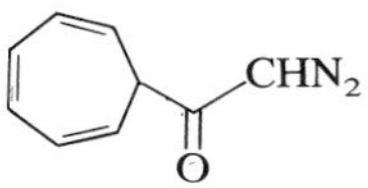 CHN_2, O	$CuSO_4$, benzene–hexane (1:3), reflux	 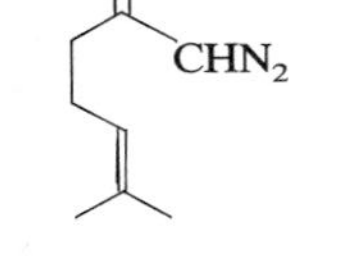	—	2

TABLE I. Intramolecular Addition of Diazoketones to Olefins (*Continued*)

Reactant	Conditions	Product(s)	Yield (%)	Refs.
C_9 (*Contd.*)				
[structure]	Cu powder, tetrahydrofuran, reflux	[structure]	54	142, 169
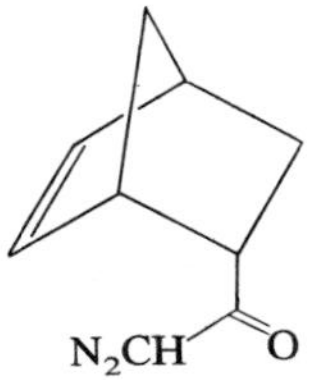 [structure]	Cu powder, tetrahydrofuran, reflux	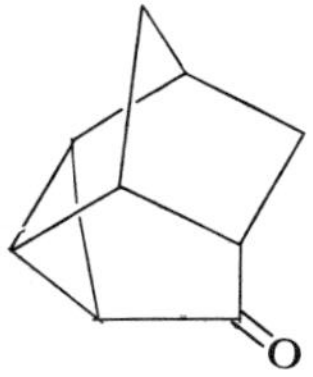 [structure]	—	170
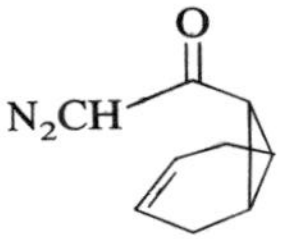 [structure]	Cu powder, hexane, reflux	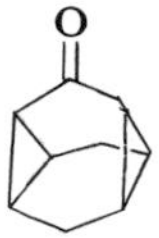 [structure]	45	171, 141
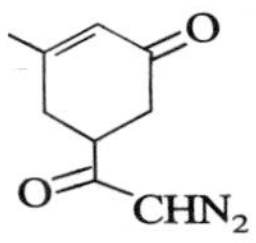 [structure]	$CuSO_4$, cyclohexane, reflux	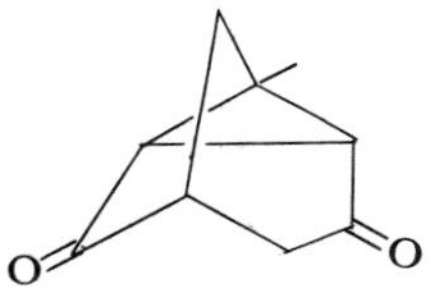 [structure]	38	172
	$CuSO_4$, dioxane, reflux	"	38	73

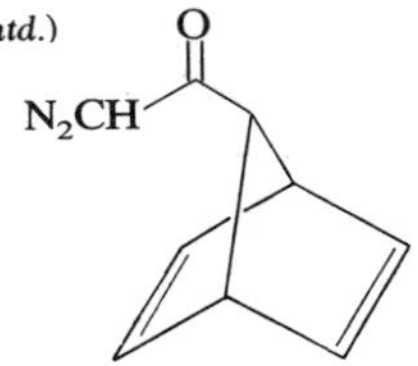

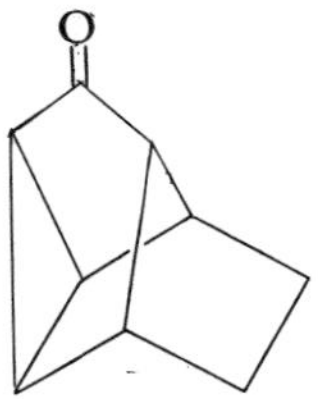

Reactant	Conditions	Product(s) and Yield(s) (%)			Yield (%)	Refs.
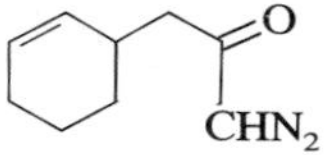 CHN_2, O	$CuSO_4$, cyclohexane, reflux	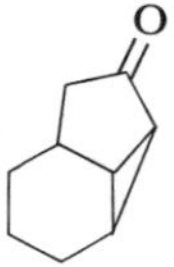 O			37	159
	$CuSO_4$, hexane, reflux	”			39	140
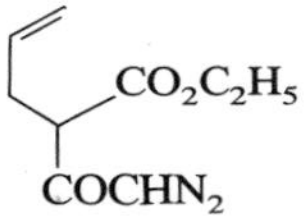 $CO_2C_2H_5$, $COCHN_2$	Di-μ-dichloro-π-allylpalladium, benzene	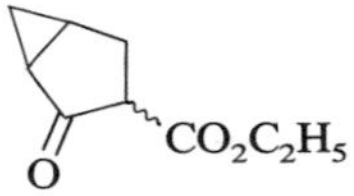 $CO_2C_2H_5$, O (53)	+	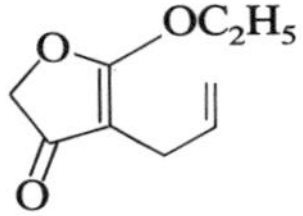 OC_2H_5, O, O (3)		146a
	$Pd(OAc)_2$, benzene	”	+	” (<1)		146a
	Bis(benzoylacetonato)palladium benzene, 80°	”	+	” (<1)		146a
	Bis(benzoylacetonato)copper, benzene, 80°	” (50)	+	” (13)		146a
	$(C_2H_5O)_3P{\cdot}CuCl$, benzene	” (3)	+	” (32)		146a
	$(CH_3O)_3P{\cdot}CuI$, benzene	” (3)	+	” (35)		146a
	$Rh_2(OAc)_4$, benzene	” (1)	+	” (58)		146a
	Benzene, 80°	” (15)	+	” (54)		146a
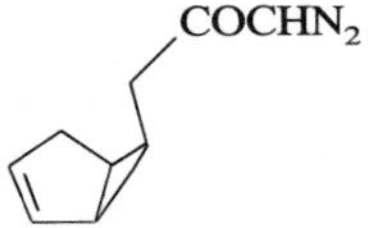 $COCHN_2$	Cu_2I_2, tetrahydrofuran	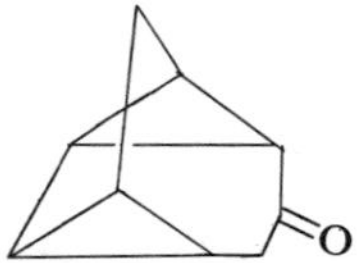 O			—	146b

TABLE I. Intramolecular Addition of Diazoketones to Olefins (*Continued*)

Reactant	Conditions	Product(s)	Yield (%)	Refs.
C_9 (*Contd.*)				
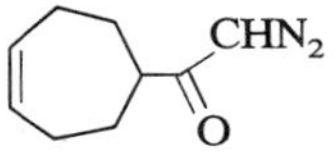	CuCl, cyclohexane, reflux	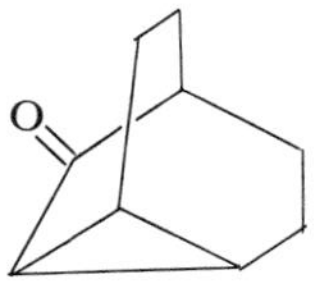	50	163
	$CuSO_4$, hexane, reflux	”	55	140
	Cu powder, hexane, reflux	”	47	140
	CuI, hexane or acetontrile–hexane (1 : 1), reflux	”	—	140
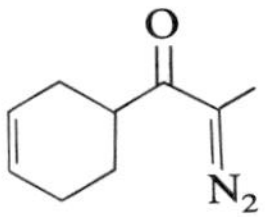	$CuSO_4$, cyclohexane, reflux	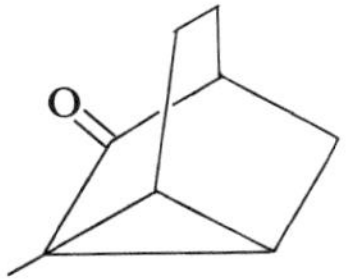	—	173
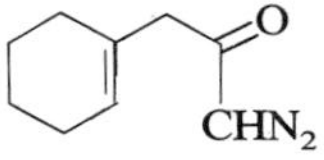	$CuSO_4$, cyclohexane, methanol, reflux	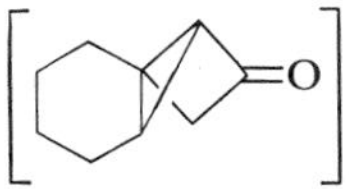	—	59
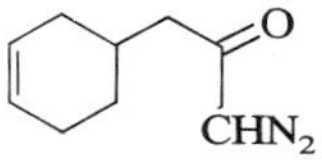	Cu powder, hexane, reflux	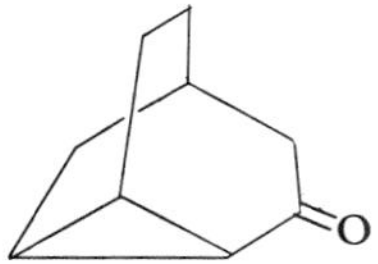	47	174

Reactant	Conditions	Product	Yield (%)	Ref.
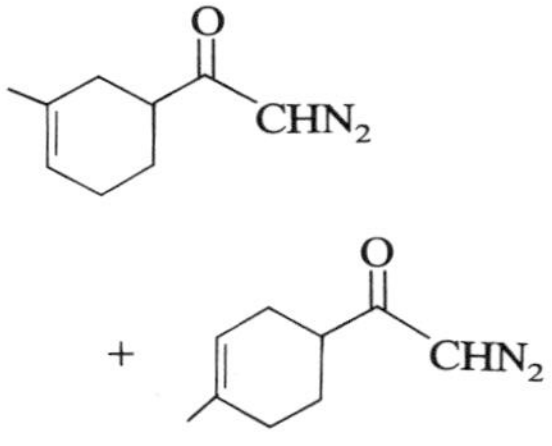	Cu powder, hexane, reflux	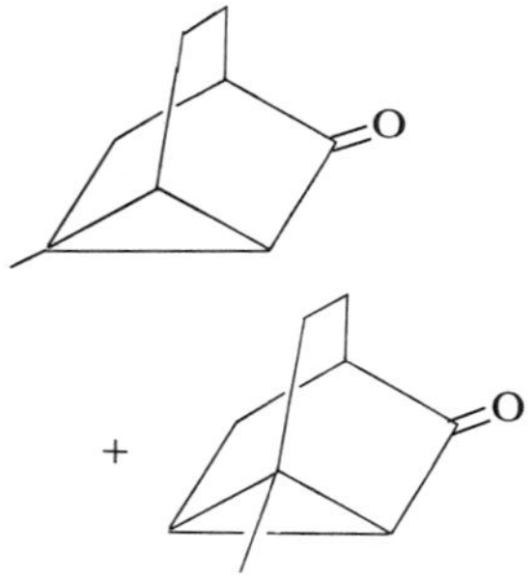	—	175
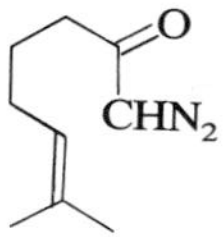	Cu powder, $CuSO_4$, cyclohexane, reflux	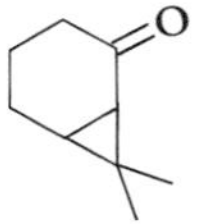	44	176
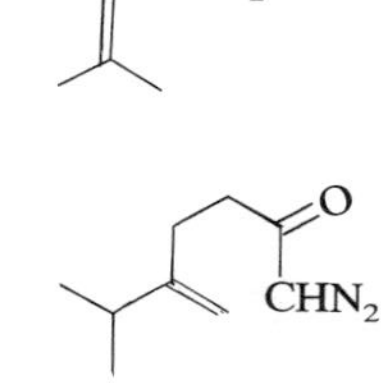	Cu, $CuSO_4$, cyclohexane, reflux	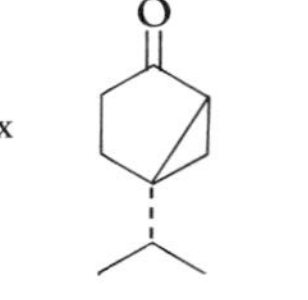	38	177
	$CuSO_4$, cyclohexane, reflux	”	—	6
	Cu bronze, cyclohexane, reflux	”	—	160
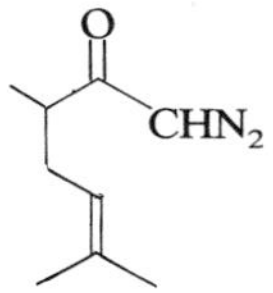	Cu powder, cyclohexane, reflux	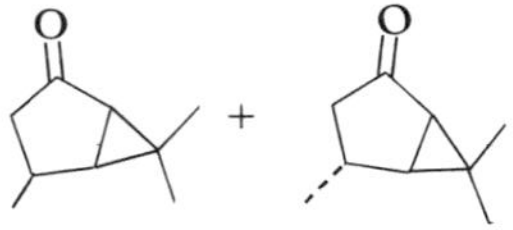 (2:3)	70	167, 168

TABLE I. INTRAMOLECULAR ADDITION OF DIAZOKETONES TO OLEFINS (*Continued*)

Reactant	Conditions	Product(s)	Yield (%)	Refs.
C_{10}				
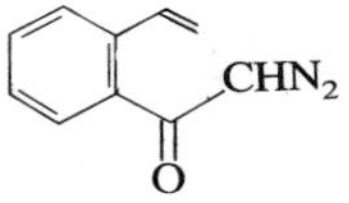	$CuSO_4$, benzene, reflux	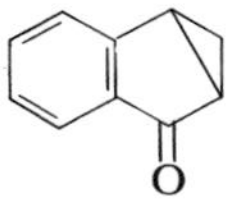	41.5	56
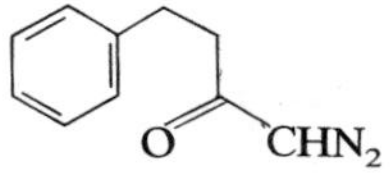	Cu powder, decalin, reflux	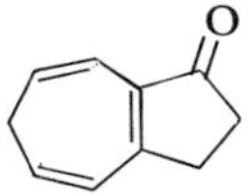	15, 20	85, 178
	CuCl, benzene, reflux	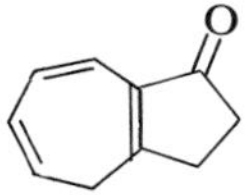	40–50	87
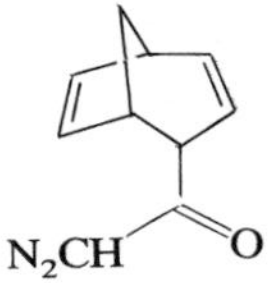	Cu bronze, cyclohexane or heptane, reflux	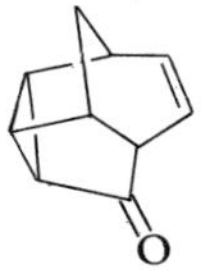	5	179
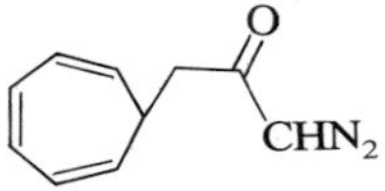	Cu powder, tetrahydrofuran, reflux		38	2
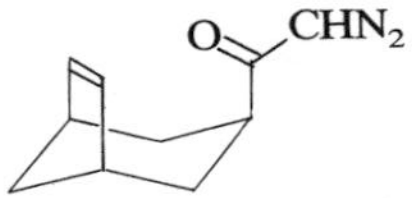	$CuSO_4$, tetrahydrofuran, reflux	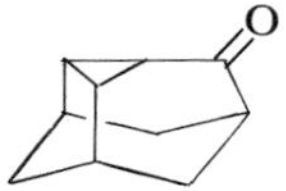	27	180, 181

Substrate	Conditions	Product	Yield (%)	Refs.
O, CHN_2	Cu bronze, cyclohexane, reflux	O	78	3, 182
CH_3O_2C, CHN_2, O	$CuSO_4$, dioxane, reflux	CO_2CH_3, O	61	73
CHN_2, O	$CuSO_4$, tetrahydrofuran, reflux	O	—	183
O, CHN_2	Cu powder, cyclohexane, reflux	O	76	85, 178
CHN_2, O	$CuSO_4$, hexane, reflux	O	47	140

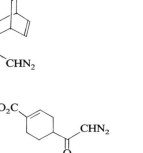

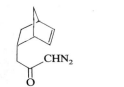

TABLE I. INTRAMOLECULAR ADDITION OF DIAZOKETONES TO OLEFINS (*Continued*)

Reactant	Conditions	Product(s)	Yield (%)	Refs.
C_{10} (*Contd.*) 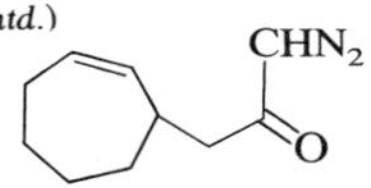 CHN$_2$ O	$CuSO_4$, hexane, reflux	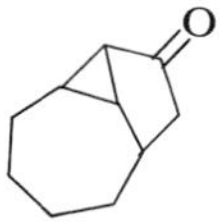 O	36	140, 159
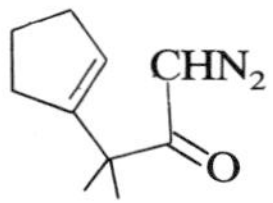 CHN$_2$ O	$CuSO_4$, cyclohexane, methanol, reflux	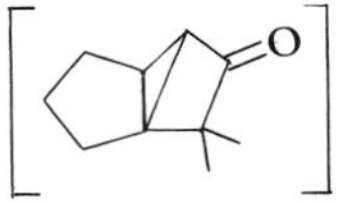 O	—	59
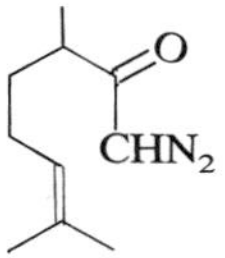 O CHN$_2$	Cu powder, $CuSO_4$, cyclohexane, reflux	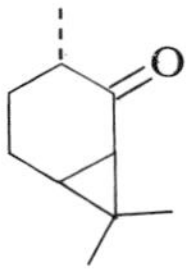 O	28	176
	Cu dust, cyclohexane, reflux	"	58	45
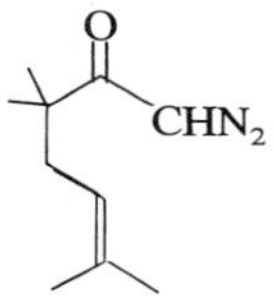 O CHN$_2$	Cu powder, cyclohexane, reflux	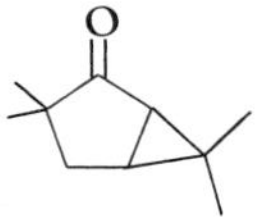 O	70	167, 184
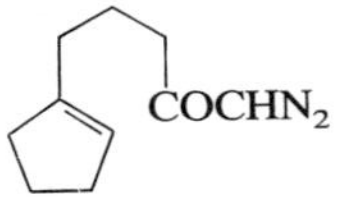 COCHN$_2$	Cu bronze, cyclohexane, reflux	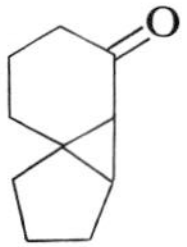 O	—	147

Reactant	Conditions	Product(s)	Yield (%)	Refs.
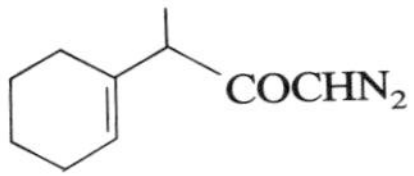 $COCHN_2$	$CuSO_4$ or Cu(acetylacetonate)$_2$, cyclohexane, reflux	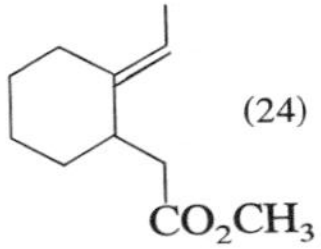 CO_2CH_3 (24) 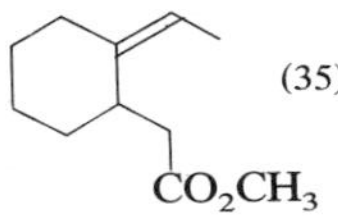 + CO_2CH_3 (35)		60a
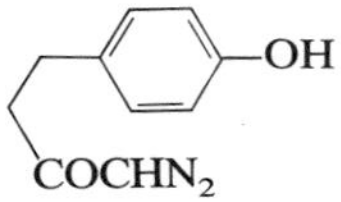 OH, $COCHN_2$	CuCl, tetrahydrofuran (0.3 *M*), 4.5 hr	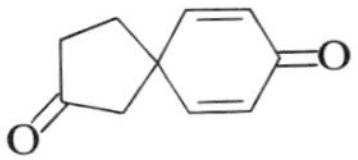 O, O	61	60b
	CuCl, tetrahydrofuran (0.16 *M*), 4.5 hr	"	79	60b
	CuCl, benzene (0.16 *M*), 5.5 hr	"	80	60b
	CuI, benzene (0.16 *M*), 5.5 hr	"	55	60b
	CuCl, benzene, 20 min	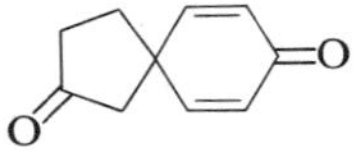 (34) + 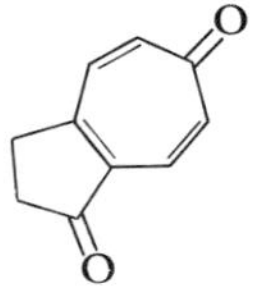 (8)		60b
C_{11}				
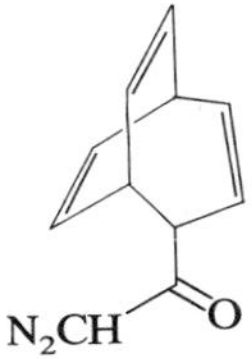 N_2CH, O	Cu bronze, benzene, 80°	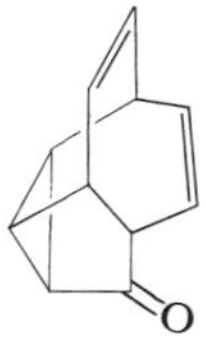 O	43	185

TABLE I. Intramolecular Addition of Diazoketones to Olefins (*Continued*)

Reactant	Conditions	Product(s)	Yield (%)	Refs.
C_{11} (*Contd.*) 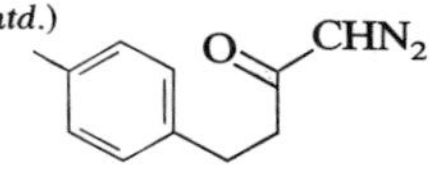 CHN_2, O	CuCl, benzene, reflux	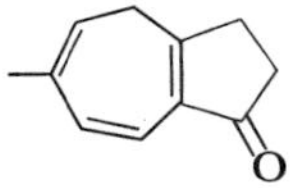 O	—	87
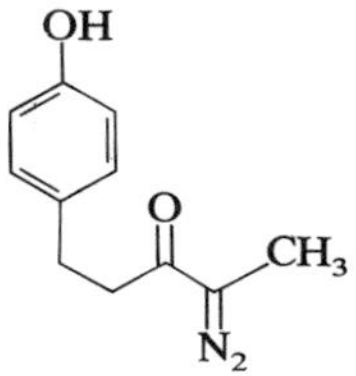 OH, O, CH_3, N_2	CuCl, tetrahydrofuran (0.16 *M*), 2 hr	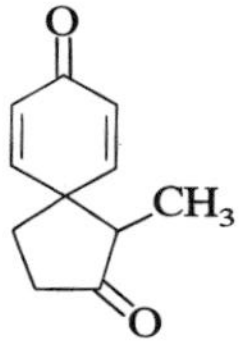 O, CH_3, O	60	60b
	CuCl, benzene (0.16 *M*), 2 hr		90	60b
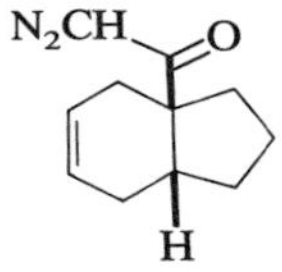 N_2CH, O, H	Cu bronze, cyclohexane, reflux	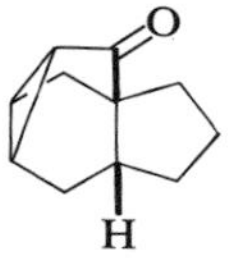 O, H	41	104
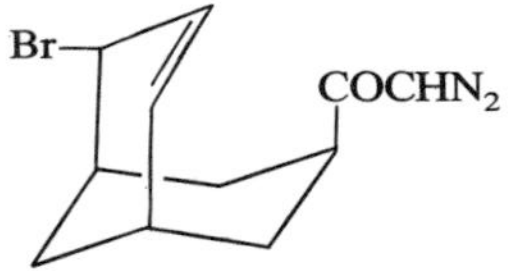 Br, $COCHN_2$	Cu bronze, cyclohexane, reflux	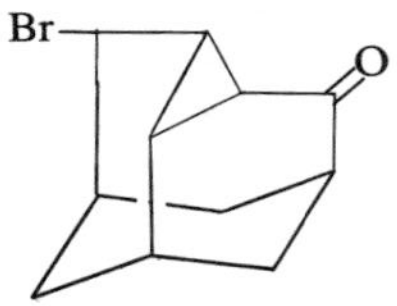 Br, O	43	186

Reactant	Conditions	Product	Yield (%)	Ref.
CH_3O_2C, CHN_2	$CuSO_4$, dioxane, reflux	CO_2CH_3	50	 73
CHN_2	$CuSO_4$, tetrahydrofuran, reflux		51	188
	Cu powder, cyclohexane, reflux	”	55	189
n-C_3H_7, n-C_3H_7, CHN_2	CH_3OH, $h\nu$,	n-C_3H_7, C_3H_7-n	10–15	42, 43
CHN_2	$CuSO_4$, cyclohexane, CH_3OH, reflux	[ ]	—	59
$COCHN_2$	$CuSO_4$, benzene, reflux		15	149

TABLE I. INTRAMOLECULAR ADDITION OF DIAZOKETONES TO OLEFINS (*Continued*)

Reactant	Conditions	Product(s)	Yield (%)	Refs.
C_{11} (*Contd.*)				
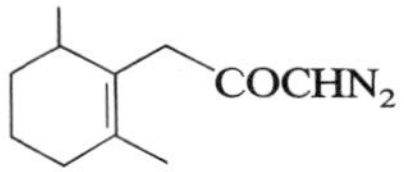 $COCHN_2$	$CuSO_4$ or Cu(acetylacetonate)$_2$, cyclohexane, 1% CH_3OH, reflux	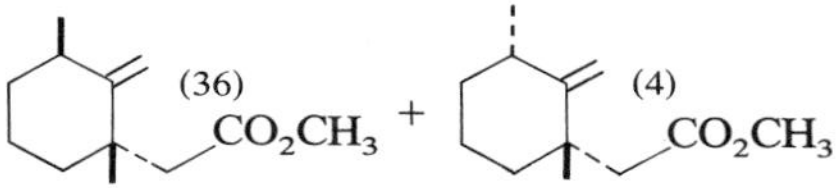 (36) CO_2CH_3 + (4) CO_2CH_3		60
 $COCHN_2$	Cu bronze, cyclohexane, reflux	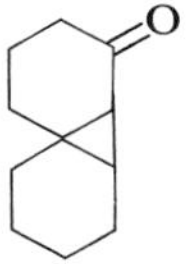 O	—	147
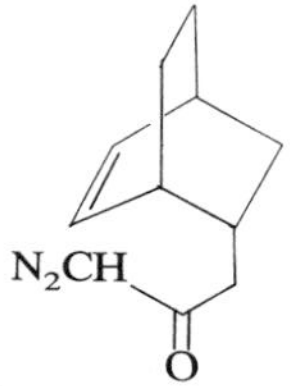 N_2CH O	$CuSO_4$, hexane, reflux	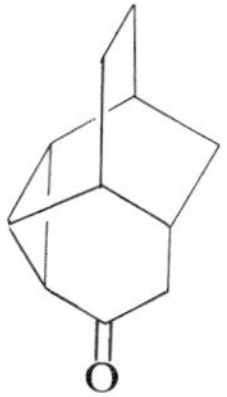 O	57	187
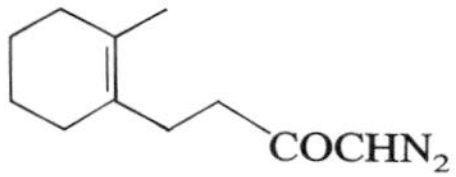 $COCHN_2$	Cu powder, cyclohexane, reflux	 O	29	56
	Cu bronze, cyclohexane, reflux	"	47	190
	$CuSO_4$, cyclohexane, reflux	"	35	190

Formula	Reactant	Conditions	Product	Yield (%)	Refs.
	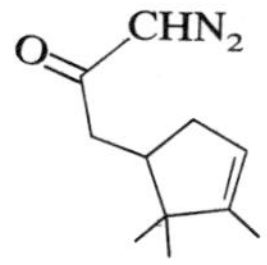 CHN_2	Cu powder, tetrahydrofuran, reflux	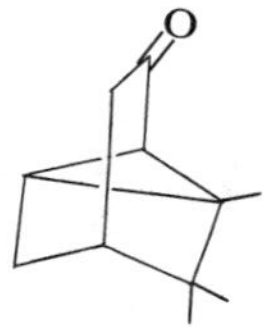	—	191
	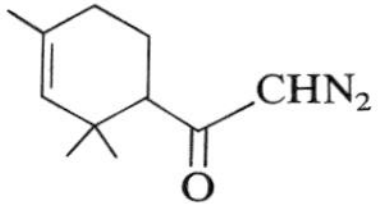 CHN_2	Cu catalysis	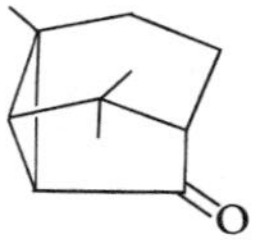	65	118a
	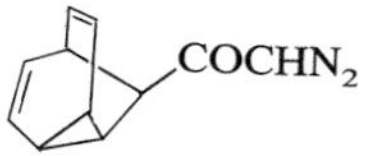 $COCHN_2$	Cu, tetrahydrofuran, reflux	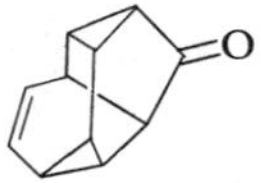	73	118b
C_{12}	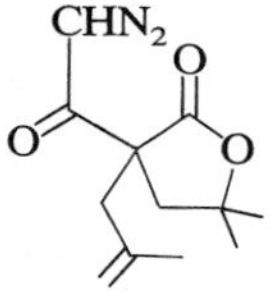 CHN_2	Di-μ-dichloro-π-allylpalladium, $(C_2H_5)_2O$, low temperature	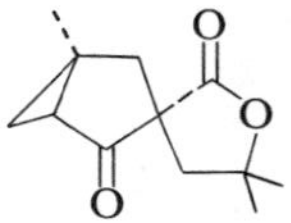	*ca.* 15	134
	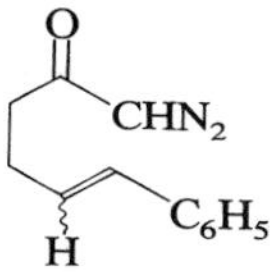 CHN_2, C_6H_5, H	$CuSO_4$, cyclohexane, reflux	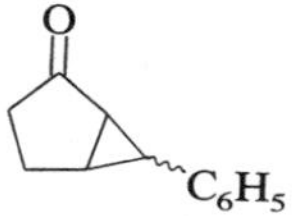 C_6H_5	59	56

TABLE I. INTRAMOLECULAR ADDITION OF DIAZOKETONES TO OLEFINS (*Continued*)

Reactant	Conditions	Product(s)	Yield (%)	Refs.
C_{12} (*Contd*)				
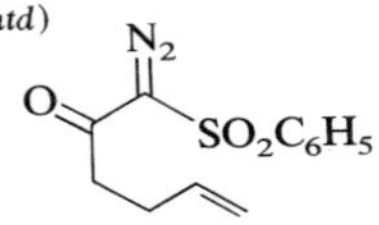 N_2, $SO_2C_6H_5$	*n*-Decane, 174°, 10 min	 $SO_2C_6H_5$	34	151a
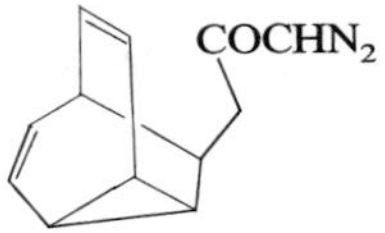 $COCHN_2$	Cu, reflux	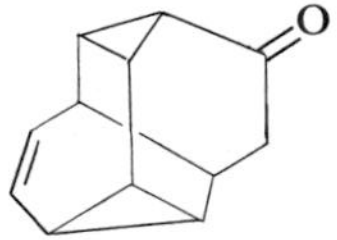	—	118b
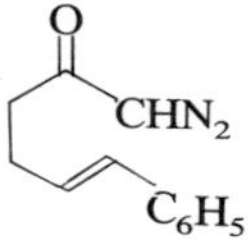 CHN_2, C_6H_5	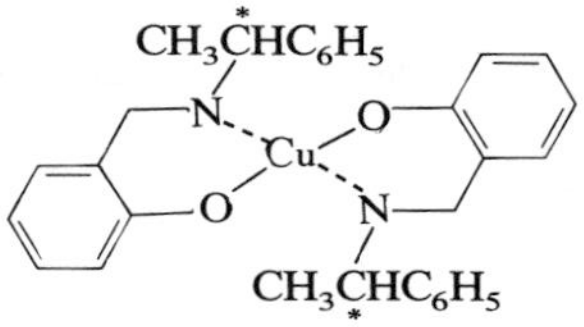 $CH_3\overset{*}{C}HC_6H_5$, N, Cu, O, O, N, $CH_3\underset{*}{C}HC_6H_5$	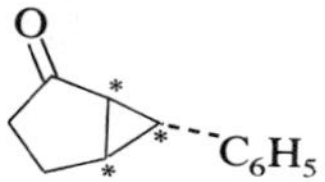 C_6H_5	64	29
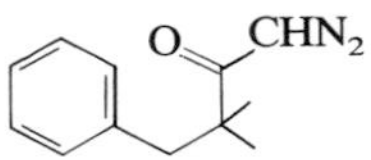 CHN_2	Cu powder, cyclohexane, reflux	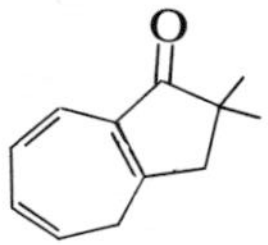	42	86, 178
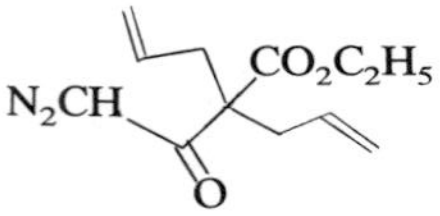 N_2CH, $CO_2C_2H_5$	 PdCl/2, ether	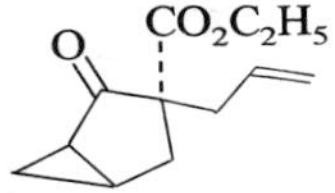 $CO_2C_2H_5$	—	192

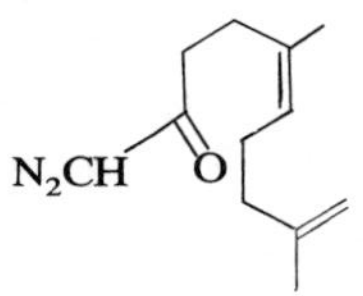 N_2CH O	Cu bronze, cyclohexane, reflux	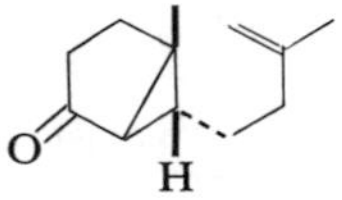 O H	—	80
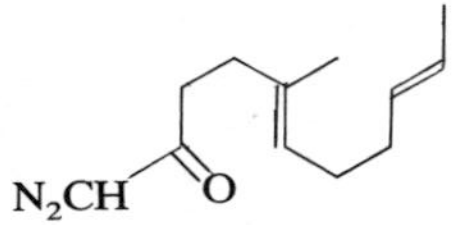 N_2CH O	”	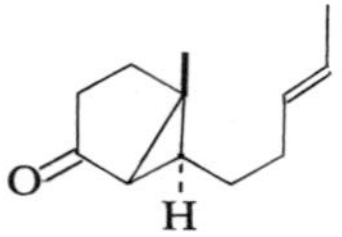 O H	80	81, 160
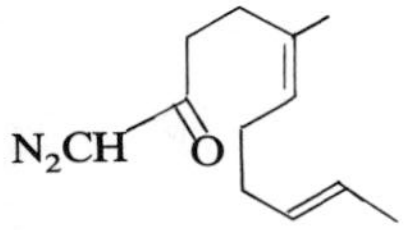 N_2CH O	Cu bronze, cyclohexane, reflux	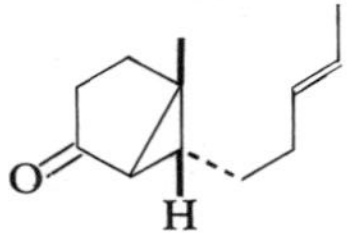 O H	—	82
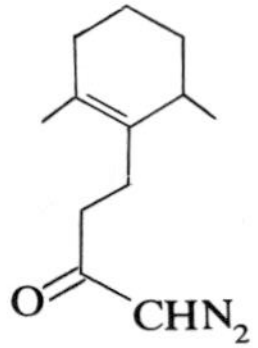 O CHN_2	Cu, cyclohexane, reflux	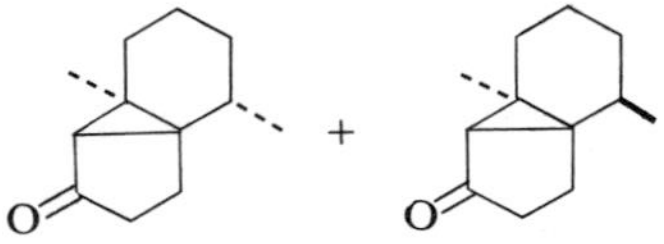 O + O (9 : 1)	44	84
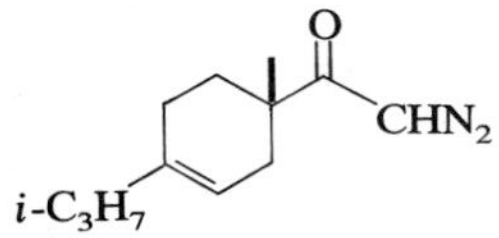 i-C_3H_7 O CHN_2	Cu powder, cyclohexane, reflux	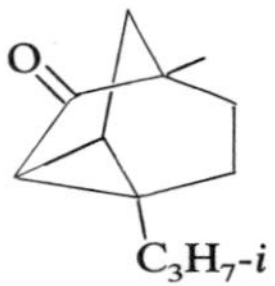 O C_3H_7-i	97	123

TABLE I. INTRAMOLECULAR ADDITION OF DIAZOKETONES TO OLEFINS (*Continued*)

Reactant	Conditions	Product(s)	Yield (%)	Refs.
C_{12} (*Contd.*)				
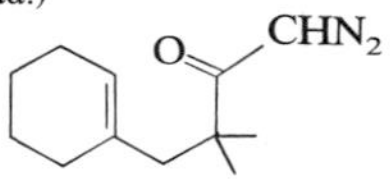 CHN_2, O	Cu powder, cyclohexane, reflux	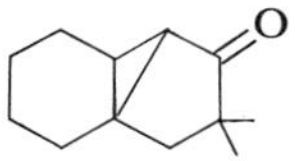 O	85	86, 178
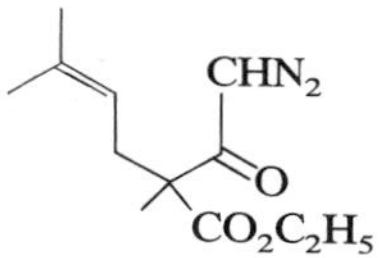 CHN_2, O, $CO_2C_2H_5$	Cu powder, cyclohexane, reflux	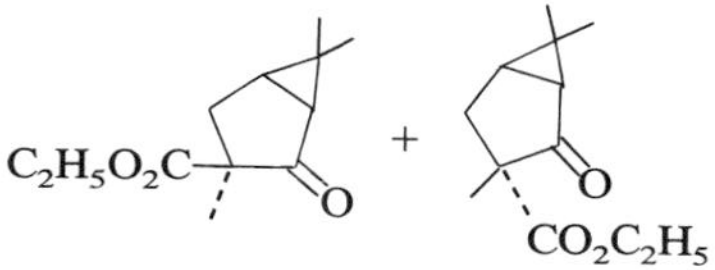 $C_2H_5O_2C$, O + O, $CO_2C_2H_5$	48	193
C_{13}				
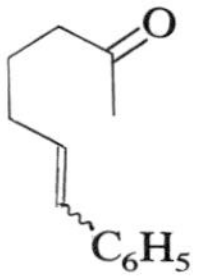 O, C_6H_5	$CuSO_4$, cyclohexane, reflux	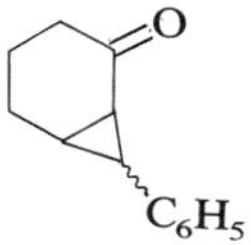 O, C_6H_5	30	56
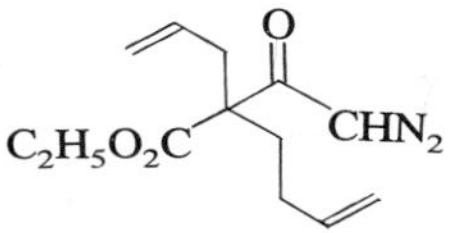 O, $C_2H_5O_2C$, CHN_2	PdCl/2, ether	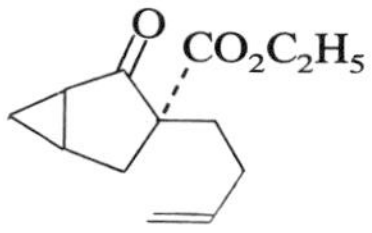 O, $CO_2C_2H_5$	—	192
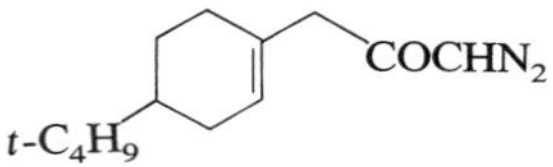 t-C_4H_9, $COCHN_2$	$CuSO_4$ or Cu(acetylacetonate)$_2$, cyclohexane, 1% CH_3OH, reflux	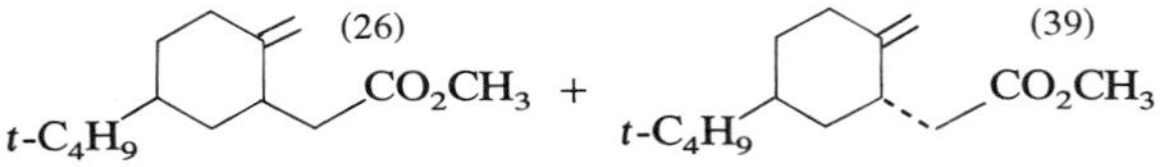 t-C_4H_9, CO_2CH_3 (26) + t-C_4H_9, CO_2CH_3 (39)		60a

C_{14}	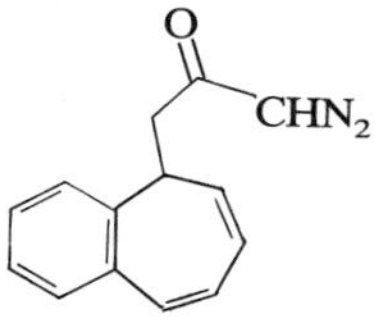	Cu	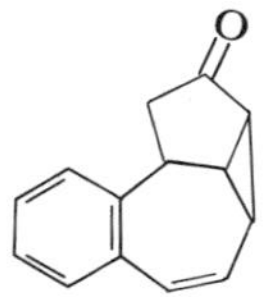	—	194
	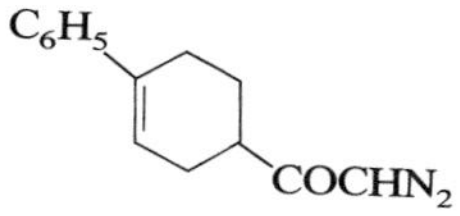	Activated Cu_2O, cyclohexane, reflux	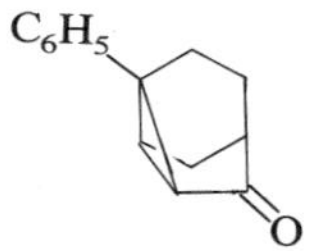	80	152
		Activated Cu_2O, cyclohexane, reflux	”	70	152
		$CuSO_4$, cyclohexane, reflux	”	57	152
	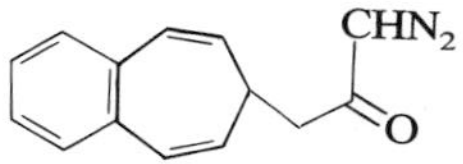	Cu	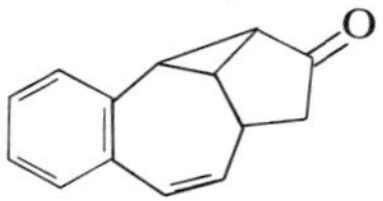	—	194
	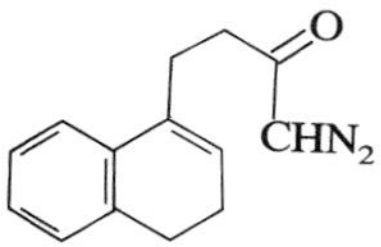	Cu powder, cyclohexane, reflux	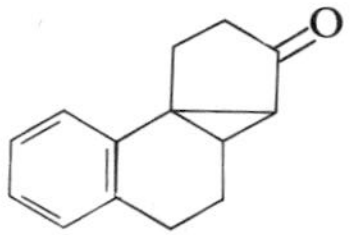	22	56

TABLE I. INTRAMOLECULAR ADDITION OF DIAZOKETONES TO OLEFINS (*Continued*)

Reactant	Conditions	Product(s)	Yield (%)	Refs.
C_{14} (*Contd.*)				
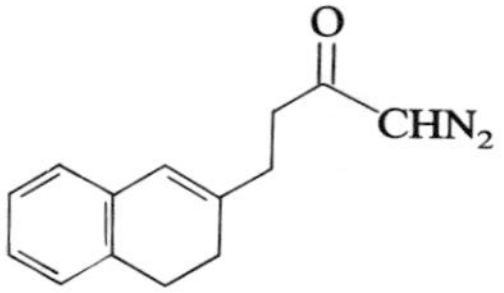 CHN_2	Cu bronze, cyclohexane, reflux	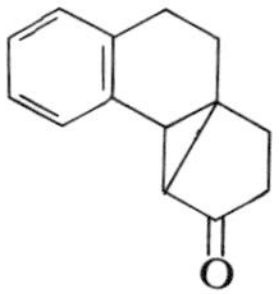	23	70
	Cu bronze, cyclohexane, reflux	”	40	74
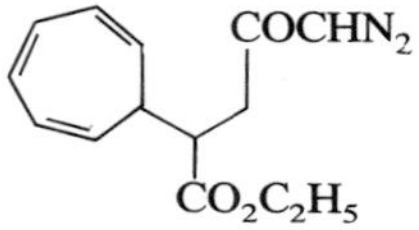 $COCHN_2$, $CO_2C_2H_5$	$CuSO_4$, toluene, reflux	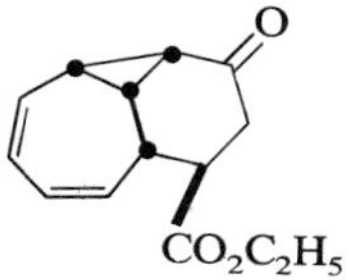 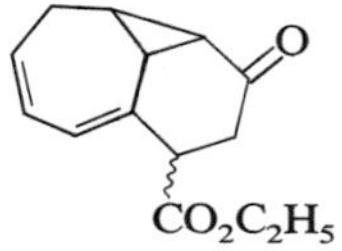 $CO_2C_2H_5$ (23) + $CO_2C_2H_5$ (21)		149
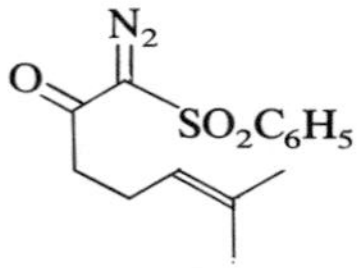 N_2, $SO_2C_6H_5$	*n*-Decane, 174°, 10 min	 $SO_2C_6H_5$	44	151a

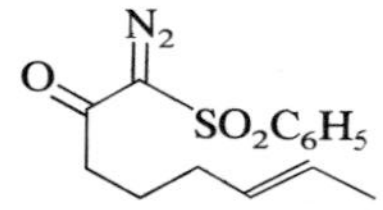	n-Decane, 174°, 10 min	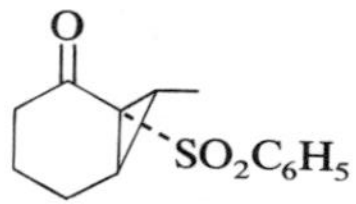	10	151
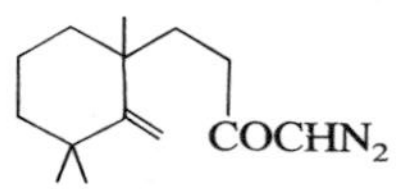	Cu catalysis	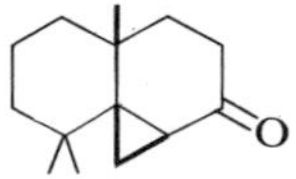	—	62a
	Cu powder, $CuSO_4$, cyclohexane, reflux	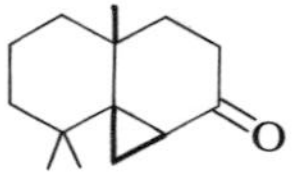	47	62b
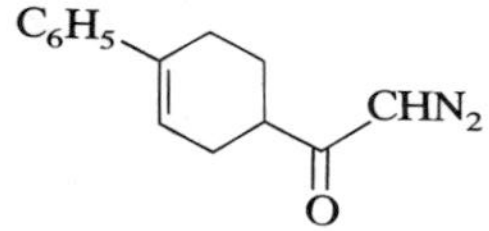	Activated CuO, cyclohexane or cyclohexane/tetrahydrofuran (7:3), $h\nu$	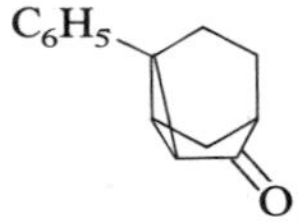	80	75
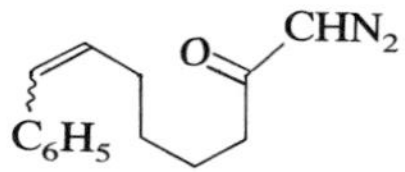	$CuSO_4$, cyclohexane, reflux	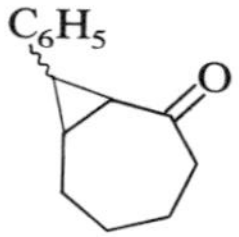	(Very low)	56
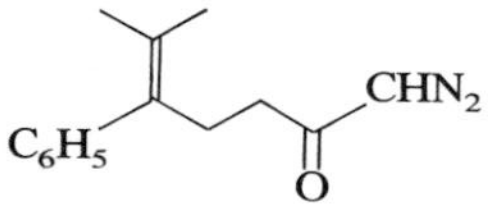	Cu bronze, benzene	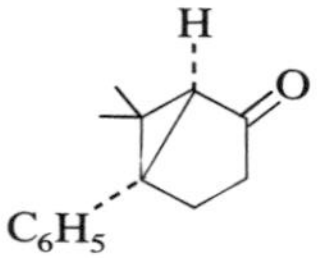	80	195
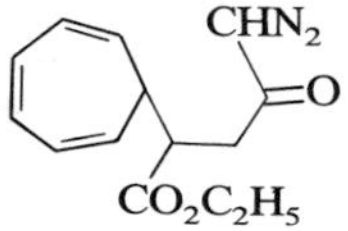	$CuSO_4$	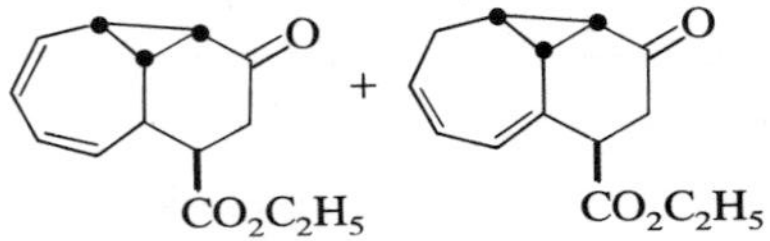 +		196

TABLE I. INTRAMOLECULAR ADDITION OF DIAZOKETONES TO OLEFINS (*Continued*)

Reactant	Conditions	Product(s)	Yield (%)	Refs.
C_{14} (*Contd.*) 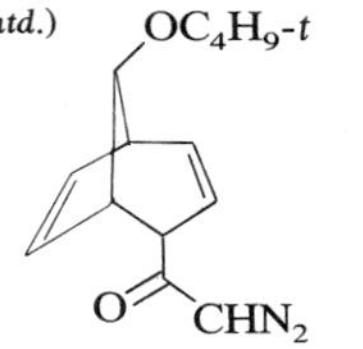 OC_4H_9-*t*; O; CHN_2	Cu bronze, cyclohexane, reflux	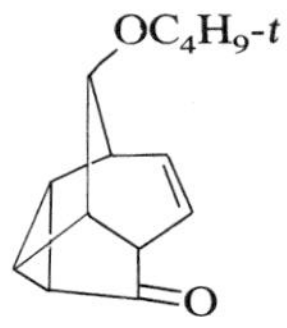 OC_4H_9-*t*; O	50	179
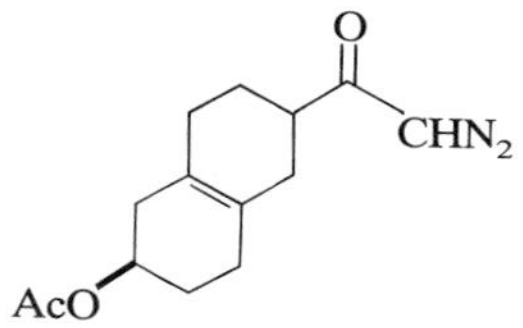 O; CHN_2; AcO	Cu powder, cyclohexane, reflux	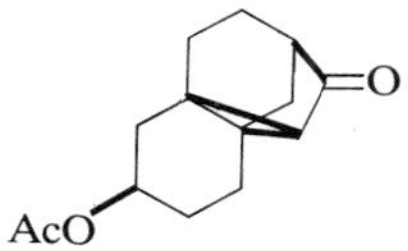 O; AcO	78	78
	Cu_2O, cyclohexane, reflux	"	(Trace)	78
	$CuSO_4$, cyclohexane, reflux	"	(45–50)	78
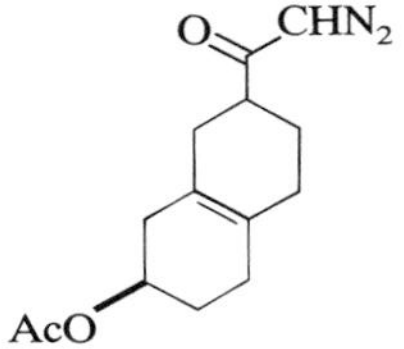 O; CHN_2; AcO	Cu powder, cyclohexane, reflux	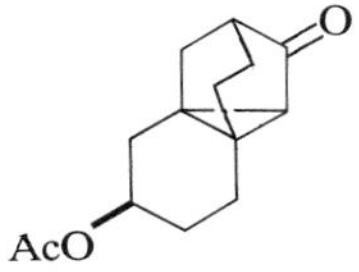 O; AcO	62	78
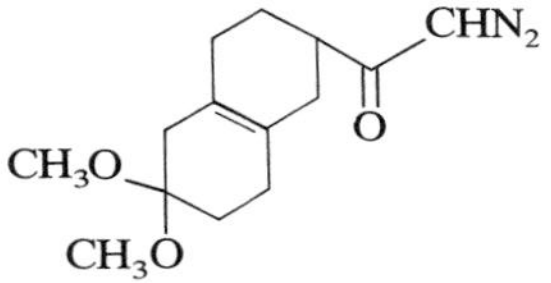 CHN_2; O; CH_3O; CH_3O	Cu powder, cyclohexane, reflux	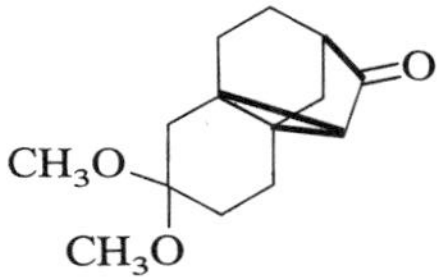 O; CH_3O; CH_3O	74	78

Reactant	Conditions	Product	Yield (%)	Refs.
	Cu_2O, cyclohexane, reflux	”	(Traces)	78
	$CuSO_4$, cyclohexane, reflux	”	(45–50)	76, 78
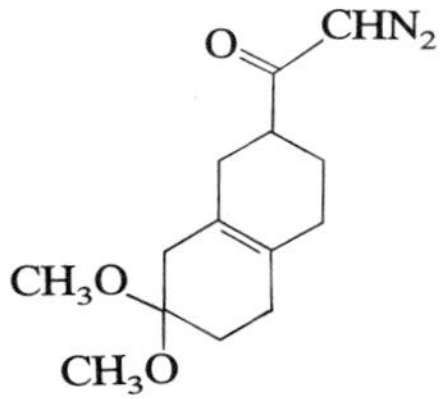 CH_3O, CH_3O, O, CHN_2	Cu powder, cyclohexane, reflux	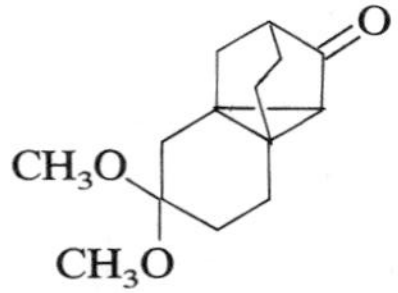 CH_3O, CH_3O, O	42	78
	$CuSO_4$, cyclohexane, reflux	”	—	76
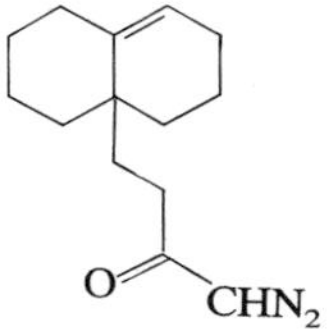 O, CHN_2	$CuSO_4$, cyclohexane, reflux	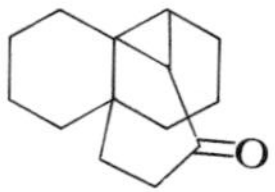 O	60	61
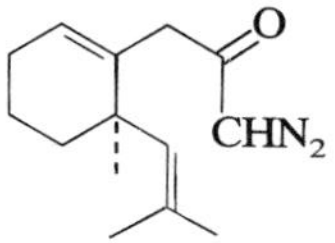 O, CHN_2	$CuSO_4$, cyclohexane, reflux	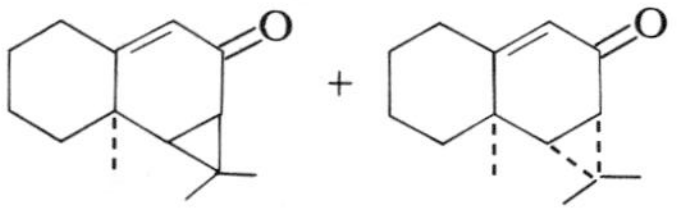 O + O (2:1)	—	197
	1. $CuSO_4$, cyclohexane, reflux 2. NaOH, CH_3OH, heat	” + ”	15	198

TABLE I. INTRAMOLECULAR ADDITION OF DIAZOKETONES TO OLEFINS (*Continued*)

Reactant	Conditions	Product(s)	Yield (%)	Refs.
C_{14} (*Contd.*) CHN_2	Cu, benzene	+ (1:9)	40	77
	Cu, cyclohexane, reflux	" + "	31	84
$C_2H_5O_2C$ CHN_2	Cu	$C_2H_5O_2C$	16	118
$CO_2C_2H_5$ CHN_2	PdCl/2, ether	$CO_2C_2H_5$	25	192
CHN_2	$CuSO_4$, cyclohexane, reflux		66	7, 11, 65

Substrate	Conditions	Product	Yield (%)	Ref.
	Cu bronze, cyclohexane, reflux	”	55	8
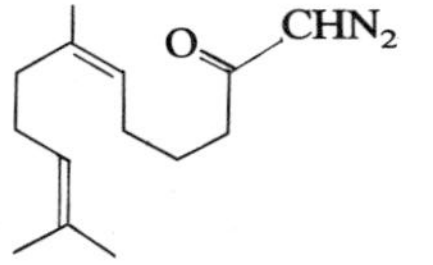 O, CHN_2	$CuSO_4$, cyclohexane, reflux	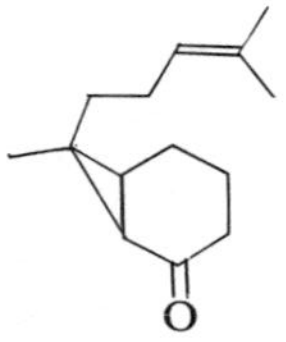 O	65	11
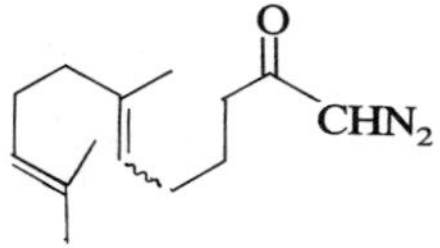 O, CHN_2	Cu powder, $CuSO_4$, cyclohexane, reflux	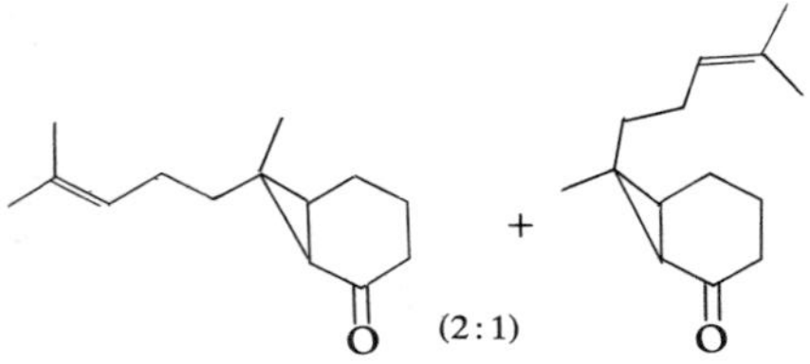 O + O (2:1)	84	10, 12
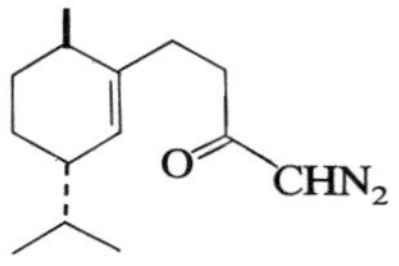 O, CHN_2	$CuSO_4$, cyclohexane, reflux	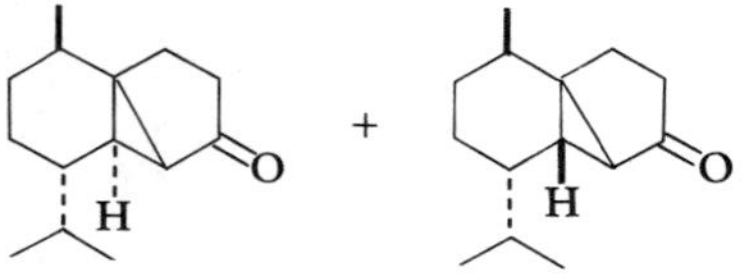 H, O + H, O (3:5)	78	199
	Cu powder, cyclohexane, reflux	” (1:1)	24	200
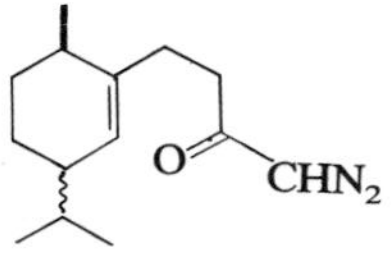 O, CHN_2	$CuSO_4$, cyclohexane, reflux	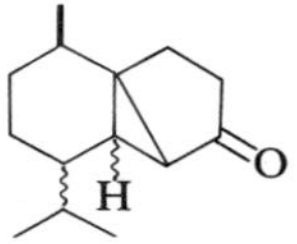 H, O	High	201
	$CuSO_4$, *n*-hexane, reflux		—	202

TABLE I. Intramolecular Addition of Diazoketones to Olefins (*Continued*)

Reactant	Conditions	Product(s)	Yield (%)	Refs.
C_{14} (*Contd.*)				
O, CHN_2	$CuSO_4$, Cu powder, cyclohexane, reflux	O	42.5	47
C_{15}				
C_6H_5, O, CHN_2	Activated CuO, cyclohexane or cyclohexane–tetrahydrofuran, (7:3), *hν*	C_6H_5, O	88	75
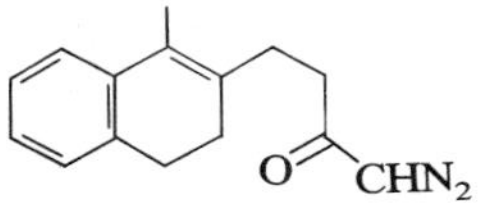 O, CHN_2	Cu bronze, cyclohexane, reflux	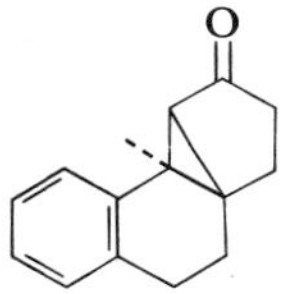 O	40	70, 74
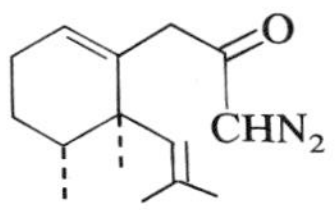 O, CHN_2	$CuSO_4$, cyclohexane, reflux	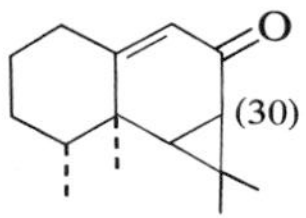 O (30) + O (15)		46
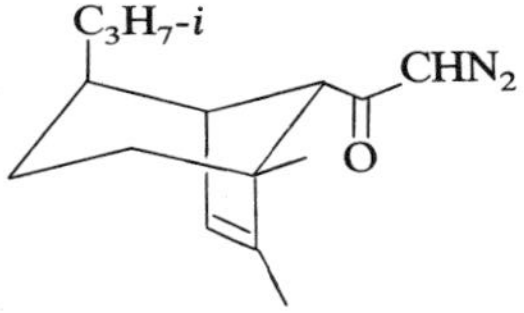 C_3H_7-*i*, CHN_2, O	$CuSO_4$, cyclohexane, reflux	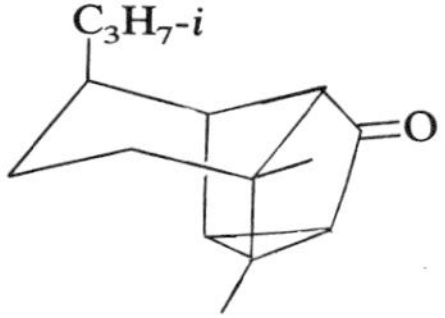 C_3H_7-*i*, O	8	203

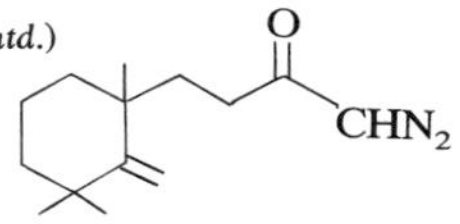

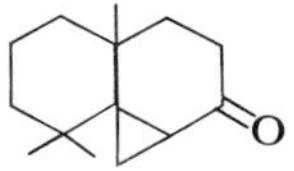

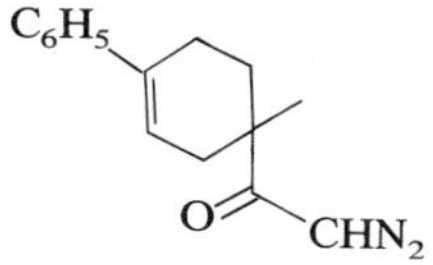

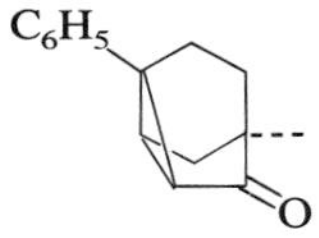

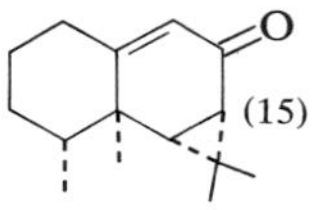

Diazo compound	Conditions	Product	Yield (%)	Refs.
N_2CH–CO– (structure)	Cu powder, tetrahydrofuran, reflux	(structure)	33	63, 64
CHN_2 (structure)	Cu powder, tetrahydrofuran	(structure) (1:1)	54	67, 68
	$CuSO_4$, cyclohexane, reflux	”	10	69
C_6H_5–, $COCHN_2$ (structure)	Activated Cu_2O, cyclohexane, $h\nu$, reflux	C_6H_5– (structure)	88	152
p-$CH_3OC_6H_4$–, $COCHN_2$ (structure)	Activated Cu_2O, cyclohexane/tetrahydrofuran (3:1), $h\nu$, reflux	p-$CH_3OC_6H_4$– (structure)	80	152
$SO_2C_6H_5$, N_2 (structure)	Cu, 80°, 10 hr, n-heptane	$SO_2C_6H_5$ (structure)	14	151a
	Cu, 174°, 10 min, n-decane	”	19	151a
	Cu, 174°, 10 min, n-decane	”	21	151a
	Cu(acetylacetonate)$_2$, 174°	”	22	151a

TABLE I. Intramolecular Addition of Diazoketones to Olefins (*Continued*)

Reactant	Conditions	Product(s)	Yield (%)	Refs.
C_{15} (*Contd.*)				
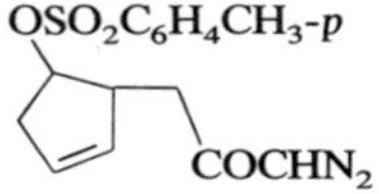 $OSO_2C_6H_4CH_3$-*p*; $COCHN_2$	Cu(acetylacetonate)$_2$	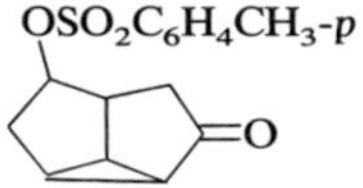 $OSO_2C_6H_4CH_3$-*p*; O	54	151b
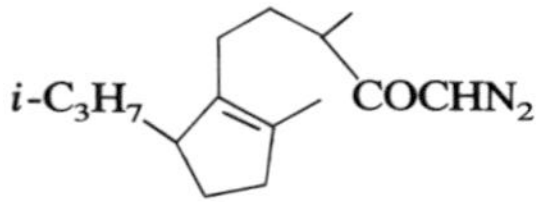 *i*-C_3H_7; $COCHN_2$	Cu powder, cyclohexane, reflux	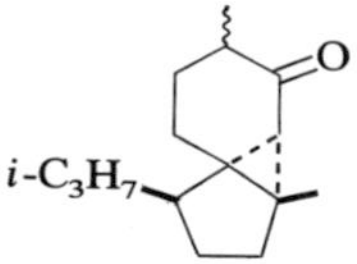 O; *i*-C_3H_7	38	153
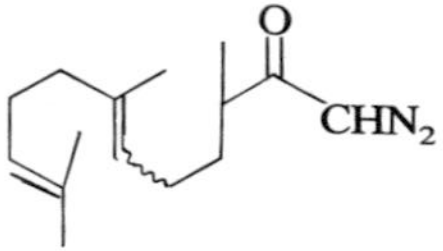 O; CHN_2	Cu powder, $CuSO_4$, cyclohexane, reflux	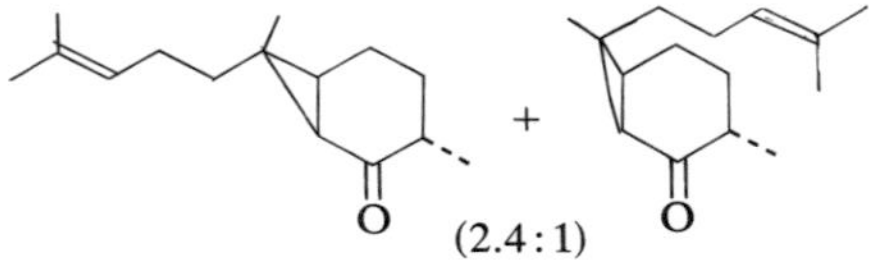 O + O (2.4 : 1)	59	12, 66
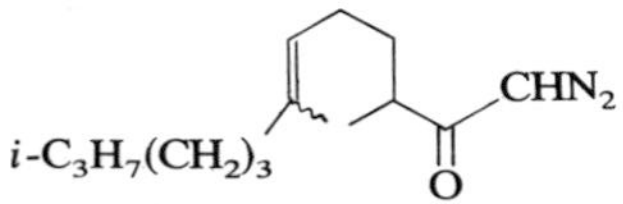 *i*-$C_3H_7(CH_2)_3$; CHN_2; O	Cu powder, $CuSO_4$, cyclohexane, reflux	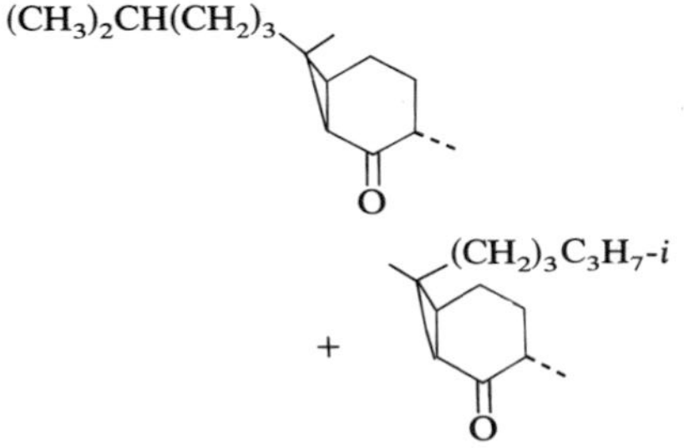 $(CH_3)_2CH(CH_2)_3$; O + $(CH_2)_3C_3H_7$-*i*; O (1 : 1)	—	176

C_{16}

Reactant	Conditions	Product	Yield (%)	Refs.
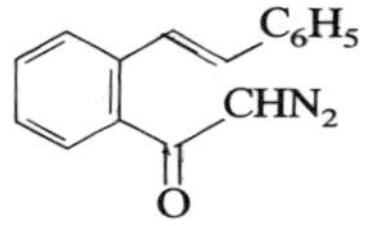 C_6H_5 CHN_2 O	CuCl, cyclohexane	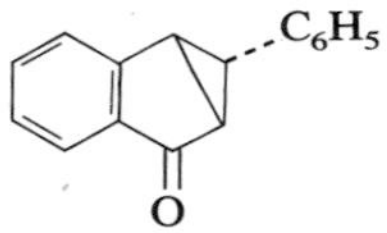 C_6H_5 O	>70	204
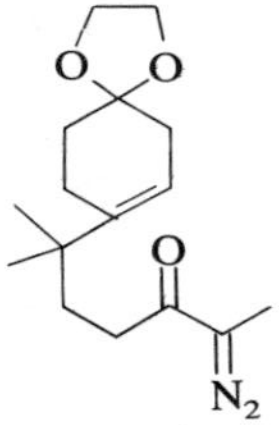 O O O N_2	Bis(N-*n*-propylsalicylidene-aminate) Cu(II), cyclohexane, 75–80°	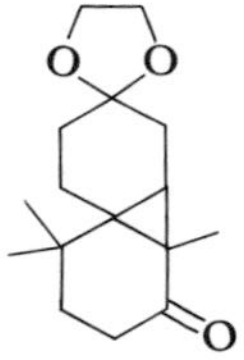 O O O	45	13
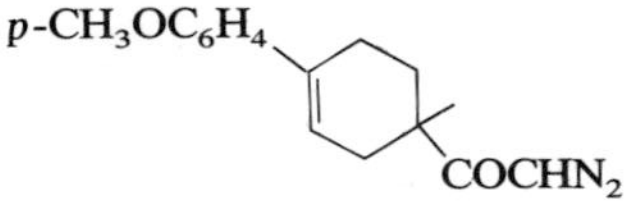 *p*-$CH_3OC_6H_4$ $COCHN_2$	Activated Cu_2O, cyclohexane tetrahydrofuran (3:1), $h\nu$, reflux	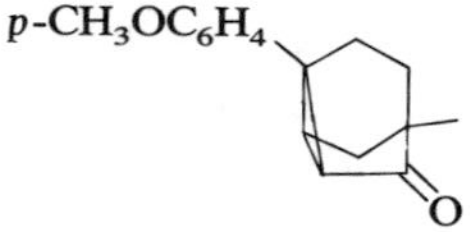 *p*-$CH_3OC_6H_4$ O	87	152
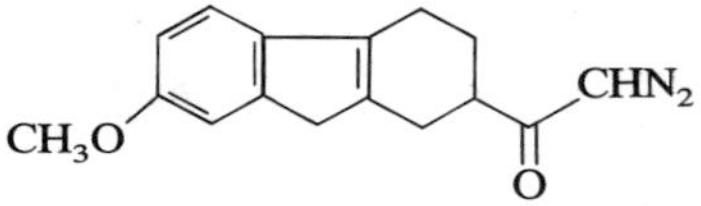 CH_3O CHN_2 O	$CuSO_4$, tetrahydrofuran, reflux	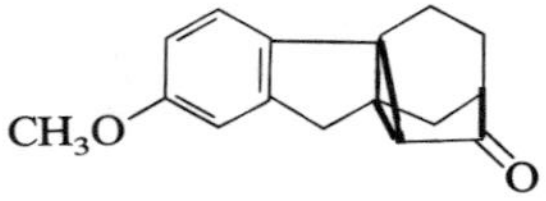 CH_3O O	27	72
	Cu bronze, cyclohexane, reflux	"	24	72
	Activated CuO, cyclohexane or cyclohexane : tetrahydrofuran (7:3), $h\nu$	"	50	75
	Cu bronze	"	—	71

TABLE I. Intramolecular Addition of Diazoketones to Olefins (*Continued*)

Reactant	Conditions	Product(s)	Yield (%)	Refs.
C_{16} (*Contd.*)				
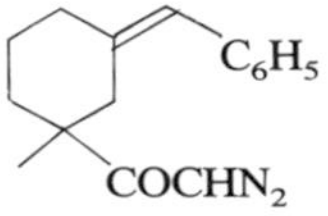 C_6H_5, $COCHN_2$	Activated Cu_2O, $h\nu$, cyclohexane : tetrahydrofuran	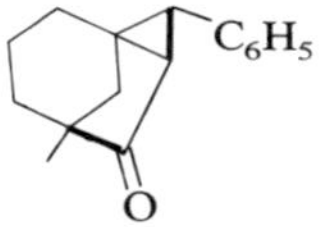 C_6H_5, O	—	154
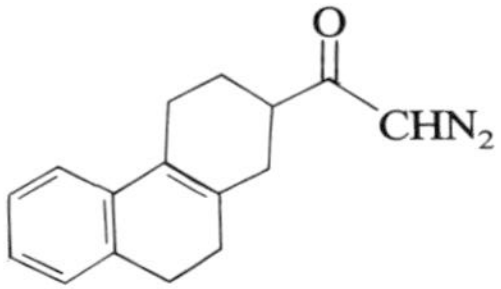 O, CHN_2	Cu bronze, cyclohexane, reflux	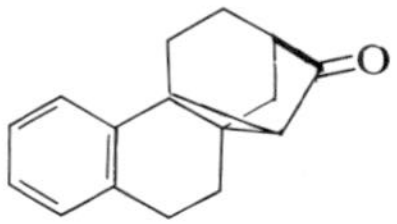 O	51	71, 72
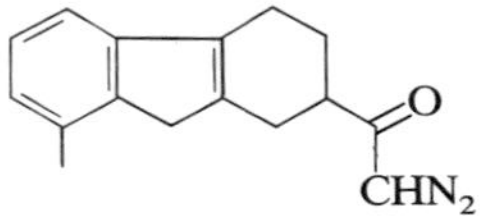 O, CHN_2	Activated CuO, cyclohexane or cyclohexane–tetrahydrofuran (7 : 3), $h\nu$	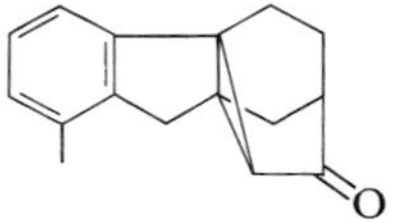 O	59	75
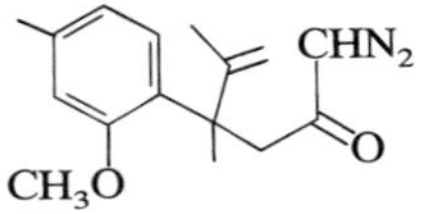 CHN_2, O, CH_3O	$CuSO_4$, n-hexane, reflux	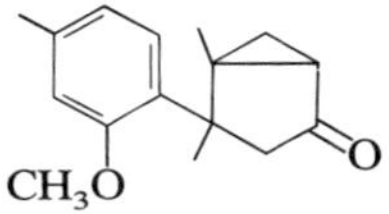 O, CH_3O	65	205
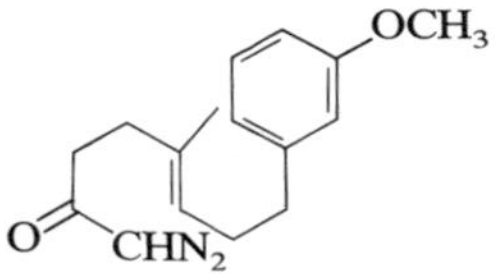 OCH_3, O, CHN_2	Cu bronze, cyclohexane, reflux	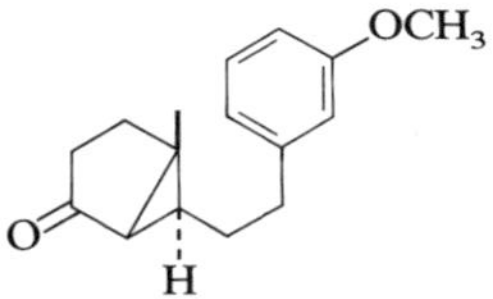 OCH_3, O, H	—	83a

		Cu bronze, cyclohexane, reflux		—	83a
		Cu, benzene, reflux		61	83b
		Cu, benzene, reflux		40	83b
C_{17}					
		Tetrahydrofuran, $h\nu$		10–15	42, 43

TABLE I. INTRAMOLECULAR ADDITION OF DIAZOKETONES TO OLEFINS (*Continued*)

Reactant	Conditions	Product(s)	Yield (%)	Refs.
C_{17} (*Contd.*)				
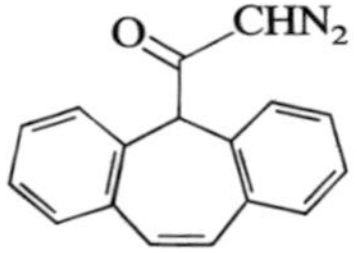 CHN_2, O	CuCl, cyclohexane, reflux or Cu powder, benzene, reflux	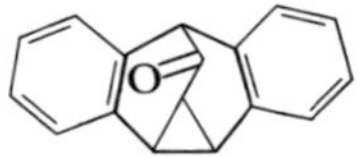 O	30	206
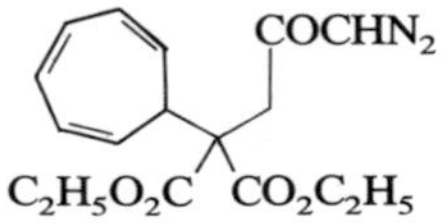 $COCHN_2$, $C_2H_5O_2C$ $CO_2C_2H_5$	$CuSO_4$, benzene, reflux	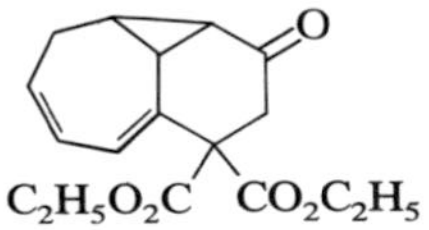 O, $C_2H_5O_2C$ $CO_2C_2H_5$	44	149
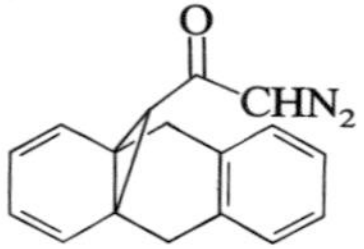 O, CHN_2	Cu	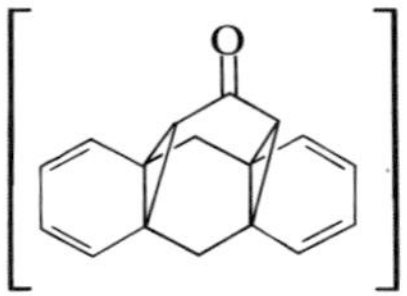 O	—	4
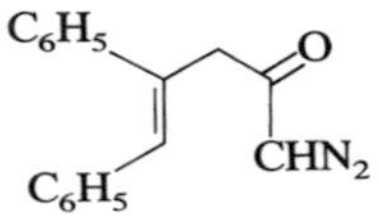 C_6H_5, O, C_6H_5, CHN_2	Cu bronze, cyclohexane, reflux	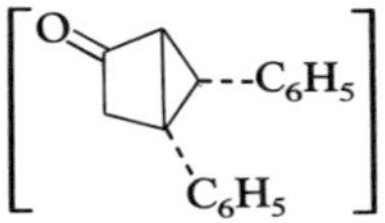 O, C_6H_5, C_6H_5	—	58
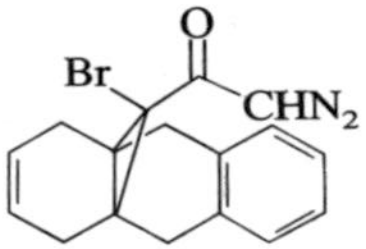 O, Br, CHN_2	Cu, cyclohexane, reflux	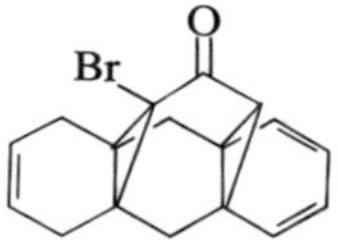 O, Br	*ca.* 50	5

	Reactant	Conditions	Product	Yield (%)	Refs.
	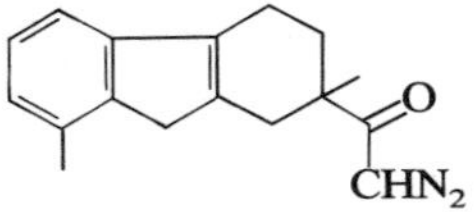	Activated CuO, cyclohexane or cyclohexane : tetrahydrofuran (7 : 3), *hν*	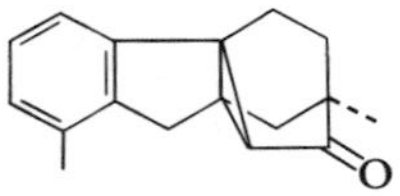	63	75
	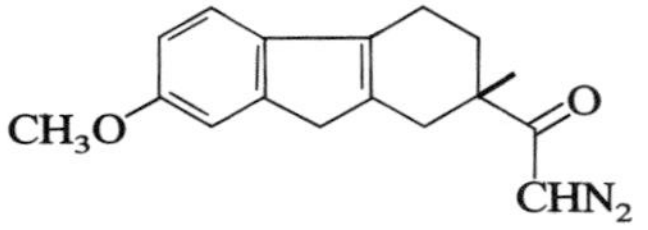	Activated CuO, cyclohexane or cyclohexane : tetrahydrofuran (7 : 3), *hν*	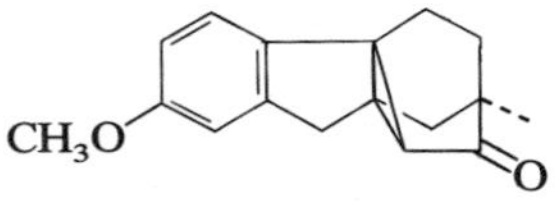	76	75
	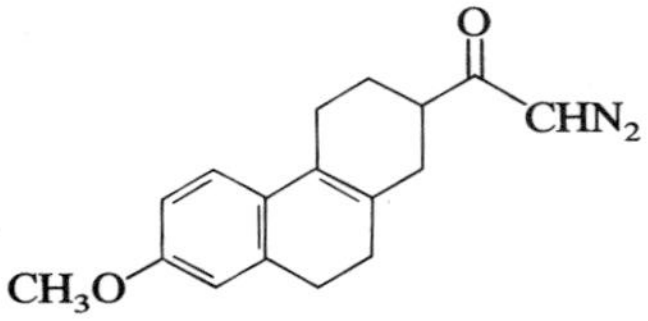	Activated CuO, cyclohexane or cyclohexane : tetrahydrofuran (7 : 3), *hν*	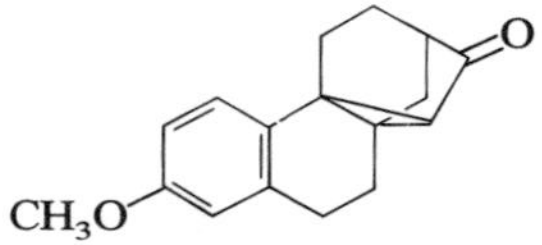	72	75
		Cu bronze, cyclohexane, reflux	”	41	72
	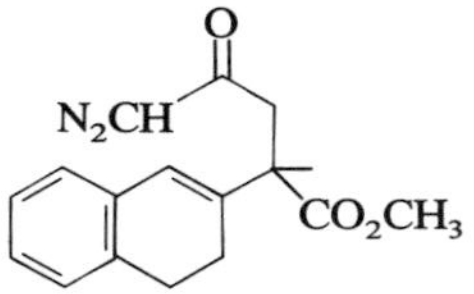	Cu bronze, cyclohexane, reflux	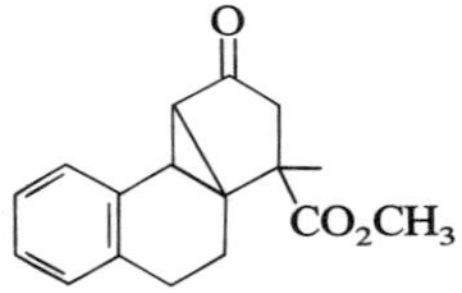	35	74
C_{18}					
		Cu		—	207

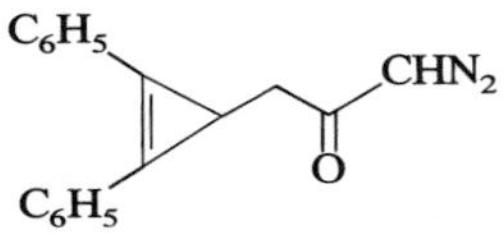

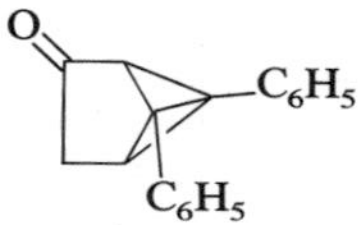

TABLE I. INTRAMOLECULAR ADDITION OF DIAZOKETONES TO OLEFINS (*Continued*)

Reactant	Conditions	Product(s)	Yield (%)	Refs.
C_{18} (*Contd.*)				
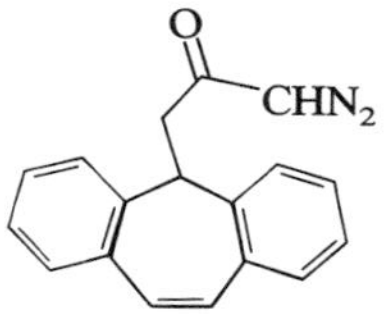 CHN$_2$		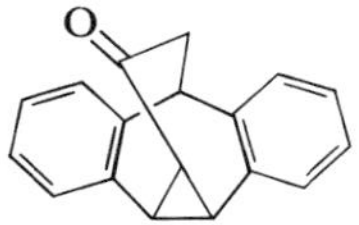	15	208
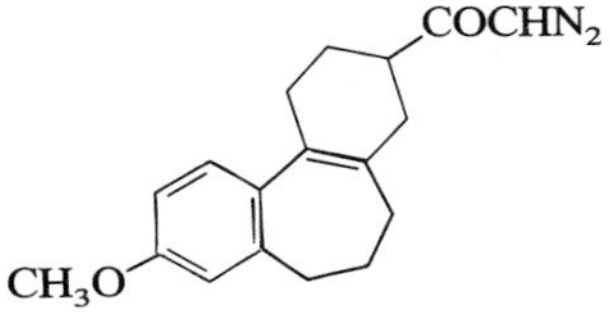 CH$_3$O, COCHN$_2$	Activated Cu_2O, cyclohexane: tetrahydrofuran, *hν*	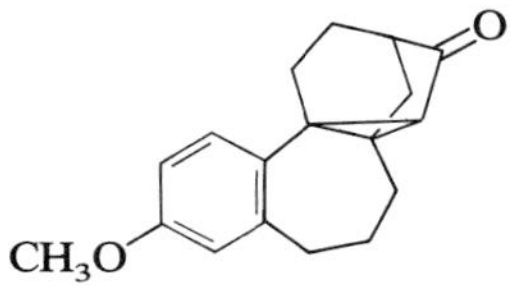 CH$_3$O	50	155
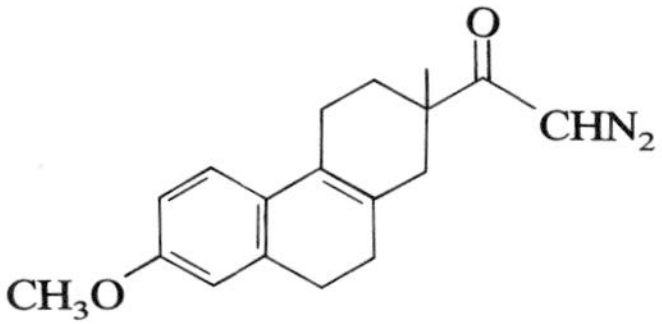 CH$_3$O, CHN$_2$	Activated CuO, cyclohexane or cyclohexane: tetrahydrofuran (7:3), *hν*	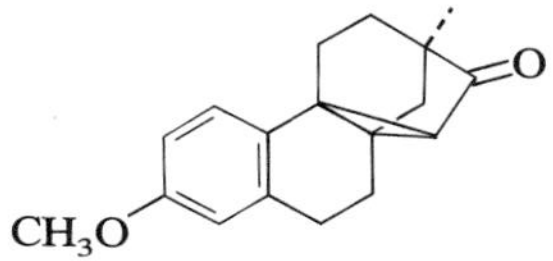 CH$_3$O	87	75
	Cu bronze, cyclohexane, reflux	"	55	72
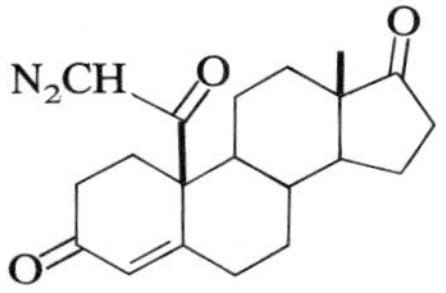 N$_2$CH	Cu powder, benzene, reflux	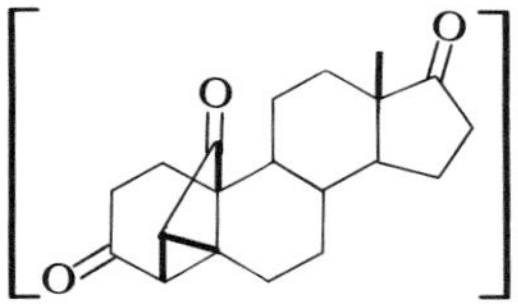	—	209

	Diazo compound	Conditions	Product	Yield (%)	Ref.
	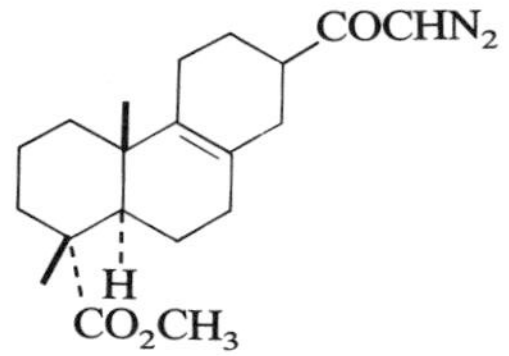	$CuSO_4$, cyclohexane, reflux	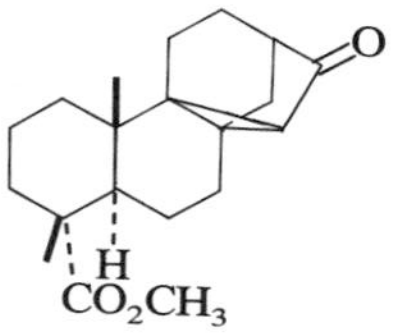	—	14
C_{21}					
		Cu powder, cyclohexane, reflux		46	156
C_{24}					
	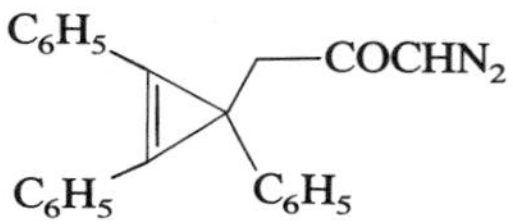	Cu powder, benzene, reflux	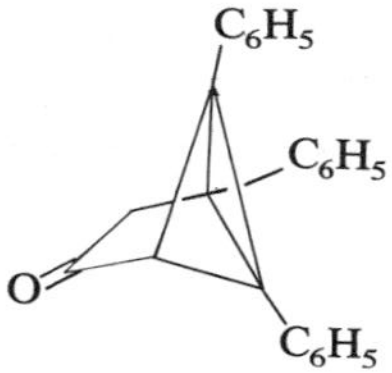	57	143, 210
	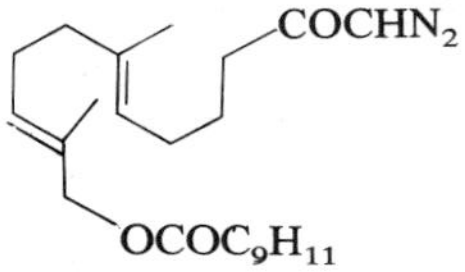	$CuSO_4$, cyclohexane, reflux	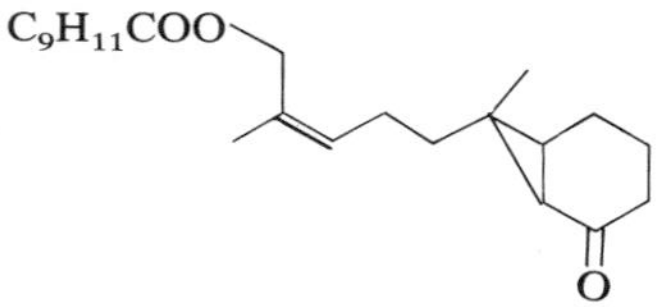	58	9

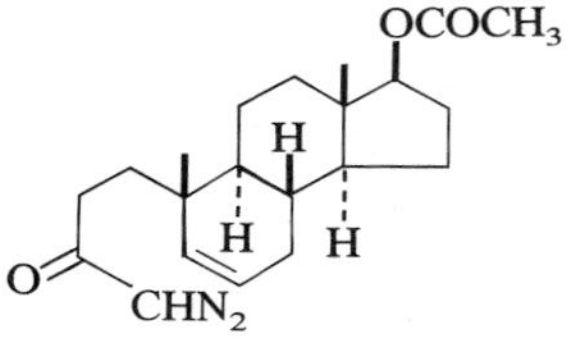

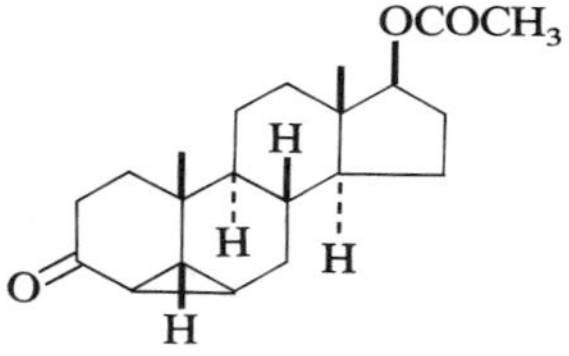

TABLE I. INTRAMOLECULAR ADDITION OF DIAZOKETONES TO OLEFINS (*Continued*)

Reactant	Conditions	Product(s)	Yield (%)	Refs.
C_{25}				
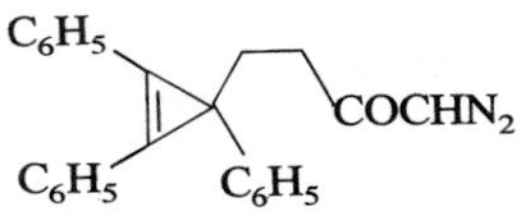 C_6H_5, C_6H_5, C_6H_5, $COCHN_2$	Cu powder, reflux	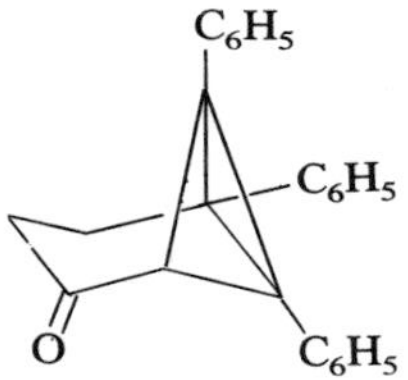 C_6H_5, C_6H_5, C_6H_5, O	42	157
C_{27}				
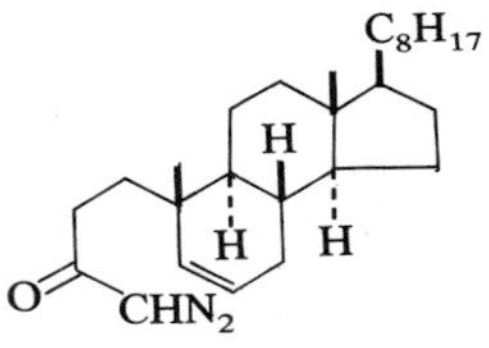 C_8H_{17}, H, H, H, O, CHN_2	Cu powder, cyclohexane, reflux	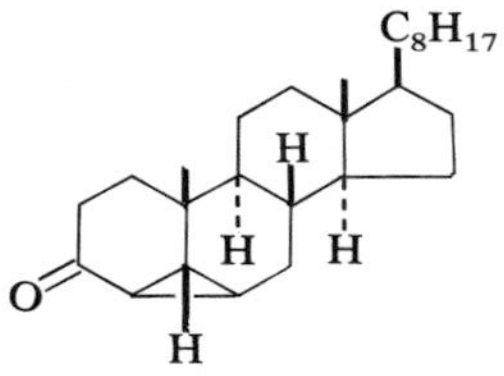 C_8H_{17}, H, H, H, H, O	37	156

Note: References 145–230 are on pp. 474–475.

TABLE II. INTRAMOLECULAR ADDITION OF DIAZOESTERS TO OLEFINS

Reactant	Conditions	Product(s)	Yield (%)	Refs.
C_5 O CHN$_2$ O	$CuSO_4$, cyclohexane, reflux	O O	37	35
	$CuCl_2$, cyclohexane, reflux	,,	34	35
	CuCl, cyclohexane, reflux	,,	28	35
	Cu powder, cyclohexane, reflux	,,	38	35
	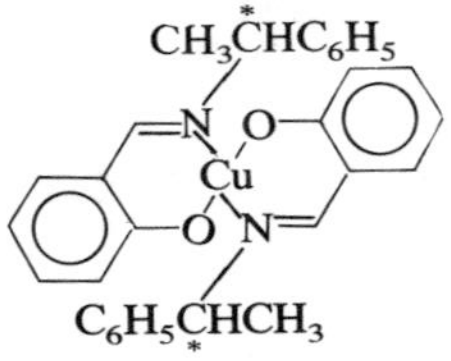 $CH_3\overset{*}{C}HC_6H_5$ N O Cu O N $C_6H_5\underset{*}{C}HCH_3$	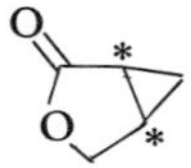 O O * *	42	29
	Cu powder, *n*-octane, reflux	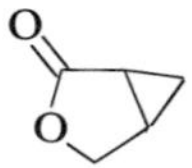 O O	50	89, 90
	Cu dust, benzene, reflux	,,	40–70	211
C_6 O CHN$_2$ O	Cu dust, benzene, reflux	O O	30	211
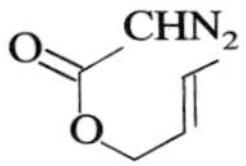 O CHN$_2$ O	Cu_2O, cyclohexane, reflux	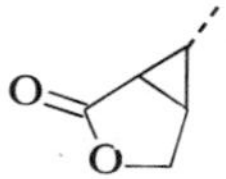 O O	39	91

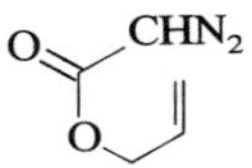

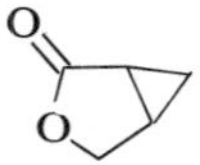

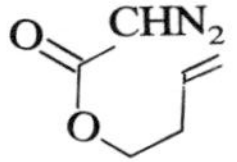

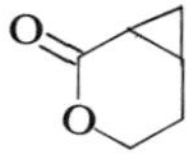

TABLE II. INTRAMOLECULAR ADDITION OF DIAZOESTERS TO OLEFINS (*Continued*)

Reactant	Conditions	Product(s)	Yield (%)	Refs.
C_7				
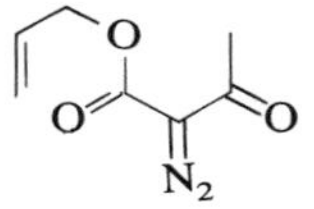 O, O, O, N_2	Cu powder, *n*-octane, reflux	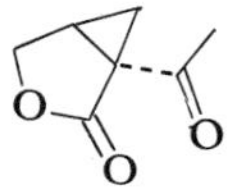 O, O, O	30	89, 90
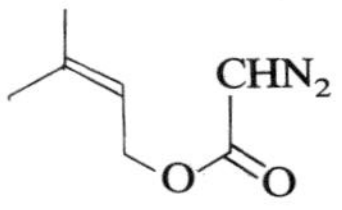 CHN_2, O, O	Cu or CuBr	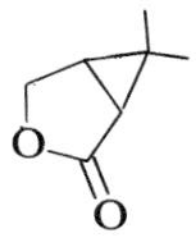 O, O	50	89, 90
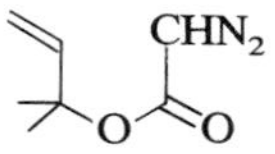 CHN_2, O, O	Cu powder, *n*-octane, reflux	 O, O	50	89, 90
C_8				
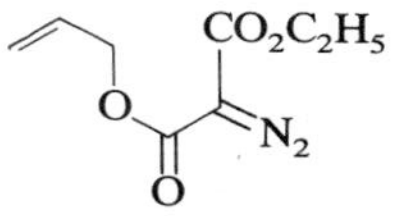 $CO_2C_2H_5$, O, N_2, O	Cu powder, *n*-octane, reflux	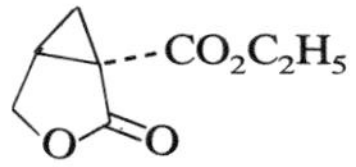 $CO_2C_2H_5$, O, O	55	89
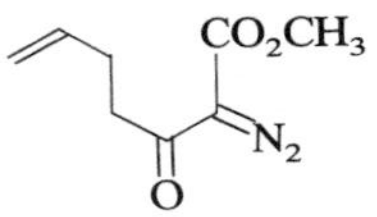 CO_2CH_3, N_2, O	$CuSO_4$, benzene, reflux	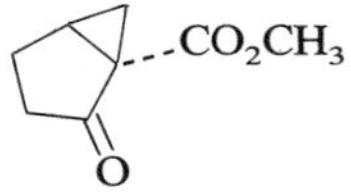 CO_2CH_3, O	69	145

Reactant	Conditions	Product	Yield (%)	Ref.
C_9				
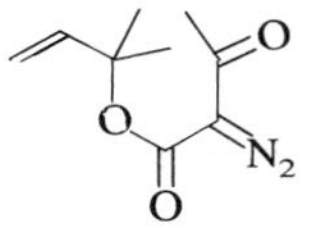 (structure: N_2, O, O, O)	Cu powder, *n*-octane, reflux	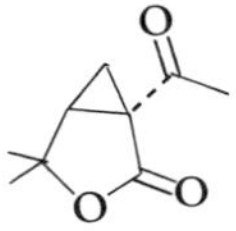 (structure: O, O, O)	30	89, 90
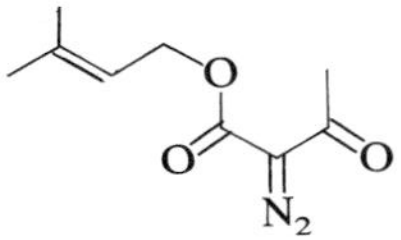 (structure: O, O, O, N_2)	Cu powder, *n*-octane, reflux	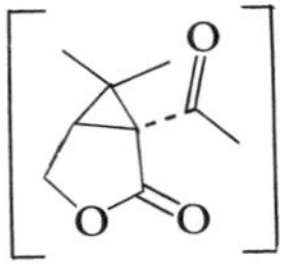 [(structure: O, O, O)]	—	89
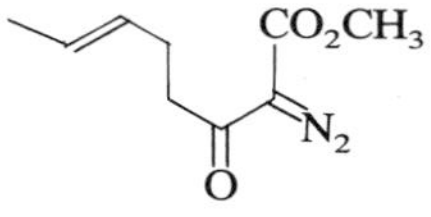 (structure: CO_2CH_3, N_2, O)	$CuSO_4$, benzene, reflux	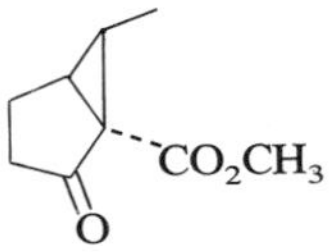 (structure: CO_2CH_3, O)	58	150
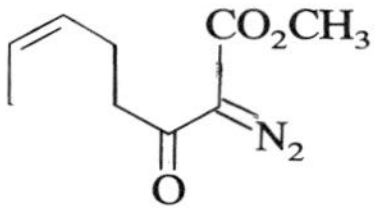 (structure: CO_2CH_3, N_2, O)	$CuSO_4$, benzene, reflux	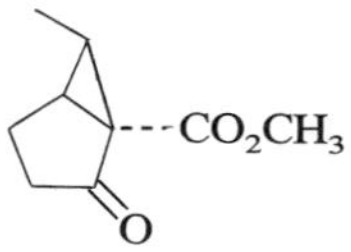 (structure: CO_2CH_3, O)	60	150
C_{10}				
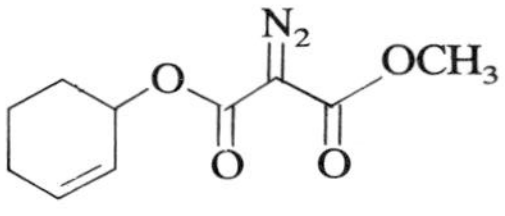 (structure: N_2, OCH_3, O, O, O)	$(CH_3O)_3P{\cdot}CuI$, benzene, reflux	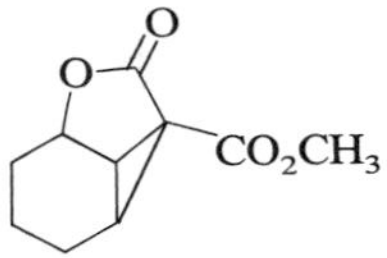 (structure: O, O, CO_2CH_3)	65	95, 96

TABLE II. INTRAMOLECULAR ADDITION OF DIAZOESTERS TO OLEFINS (*Continued*)

Reactant	Conditions	Product(s)	Yield (%)	Refs.
C_{10} (*Contd.*)				
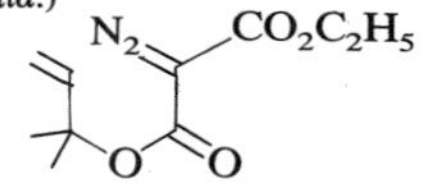 N_2, $CO_2C_2H_5$, O, O	Cu powder, *n*-octane, reflux	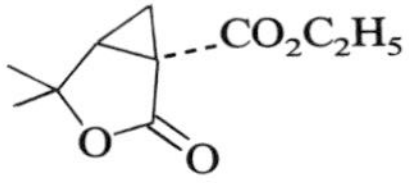 $CO_2C_2H_5$, O, O	50	89
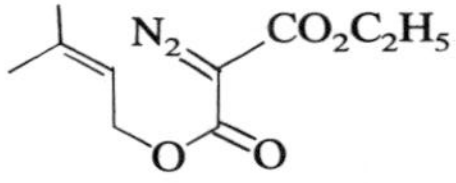 N_2, $CO_2C_2H_5$, O, O	Cu powder, *n*-octane, reflux	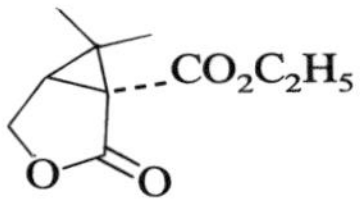 $CO_2C_2H_5$, O, O	50	89
C_{11}				
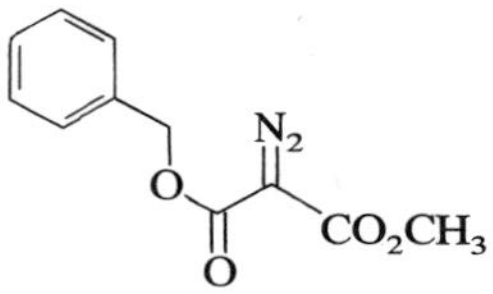 N_2, O, O, CO_2CH_3	Cu powder, *n*-octane, reflux, or benzene, 130°	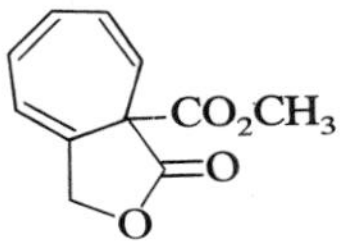 CO_2CH_3, O, O	20	92, 93
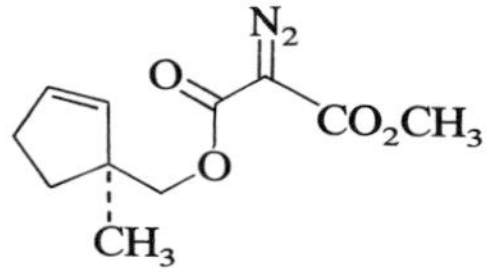 N_2, O, O, CO_2CH_3, CH_3	Cu, *n*-octane, reflux	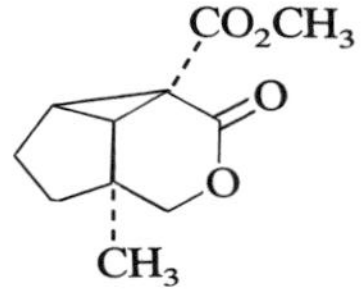 CO_2CH_3, O, O, CH_3	67	212
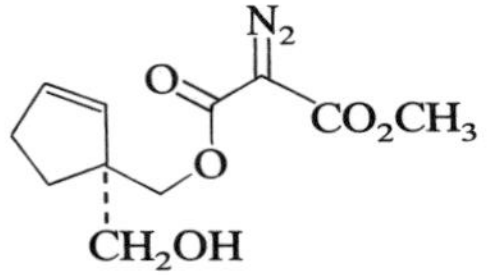 N_2, O, O, CO_2CH_3, CH_2OH	Cu, *n*-octane, reflux	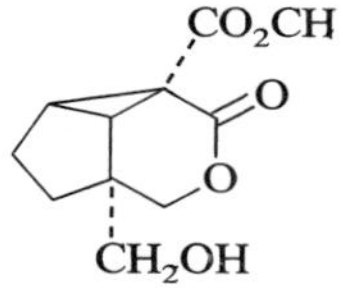 CO_2CH, O, O, CH_2OH	52	212

Reactant	Conditions	Product	Yield (%)	Refs.
CO_2CH_3, N_2, O	$CuSO_4$, benzene, reflux	CO_2CH_3, O	64	150
N_2CH, O	Cu_2O, cyclohexane, reflux	O	21	91
C_{12} N_2, CO_2CH_3, O	Cu powder, *n*-octane, reflux	CO_2CH_3, O	19	92, 93a
$(CH_3)_2C{=}CH(CH_2)_2$, CH_3, $OCOCHN_2$	Cu(>10 eq), cyclohexane, reflux	$(CH_3)_2C{=}CH(CH_2)_2$, CH_3, H, H, O	27	148
	Cu_2O(>10 eq),	"	33	148
	$Cu(acac)_2$ (>10 eq), toluene, reflux	"	22	148
	$Cu(acac)_2$ (<0.1 eq), toluene, reflux	"	76	148

TABLE II. Intramolecular Addition of Diazoesters to Olefins (*Continued*)

Reactant	Conditions	Product(s)	Yield (%)	Refs.
C_{12} (*Contd.*)				
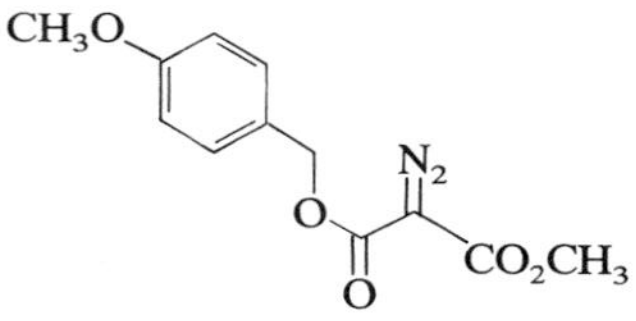 CH_3O, N_2, CO_2CH_3	Cu powder, *n*-octane, reflux	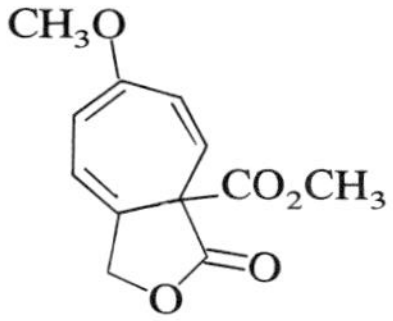 CH_3O, CO_2CH_3	60	92, 93a
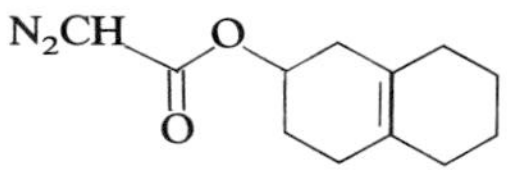 N_2CH	Cu_2O, cyclohexane, reflux	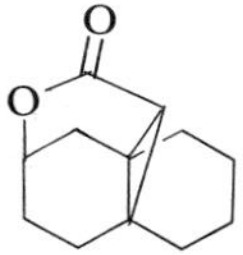	<23	91, 211
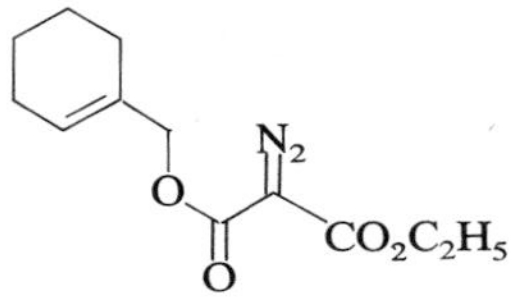 N_2, $CO_2C_2H_5$	$(CH_3O)_3P{\cdot}CuI$, toluene, reflux	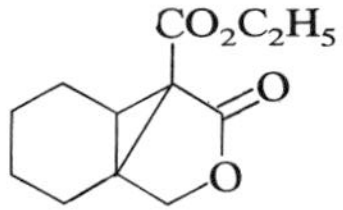 $CO_2C_2H_5$	72	97
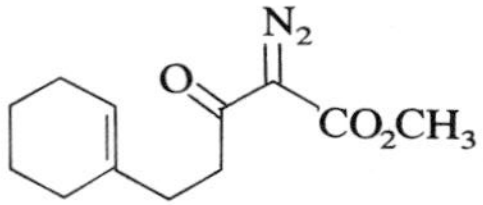 N_2, CO_2CH_3	$(CH_3O)_3P{\cdot}CuI$, toluene, reflux	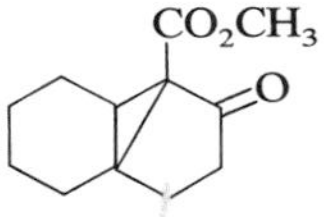 CO_2CH_3	70	97

	Reactant	Conditions	Product	Yield (%)	Refs.
C_{13}	C_6H_5, N_2	Cu, Cu_2O, $CuSO_4$, toluene, reflux	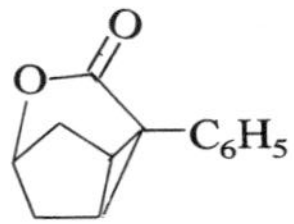 C_6H_5	—	213
	N_2, CO_2CH_3	Cu powder, octane, reflux	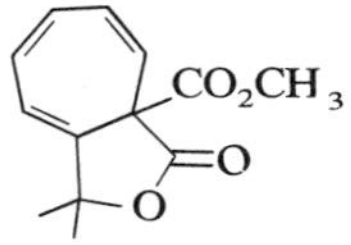 CO_2CH_3	45	92, 93a
	OCH_3, CH_3O_2C, N_2	Cu(acetylacetonate)$_2$, toluene, reflux	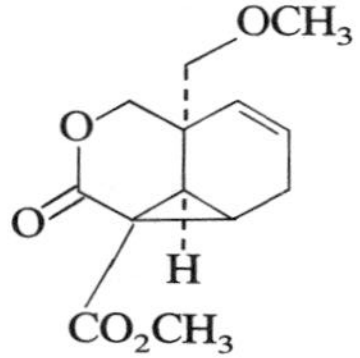 OCH_3, H, CO_2CH_3	71	93b
C_{14}	CO_2CH_3, N_2, $C_5H_9O_2$	Cu powder, xylene, reflux	CO_2CH_3, $C_5H_9O_2$	50	95

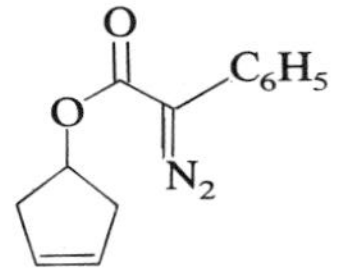

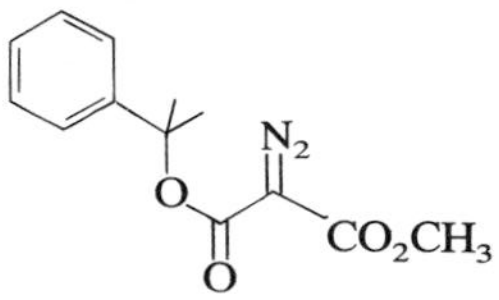

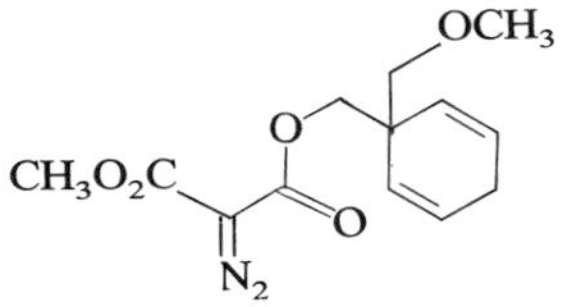

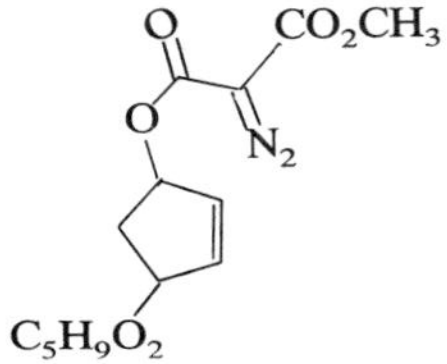

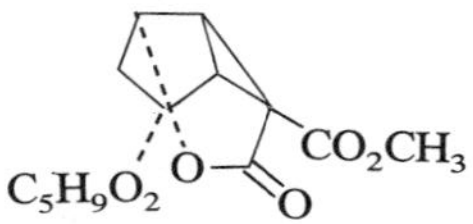

TABLE II. INTRAMOLECULAR ADDITION OF DIAZOESTERS TO OLEFINS (*Continued*)

Reactant	Conditions	Product(s)	Yield (%)	Refs.
C_{14}				
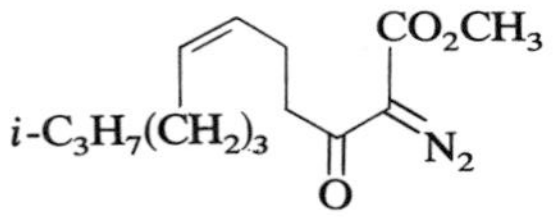 i-$C_3H_7(CH_2)_3$, CO_2CH_3, N_2, O	Cu bronze, toluene, reflux	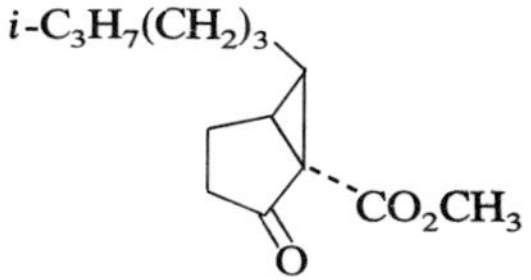 i-$C_3H_7(CH_2)_3$, CO_2CH_3, O	73–80	98
C_{15} 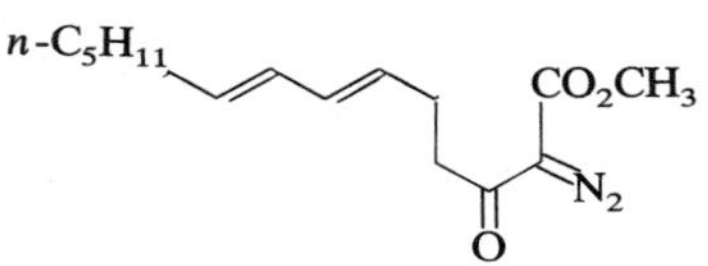 n-C_5H_{11}, CO_2CH_3, N_2, O	Cu(acetylacetonate)$_2$, benzene, reflux	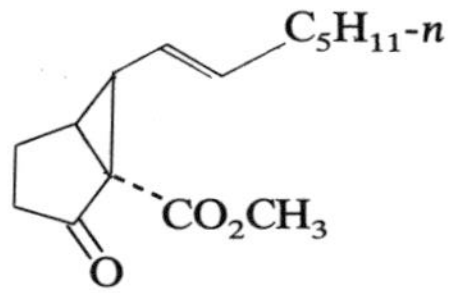 C_5H_{11}-n, CO_2CH_3, O	63	100
C_{16} 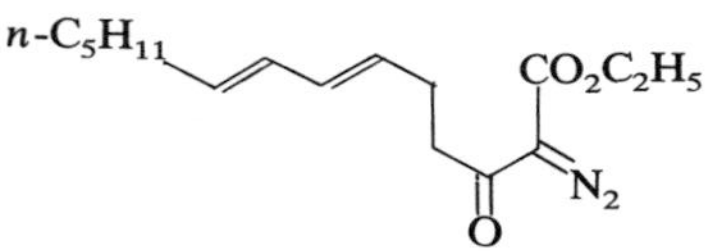 n-C_5H_{11}, $CO_2C_2H_5$, N_2, O	Cu bronze, toluene, reflux	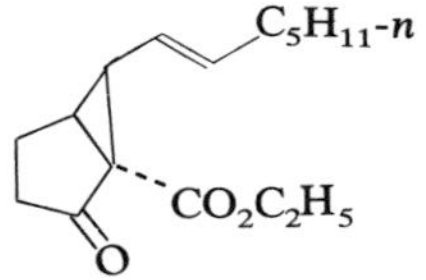 C_5H_{11}-n, $CO_2C_2H_5$, O	50	99
C_{17}				
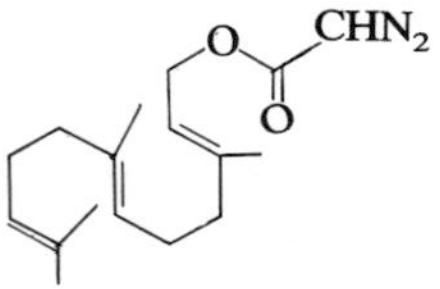 O, CHN_2, O	Cu (>10 eq) or Cu(acac)$_2$(>10 eq), toluene, reflux	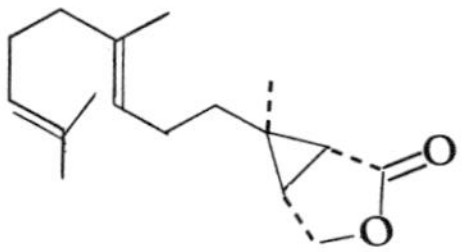 O, O	20	94, 148
	Cu(acac)$_2$(<0.1 eq), toluene, reflux	"	71	148

	Reactant	Conditions	Product(s)	Yield (%)	Ref.
C_{18}					
	CO_2CH_3, N_2, O, $NC_8H_4O_2$	Cu bronze, toluene, reflux	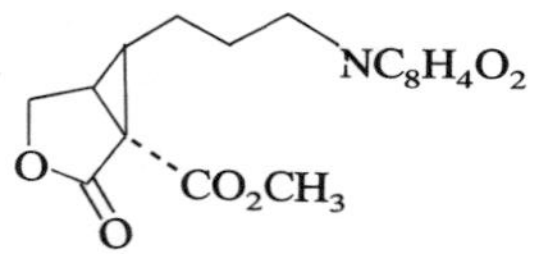 $NC_8H_4O_2$, O, CO_2CH_3	65	214a
	$C_8H_4O_2N$, CO_2CH_3, N_2, O	Cu bronze, toluene, reflux	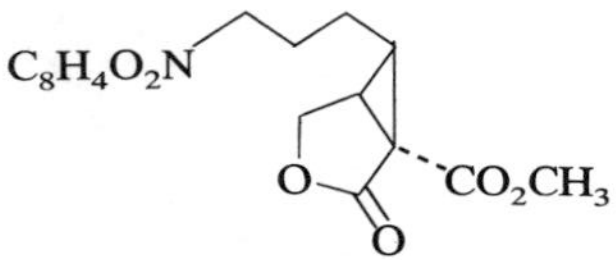 $C_8H_4O_2N$, O, CO_2CH_3	81	214a
C_{26}					
	N_2, O, CO_2CH_3, OCH_3, CH_3O, CH_3, $NC_8H_4O_2$, OCH_3	Cu bronze, toluene, reflux	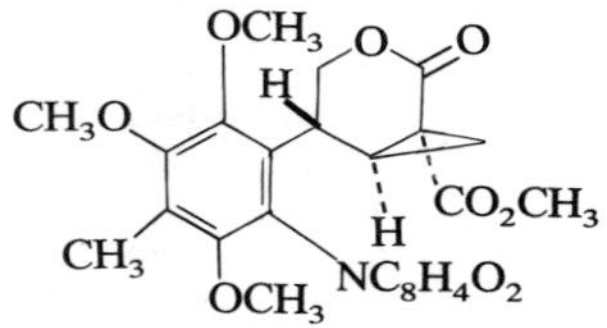 OCH_3, O, H, CH_3O, CO_2CH_3, H, CH_3, OCH_3, $NC_8H_4O_2$ (25) 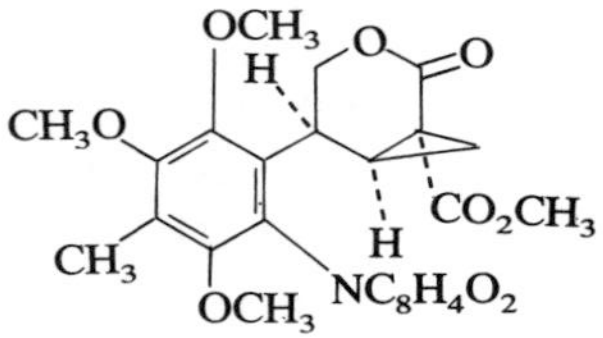 + OCH_3, O, H, CH_3O, CO_2CH_3, H, CH_3, OCH_3, $NC_8H_4O_2$ (5)		214b

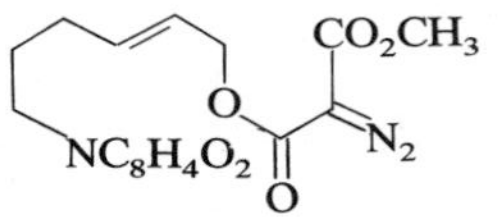

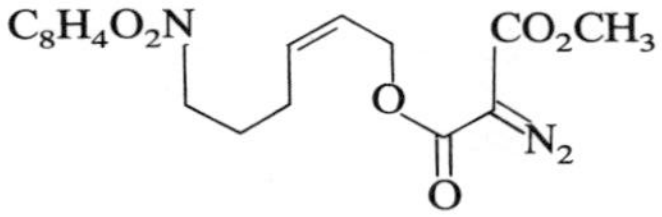

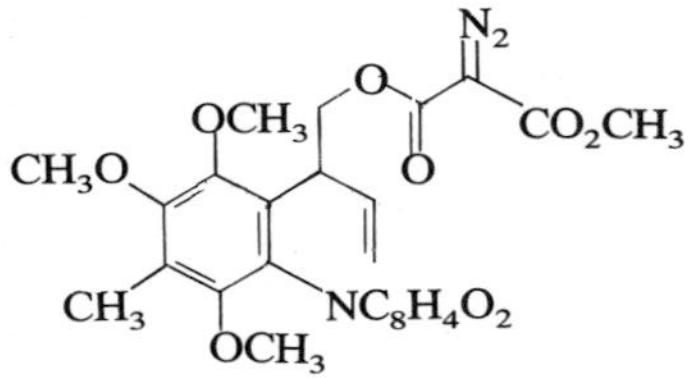

Note: References 145–230 are on pp. 474–475.

TABLE III. Intramolecular Insertion of Diazocarbonyl Compounds into C–H Bonds

Reactant	Conditions	Product(s)	Yield (%)	Refs.
C_6				
$(C_2H_5)_2NCOCHN_2$	Dioxane, $h\nu$	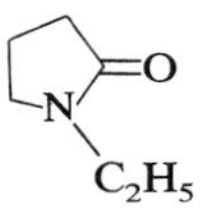 C_2H_5 (43) 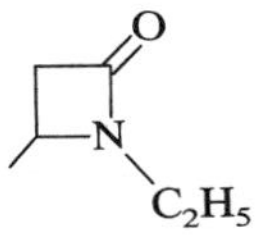 + C_2H_5 (57)		35, 36
	Methanol, $h\nu$	" (5) + " (43)		35, 36
	Cyclopentane, $h\nu$	" (47) + " (53)		36
	Ethyl acetate, $h\nu$	" (37) + " (63)		36
	Methylene chloride, $h\nu$	" (38) + " (62)		36
	Acetone, $h\nu$	" (40) + " (60)		36
	Acetonitrile, $h\nu$	" (31) + " (69)		36
	Cyclopentane–methanol (9:1), $h\nu$	" (19) + " (81)		36
	Dioxane, LiBr (1%), $h\nu$	" (34) + " (66)		36
	Acetone, LiBr (3.5%), $h\nu$	" (13) + " (87)		36
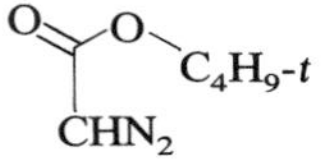 C_4H_9-t, CHN_2	Cyclohexane, $h\nu$	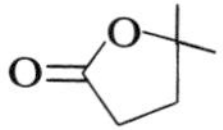	(*ca.* 4)	37
C_7				
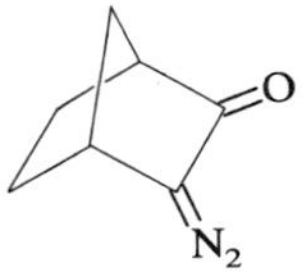 N_2	Cu powder, benzene, reflux	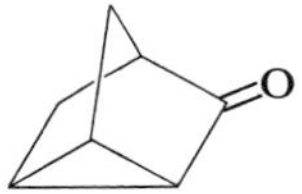 (60)		85
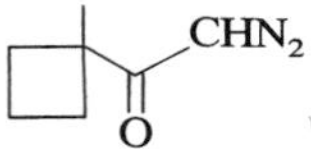 CHN_2	$CuSO_4$, cyclohexane, reflux	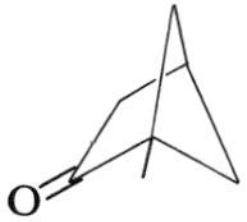	—	105

$N_2{=}C(CO_2C_2H_5)_2$	$C_6H_5C(=S)C_6H_5$, cyclohexane, $h\nu$	[O, O, $CO_2C_2H_5$]	—	215
N, CHN_2, O	CH_2Cl_2, $h\nu$	O, N	—	50
O, O, CHN_2	Cyclohexane, $h\nu$	O, O (<2) + O, O (<2)		38
C_8				
O, N_2	Cu powder, benzene, reflux	O	5	102
O, N, CHN_2	Dioxane, $h\nu$	O, N	23	37
O, N_2	Petroleum ether (60–90°), CuO, heat	O	31	108

TABLE III. INTRAMOLECULAR INSERTION OF DIAZOCARBONYL COMPOUNDS INTO C–H BONDS (*Continued*)

Reactant	Conditions	Product(s)	Yield (%)	Refs.
C_8 (*Contd.*)				
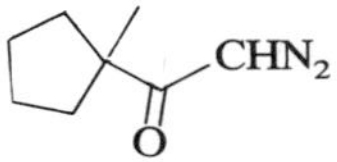 1-methylcyclopentyl–$COCHN_2$	$CuSO_4$, cyclohexane, reflux	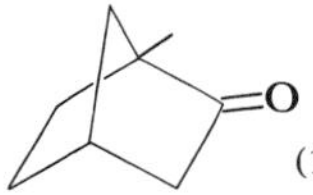 (19) 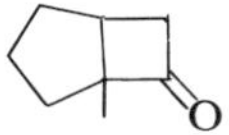 + (3)		105
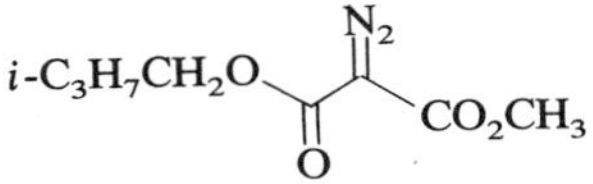 i-$C_3H_7CH_2OCOC(N_2)CO_2CH_3$	Cu powder or Cu salts, benzene, reflux	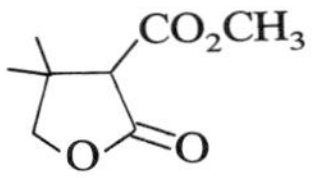 lactone with CO_2CH_3	14–30	39
	Cu powder or Cu salts, chlorobenzene, reflux	"	62	39
	Freon, tetrahydrofuran, $h\nu$	"	Trace	114
	CCl_4, $h\nu$	"	Trace	114
	Benzene, 140°	"	4	114
	Cu, n-octane, 120°	"	Trace	114
	CuCN, benzene, 140°	"	14	114
	Cu, benzene, 140°	"	30	114
	Cu, chlorobenzene, 131°	"	62	114
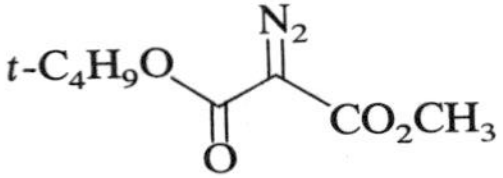 t-$C_4H_9OCOC(N_2)CO_2CH_3$	$CuSO_4$, chlorobenzene, reflux	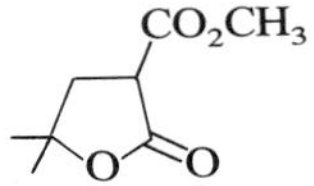 lactone with CO_2CH_3	11	39
	Cu powder, chlorobenzene, reflux	"	11	114
	CCl_4, $h\nu$	"	10	114
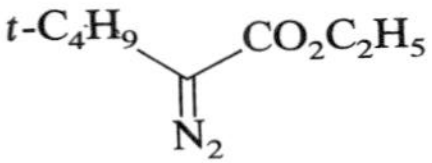 t-$C_4H_9C(N_2)CO_2C_2H_5$	Cyclopentane, $h\nu$	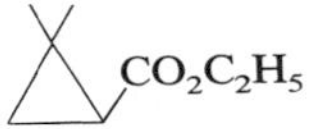 cyclopropane–$CO_2C_2H_5$	35	132a

C_9

Reactant	Conditions	Product(s)	Yield (%)	Refs.
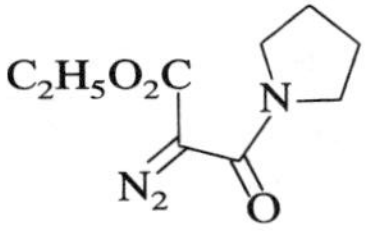 $C_2H_5O_2C$, N_2, N, O	CCl_4, $h\nu$	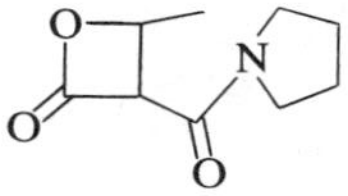 O, O, O, N	—	51
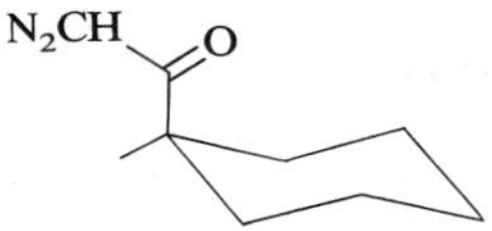 N_2CH, O	Silver benzoate, $(C_2H_5)_3N$, CH_3OH, heat	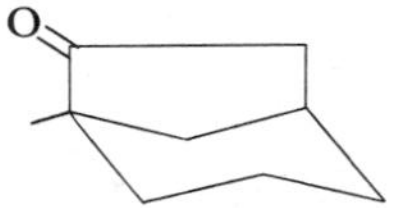 O	8	216
	$CuSO_4$, cyclohexane	”	48	216
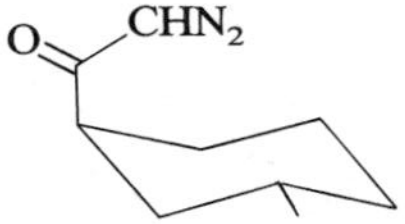 O, CHN_2	$CuSO_4$, cyclohexane, reflux	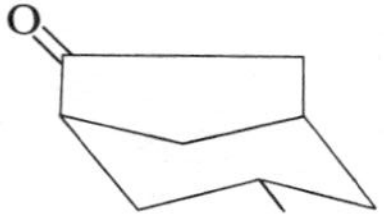 O (2) 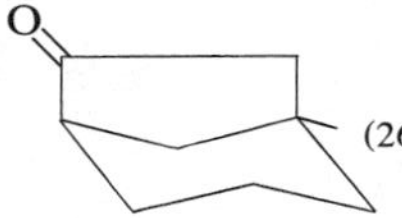 + O (26)		216
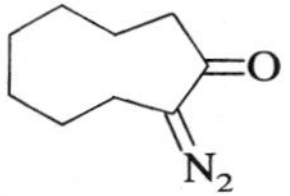 O, N_2	CuO, petroleum ether (60–90°), heat	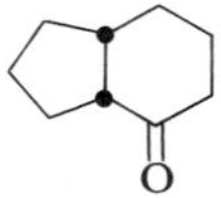 O	52	108
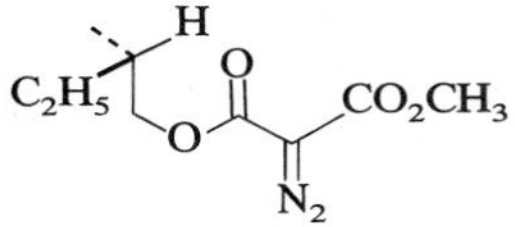 H, C_2H_5, O, O, CO_2CH_3, N_2	$CuSO_4$, chlorobenzene, reflux	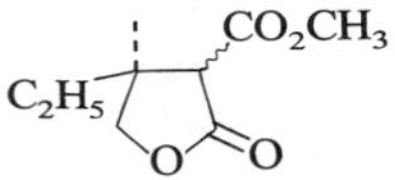 C_2H_5, CO_2CH_3, O, O	47.5[a]	39
	Cu or $CuSO_4$, chlorobenzene, reflux	”	32	114

TABLE III. INTRAMOLECULAR INSERTION OF DIAZOCARBONYL COMPOUNDS INTO C–H BONDS (*Continued*)

Reactant	Conditions	Product(s)	Yield (%)	Refs.
C_9 (*Contd.*)				
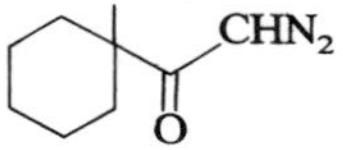 CHN_2, O	Ag_2O, methanol	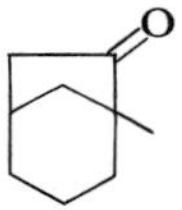 O	6	105
	$CuSO_4$, cyclohexane, reflux	”	62	105
C_{10}				
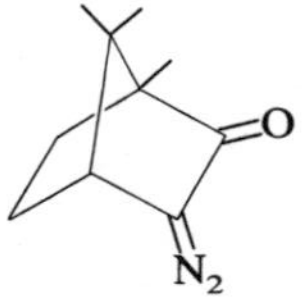 O, N_2	Cu, ethanol	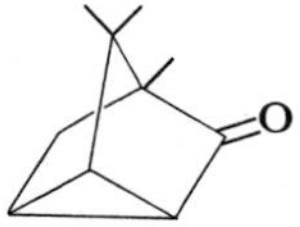 O	91	101
	Cu, cyclohexene	”	94	101
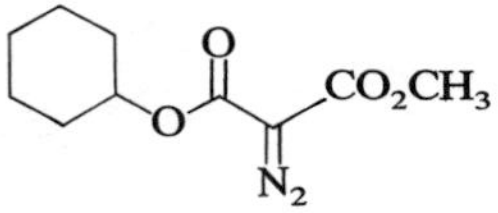 O, CO_2CH_3, O, N_2	$CuSO_4$, chlorobenzene, reflux	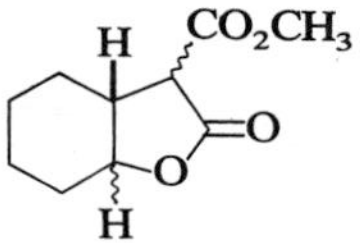 H, CO_2CH_3, O, O, H (5) 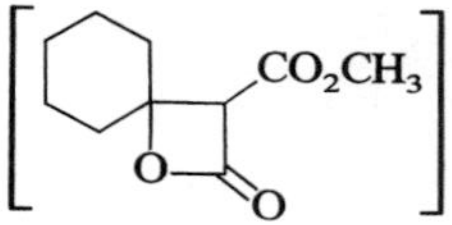 + [CO_2CH_3, O, O]	(*ca.* 30)	39
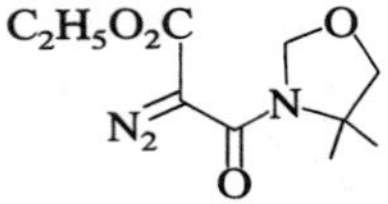 $C_2H_5O_2C$, O, N_2, N, O	CCl_4, $h\nu$	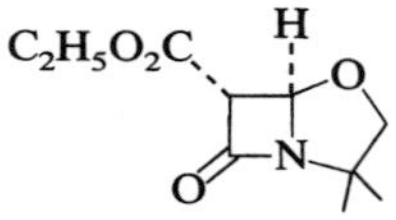 $C_2H_5O_2C$, H, O, O, N (*ca.* 55)		55

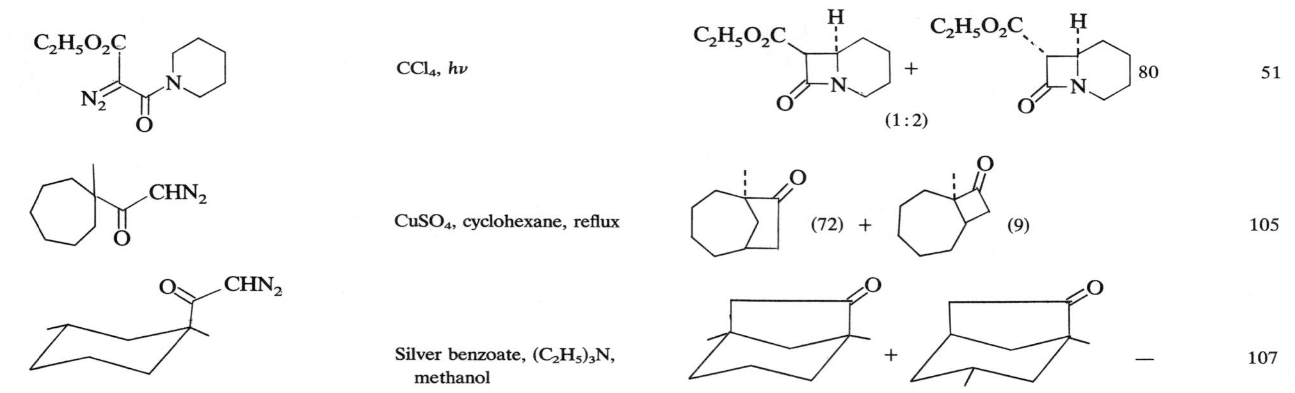

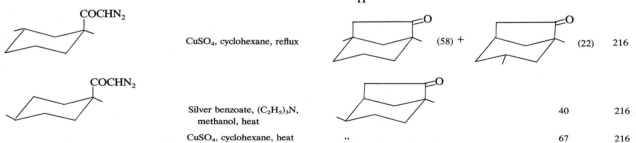

Reactant	Conditions	Product(s)	Yield (%)	Ref.
	CCl_4, *hν*	$C_2H_5O_2C$, H, O, N + $C_2H_5O_2C$, H, O, N (1:2)	80	51
	$CuSO_4$, cyclohexane, reflux	O (72) + O (9)		105
	Silver benzoate, $(C_2H_5)_3N$, methanol	O + O (2:1)	—	107
	CuO, petroleum ether (60–90°), heat	H, O, H	13	108
	$CuSO_4$, cyclohexane, reflux	O (58) + O (22)		216
	Silver benzoate, $(C_2H_5)_3N$, methanol, heat	O	40	216
	$CuSO_4$, cyclohexane, heat	,,	67	216

TABLE III. Intramolecular Insertion of Diazocarbonyl Compounds into C–H Bonds (*Continued*)

Reactant	Conditions	Product(s)	Yield (%)	Refs.
C_{11}				
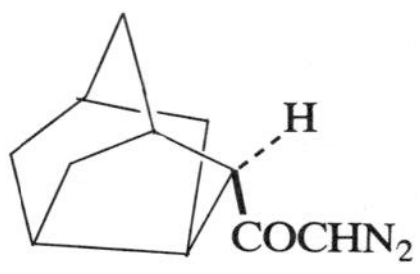 H, $COCHN_2$	Cu, toluene, reflux	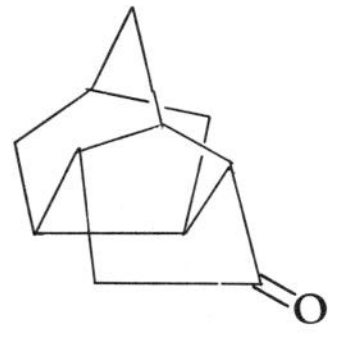 O	51	217a
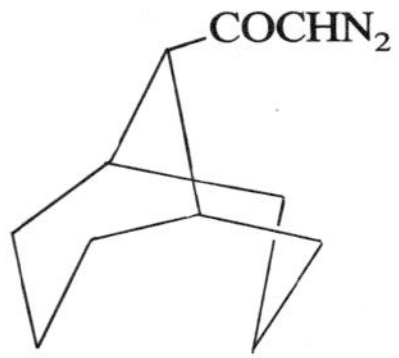 $COCHN_2$	$CuSO_4$, toluene, reflux	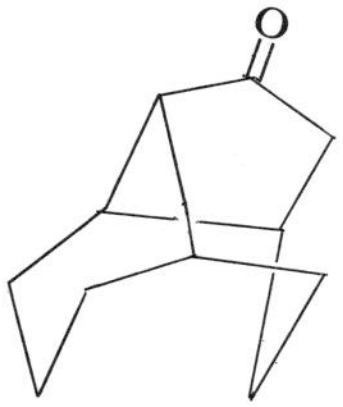 O	74	217b
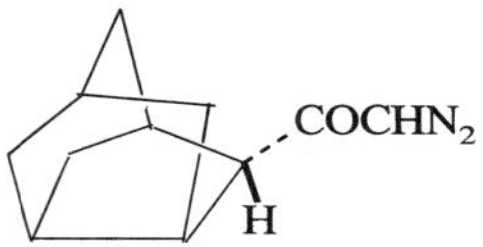 $COCHN_2$, H	Cu, toluene, reflux	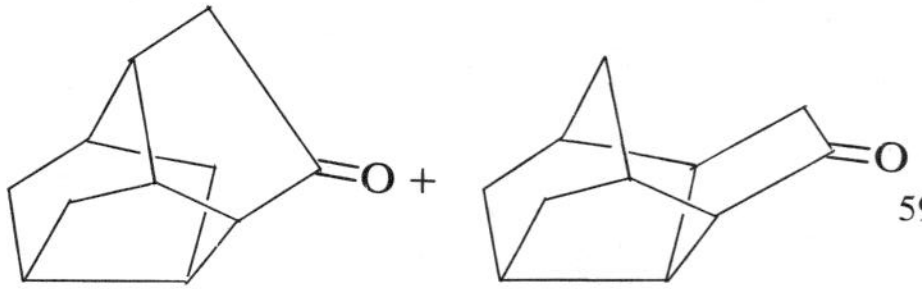 O + O	59	217a
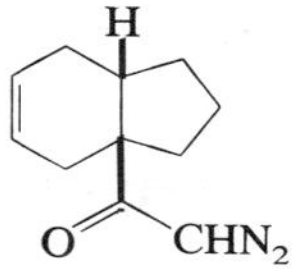 H, O, CHN_2	Silver benzoate, $(C_2H_5)_3N$, methanol	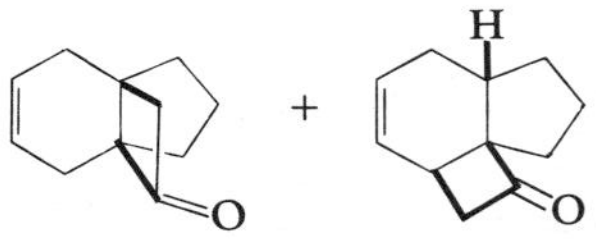 O + H, O		104

	Reactant	Conditions	Product	Yield (%)	Ref.
	t-$C_4H_9O_2C$, N_2, O, N	CCl_4, $h\nu$		—	51
C_{12}	C_6H_5, N_2, O, N	Methylene chloride, $h\nu$	C_6H_5, H, O, N	—	50
	C_4H_9-t, CHN_2, O	Cu powder, benzene	O	12	218
	t-$C_4H_9O_2C$, N_2, O, N	CCl_4, $h\nu$	t-$C_4H_9O_2C$, H, O, N	14	51
	t-C_4H_9, O, CHN_2	$CuSO_4$, cyclohexane, heat	t-C_4H_9, O	16	216

TABLE III. INTRAMOLECULAR INSERTION OF DIAZOCARBONYL COMPOUNDS INTO C–H BONDS (*Continued*)

Reactant	Conditions	Product(s)	Yield (%)	Refs.
C_{12} (*Contd.*)				
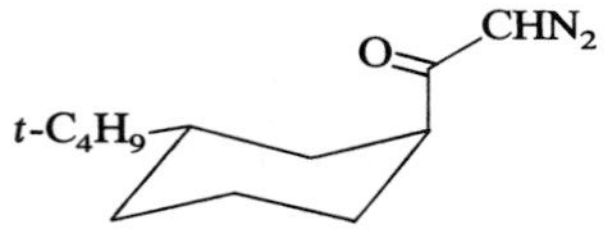 t-C_4H_9, CHN_2, O	$CuSO_4$, cyclohexane, heat	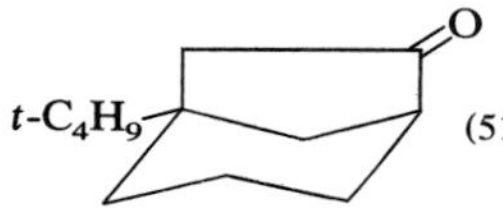 t-C_4H_9, O (51) 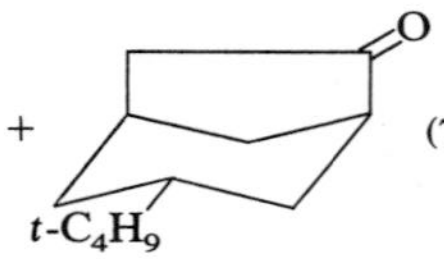 + t-C_4H_9, O	(7)	216
C_{13}				
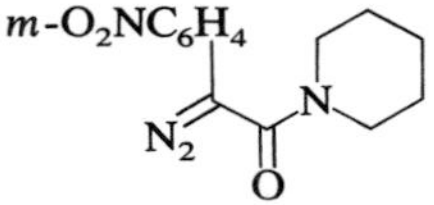 m-$O_2NC_6H_4$, N_2, O, N	Dioxane, reflux	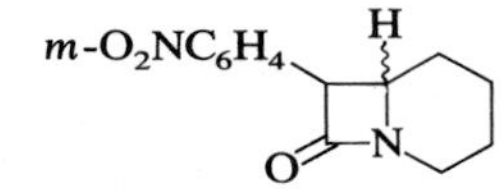 m-$O_2NC_6H_4$, H, O, N	32	50
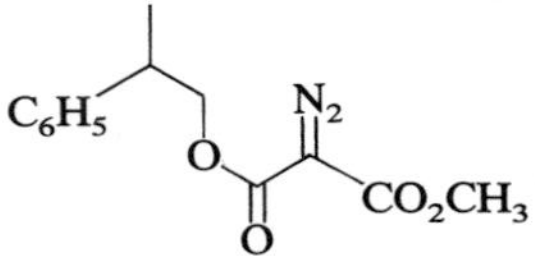 C_6H_5, N_2, O, CO_2CH_3	$CuSO_4$, chlorobenzene, reflux	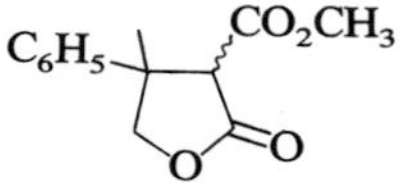 C_6H_5, CO_2CH_3, O	8.5	114
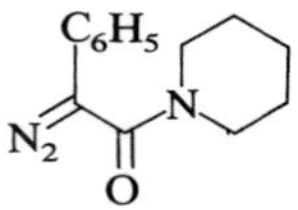 C_6H_5, N_2, O, N	Dioxane, reflux	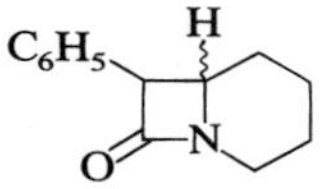 C_6H_5, H, O, N	39.5	50
	Methylene chloride, $h\nu$ or heat	"	50	49

Reactant	Conditions	Product(s)	Yield (%)	Refs.
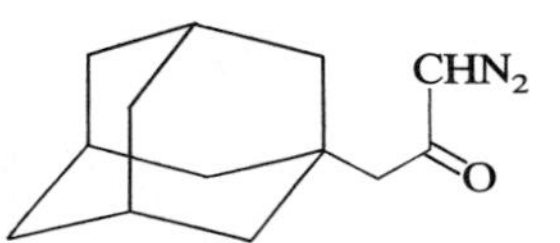	$CuSO_4$ or CuO, toluene, reflux		70–80	103
C_{14}				
	n-Decane, 174°, 10 min		12	151a
C_{15}				
	Cu, 80°, 10 hr, *n*-heptane		26	151a
	Cu, 174°, 10 min, *n*-decane	"	12	151a
	n-Decane, 174°, 10 min	"	13	151a
	Cu(acetylacetonate)$_2$, 174°, 10 min, *n*-decane	"	23	151a
	Methylene chloride, 10°, $h\nu$ (also thermal)		—	49

TABLE III. INTRAMOLECULAR INSERTION OF DIAZOCARBONYL COMPOUNDS INTO C–H BONDS (*Continued*)

Reactant	Conditions	Product(s)	Yield (%)	Refs.
C_{15} (*Contd.*)				
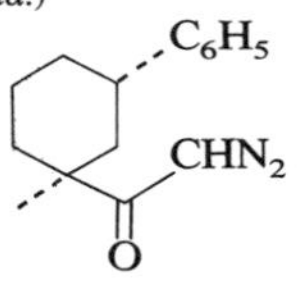 C_6H_5, CHN_2, O	Cu_2O, cyclohexane, $h\nu$, reflux	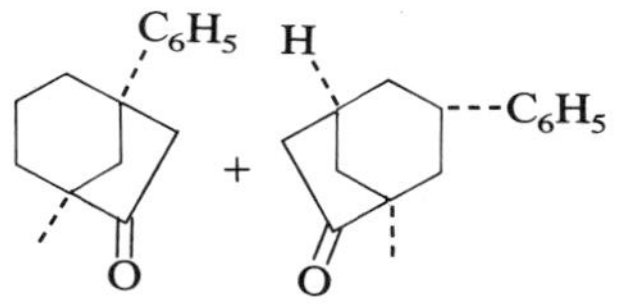 C_6H_5, O + H, C_6H_5, O (9:2)	55–60	113
	$CuSO_4$, cyclohexane, reflux	"	65–70	113
C^{16}				
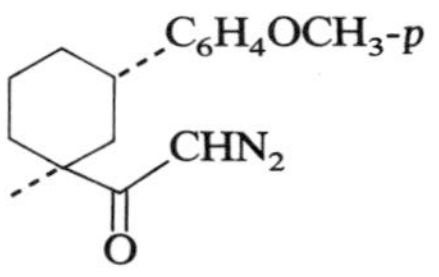 $C_6H_4OCH_3$-*p*, CHN_2, O	Cu_2O, cyclohexane, $h\nu$, reflux	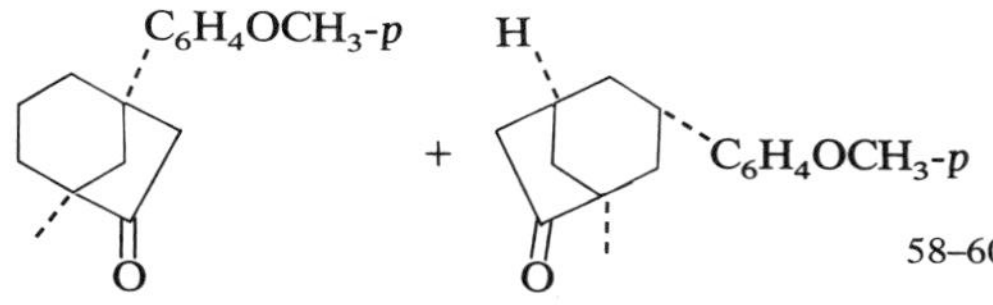 $C_6H_4OCH_3$-*p*, O + H, $C_6H_4OCH_3$-*p*, O (9:2)	58–60	113
	$CuSO_4$, cyclohexane, reflux	"	75	113
C_{17}				
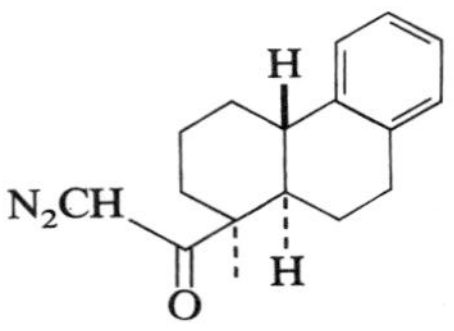 H, N_2CH, O, H	Cu_2O, cyclohexane–tetrahydrofuran (1:1), $h\nu$, reflux	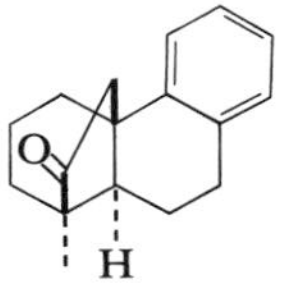 O, H	55	113, 219

Reactant	Conditions	Product(s)	Yield (%)	Refs.
	$CuSO_4$, cyclohexane, tetrahydrofuran, reflux	”	54	111, 219
CHN_2 (O)	$CuSO_4$, tetrahydrofuran, heat	O	53	111
	$CuSO_4$, tetrahydrofuran, reflux	”	52.5	220
	Cu_2O, cyclohexane, *hν*, reflux	”	58	220
$COCHN_2$	Cu_2O, cyclohexane, *hν*, reflux	O	23	220
C_{18}				
$t\text{-}C_4H_9O_2C$, N_2, S, N, O, H, $CO_2CH_2C_6H_5$	CCl_4, *hν*	$t\text{-}C_4H_9O_2C$, S, N, O, H, $CO_2CH_2C_6H_5$ (50) + O, O, O, S, N, H, $CO_2CH_2C_6H_5$ (−)		51

TABLE III. Intramolecular Insertion of Diazocarbonyl Compounds into C–H Bonds (*Continued*)

Reactant	Conditions	Product(s)	Yield (%)	Refs.
C_{21} (*Contd.*)				
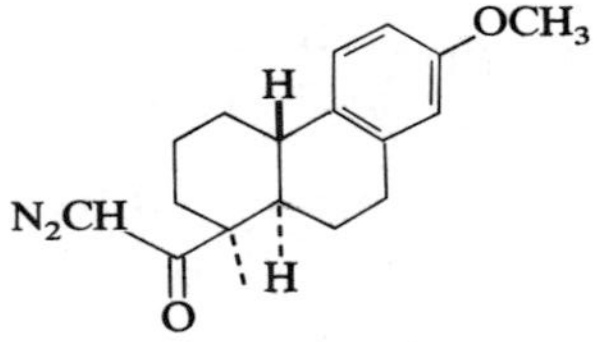 N_2CH … OCH_3	$CuSO_4$, cyclohexane, tetrahydrofuran, reflux	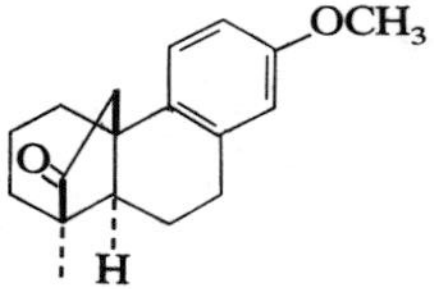 OCH_3 … H	38	111, 219
	Cu_2O, cyclohexane–tetrahydrofuran (1:1), $h\nu$, reflux	"	55	113, 219
C_{20}				
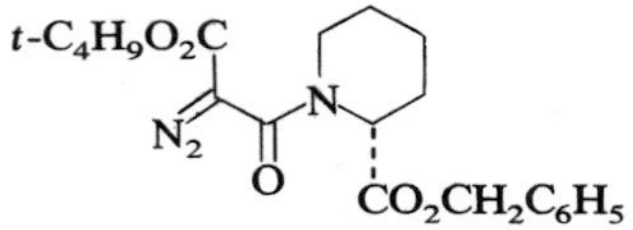 t-$C_4H_9O_2C$ … N_2 … $CO_2CH_2C_6H_5$	CCl_4, $h\nu$	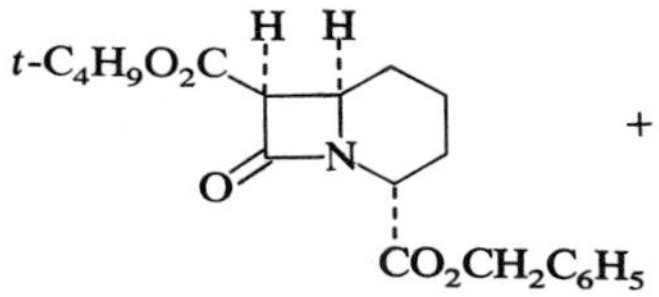 t-$C_4H_9O_2C$ … $CO_2CH_2C_6H_5$ (8) 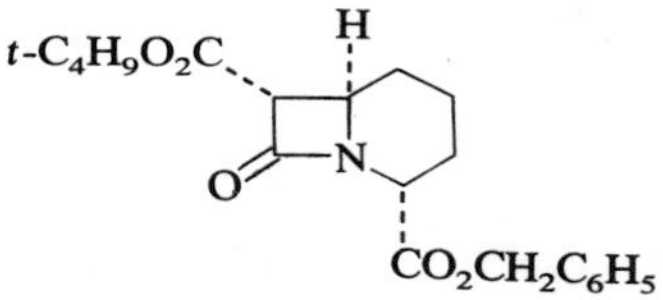 + t-$C_4H_9O_2C$ … $CO_2CH_2C_6H_5$ (11)		52, 53

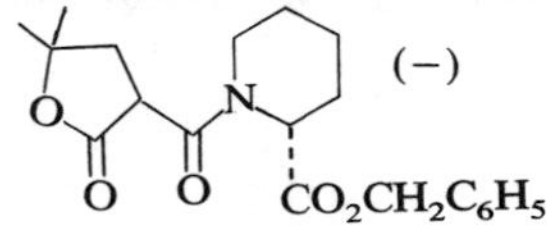

Formula	Reactant	Conditions	Product(s)	Yield (%)	Ref.
			+ (−) $CO_2CH_2C_6H_5$		
C_{21}	CHN$_2$, H, H, H	Cu_2O, cyclohexane, $h\nu$, reflux	H, H, O	95	110
	$C_6H_5C(CH_3)_2O$, O, N_2, O, N, S, $CO_2C_4H_9$-t	Benzene, $h\nu$	$C_6H_5C(CH_3)_2O_2C$, H, S, N, O, $CO_2C_4H_9$-t + $C_6H_5C(CH_3)_2O_2C$, H, S, N, O, $CO_2C_4H_9$-t (7:3)		54

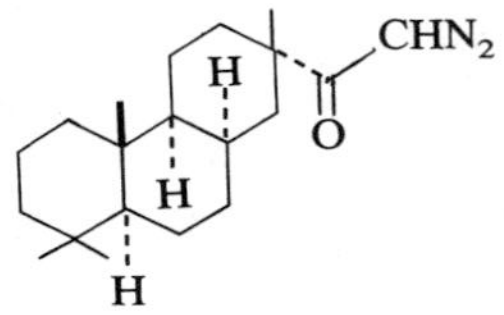

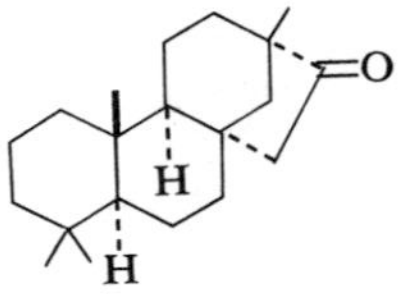

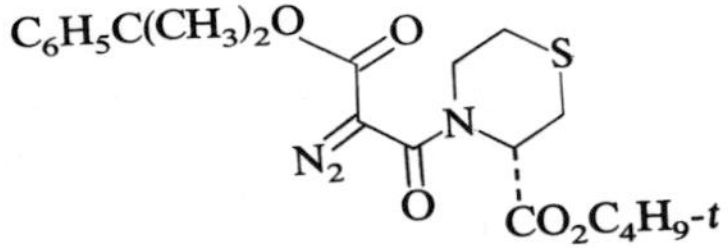

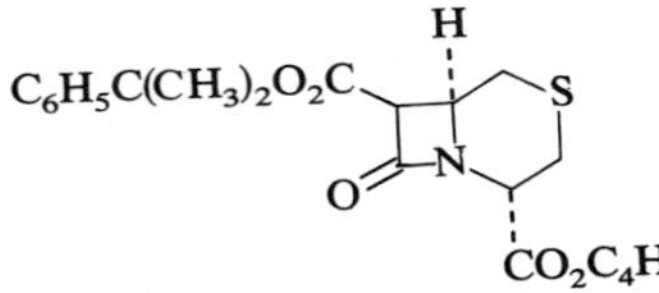

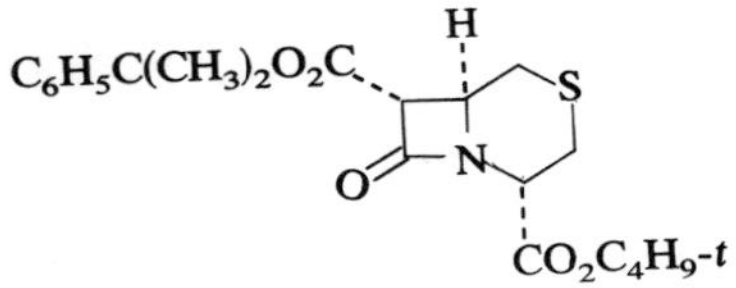

TABLE III. INTRAMOLECULAR INSERTION OF DIAZOCARBONYL COMPOUNDS INTO C–H BONDS (*Continued*)

Reactant	Conditions	Product(s)	Yield (%)	Refs.
C_{21} (*Contd.*)				
C_3H_7-*i*; H; O; N_2CH	Ag_2O, methanol	C_3H_7-*i*; H; O	22	105
	Methanol, *hν*	"	—	105
	$CuSO_4$, cyclohexane, reflux	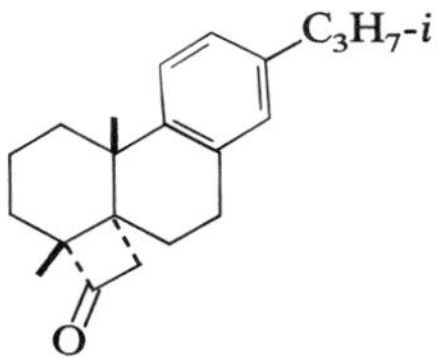 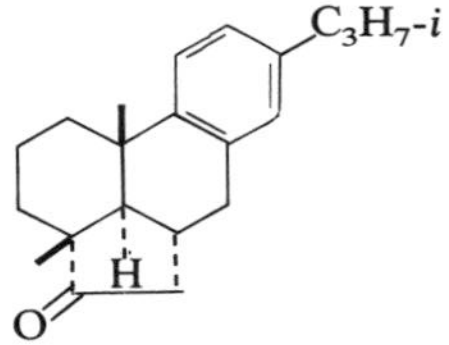 C_3H_7-*i*; O + C_3H_7-*i*; H; O		105
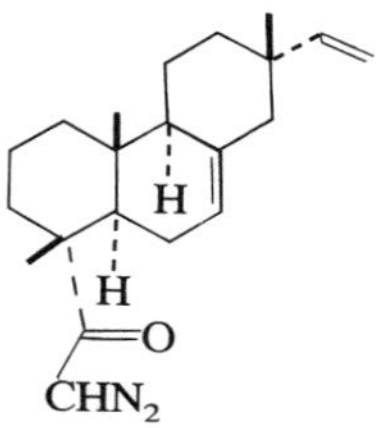 H; H; O; CHN_2	Ag_2O, benzene, heat	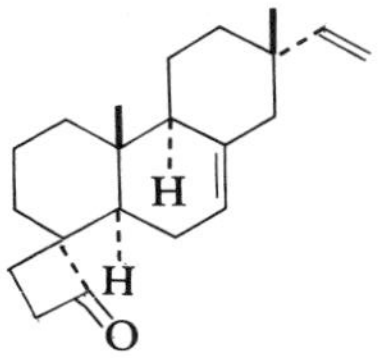 H; H; O	—	106

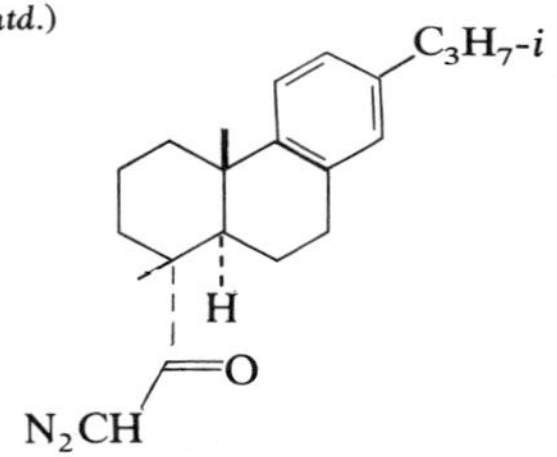

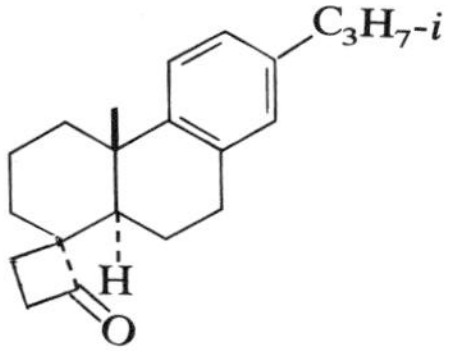

Reactant	Conditions	Product(s)	Yield (%)	Refs.
	CuO, toluene, heat		—	41a
C_{21}				
	$CuSO_4$, benzene, $h\nu$		40	41b
C_{23}				
	Cu_2O, cyclohexane, heat		—	109

[a] Insertion occurs with retention of configuration.

Note: References 145–230 are on pp. 474–475.

TABLE IV. ACID-CATALYZED CYCLIZATIONS OF DIAZOKETONES

Reactant	Conditions	Product(s)	Yield (%)	Refs.
C_5				
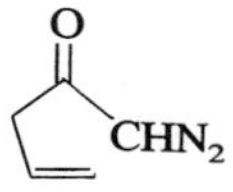 CHN_2	1. $BF_3 \cdot (C_2H_5)_2O$, CH_3NO_2 2. 10% aq HCl, reflux	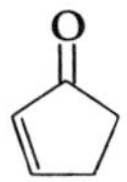 O	13	131
C_6				
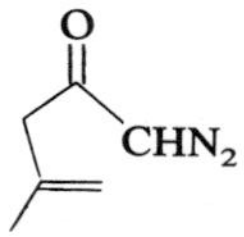 CHN_2	1. $BF_3 \cdot (C_2H_5)_2O$, CH_3NO_2 2. 10% aq HCl, reflux	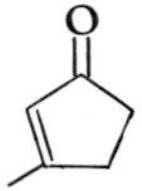 O	40–65	131
C_7				
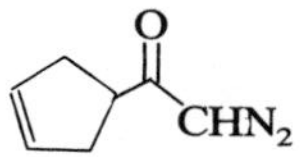 CHN_2	$HClO_4$, dioxane–H_2O (3:2)	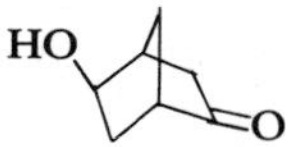 HO, O	60	117
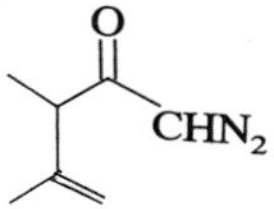 CHN_2	1. $BF_3 \cdot (C_2H_5)_2O$, CH_3NO_2 2. 10% aq HCl, reflux	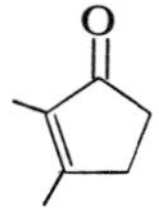 O	40–65	131
C_8				
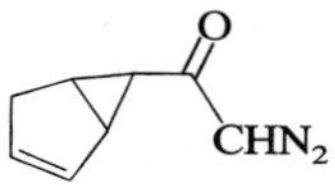 CHN_2	AcOH, 40°	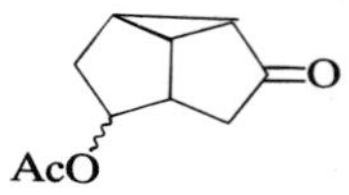 AcO, O	70–80	129

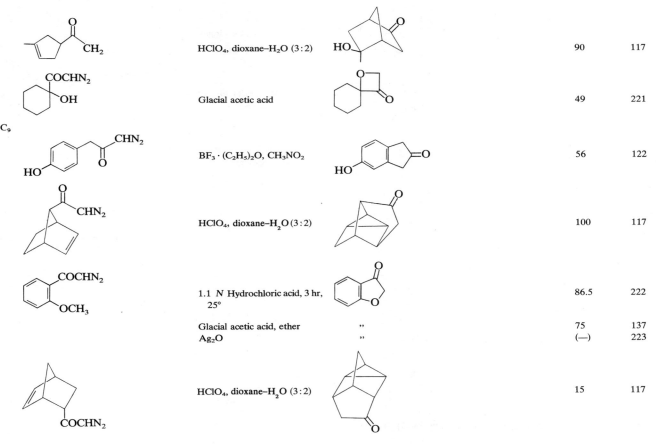

	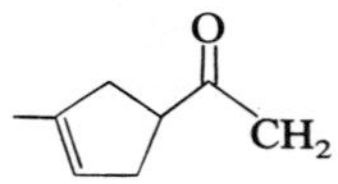	$HClO_4$, dioxane–H_2O (3:2)	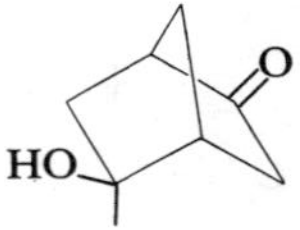	90	117
	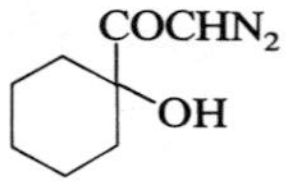	Glacial acetic acid	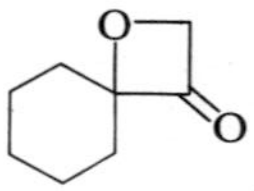	49	221
C_9	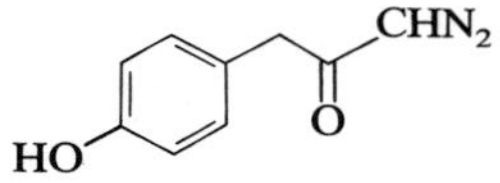	$BF_3 \cdot (C_2H_5)_2O$, CH_3NO_2	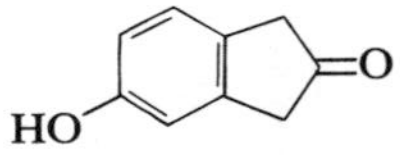	56	122
	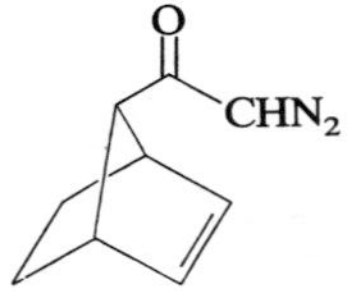	$HClO_4$, dioxane–H_2O (3:2)	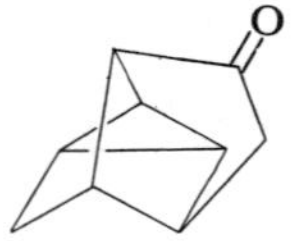	100	117
	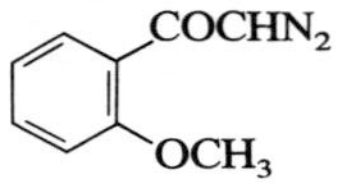	1.1 *N* Hydrochloric acid, 3 hr, 25°	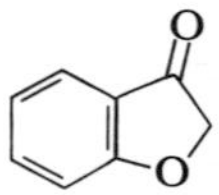	86.5	222
		Glacial acetic acid, ether	”	75	137
		Ag_2O	”	(—)	223
		$HClO_4$, dioxane–H_2O (3:2)	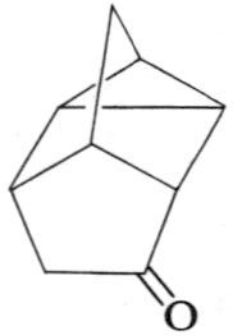	15	117

TABLE IV. Acid-Catalyzed Cyclizations of Diazoketones (*Continued*)

Reactant	Conditions	Product(s)	Yield (%)	Refs.
C_9 (*Contd.*)				
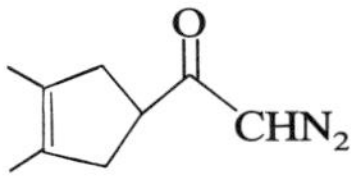 O, CHN_2	$HClO_4$, dioxane–H_2O (3:2)	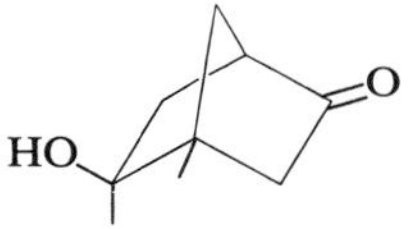 HO, O	100	117
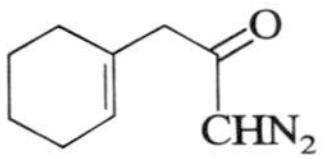 O, CHN_2	1. $BF_3 \cdot (C_2H_5)_2O$, CH_3NO_2 2. 10% aq HCl, reflux	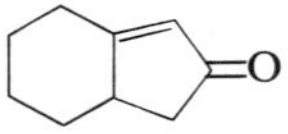 O	50–68	130
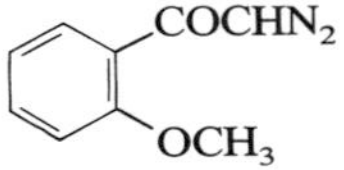 $COCHN_2$, OCH_3	Glacial acetic acid, 50°, 10% H_2SO_4	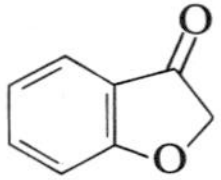 O, O	(—)	137
	Formic acid, 0–5°	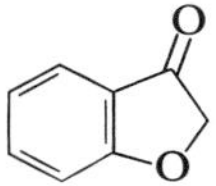 O, O	(—)	223
C_{10}				
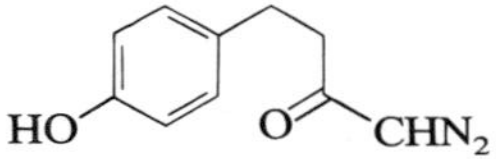 HO, O, CHN_2	$BF_3 \cdot (C_2H_5)_2O$, CH_3NO_2	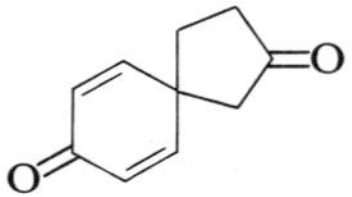 O, O	67	122
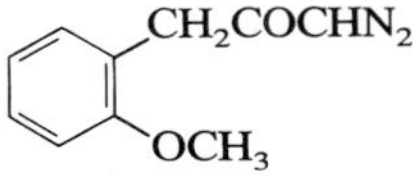 CH_2COCHN_2, OCH_3	Sulfuric acid, ether	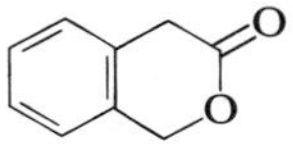 O, O	4	224

Reactant	Conditions	Product(s) and Yield(s) (%)		Refs.
	Sulfuric acid, dioxane	”	15	224
	BF_3, ether	”	35	224
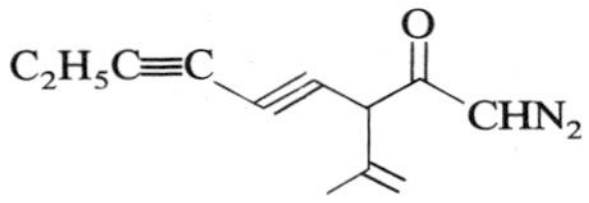 $C_2H_5C{\equiv}C$ … CHN_2	1. $BF_3 \cdot (C_2H_5)_2O$, CH_3NO_2 2. 10% aq HCl, reflux	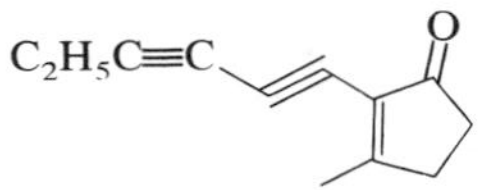 $C_2H_5C{\equiv}C$	40–65	131
C_{11}				
CHN_2, OCH_3	CF_3CO_2H	OCH_3	20	127
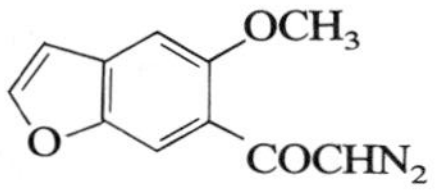 OCH_3, $COCHN_2$	50% Formic acid or dilute alcoholic H_2SO_4	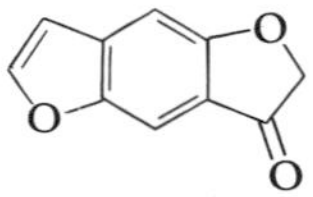	35	225
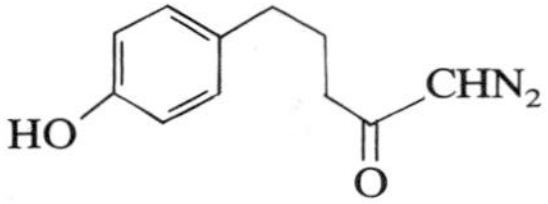 HO, CHN_2	$BF_3 \cdot (C_2H_5)_2O$, CH_3NO_2	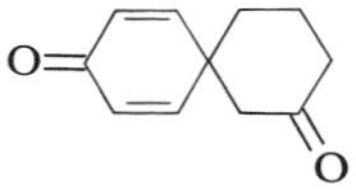	11	122
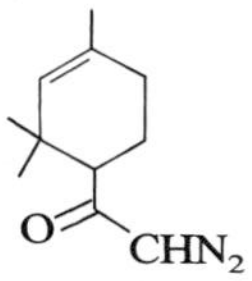 CHN_2	$BF_3 \cdot (C_2H_5)_2O$, $ClCH_2CH_2Cl$	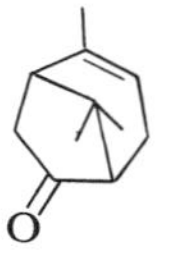 (56) 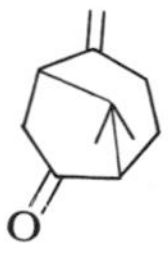 + (6)		118a
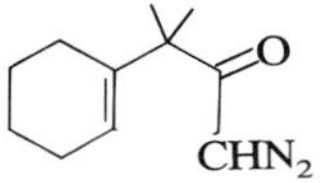 CHN_2	1. $BF_3 \cdot (C_2H_5)_2O$, CH_3NO_2 2. 10% aq HCl, reflux	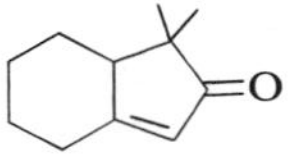	50–68	130

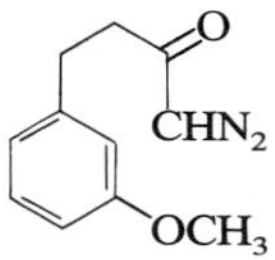

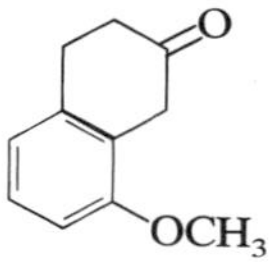

TABLE IV. ACID-CATALYZED CYCLIZATIONS OF DIAZOKETONES (*Continued*)

Reactant	Conditions	Product(s)	Yield (%)	Refs.
C_{11} (*Contd.*)				
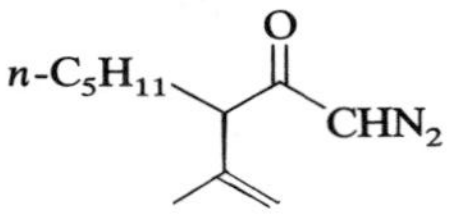 n-C_5H_{11}, O, CHN_2	1. $BF_3 \cdot (C_2H_5)_2O$, CH_3NO_2 2. 10% aq HCl, reflux	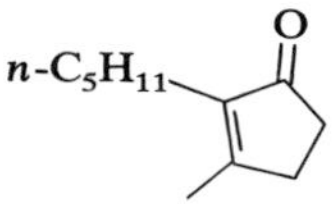 n-C_5H_{11}, O	40–65	131
C_{12}				
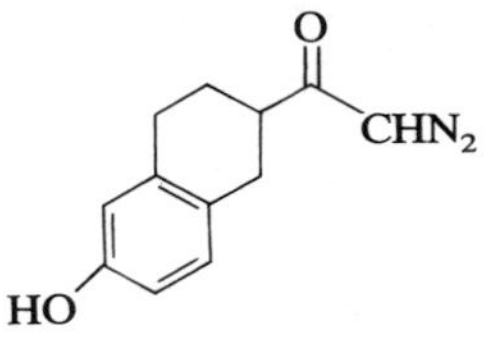 O, CHN_2, HO	HBF_4, CH_3NO_2	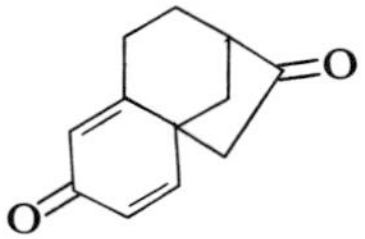 O, O	70–74	119, 126
	CF_3CO_2H, −20°	”	100	119, 126
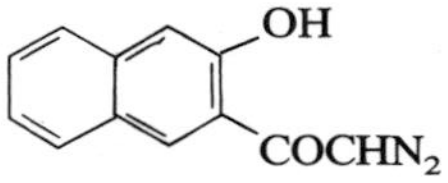 OH, $COCHN_2$	HCl, ethanol, heat	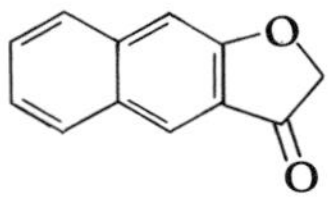 O, O	64	226
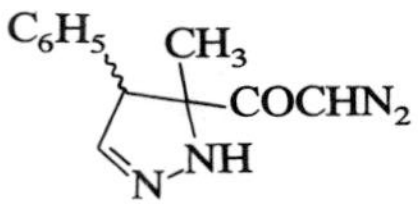 C_6H_5, CH_3, $COCHN_2$, NH, N	Glacial acetic acid	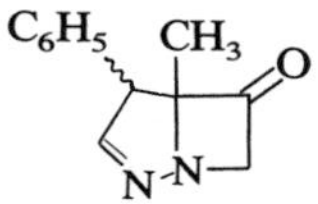 C_6H_5, CH_3, O, N, N	(—)	227
	HCl, methanol	”	(—)	227

Reactant	Conditions	Product(s) and Ratio	Yield (%)	Refs.
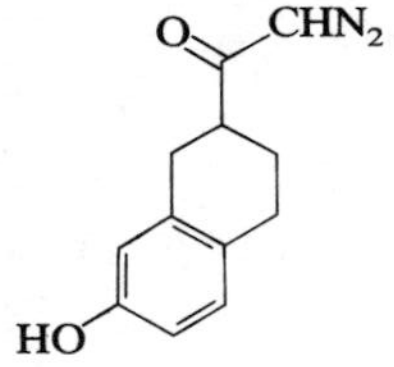 CHN_2 ketone (HO-tetralin)	$BF_3 \cdot (C_2H_5)_2O$, CH_3NO_2, room temperature	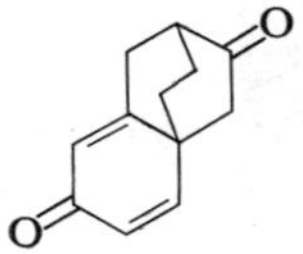	32	120
	CF_3CO_2H	"	48	126
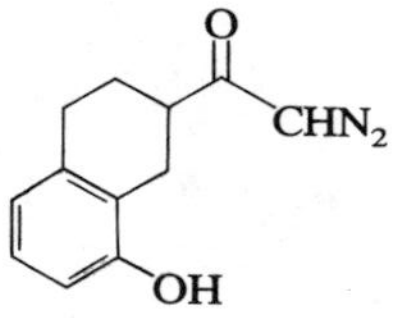 CHN_2 ketone (OH-tetralin)	CF_3CO_2H, CH_2Cl_2, 0°	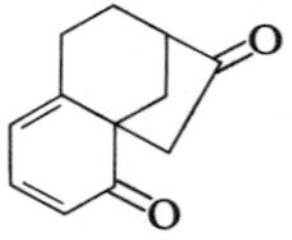	96	127
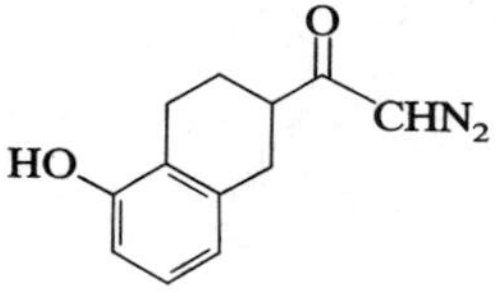 CHN_2 ketone (HO-tetralin)	CF_3CO_2H, CH_2Cl_2, 0°	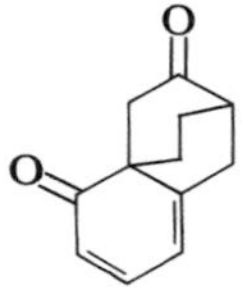	74	127
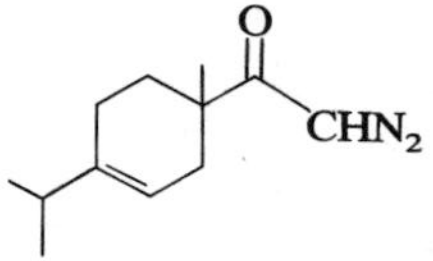 CHN_2 ketone	$BF_3 \cdot (C_2H_5)_2O$, CH_3NO_2, 0°	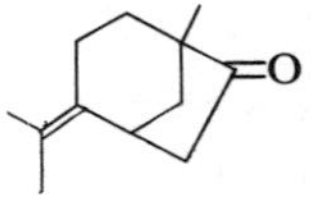 (80) 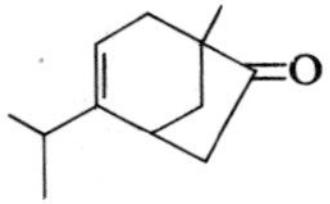 + (20)		123
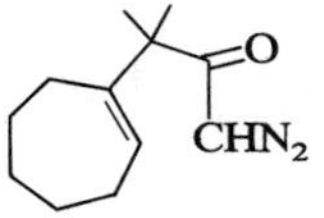 CHN_2 ketone	1. $BF_3 \cdot (C_2H_5)_2O$, CH_3NO_2 2. 10%, aq HCl, reflux	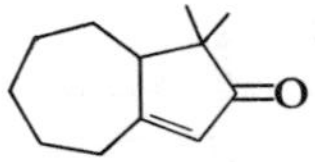	50–68	130

TABLE IV. ACID-CATALYZED CYCLIZATIONS OF DIAZOKETONES (*Continued*)

Reactant	Conditions	Product(s)	Yield (%)	Refs.
C_{13}				
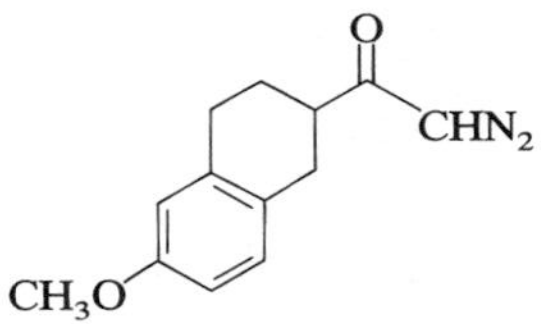 CH_3O, CHN_2, O	CF_3CO_2H	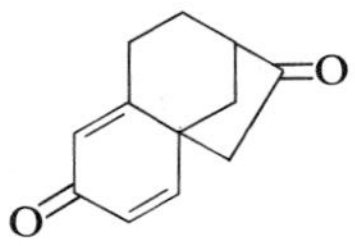 O, O	86	119, 126
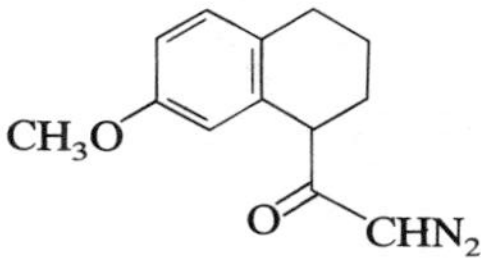 CH_3O, CHN_2, O	CF_3CO_2H, CH_2Cl_2, −20°	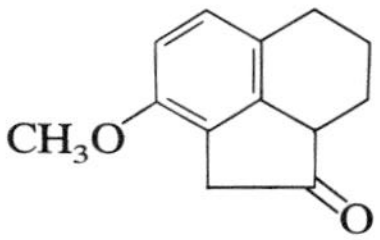 CH_3O, O (33) 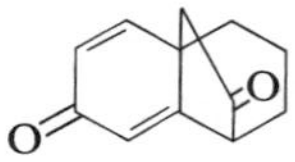 + O, O		126
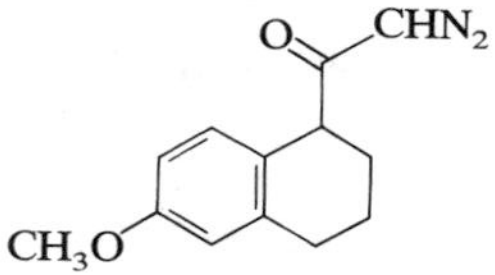 CH_3O, CHN_2, O	CF_3CO_2H, CH_2Cl_2, −20°	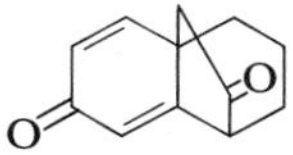 O, O	58	126
C_{14}				
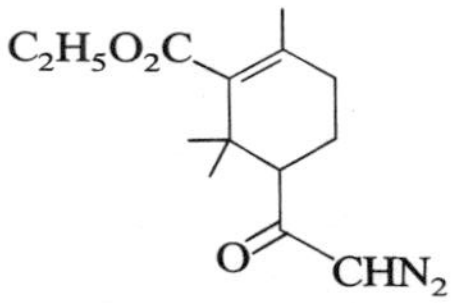 $C_2H_5O_2C$, CHN_2, O	$BF_3 \cdot (C_2H_5)_2O$, $ClCH_2CH_2Cl$	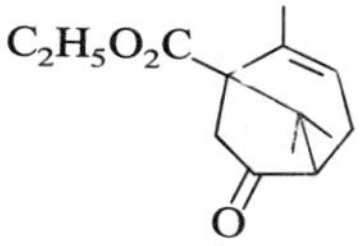 $C_2H_5O_2C$, O (30) 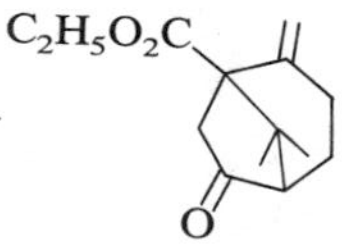 + $C_2H_5O_2C$, O (3)		118

Reactant	Conditions	Product(s)	Yield (%)	Ref.
C_6H_5, $COCHN_2$	$BF_3 \cdot (C_2H_5)_2O$, CH_2Cl_2	O, C_6H_5	74	152
	50% aq HBF_4, CH_3NO_2	O, C_6H_5 (30) + O, C_6H_5, OH	(21)	152
$COCHN_2$	10% H_2SO_4 solution in acetic acid	O	80	115
$COCHN_2$	$BF_3 \cdot (C_2H_5)_2O$, CH_3NO_2	O	21	121
$COCHN_2$, O, O, $OCOC_2H_5$	Aq HCl, 0–5°	O, O, O, O	61.5	228

TABLE IV. ACID-CATALYZED CYCLIZATIONS OF DIAZOKETONES (*Continued*)

Reactant	Conditions	Product(s)	Yield (%)	Refs.
C_{15}				
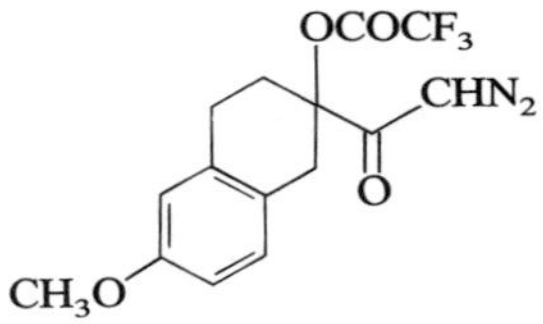 $OCOCF_3$, CHN_2, O, CH_3O	CF_3CO_2H, CH_2Cl_2, −20°	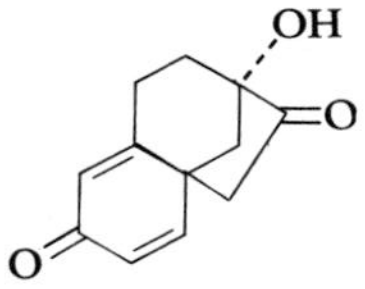 OH, O, O	70	119, 126
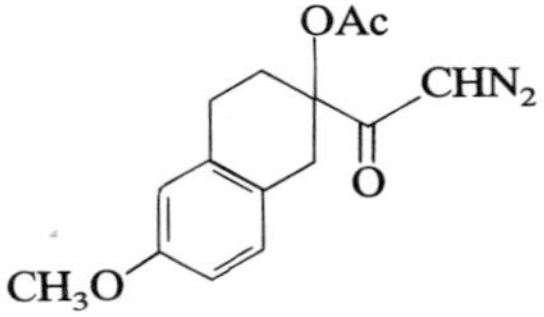 OAc, CHN_2, O, CH_3O	—	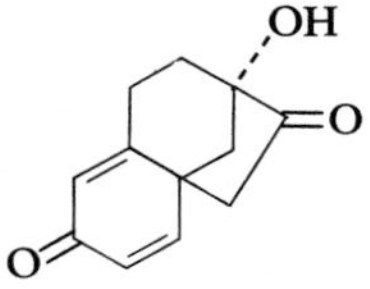 OH, O, O	16	119
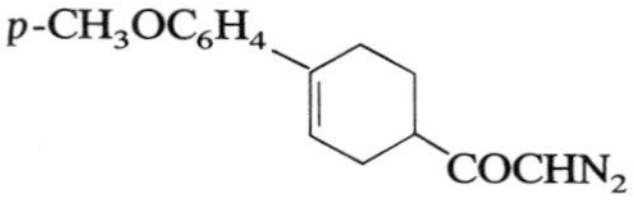 p-$CH_3OC_6H_4$, $COCHN_2$	$BF_3 \cdot (C_2H_5)_2O$, CH_2Cl_2	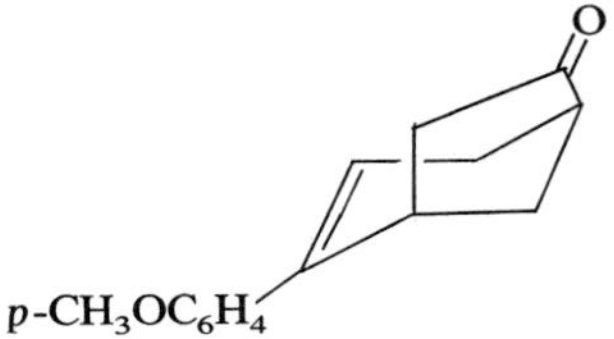 O, p-$CH_3OC_6H_4$	82	152
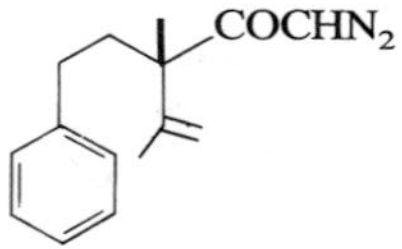 $COCHN_2$	$BF_3 \cdot (C_2H_5)_2O$, CH_3NO_2	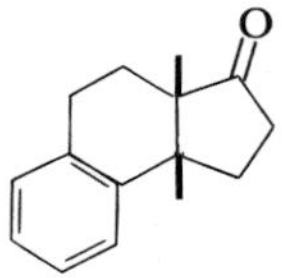 O (46) 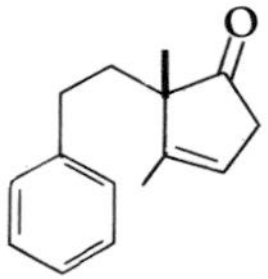 + O (10)		121

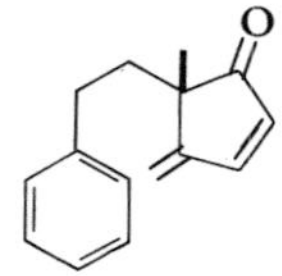

Reactant	Conditions	Product(s) and Yield(s) (%)		Refs.
		+ 2-methyl-2-(2-phenylethyl)-3-methylenecyclopent-4-en-1-one (2)		
4-C_6H_5-1-methylcyclohex-3-enyl $COCHN_2$	$BF_3 \cdot (C_2H_5)_2O$, CH_2Cl_2	C_6H_5-substituted methylbicyclooctenone	80	152
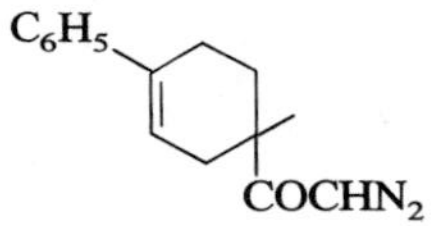 4-C_6H_5-1-methylcyclohex-3-enyl $COCHN_2$	50% Aq HBF_4	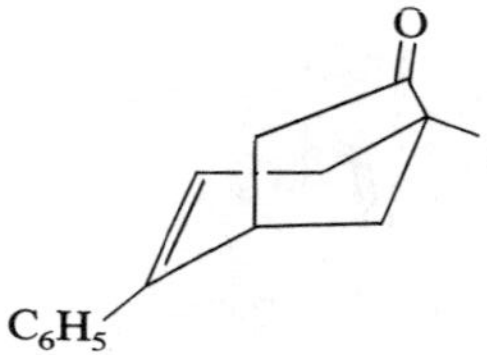 C_6H_5-substituted methylbicyclooctenone (26) 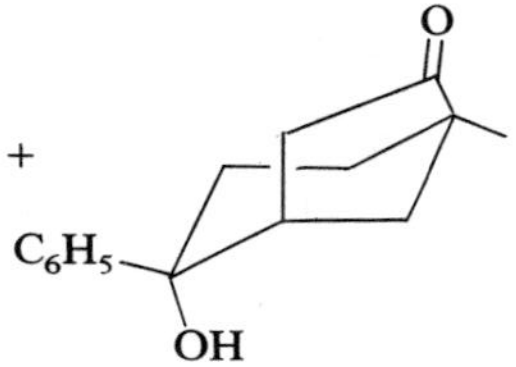 + C_6H_5, OH-substituted methylbicyclooctanone (18)		152
C_{16}				
CH_3O-substituted tetrahydrofluorenyl $COCHN_2$	$BF_3 \cdot (C_2H_5)_2O$, CH_2Cl_2, −10–5°	CH_3O-substituted bridged ketone	45	72
	HBF_4, CH_3NO_2	"	97	119
	CF_3CO_2H, CH_2Cl_2, 0°	"	100	125

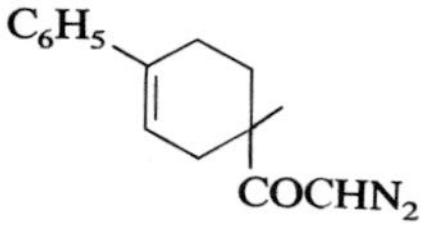

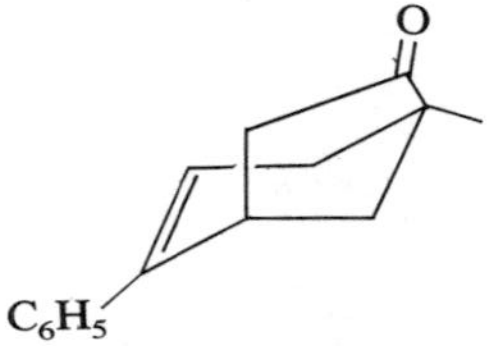

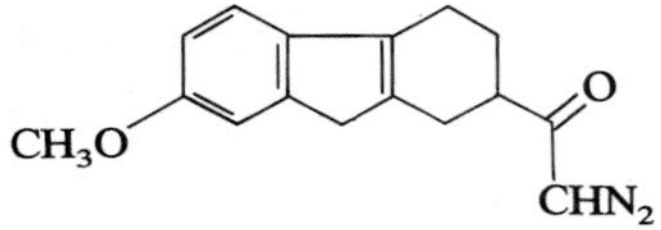

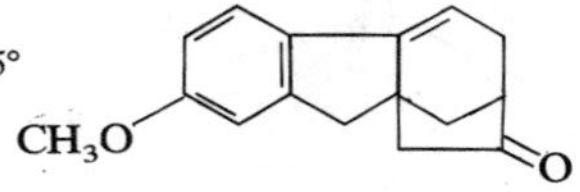

TABLE IV. Acid-Catalyzed Cyclizations of Diazoketones (*Continued*)

Reactant	Conditions	Product(s)	Yield (%)	Refs.
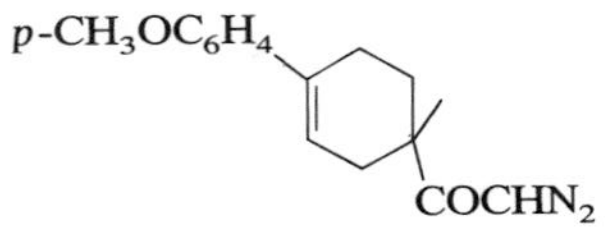 p-$CH_3OC_6H_4$, $COCHN_2$	$BF_3 \cdot (C_2H_5)_2O$, CH_2Cl_2	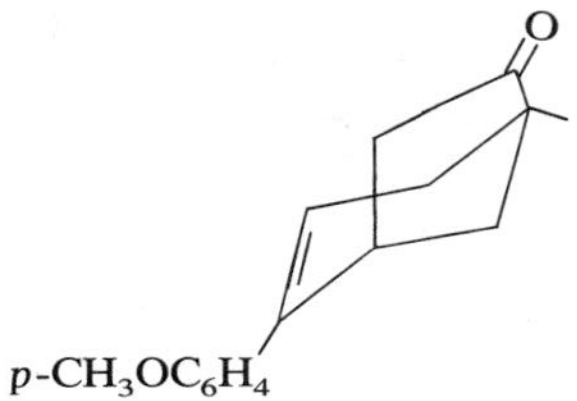 p-$CH_3OC_6H_4$, O	85	152
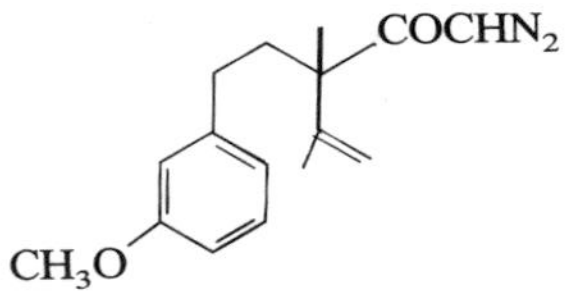 CH_3O, $COCHN_2$	$BF_3 \cdot (C_2H_5)_2O$, CH_3NO_2	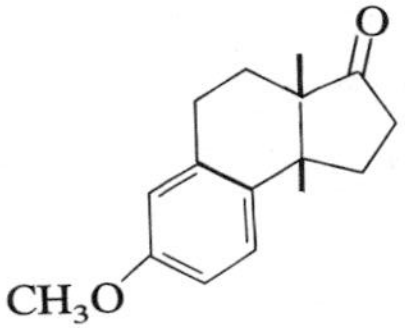 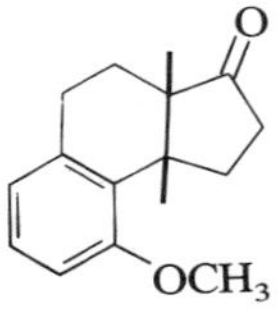 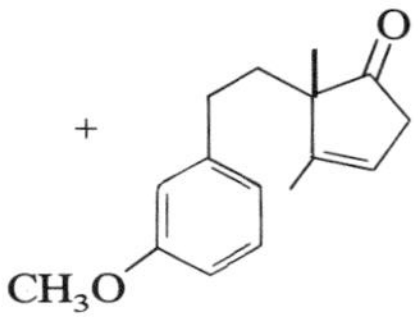 CH_3O, O (31) + O, OCH_3 (12) + CH_3O, O (10)	(31), (12), (10)	121
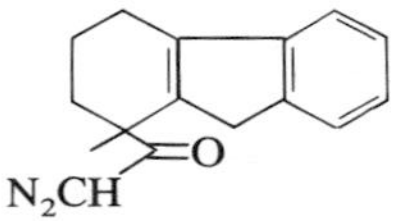 N_2CH, O	$HClO_4$ or HI	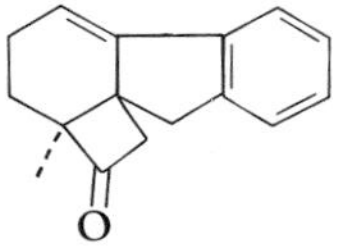 O	50–55	128

	HBF_4, CH_3NO_2		78	119
	70% Aq $HClO_4$ or 48% aq HBF_4 or conc H_2SO_4, $CHCl_3$		80–99	128
	57% HI, $CHCl_3$	"	60	128
	$BF_3 \cdot (C_2H_5)_2O$, $ClCH_2CH_2Cl$, 0°		43	72
	HBF_4, CH_3NO_2	"	97	119, 124
	CF_3CO_2H, 0°	"	100	124
C_{18}				
	1. CF_3CO_2H, CH_2Cl_2, 0° 2. Hydrolysis		62	125

TABLE IV. Acid-Catalyzed Cyclizations of Diazoketones (*Continued*)

Reactant	Conditions	Product(s)	Yield (%)	Refs.
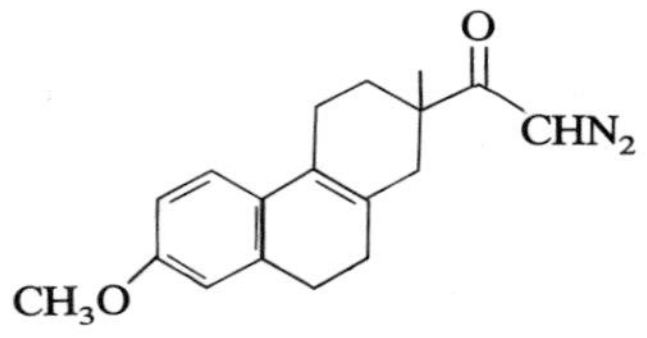 CH_3O, O, CHN_2	$BF_3 \cdot (C_2H_5)_2O$, CH_2Cl_2, 0°	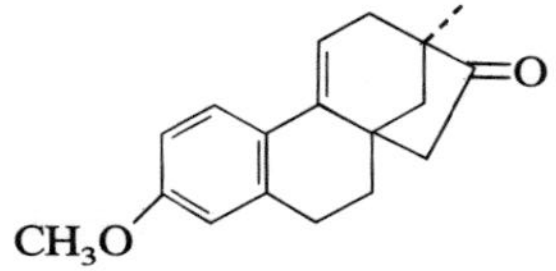 CH_3O, O	41	72
	CF_3CO_2H, 0°	"	100	124
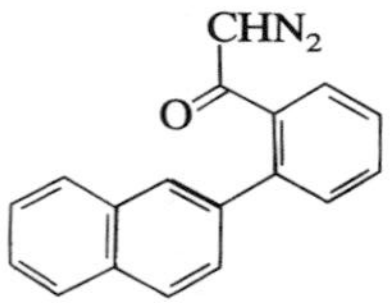 CHN_2, O	10% Soln H_2SO_4 in acetic acid, room temp	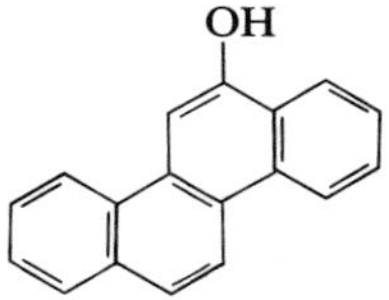 OH	(—)	116
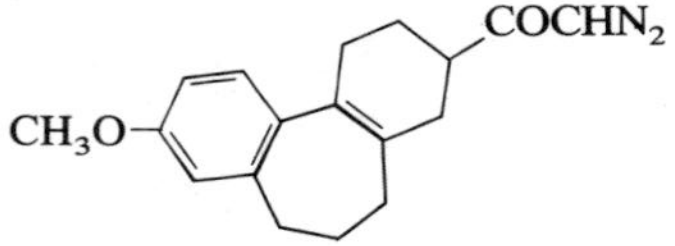 CH_3O, $COCHN_2$	$BF_3 \cdot (C_2H_5)_2O$, CH_2Cl_2	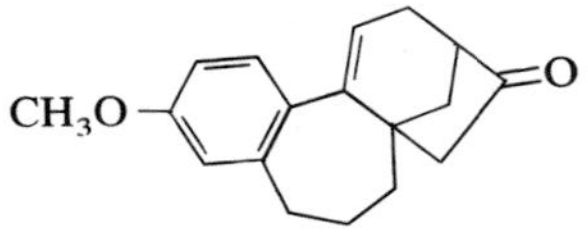 CH_3O, O	50	155
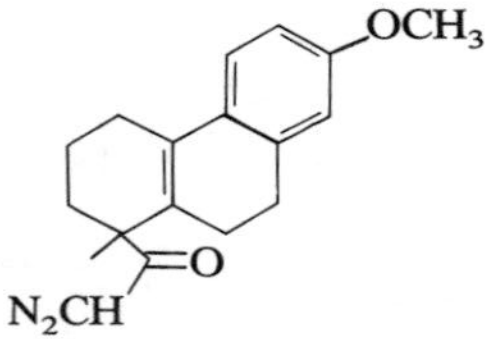 OCH_3, O, N_2CH	70% Aq $HClO_4$ or 57% HI, $CHCl_3$	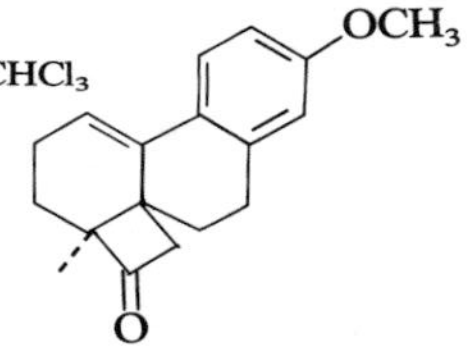 OCH_3, O	50–70	128

	Reactant	Conditions	Product	Yield (%)	Refs.
C_{19}	CH_3O; $OCOCF_3$; CHN_2; O	1. CF_3CO_2H, 0° 2. Hydrolysis	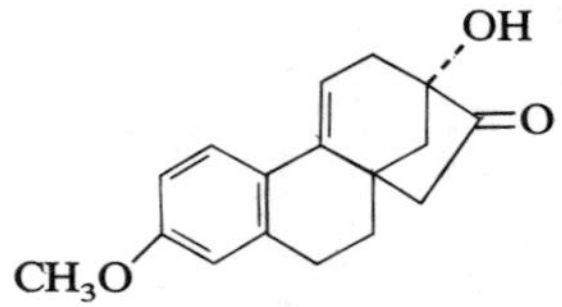 CH_3O; OH; O	60	124
C_{21}	HO; $COCHN_2$; OH; O	Phosphoric acid, dioxane	HO; O; O; O	(—)	229
C_{22}	HO; $COCHN_2$; N; N	Glacial acetic acid, 25°	HO; O; N; N	78	230
C_{24}	AcO; $COCHN_2$; N; N	Glacial acetic acid, heat	AcO; O; N; N	82	230

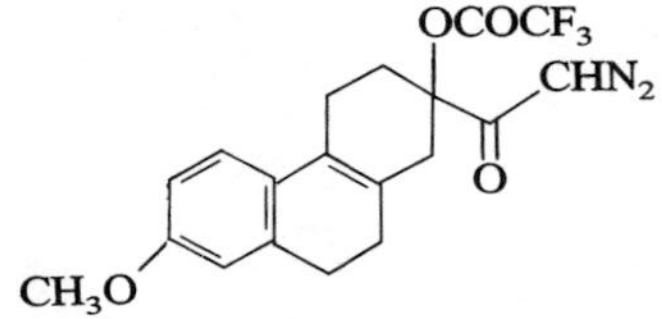

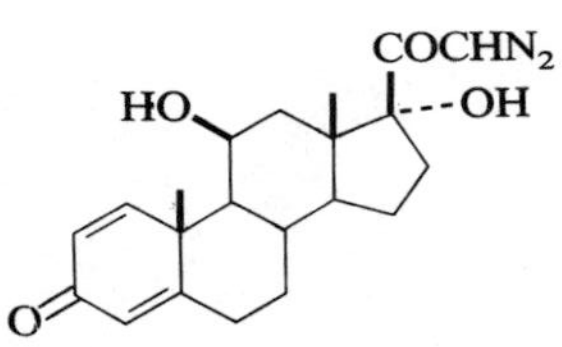

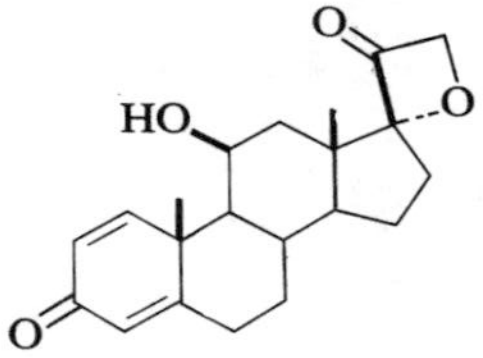

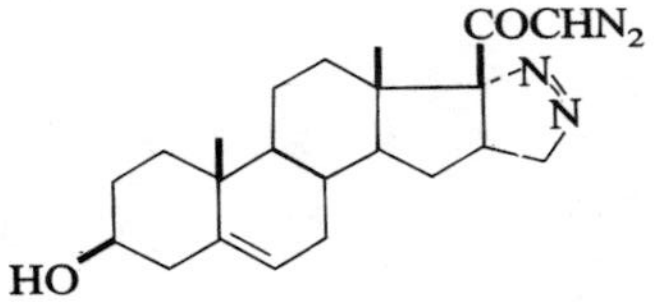

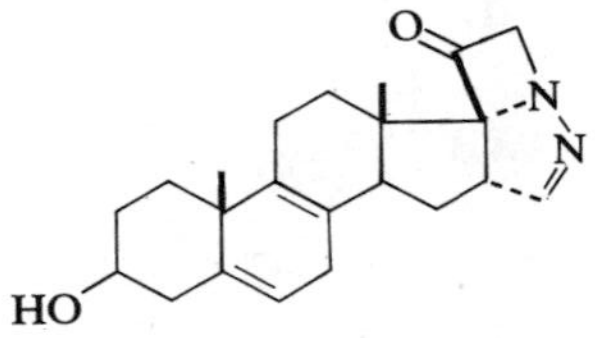

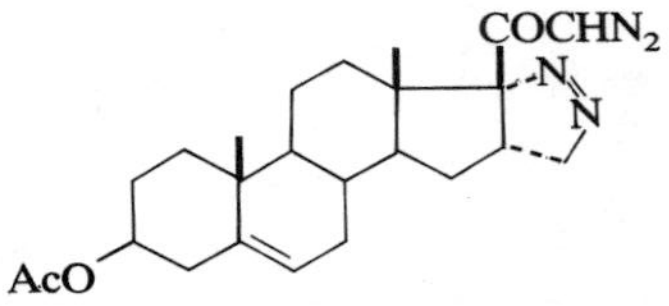

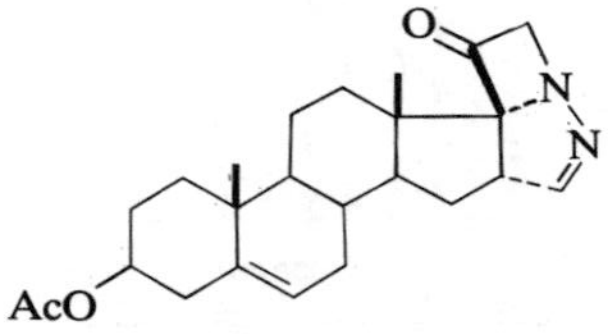

Note: References 145–230 are on pp. 474–475.

TABLE V. Miscellaneous Reactions

Reactant	Conditions	Product(s)	Yield (%)	Refs.
C_8				
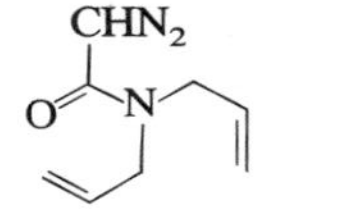 CHN_2, O, N	Dioxane, $h\nu$	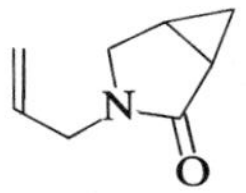 N, O	77	37
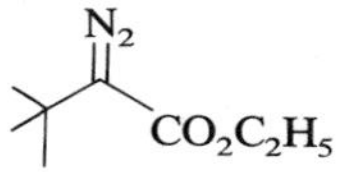 N_2, $CO_2C_2H_5$	Cyclopentane, $h\nu$	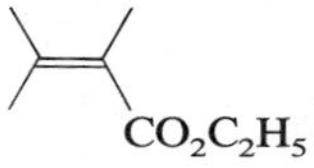 $CO_2C_2H_5$	65	132a
C_9				
 $COCHN_2$	Acetone, 1 N H_2SO_4 (4:1 v/v)	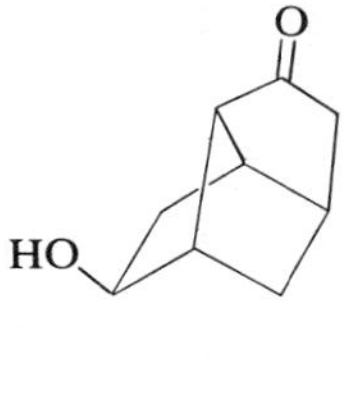 O, HO (47) 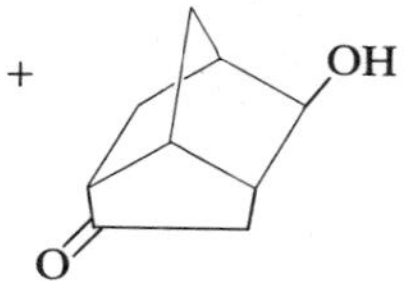 + O, OH (28) 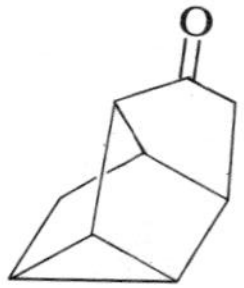 + O (25)		132b

		,,	,,	,,		
 C_{10} $COCHN_2$	Acetone, 1 N H_2SO_4 (4 : 1 v/v)	(3)	(23)	(1)		132b
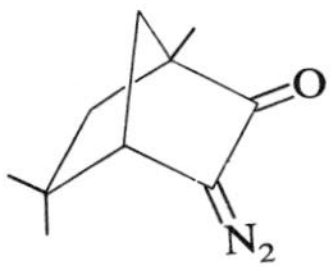 N_2	Cu bronze, benzene, reflux	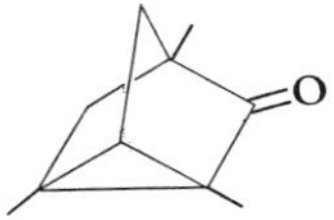			53	101
	Cyclohexene, 130°		,,		43[a]	101
C_{11} N_2CH CHN_2	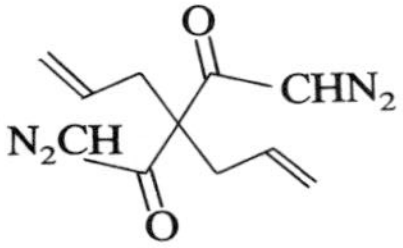 $CuSO_4$, dioxane, 100°	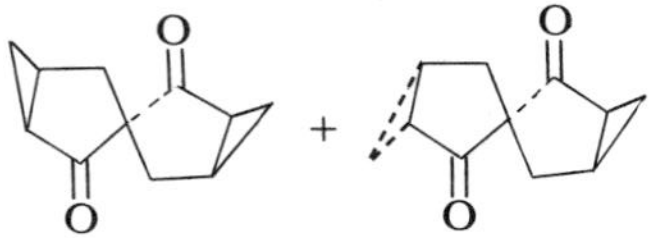 + (2 : 1)			15	133
	Di-μ-dichloro-π-allyl-palladium, $(C_2H_5)_2O$, 5–15°	,,	(6 : 1)	,,	22	133
	Di-μ-dichloro-π-allyl-palladium, tetrahydrofuran, 5–15°	,,	(6 : 1)	,,	24	133
	Di-μ-dichloro-π-allyl-palladium, dioxane, 5–15°	,,	(6 : 1)	,,	35	133
	Di-μ-dichloro-π-allyl-palladium, benzene, 5–15°	,,	(3 : 1)	,,	23	133

TABLE V. Miscellaneous Reactions (*Continued*)

Reactant	Conditions	Product(s)	Yield (%)	Refs.
C_{12}				
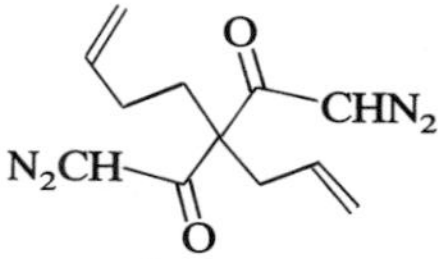 N_2CH, O, O, CHN_2	Di-μ-dichloro-π-allyl-palladium, $(C_2H_5)_2O$, low temp	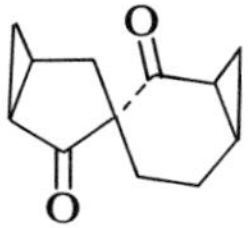 O, O	20	134
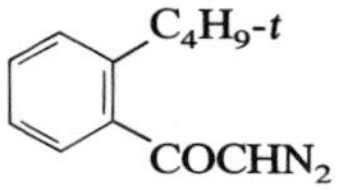 C_4H_9-*t*, $COCHN_2$	Cu powder, DMSO, room temp	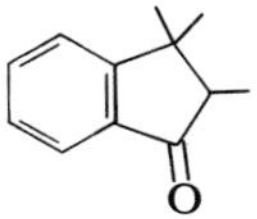 O	1	218
C_{13}				
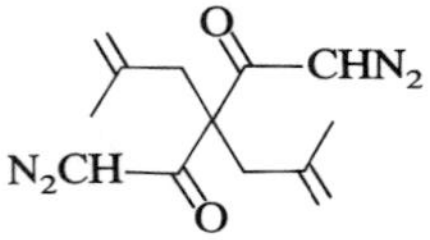 N_2CH, O, O, CHN_2	Di-μ-dichloro-π-allyl-palladium, $(C_2H_5)_2O$, low temp	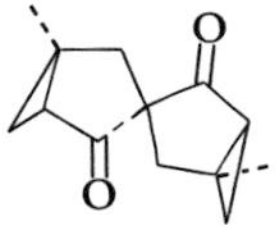 O, O	15	134
C_{16}				
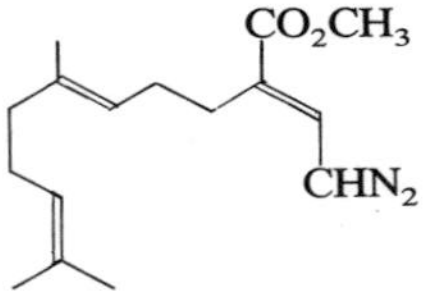 CO_2CH_3, CHN_2	CuI, tetrahydrofuran, 35°	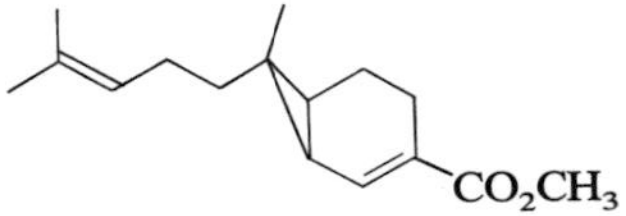 CO_2CH_3	50	48

Reactant	Conditions	Product	Yield (%)	Refs.
C_{17}				
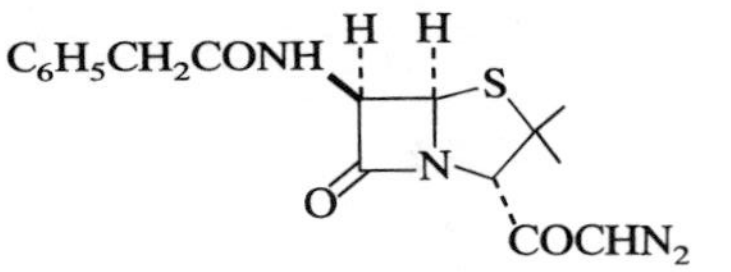 $C_6H_5CH_2CONH$ (penam, $COCHN_2$)	$Cu(acac)_2$, tetrahydrofuran, reflux	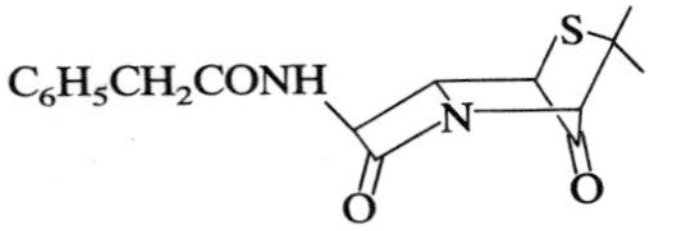 $C_6H_5CH_2CONH$ (tricyclic diketone)	90	136
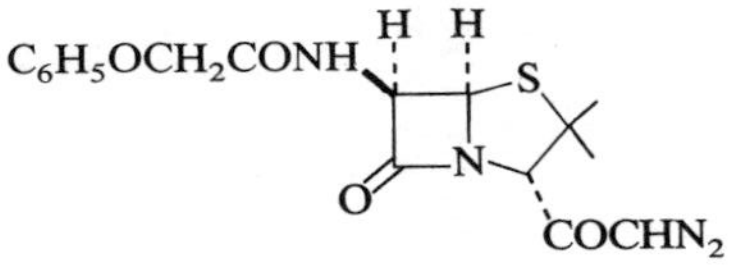 $C_6H_5OCH_2CONH$ (penam, $COCHN_2$)	$Cu(acac)_2$, tetrahydrofuran, reflux	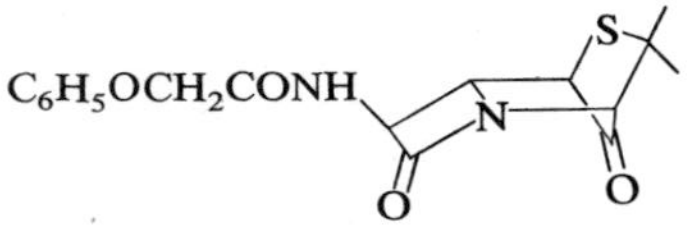 $C_6H_5OCH_2CONH$ (tricyclic diketone)	80	136
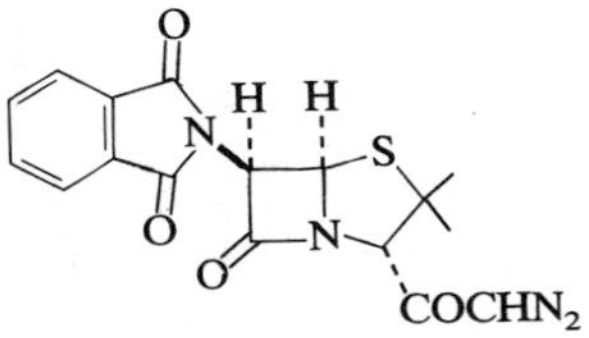 Phthalimido penam, $COCHN_2$	$Cu(acac)_2$, tetrahydrofuran, reflux	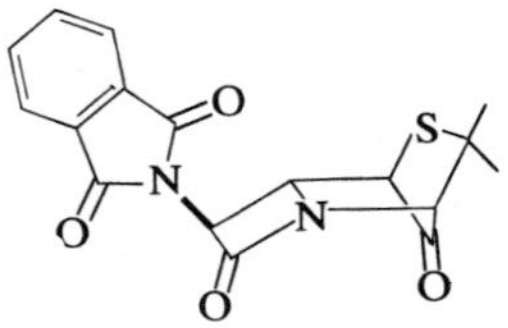 Phthalimido tricyclic diketone	43	136
C_{21}				
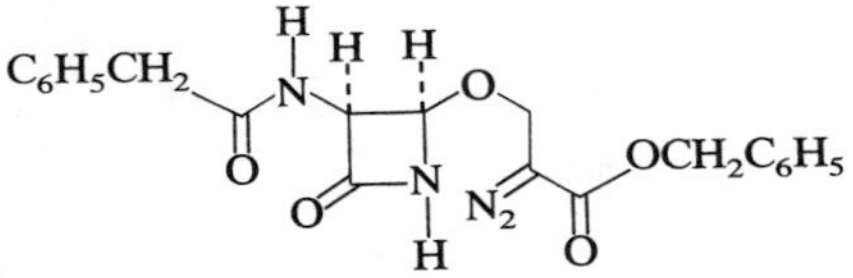 $C_6H_5CH_2CONH$ azetidinone, $OCH_2C(=N_2)CO_2CH_2C_6H_5$	Rhodium acetate, benzene, 3 hr	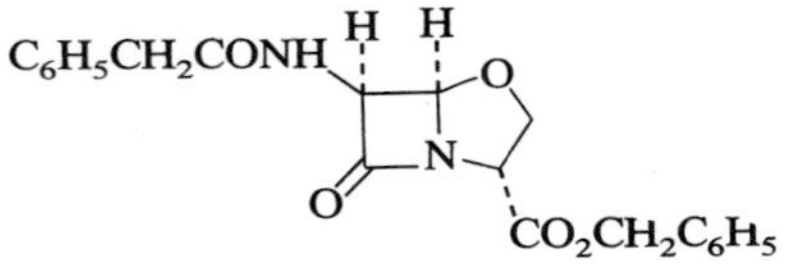 $C_6H_5CH_2CONH$ oxapenam, $CO_2CH_2C_6H_5$	20	135

[a] No norcarane derivative was formed.

Note: References 145–230 are on pp. 474–475.

REFERENCES TO TABLES I–V

[145] K. Kondo, E. Hiro, and D. Tunemoto, *Tetrahedron Lett.*, **1976,** 4489.

[146] (a) S. Bien, A. Gillon, and S. Kohen, *J. Chem. Soc., Perkin Trans. I,* **1976,** 489; (b) R. D. Miller and D. L. Dolce, *Tetrahedron Lett.*, **1977,** 3329.

[147] J. F. Ruppert and J. D. White, *J. Chem. Soc., Chem. Commun.*, **1976,** 976.

[148] W. H. Robinson, Ph.D. Dissertation, University of Illinois, 1972 [*Diss. Abstr. Int., B,* **34,** 1045 (1973)].

[149] E. Vedejs, W. R. Wilber, and R. Twieg, *J. Org. Chem.*, **42,** 401 (1977).

[150] D. Tunemoto, N. Araki, and K. Kondo, *Tetrahedron Lett.*, **1977,** 109.

[151] (a) I. Kuwajima, Y. Higuchi, H. Iwasawa, and T. Sato, *Chem. Lett. (Jpn.)*, **1976,** 1271; (b) M. Rull, F. Serratosa, and J. Vilarrasa, *Tetrahedron Lett.*, **1977,** 4549.

[152] U. R. Ghatak, S. K. Alam, P. C. Chakraborti, and B. C. Ranu, *J. Chem. Soc., Perkin Trans. I,* **1976,** 1669.

[153] J. F. Ruppert, M. A. Avery, and J. D. White, *J. Chem. Soc., Chem. Commun.*, **1976,** 978.

[154] A. Sarkar, S. Chatterjee, and P. C. Dutta, *Tetrahedron Lett.*, **1976,** 4633.

[155] U. R. Ghatak and J. K. Ray, *J. Chem. Soc., Perkin Trans. I,* **1977,** 518.

[156] F. T. Bond, W. Weyler, B. Brunner, and J. E. Stemke, *J. Med. Chem.*, **19,** 255 (1976).

[157] A. Monahan and D. Lewis, *J. Chem. Soc., Perkin Trans. I,* **1977,** 60.

[158] G. A. Russell, P. R. Whittle, and J. McDonnell, *J. Am. Chem. Soc.*, **89,** 5515 (1967).

[159] G. A. Russell, J. J. McDonnell, P. R. Whittle, R. S. Givens, and R. C. Keske, *J. Am. Chem. Soc.*, **93,** 1452 (1971).

[160] M. Marx, *Diss. Abstr.*, **27,** 4266–B (1967).

[161] P. K. Freemann and D. G. Kuper, *Chem. Ind. (London)*, **1965,** 424.

[162] J. Meinwald and G. H. Wahl, Jr., *Chem. Ind. (London)*, **1965,** 425.

[163] V. Ioan, M. Popovici, E. Mosanu, M. Eliean, and C. D. Nenitzescu, *Rev. Roum. Chim.*, **10,** 185 (1965) [*C.A.*, **63,** 4181f (1965)].

[164] S. A. Monti, D. J. Bucheck, and J. C. Shepard, *J. Org. Chem.*, **34,** 3080 (1969).

[165] G. A. Russell, J. J. McDonnell, and P. R. Whittle, *J. Am. Chem. Soc.*, **89,** 5516 (1967).

[166] W. G. Dauben and W. M. Welch, *Tetrahedron Lett.*, **1971,** 4531.

[167] S. Julia, M. Julia, and G. Linstrumelle, *Bull. Soc. Chim. Fr.*, **1964** 2693.

[168] S. Julia and G. Linstrumelle, *Bull. Soc. Chim. Fr.*, **1966,** 3490.

[169] R. M. Coates and J. L. Kirkpatrick, *J. Am. Chem. Soc.*, **90,** 4162 (1968).

[170] A. Nickon, H. Kwasnik, T. Swartz, R. O. Williams, and J. B. DiGiorgio, *J. Am. Chem. Soc.*, **87,** 1615 (1965).

[171] U. Biethan, U. v. Gizycki, and H. Musso, *Tetrahedron Lett.*, **1965,** 1477.

[172] D. Becker and H. J. E. Loewenthal, *J. Chem. Soc., Chem. Commun.*, **1965,** 149.

[173] C. J. V. Scanio and D. L. Lickei, *Tetrahedron Lett.*, **1972,** 1363.

[174] H. Klusacek and H. Musso, *Chem. Ber.*, **103,** 3066 (1970).

[175] J. A. Berson, D. S. Donald, and W. J. Libbey, *J. Am. Chem. Soc.*, **91,** 5580 (1969).

[176] K. Mori and M. Matsui, *Tetrahedron*, **25,** 5013 (1969).

[177] K. Mori, M. Ohki, and M. Matsui, *Tetrahedron*, **26,** 2821 (1970).

[178] S. Julia, A. Costantino, and G. Linstrumelle, *C. R. Acad. Sci. Paris, Ser. C,* **264,** 407 (1967).

[179] G. W. Klumpp, G. Ellen, J. Japenga, and G. M. de Hoog, *Tetrahedron Lett.*, **1972,** 1741.

[180] J. E. Baldwin and W. D. Fogelsong, *J. Am. Chem. Soc.*, **90,** 4303 (1968).

[181] J. E. Baldwin and W. D. Fogelsong, *Tetrahedron Lett.*, **1966,** 4089.

[182] M. Tichý and J. Sicher, *Coll. Czech. Chem. Commun.* **1972,** 3106 [*C.A.*, **78,** 3779g (1973)].

[183] B. Boyer, P. Dubreuil, G. Lamaty, and J. P. Roque, *Tetrahedron Lett.*, **1974,** 2919.

[184] S. Julia, M. Julia, and G. Linstrumelle, *Bull. Soc. Chim. Fr.*, **1966,** 3499.

[185] E. Vedejs and R. A. Shepherd, *J. Org. Chem.*, **41,** 742 (1976).

[186] D. P. G. Hamon, G. F. Taylor, and R. N. Young, *Tetrahedron Lett.*, **1975,** 1623.

[187] A. Krantz and C. Y. Lin, *J. Am. Chem. Soc.*, **95,** 5662 (1973).

[188] R. K. Murray, K. A. Babiak, and T. K. Morgan, Jr., *J. Org. Chem.*, **40,** 2463 (1975).

[189] D. P. G. Hamon and G. F. Taylor, *Aust. J. Chem.*, **28,** 2255 (1975).

[190] H. E. Zimmerman, R. G. Lewis, J. J. McCullough, A. Padwa, S. W. Staley, and M. Semmelhack, *J. Am. Chem. Soc.*, **88,** 1965 (1966).

[191] Y. Bessière-Chrètien and M. M. El Gaïed, *Bull, Soc. Chim. Fr.*, **1971,** 2189.
[192] S. Bien and D. Ovadia, *J. Org. Chem.*, **39,** 2258 (1974).
[193] G. Linstrumelle and S. Julia, *Bull. Soc. Chim. Fr.*, **1966,** 3507.
[194] E. Vedejs, R. A. Shepherd, and R. P. Steiner, *J. Am. Chem. Soc.*, **92,** 2158 (1970).
[195] H. E. Zimmerman and L. M. Tolbert, *J. Am. Chem. Soc.*, **97,** 5497 (1975).
[196] E. Vedejs and W. R. Wilber, *Tetrahedron Lett.*, **1975,** 2679.
[197] E. Piers, W. de Waal, and R. W. Britton, *Can. J. Chem.*, **47,** 4299 (1969).
[198] E. Piers, W. de Waal, and R. W. Britton, *J. Chem. Soc., Chem. Commun.*, **1968,** 188.
[199] E. Piers, R. W. Britton, and W. de Waal, *Can. J. Chem.*, **49,** 12 (1971).
[200] A. Tanaka, H. Uda, and A. Yoshikoshi, *J. Chem. Soc., Chem. Commun.*, **1969,** 308.
[201] E. Piers, R. W. Britton, and W. de Waal, *Tetrahedron Lett.*, **1969,** 1251.
[202] O. P. Vig, M. S. Bhatia, A. K. Verma, and K. L. Matta, *J. Indian Chem. Soc.*, **47,** 277 (1970).
[203] E. Piers, M. B. Geraghty, R. D. Smillie, and M. Soucy, *Can. J. Chem.*, **53,** 2849 (1975).
[204] M. Popovici, V. Ioan, M. Elian, and C. D. Nenitzescu, *Rev. Roum. Chim.*, **12,** 583 (1967) [*C.A.*, **68,** 29448s (1968)].
[205] O. P. Vig, M. S. Bhatia, I. R. Trehan, and K. L. Matta, *J. Indian Chem. Soc.*, **47,** 282 (1970).
[206] V. Ioan, M. Popovici, and C. D. Nenitzescu, *Tetrahedron Lett.*, **1965,** 3383.
[207] S. Masamune and N. T. Castellucci, *Proc. Chem. Soc. (London)*, **1964,** 298.
[208] M. Baciu, M. Elian, E. Cioranescu, and C. D. Nenitzescu, *Tetrahedron Lett.*, **1972,** 2573.
[209] J. L. Mateos and V. M. Coronado, *Rev. Latinoamer. Quim.*, **6,** 72 (1975) [*C.A.*, **83,** 147641g (1976)].
[210] A. Small, *J. Am. Chem. Soc.*, **86,** 2091 (1964).
[211] L. Solomon, *Diss. Abstr.*, **26,** 101 (1965).
[212] N. Nakamura and K. Sakai, *Tetrahedron Lett.*, **1976,** 2049.
[213] T. M. Brennan and R. Hill, *J. Am. Chem. Soc.*, **90,** 5614 (1968).
[214] (a) S. Danishefsky, R. McKee, and R. K. Singh, *J. Am. Chem. Soc.*, **99,** 4783 (1977); (b) S. Danishefsky and R. Doehner, *Tetrahedron Lett.*, **1977,** 3031.
[215] J. A. Kaufman and S. J. Weininger, *J. Chem. Soc., Chem. Commun.*, **1969,** 593.
[216] W. C. Agosta and S. Wolf, *J. Org. Chem.*, **40,** 1027 (1975).
[217] (a) S. A. Godleski, P. v. R. Schleyer, and E. Osawa, *J. Chem. Soc., Chem. Commun.*, **1976,** 38; (b) Y. Inamoto, K. Aigami, N. Takaishi, Y. Fujikura, K. Tsuchihashi, and H. Ikeda, *J. Org. Chem.*, **42,** 3833 (1977).
[218] R. T. Landsbury and J. G. Colson, *Chem. Ind. (London)*, **1962,** 821.
[219] U. R. Ghatak and S. Chakrabarty, *J. Org. Chem.*, **41,** 1089 (1976).
[220] U. R. Ghatak, J. K. Ray, and S. Chakrabarty, *J. Chem. Soc., Perkin Trans. I*, **1976,** 1975.
[221] J. R. Marshall and J. Walker, *J. Chem. Soc.*, **1952,** 467.
[222] A. K. Bose and P. Yates, *J. Am. Chem. Soc.*, **74,** 4703 (1952).
[223] P. Pfeiffer and E. Enders, *Chem. Ber.*, **84,** 247 (1951).
[224] H. E. Sheffer and J. A. Moore, *J. Org. Chem.*, **28,** 129 (1963).
[225] A. Seetharamiah, *J. Chem. Soc.*, **1948,** 894.
[226] H. Krzikalla and B. Eistert, *J. Prakt. Chem.*, **143,** 50 (1935).
[227] J. A. Moore and R. W. Medeiros, *J. Am. Chem. Soc.*, **81,** 6026 (1959).
[228] F. W. Bruchhausen and H. Hoffmann, *Chem. Ber.*, **74,** 1584 (1941).
[229] R. Hirschmann, *Chem. Ind. (London)*, **1958,** 1259.
[230] J. A. Moore, W. F. Holton, and E. L. Wittle, *J. Am. Chem. Soc.*, **84,** 390 (1962).

AUTHOR INDEX, VOLUMES 1–26

CHAPTER AND TOPIC INDEX, VOLUMES 1–26

Many chapters contain brief discussions of reactions and comparisons of alternative synthetic methods which are related to the reaction that is the subject of the chapter. These related reactions and alternative methods are not usually listed in this index. In this index the volume number is in BOLDFACE, the chapter number in ordinary type.

SUBJECT INDEX, VOLUME 26

Since the table of contents provides a quite complete index, only those items not readily found from the contents pages are listed here. Numbers in BOLDFACE refer to experimental procedures.